Airplane Design

Part IV: Layout of Landing Gear and Systems

Dr. Jan Roskam
Ackers Distinguished Professor of Aerospace Engineering
The University of Kansas, Lawrence

2010

DARcorporation
Design • Analysis • Research

1440 Wakarusa Drive, Suite 500 • Lawrence, Kansas 66049, U.S.A.

PUBLISHED BY
Design, Analysis and Research Corporation (*DARcorporation*)
1440 Wakarusa Drive, Suite 500
Lawrence, Kansas 66049
U.S.A.
Phone: (785) 832-0434
Fax: (785) 832-0524
e-mail: info@darcorp.com
http://www.darcorp.com

Library of Congress Catalog Card Number: 97-68580

ISBN 978-1-884885-53-2

In all countries, sold and distributed by
Design, Analysis and Research Corporation
1440 Wakarusa Drive, Suite 500
Lawrence, Kansas 66049
U.S.A.

Printed in the United States of America
First Printing, 1986
Second Printing, 1989
Third Printing, 2000
Fourth Printing, 2004
Fifth Printing, 2007
Sixth Printing, 2010

TABLE OF CONTENTS

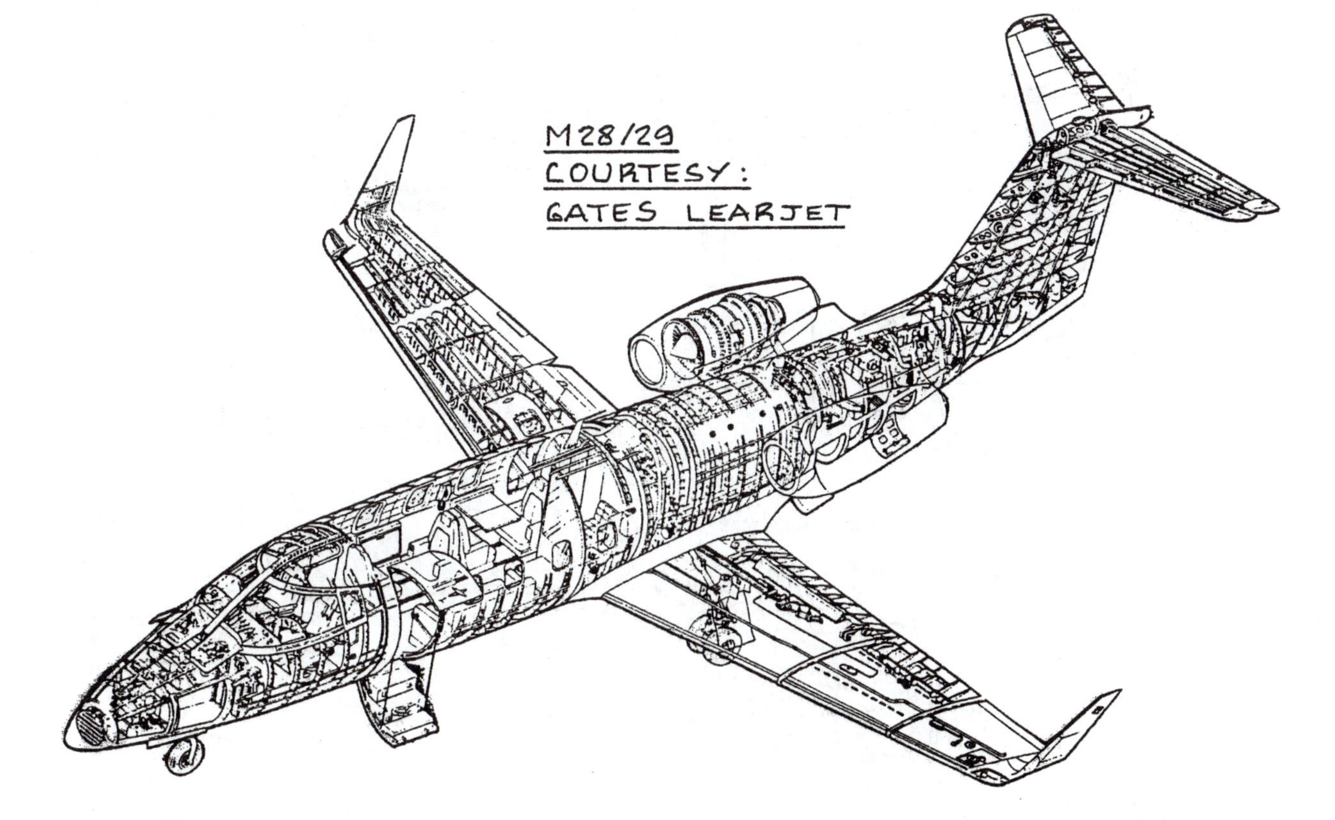
M28/29
COURTESY:
GATES LEARJET

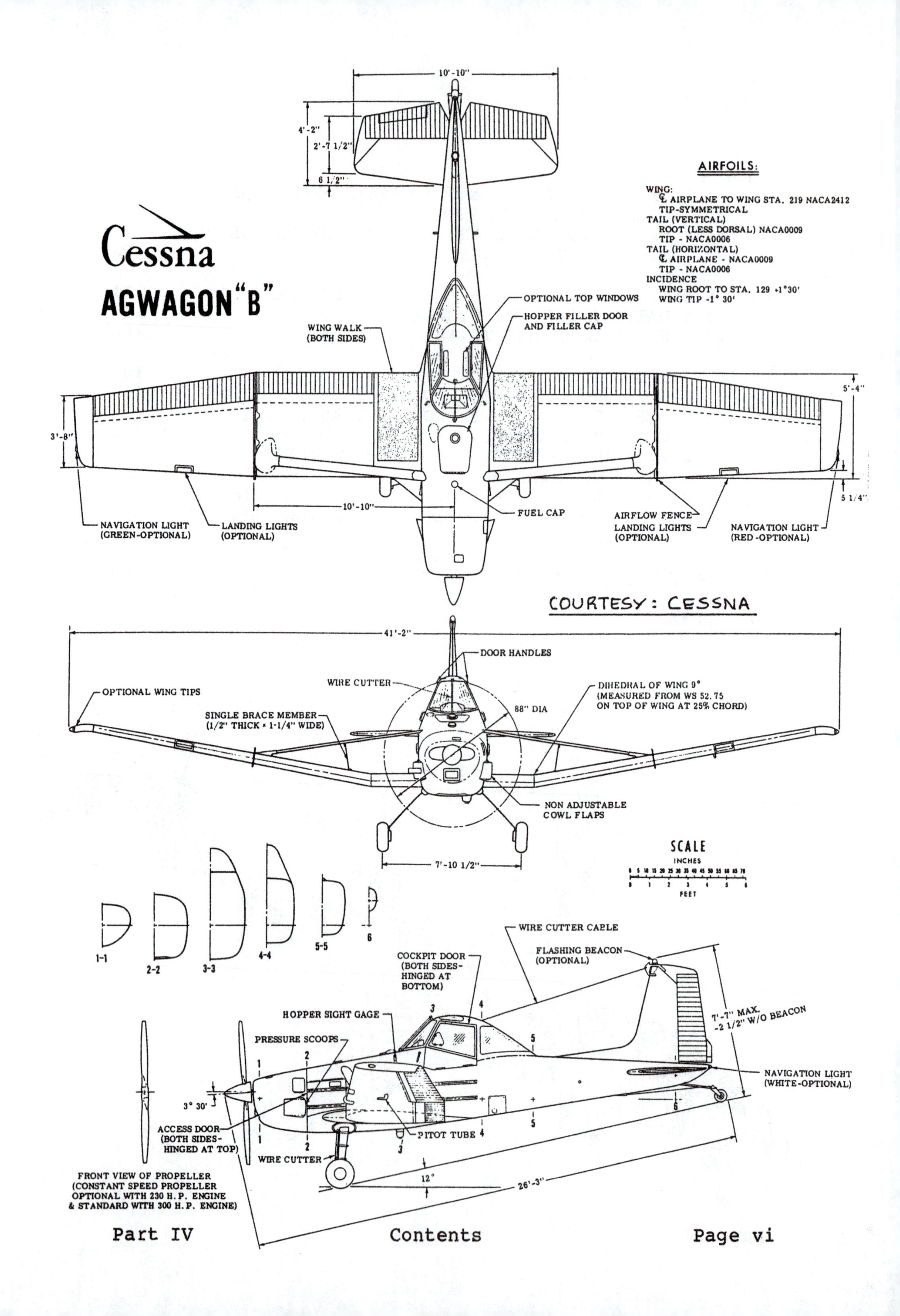
Cessna
AGWAGON "B"
AIRFOILS:
WING:
℄ AIRPLANE TO WING STA. 219 NACA2412
TIP-SYMMETRICAL
TAIL (VERTICAL)
ROOT (LESS DORSAL) NACA0009
TIP - NACA0006
TAIL (HORIZONTAL)
℄ AIRPLANE - NACA0009
TIP - NACA0006
INCIDENCE
WING ROOT TO STA. 129 +1°30'
WING TIP -1° 30'
10'-10"
4'-2"
2'-7 1/2"
6 1/2"
OPTIONAL TOP WINDOWS
HOPPER FILLER DOOR AND FILLER CAP
WING WALK (BOTH SIDES)
5'-4"
3'-8"
5 1/4"
10'-10"
FUEL CAP
NAVIGATION LIGHT (GREEN-OPTIONAL)
LANDING LIGHTS (OPTIONAL)
AIRFLOW FENCE
LANDING LIGHTS (OPTIONAL)
NAVIGATION LIGHT (RED-OPTIONAL)
COURTESY: CESSNA
41'-2"
DOOR HANDLES
WIRE CUTTER
OPTIONAL WING TIPS
DIHEDRAL OF WING 9° (MEASURED FROM WS 52.75 ON TOP OF WING AT 25% CHORD)
SINGLE BRACE MEMBER (1/2" THICK x 1-1/4" WIDE)
88" DIA
NON ADJUSTABLE COWL FLAPS
7'-10 1/2"
SCALE
INCHES
FEET
1-1
2-2
3-3
4-4
5-5
6
WIRE CUTTER CABLE
FLASHING BEACON (OPTIONAL)
COCKPIT DOOR (BOTH SIDES-HINGED AT BOTTOM)
HOPPER SIGHT GAGE
PRESSURE SCOOPS
7'-7" MAX. -2 1/2" W/O BEACON
NAVIGATION LIGHT (WHITE-OPTIONAL)
3° 30'
ACCESS DOOR (BOTH SIDES-HINGED AT TOP)
PITOT TUBE
WIRE CUTTER
12°
26'-3"
FRONT VIEW OF PROPELLER (CONSTANT SPEED PROPELLER OPTIONAL WITH 230 H.P. ENGINE & STANDARD WITH 300 H.P. ENGINE)

TABLE OF SYMBOLS

Symbol	Definition	Dimension
a_x	forward deceleration	ft/sec^2
b	wing span	ft
b_t	maximum tire width	
$\bar{c}$	wing mean geometric chord	ft
c_j	specific fuel cons.(jet)	lbs/lbs/hr
c_p	specific fuel cons.(pist)	lbs/hp/hr
C.F.	Centrifugal force	lbs
C_h	Hingemoment coefficient	----
C_{m_r}	Recoil pitching mom. coeff.	----
C_{n_r}	Recoil yawing mom. coeff.	----
C_R	Top tire clearance	in
d_s	shock absorber diameter	ft
D	Tire bead seat diameter	in
D_i	Drag in Fig.2.66a	lbs
D_o	Outside tire diameter	in
D_F	Tire flange diameter	in
D_G	Grown tire diameter	in
D_s	Tire shoulder diameter	in
D_t	Outside tire diameter	in
E_t	Kinetic energy	ftlbs
f_{dyn}	factor which multiplies tire static load to get dynamic load	-----
F	Force on landing gear	lbs
F_c	cable or push-rod control force	lbs
F_p	Pilot control force	lbs
F_r	Retraction force, also gun recoil force	lbs

g	acceleration of gravity	ft/sec^2
G	Gearing ratio	rad/ft
h_{cg}	distance from the center of gravity to the runway	ft
H	Tire section height	in
HM	Hingemoment	ftlbs
i_H	Stabilizer incidence	deg or rad
l_i	See Fig.2.66a	ft
l_m	Distance from main gear to the center of gravity	ft
l_n	Distance from the nose gear to the center of gravity	ft
m	airplane mass	slugs
n	load factor	-----
n_s	number of struts in main gear	-----
n_t	number of tires (nosegear)	-----
N_g	Landing gear load factor	-----
P	Force on landing gear	lbs
P_n	Static load on nosegear as defined in Fig.2.14	lbs
P_m	Static load on main gear strut defined in Fig.2.14	lbs
P_{TO}	Take-off power	hp
$\bar{q}$	dynamic pressure	psf
R	See Fig.2.66b	ft
s_p	cockpit control travel	in or ft
s_s	allowable shock absorber deflection	ft
s_t	allowable tire deflection	ft
s_x	tire shoulder clearance	in
S	Wing area	ft^2
S_p	Effective piston area	in^2
t	Tire deflection	in/ft
T_{TO}	Take-off thrust	lbs
V	Airplane speed	fps/kts
V_{s_1}	Stall speed, landing	mph

$V_{s_{TO}}$	Stall speed, take-off	mph
$V_{tire/max}$	Maximum allowable tire speed	mph
w_t	Touchdown rate	fps
W	Airplane weight	lbs
W (note!)	Maximum tire width, also called Tire section width	in
W_g	Landing gear weight	lbs
W_G	Grown tire width	in
W_s	Tire shoulder width	in
x_i	See Fig.2.66a	ft
x_p	See Fig.2.66b	ft
y_r	See Fig.3.7	ft
z_i	See Fig.2.66a	ft
z_p	See Fig.2.66b	ft
z_r	See Fig.3.7	ft

Greek Symbols
=============

γ	Flight path angle	deg
η_s	Shock abs. efficiency	-----
η_t	Tire abs. efficiency	-----
δ	Control deflection angle	deg or rad
μ	friction coefficient	-----
phi	gear retraction angle	deg

Subscripts
==========

ave	average
dyn	dynamic
g	ground
h	horizontal
L	Landing
m	main gear

max	maximum
n	nose gear
r	recoil
R	Rudder
TO	Take-off
v	vertical

Acronyms
========

ac	alternating current
ACLS	Air cushion landing system
AGM	Air-to-ground missile
AIM	Air intercept missile
ARM	Anti radiation missile
ASM	Air-to-surface missile
BBL	Body Buttock Line
CCW	Counterclock wise
CW	Clock wise
dc	direct current
ESWL	Equivalent Single Wheel Load
FAA	Federal Aviation Administration
FAR	Federal Aviation Regulation
FOD	Foreign Object Damage
HUD	Heads up display
ICAO	International Cicil Aviation Organization
KVA	Kilovolt-amperes
LCN	Load Classification Number
N.A.	Not available
NTSB	National Transportation and Safety Board
RCS	Radar cross section
SMTD	STOL and maneuver demonstrator
STOL	Short take-off and landing
VDC	Volts direct current
V/STOL	Vertical/Short Take-off and Landing

ACKNOWLEDGEMENT
===============

Writing a book on airplane design is impossible without the supply of a large amount of data. The author is grateful to the following companies for supplying the raw data, manuals, sketches and drawings which made the book what it is:

Aerospatiale
Beech Aircraft Corp.
The Boeing Company
British Aerospace Corp.
Cessna Aircraft Company
Fairchild Republic Co.
Gates Learjet Corporation
Piper Aircraft Corporation
General Dynamics Corporation
Grumman Aerospace Corp.
Gulfstream Aerospace Corp.
Lockheed Aircraft Corp.
McDonnell Douglas Corp.
Short Brothers and Harland Ltd.

The author wishes to specifically acknowledge the cooperation received from Boeing and from McDonnell Douglas Corporation in providing large numbers of layout drawings with permission to publish.

A significant amount of airplane design information has been accumulated by the author over many years from the following magazines:

Interavia (Swiss, monthly)
Flight International (British, weekly)
Business and Commercial Aviation (USA, monthly)
Aviation Week and Space Technology (USA, weekly)
Journal of Aircraft (USA, AIAA, monthly)

The author wishes to acknowledge the important role played by these magazines in his own development as an aeronautical engineer. Aeronautical engineering students and graduates should read these magazines regularly.

Nearly all weapons and military payload drawings in this book were drawn by Mr. Govert Tukker of Molenaarsgraaf, The Netherlands. The author is grateful to Mr. Tukker for his skill and patience in carrying out this most difficult assignment.

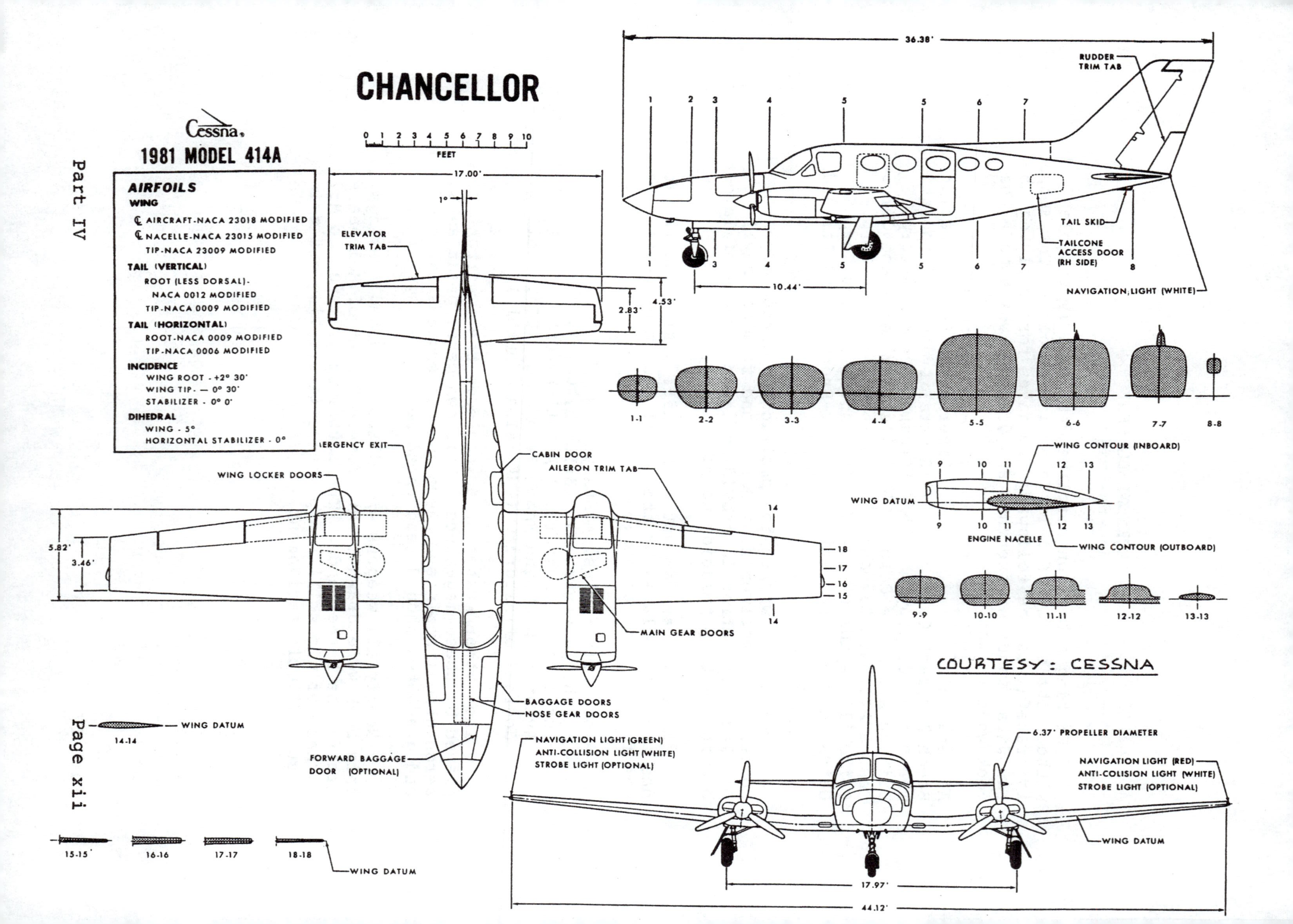
CHANCELLOR
Cessna
1981 MODEL 414A
AIRFOILS
WING
℄ AIRCRAFT-NACA 23018 MODIFIED
℄ NACELLE-NACA 23015 MODIFIED
TIP-NACA 23009 MODIFIED
TAIL (VERTICAL)
ROOT (LESS DORSAL)-
NACA 0012 MODIFIED
TIP-NACA 0009 MODIFIED
TAIL (HORIZONTAL)
ROOT-NACA 0009 MODIFIED
TIP-NACA 0006 MODIFIED
INCIDENCE
WING ROOT - +2° 30'
WING TIP - — 0° 30'
STABILIZER - 0° 0'
DIHEDRAL
WING - 5°
HORIZONTAL STABILIZER - 0°
0 1 2 3 4 5 6 7 8 9 10
FEET
17.00'
1°
ELEVATOR TRIM TAB
2.83'
4.53'
36.38'
RUDDER TRIM TAB
TAIL SKID
TAILCONE ACCESS DOOR (RH SIDE)
NAVIGATION LIGHT (WHITE)
10.44'
1-1
2-2
3-3
4-4
5-5
6-6
7-7
8-8
WING CONTOUR (INBOARD)
WING DATUM
ENGINE NACELLE
WING CONTOUR (OUTBOARD)
9-9
10-10
11-11
12-12
13-13
ERGENCY EXIT
CABIN DOOR
AILERON TRIM TAB
WING LOCKER DOORS
5.82'
3.46'
MAIN GEAR DOORS
BAGGAGE DOORS
NOSE GEAR DOORS
FORWARD BAGGAGE DOOR (OPTIONAL)
COURTESY: CESSNA
14-14
WING DATUM
15-15
16-16
17-17
18-18
WING DATUM
NAVIGATION LIGHT (GREEN)
ANTI-COLLISION LIGHT (WHITE)
STROBE LIGHT (OPTIONAL)
6.37' PROPELLER DIAMETER
NAVIGATION LIGHT (RED)
ANTI-COLISION LIGHT (WHITE)
STROBE LIGHT (OPTIONAL)
WING DATUM
17.97'
44.12'

1. INTRODUCTION

The purpose of this series of books on Airplane Design is to familiarize aerospace engineering students with the design methodology and design decision making involved in the process of designing airplanes.

The series of books is organized as follows:

PART I:	PRELIMINARY SIZING OF AIRPLANES
PART II:	PRELIMINARY CONFIGURATION DESIGN AND INTEGRATION OF THE PROPULSION SYSTEM
PART III:	LAYOUT DESIGN OF COCKPIT, FUSELAGE, WING AND EMPENNAGE: CUTAWAYS AND INBOARD PROFILES
PART IV:	LAYOUT DESIGN OF LANDING GEAR AND SYSTEMS
PART V:	COMPONENT WEIGHT ESTIMATION
PART VI:	PRELIMINARY CALCULATION OF AERODYNAMIC, THRUST AND POWER CHARACTERISTICS
PART VII:	DETERMINATION OF STABILITY, CONTROL AND PERFORMANCE CHARACTERISTICS: FAR AND MILITARY REQUIREMENTS
PART VIII:	AIRPLANE COST ESTIMATION: DESIGN, DEVELOPMENT, MANUFACTURING AND OPERATING

The purpose of PART IV is to present a systematic approach to the problem of airplane landing gear design and airplane systems design during the preliminary design phase.

Chapter 2 presents a discussion of methods employed to yield realistic layouts of landing gears. Specific problems addressed are:

a) tire selection and sizing
b) strut sizing
c) landing gear disposition in view of turnover criteria, rotation and ground handling requirements
d) retraction kinematics

Chapter 3 contains data on weapons integration problems encountered during the preliminary design of military airplanes. In laying out military airplane designs, geometric and weights data on weapons and other military payloads are needed. Such data are given in Chapter 3.

Chapter 4 contains a discussion of design considerations for primary and secondary flight control systems. Both reversible and irreversible flight control systems are addressed and many example layouts included.

The layout design of airplane systems is an important subject during the preliminary design phase. The main reason for this is the fact that many systems have a large impact on flight safety. Another reason is that early design decisions tend 'lock in' most of the life cycle cost of an airplane. It is therefore essential, that attention be given to the layout design of all systems which are important to the operation of an airplane. The following systems are covered:

Chapter 5 addresses the problem of preliminary fuel system design. A number of guidelines for 'design for safety' are included.

Chapter 6: provides an introduction to hydraulic system layout design.

Chapter 7: Gives a brief overview of design decisions involved in laying out airplane electrical systems.

Chapter 8: Environmental systems are important components of many airplanes. This chapter deals with the layout design of the pressurization system, the airconditioning systen and the oxygen system.

Chapter 9: This chapter deals with layout design problems associated with cockpit instrumentation, flight management and other avionics systems.

De-icing, anti-icing, rain removal and defog systems are covered in Chapter 10.

Chapter 11 presents examples of the layout design problems encountered in the preliminary design of emergency escape and ejection systems.

Particularly in passenger transports the design of the potable water and waste system is an important aspect of layout design. Chapter 12 covers these systems.

Chapter 13 addresses the role of safety and survivability in preliminary design thinking and in preliminary design decision making. A review of aviation safety and how it is measured is also given.

2. LANDING GEAR LAYOUT DESIGN

The purpose of this chapter is to provide methods and data to assist in preparing satisfactory landing gear layouts. The material presented here is meant to be used in conjunction with Steps 18 and 29 in p.d. sequence II of Part II. Methods consistent with p.d. sequence II are referred to as Class II methods. The Class I landing gear layout design procedure was presented in Chapter 9 of Part II.

The material is organized as follows:

2.1 Function of landing gear components
2.2 Discussion of landing gear types
2.3 Compatibility of landing gear and runway surface: determination of allowable wheel loads
2.4 Tires: types, performance, sizing and data
2.5 Strut-wheel interface, struts and shock absorbers
2.6 Brakes and braking capability
2.7 Design considerations for landing gears of carrier based airplanes
2.8 Review of landing gear layout geometry
2.9 Steering, turnradii and ground operation
2.10 Retraction kinematics
2.11 Examples of landing gear layouts
2.12 Unconventional landing gear configurations

References 1 through 5 are excellent sources for additional information on landing gear design.

2.1 FUNCTION OF LANDING GEAR COMPONENTS

There are five reasons for incorporating landing gears in airplanes:

1. To absorb landing shocks and taxiing shocks.
2. To provide ability for ground maneuvering: taxi, take-off roll, landing roll and steering.
3. To provide for braking capability.
4. To allow for airplane towing.
5. To protect the ground surface.

Landing gears must be capable of absorbing landing

and taxi loads as well as transmit part of these loads to the airframe. The magnitude of these loads depends on the type of airplane as well as on its mission. Ref.2 contains detailed discussions of landing gear loads. Three types of loads must be considered in the layout design of landing gears:

1. Vertical loads, primarily caused by non-zero touchdown rates and taxiing over rough surfaces.

2. Longitudinal loads primarily caused by 'spin-up' loads, braking loads and rolling friction loads.

3. Lateral loads primarily caused by 'crabbed landings', cross-wind taxiing and ground turning.

1. Vertical Landing Gear Loads

The magnitude of vertical landing gear loads depends on the touchdown rate. Design touchdown rates (also called sink speeds) are as follows:

FAR 23: $w_t = 4.4(W/S)_L^{1/4}$, but no less than 7 and no more than 10 fps (Derived from FAR 23.725)

FAR 25: w_t = 12 fps (FAR 25.723)

USAF: w_t = 10 fps (13 fps for trainers)

USN: w_t = 10 fps for transports

w_t = 17 fps for other non-carrier based airplanes

w_t = 22 fps for carrier based airplanes: these must contend with heaving decks.

Except for trainers and carrier based airplanes these sink speeds are hardly ever experienced: they are very conservative. A 4 fps sink rate is considered a 'hard' landing. Figure 2.1 illustrates the probabilities associated with encountering specific sink speeds.

To properly absorb the shock associated with any sink speed most landing gears contain two elements: tires and shock absorbers. Figure 2.2 shows both components in a typical landing gear layout.

The role of tires is discussed in Section 2.4.

Shock absorbers can be designed as separate elements

or they can be integrated into the gear strut. Section 2.5 presents a discussion on struts and on shock absorbers.

Figure 2.3 shows what happens to the ground reaction force during a typical landing. Note that longitudinal and lateral tire-ground forces are not considered here.

Figure 2.4 shows the interplay between spring and damping forces acting on the shock absorber. Figure 2.5 shows the same for the tire. More detailed design considerations for tires and shock absorbers (shock struts) are presented in Sections 2.4 and 2.5 respectively.

2. Longitudinal and Lateral Loads

In addition to the 'vertical' landing gear loads just mentioned, there are the longitudinal and lateral loads. Figure 2.6 illustrates these loads. The landing gear elements which resist these loads are called the drag-brace and the side-brace respectively. Figure 2.7 shows a landing gear with all elements mentioned sofar. In simple landing gears such as shown in Figure 2.8 the drag-brace and side-brace capability are all included in the main strut design. In the case of Figure 2.8 the strut is normally referred to as the 'spring-leaf' or the 'spring-tube'.

For details regarding the structural design of landing gear elements the reader should consult Ref. 2.

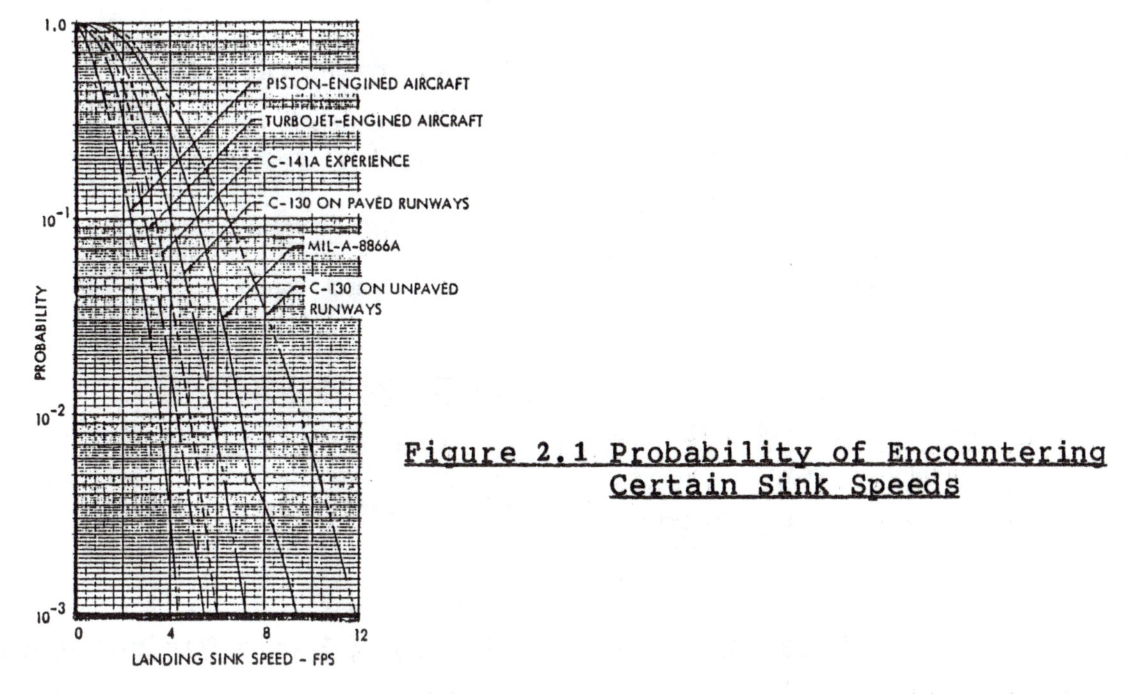

Figure 2.1 Probability of Encountering Certain Sink Speeds

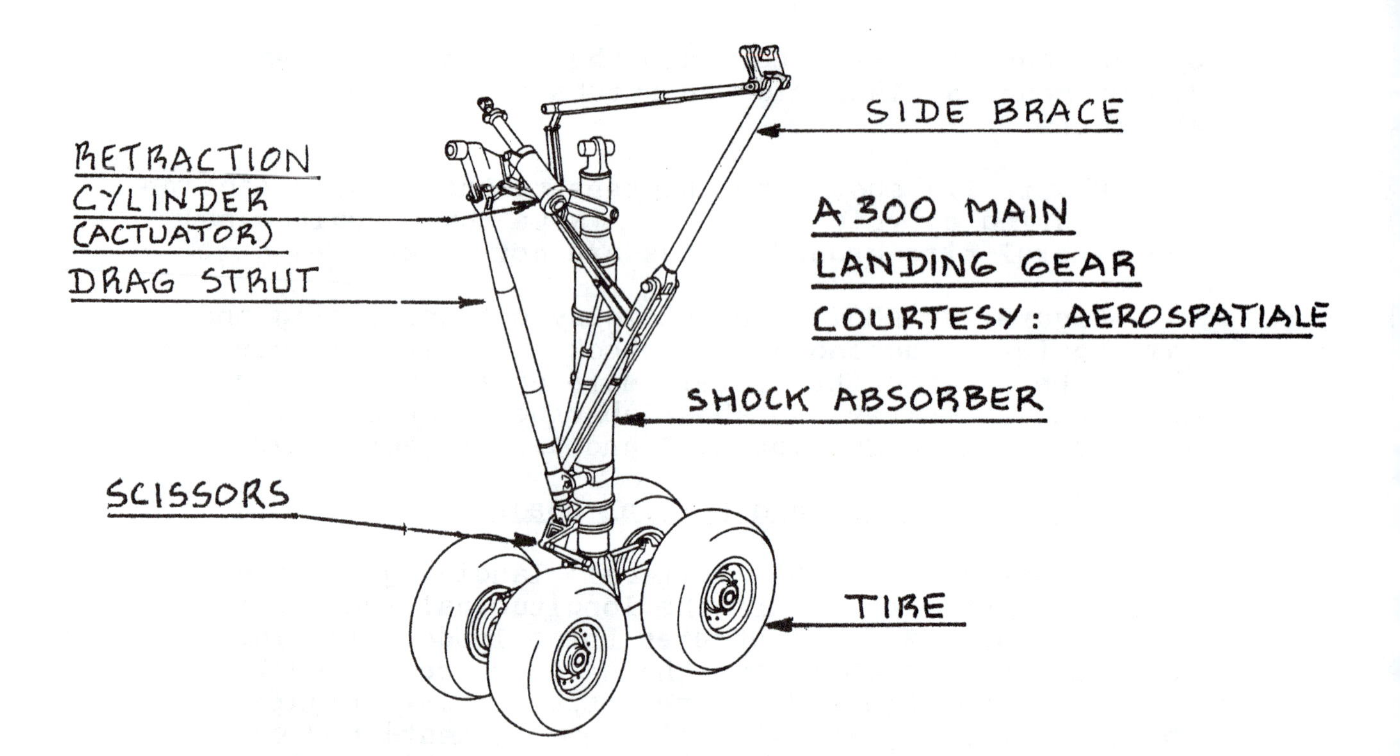

Figure 2.2 Example of Shock Absorber and Tires

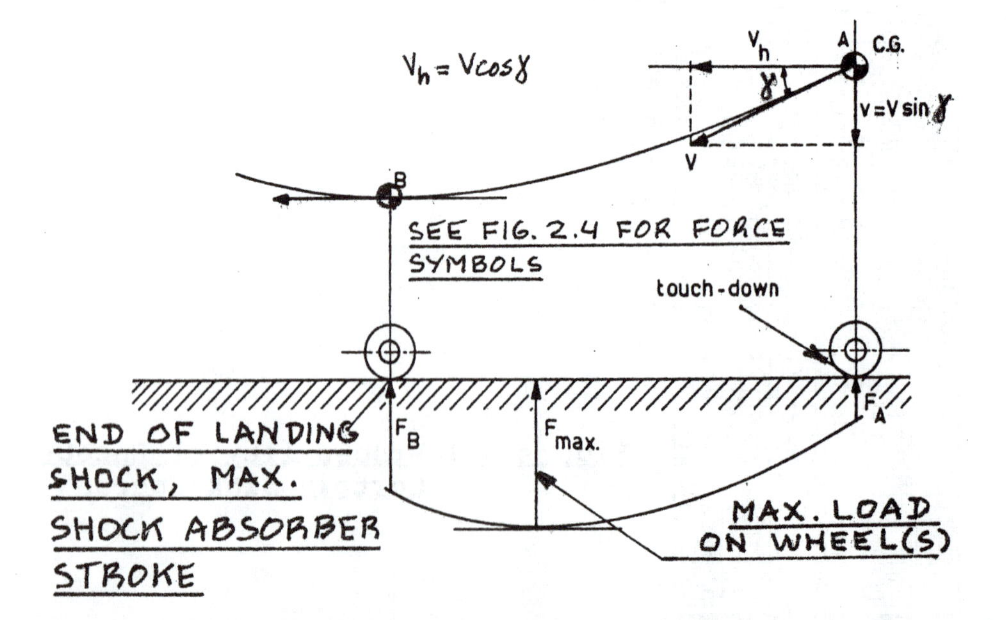

Figure 2.3 Change of Ground Reaction Force During Landing

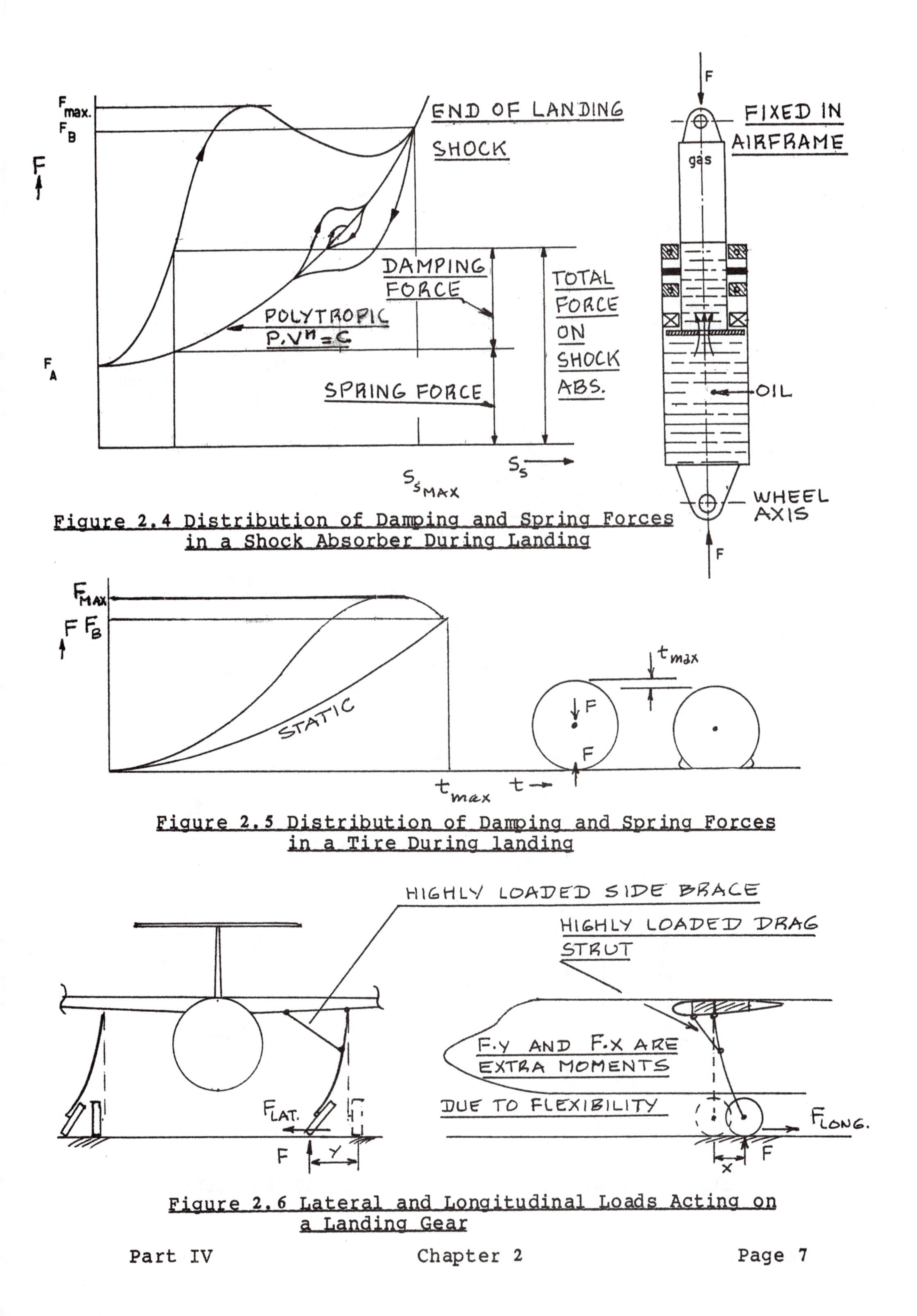

Figure 2.4 Distribution of Damping and Spring Forces in a Shock Absorber During Landing

Figure 2.5 Distribution of Damping and Spring Forces in a Tire During landing

Figure 2.6 Lateral and Longitudinal Loads Acting on a Landing Gear

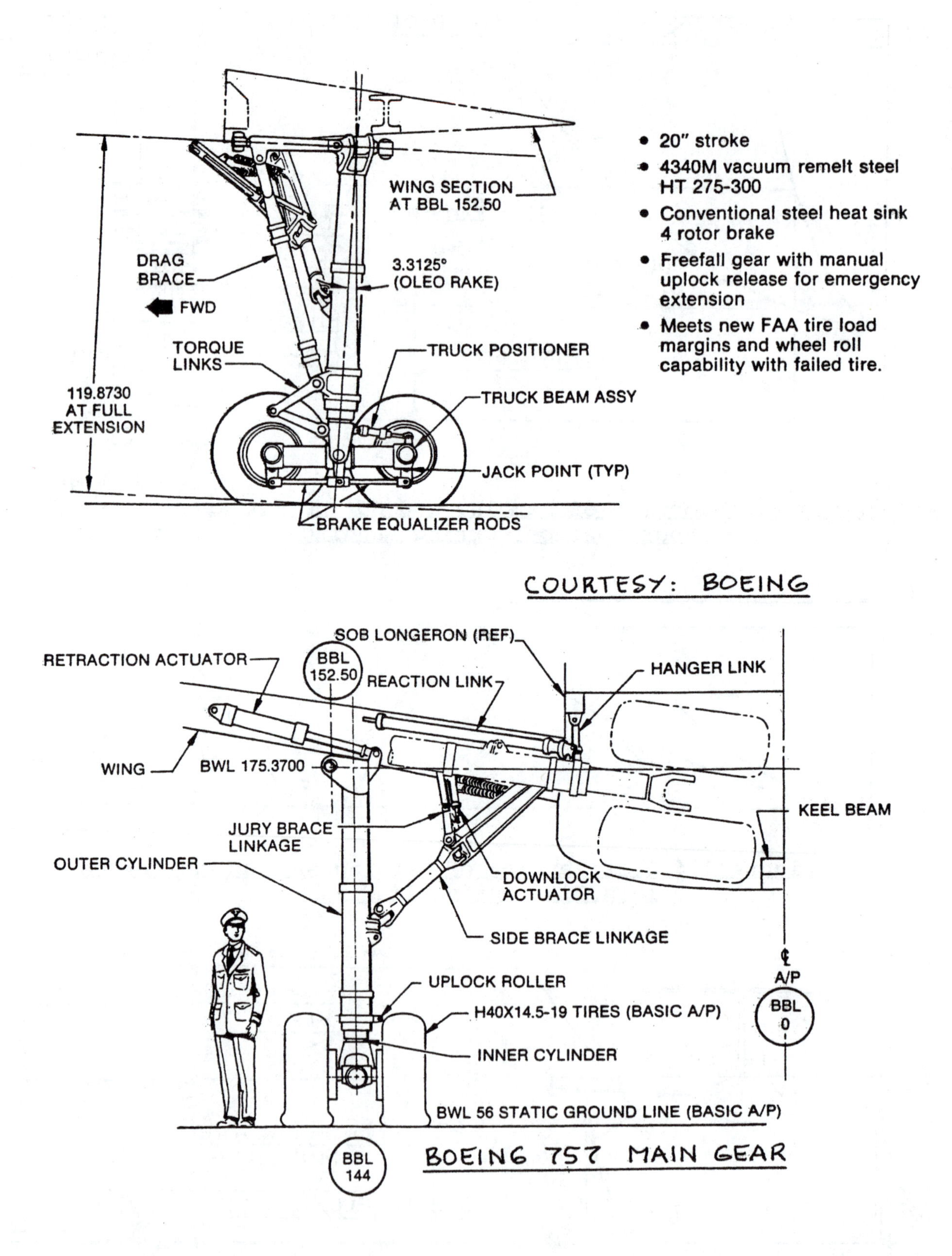

Figure 2.7 Example of a Jet Transport Landing Gear

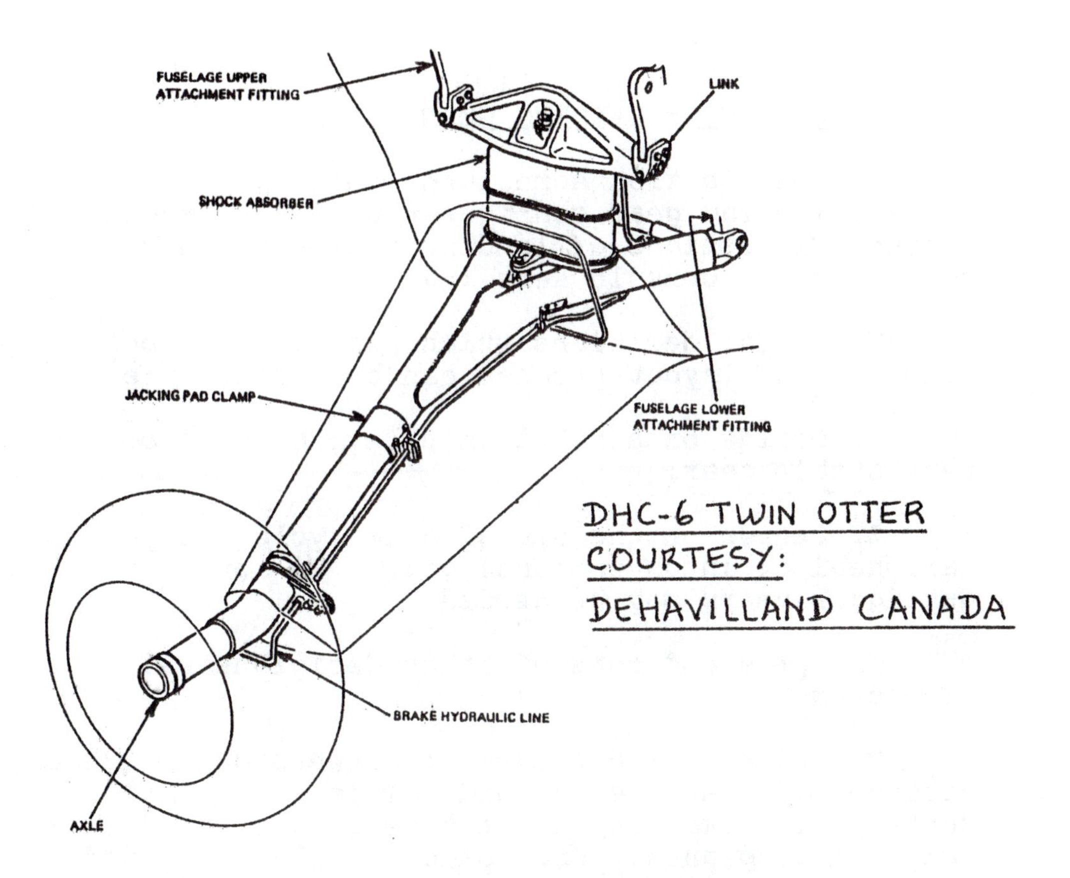

Figure 2.8 Example of a Commuter Landing Gear

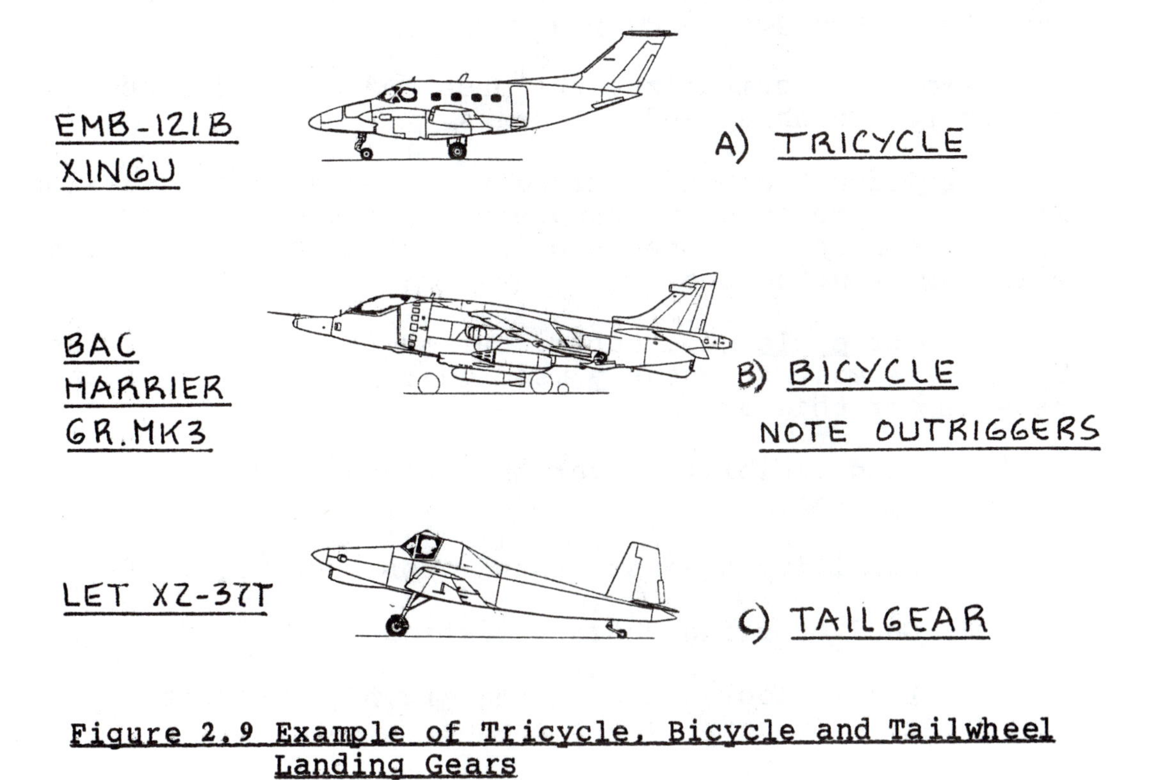

Figure 2.9 Example of Tricycle, Bicycle and Tailwheel Landing Gears

2.2 DISCUSSION OF LANDING GEAR TYPES

In this section some fundamental aspects of the overall landing gear configuration are discussed. For further discussions of landing gear configurations the reader should consult Refs 1-5.

Two major decisions which must be made before the landing gear layout process can be started are:

1. Decide on a fixed (non-retractable) or a retractable gear.

2. Decide on the use of a tricycle, bicycle, tailwheel or unconventional gear. In some cases outrigger gears may be needed.

The pros and cons of these decisions will now be discussed.

Decision 1 is a trade-off between gear induced aerodynamic drag, weight and complexity (cost!). Ref.6 contains information on the type of gear used in a wide range of airplanes. The reader should study Ref.6 to get some insight into which types of gears are used by various manufacturers for different types of airplanes. Experience indicates that airplanes with cruise speeds above 150 kts tend to use retractable landing gears because of the gear drag penalty.

Table 2.1 summarizes the pros and cons of fixed versus retractable landings gears.

Decision 2 depends strongly on the airplane mission. Figure 2.9 presents an example of each type of gear. There are pros and cons associated with each type. These will now be discussed.

The tricycle gear configuration (see Fig.2.9a) has become the most frequently used gear layout. Important reasons for this are:

1. Good visibility over the nose during ground operation.

2. Stability against groundloops: see Fig.2.10.

3. Good steering characteristics.

4. Level floor while on the ground. This is important in cargo and passenger transports.

Table 2.1 SUMMARY OF PROS AND CONS OF FIXED VERSUS RETRACTABLE LANDING GEARS

Gear Type:	Fixed	Retractable
Characteristics:		
Aerodynamic drag	High	Minimal
Weight	Low	High
Complexity and cost	Low	High
Maintenance cost	Insignificant	Significant

Table 2.2 SUMMARY OF PROS AND CONS FOR THREE GEAR TYPES

Gear Type:	Tricycle	Bicycle	Tailwheel
Characteristic:			
Groudloop behavior:	Stable	Stability depends on c.g. location	Unstable
Visibility over the nose:	Good	Good	Poor
Floor attitude on the ground:	Level	Can be level	Not level
Weight:	Medium	High	Low
Steering after touchdown:	Good	Marginal to good	Poor
Steering while taxiing:	Good	Good	Poor
Take-off rotation:	Good	Marginal to impossible	Good
Take-off procedure	Easy	Easy	Needs skill

The bi-cycle gear configuration (see Fig.2.9b) is used in cases where placement of essential components prohibits the use of either the tricycle or the tailwheel configuration. Examples are: the B52 bomber (Fig.3.28a, Part II) and the AV8B Harrier V/STOL fighter (Bottom of p.221, Part II).

In the case of the B52 bomber, a large uninterrupted bombbay was desired. This made retraction of the main gear into the middle of the fuselage impossible. The high wing layout ruled against gear retraction into the wing. The remaining option of a bicycle gear was used.

In the case of the Harrier, the entire center area of the fuselage is occupied by the Pegasus 'four-poster' engine with swiveling nozzles. This made main gear retraction into the fuselage impossible. Because of the thin wing, retraction of the main gear into the wing was also ruled out. That left the bicycle gear layout.

The reader will note that in the case of bicycle gear layouts so-called 'outrigger' gears are needed to provide lateral stability during ground operation. These ourigger gears are normally retracted into the wings (B52) or into wingtip fairings (Harrier).

An important consequence of the bicycle gear arrangement is that take-off rotation is made difficult if not impossible. The wing must therefore be set at an incidence angle governed by take-off considerations instead of by cruise drag considerations. The B52 is an example of this!

The tailwheel configuration (See Fig.2.9c) is used primarily in homebuilt airplanes and in airplanes which must operate from rather rough surfaces. Nose-gear configurations tend to become very heavy if the nosegear is designed to withstand the severe stresses which are the consequence of operating from rough fields.

Tailwheel gears are nearly always lighter than other types of gears. The following disadvantages are the reason for the demise of the tailwheel configuration in most airplanes:

1. Strong tendency to groundloop: see Fig.2.10.

2. Visibility over the nose is poor during ground operation. This results in the need to zig-zag during taxiing.

3. Steering while taxiing is compromised.

Table 2.2 summarizes the pros and cons of the tricycle, bicycle and tailwheel type landing gears. If none of these landing gear configurations are suitable an unconventional layout may be called for. Section 2.12 presents some thoughts on unconventional landing gear configurations.

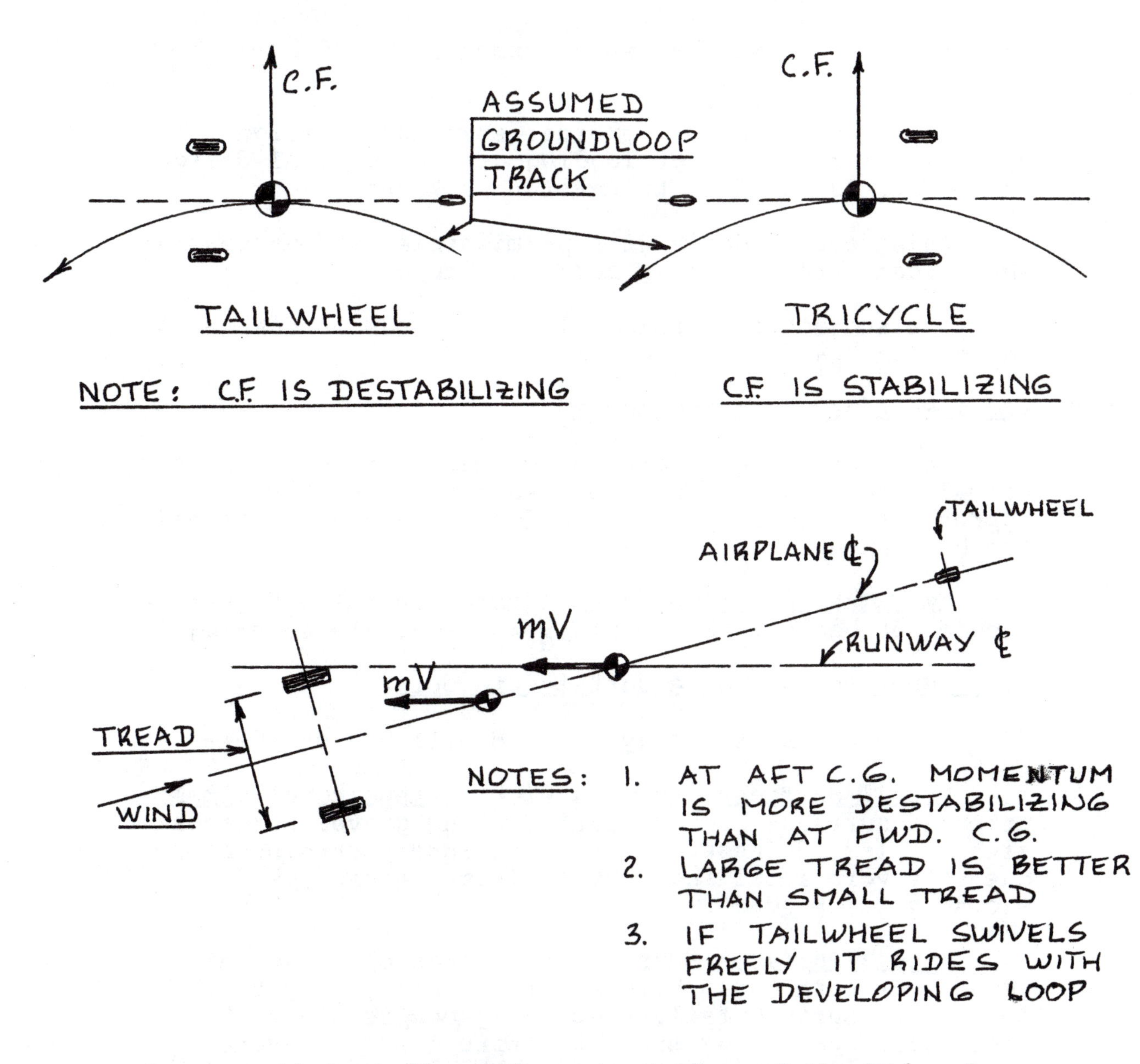

Figure 2.10 Groundloop Characteristics of the Tricycle and the Tailwheel Landing Gear Configurations

2.3 COMPATIBILITY OF LANDING GEAR AND RUNWAY SURFACE: DETERMINATION OF ALLOWABLE WHEEL LOADS

The load on each landing gear strut (also called landing gear leg) as well as the load on each tire may not exceed values which:

1. cause structural damage to the gear or to the airplane

2. cause tire damage

3. cause runway damage or excessive surface deformations

This text does not consider structural design details. For aspects of landing gear structural design the reader is referred to References 2 and 7.

This section deals with permissible landing gear and wheel loads from a runway surface viewpoint.

The subject of permissible tire loads is discussed in Section 2.4.

2.3.1 Nosegear Steering Loads

To allow for adequate nosewheel steering, a minimum normal force must act on the nose gear so that the appropriate levels of friction forces needed for steering can be generated.

IMPORTANT NOTE: The normal force on the nosegear should not be less than $0.08W_{TO}$ for adequate steering.

2.3.2 Gear Loads From A Surface Viewpoint

Three types of runway surface will be considered:

Type 1 Surfaces: Runways with unprepared or simply prepared surfaces: grassy surfaces and gravel surfaces are examples of these. Surface failure occurs normally due to severe local indentation (ruts) caused by excessive tire loads.

Type 2 Surfaces: Runways with flexible pavement: asphalt or tarmacadam. These are normally very thick surfaces. Surface failure normally occurs due to local indentation caused by excessive tire loads. Severe surface waviness may result from this.

Type 3 Surfaces: Runways with rigid pavement: concrete. These surfaces normally have about one half the thickness of flexible pavements. Failure often occurs due to corner fracture of a slab caused by excessive tire loads.

Figure 2.11 illustrates the very heavy pavement requirements imposed by modern jet transport airplanes.

2.3.2.1 Allowable gear loads for Type 1 surfaces:

To avoid gear induced surface damage for Type 1 surfaces the tire pressures should not exceed the values given in Table 2.3. Note that this table does NOT apply whenever the load per strut exceeds 10,000 lbs. If the load per strut is greater than 10,000 lbs the gear design practices associated with surface types 2 and 3 must be used.

2.3.2.2 Allowable gear loads for Types 2 and 3 surfaces:

To avoid gear induced surface damage for Type 2 and for Type 3 surfaces the so-called LCN method (LCN = Load Classification Number) is suggested. This method has been established by the ICAO (International Civil Aviation Organization). All major runways in the world have been assigned an LCN. Landing gears must be designed so that their LCN number does not exceed the lowest runway LCN from which the airplane is intended to operate.

Table 2.4 presents typical LCN design values associated with transport airplanes.

For landing gears with a SINGLE WHEEL PER STRUT the relationship between its LCN, its load per wheel and its tire pressure is illustrated in Figure 2.12. When operating from runways with a given LCN value, Fig.2.12 can be used to find the allowable combinations of load per wheel and tire pressure.

For landing gears with MULTIPLE WHEELS PER STRUT the so-called ESWL (Equivalent Single Wheel Load) must be determined first, before Figure 2.12 can be used. The definition of ESWL is as follows:

Definition: The ESWL of a group of two or more wheels equals the load of a single wheel with the same pressure and causing the same pavement stresses.

Precise methods for computing the ESWL for arbitrary

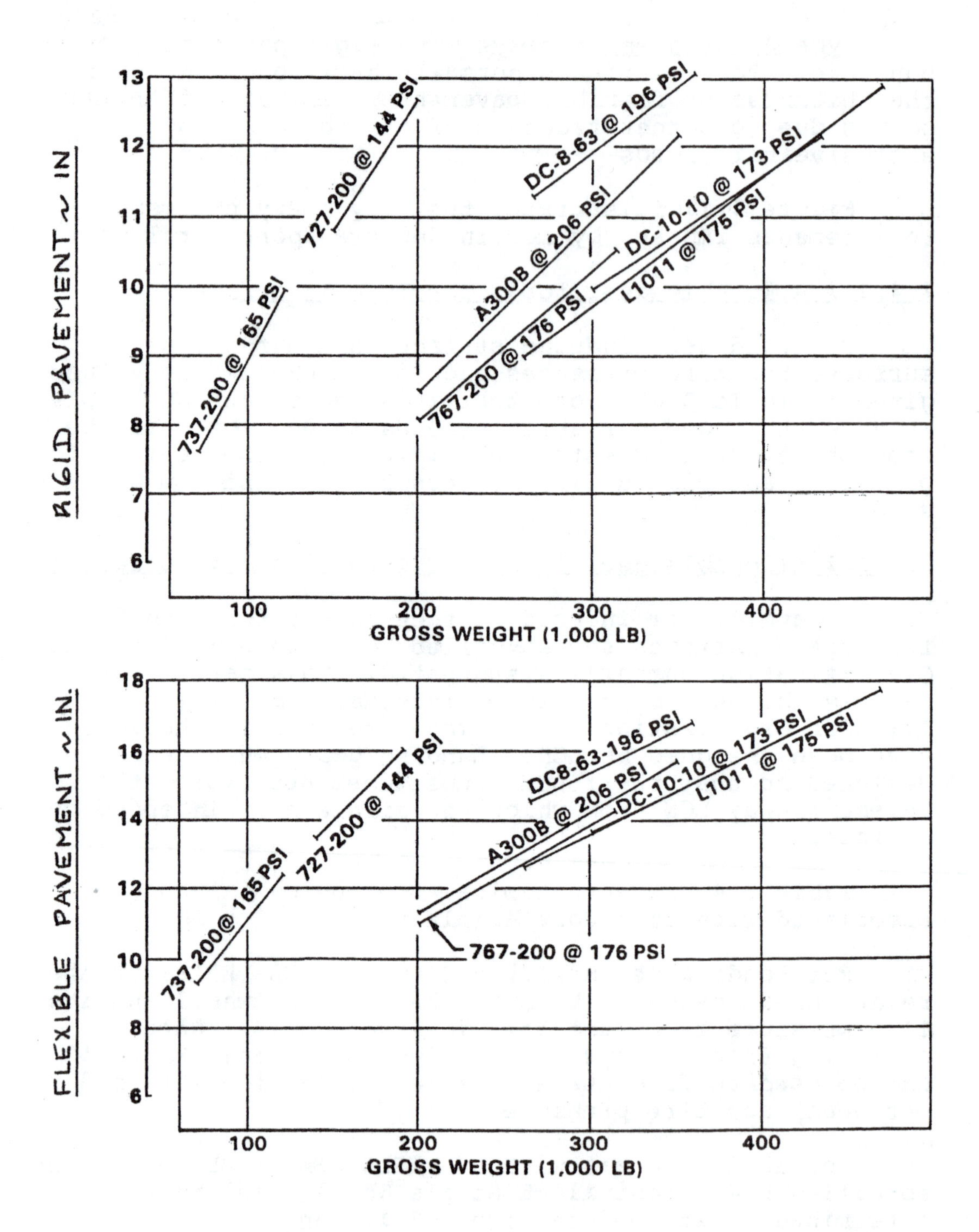

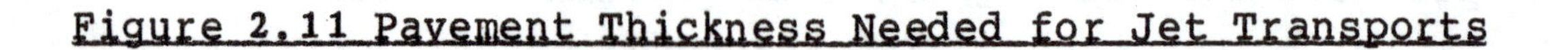
Figure 2.11 Pavement Thickness Needed for Jet Transports

Table 2.3 Recommended Tire Pressures for Various Surfaces

Description of Surface	Maximum Allowable Tire Pressure	
	kg/cm^2	psi
Soft, loose desert sand	1.8 - 2.5	25 - 35
Wet, boggy grass	2.1 - 3.2	30 - 45
Hard desert sand	2.8 - 4.2	40 - 60
Hard grass depending on the type of subsoil	3.2 - 4.2	45 - 60
Small tarmac runway with poor foundation	3.5 - 5.0	50 - 70
Small tarmac runway with good foundation	5.0 - 6.3	70 - 90
Large, well maintained concrete runways	8.5 - 14	120 - 200

Table 2.4 Tire Pressure and LCN Data for Transports

Airplane Type	W_{TO}	Tire Pressure	LCN
	lbs	psi	
Fokker F-27 Mk 500	45,000	80	19
Fokker F-28 Mk 2000	65,000	100	27
McDD DC-9/10	90,700	129	39
Boeing 737-200	110,110	162	49
Boeing 727-200	190,000	160	80
Boeing 757-200	210,000	157	50
Boeing 707-320C	300,000	180	80
McDD DC-10/10	410,000	175	88

gear configurations may be found in References 2 and 3. For preliminary design purposes the following equations give reasonable results:

For DUAL WHEEL layouts (See Fig.2.13 for examples):

$$ESWL = P_n/1.33 \text{ or } P_m/1.33 \qquad (2.1)$$

For TANDEM TWIN layouts (See Fig.2.13 for examples):

$$ESWL = P_n/2 \text{ or } P_m/2 \qquad (2.2)$$

Definitions for P_n and P_m are given in Figure 2.14.

For landing gears with more than four wheels per strut the ESWL should be determined with the methods given in References 2 and 3.

For military airplanes operating from soft fields or steelmatted fields the design procedures of Reference 2 should be used.

Tables 9.1 and 9.2 in Part II provide information on the number of struts and the number of wheels per strut used by a range of airplane types. Reference 6 contains considerably more data on this subject.

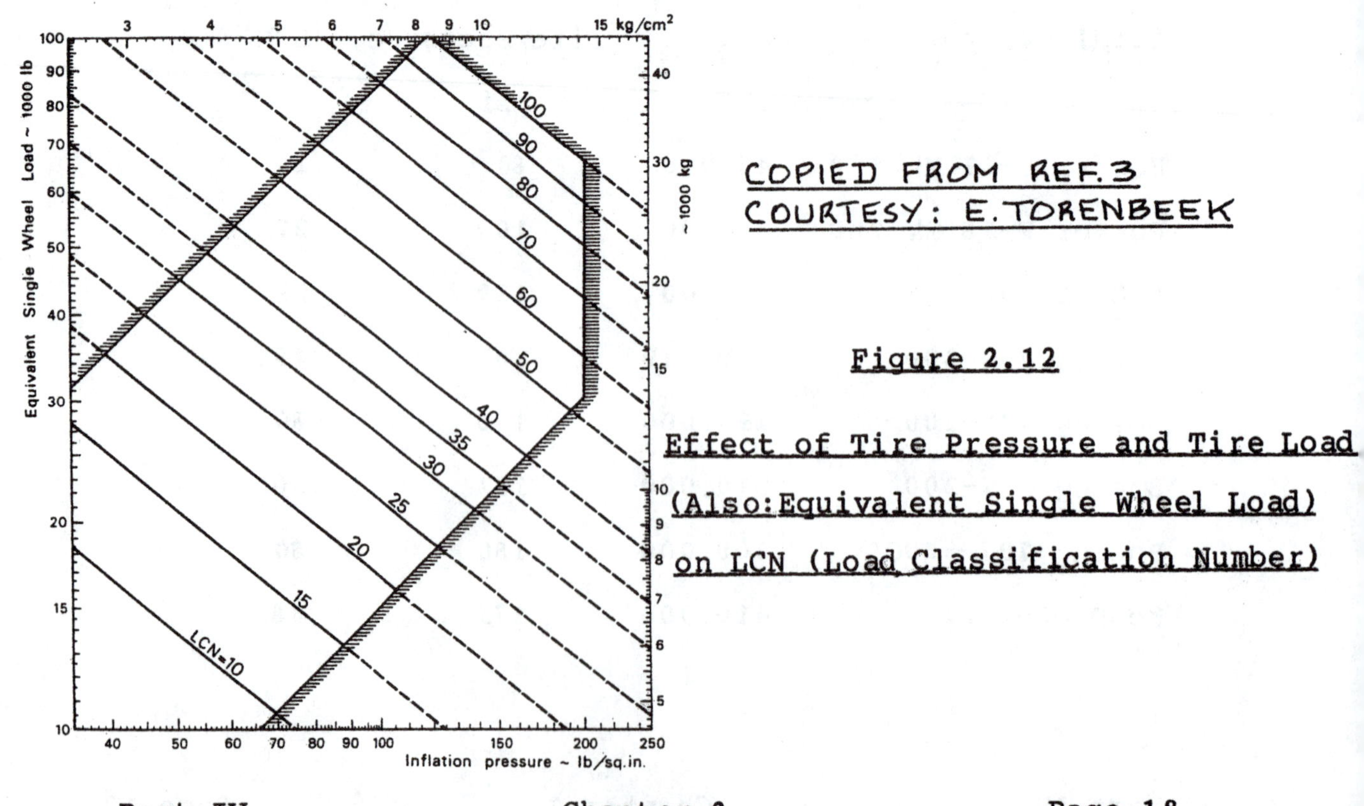

Figure 2.12

Effect of Tire Pressure and Tire Load (Also:Equivalent Single Wheel Load) on LCN (Load Classification Number)

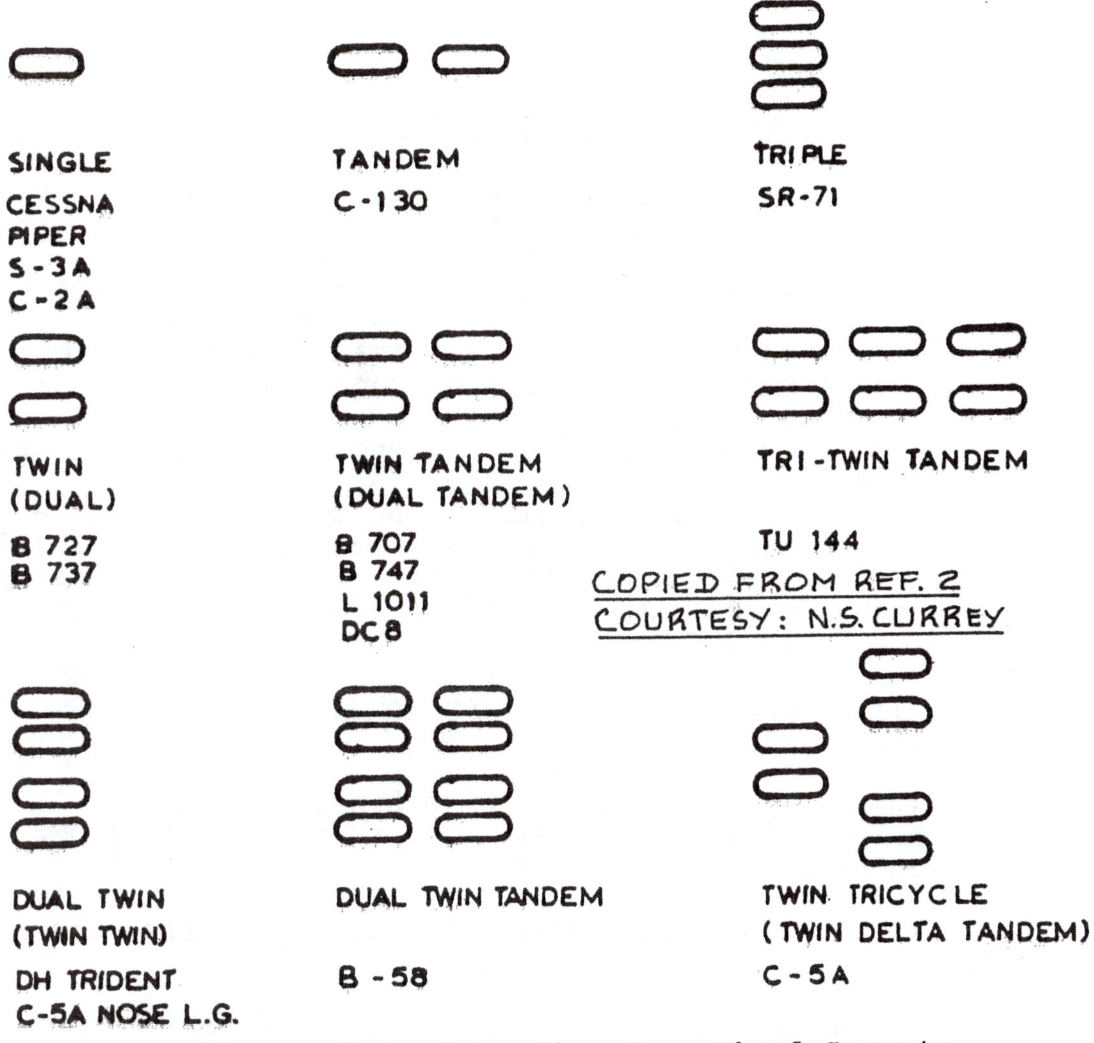

Figure 2.13 Examples of Landing Gear Wheel Layouts

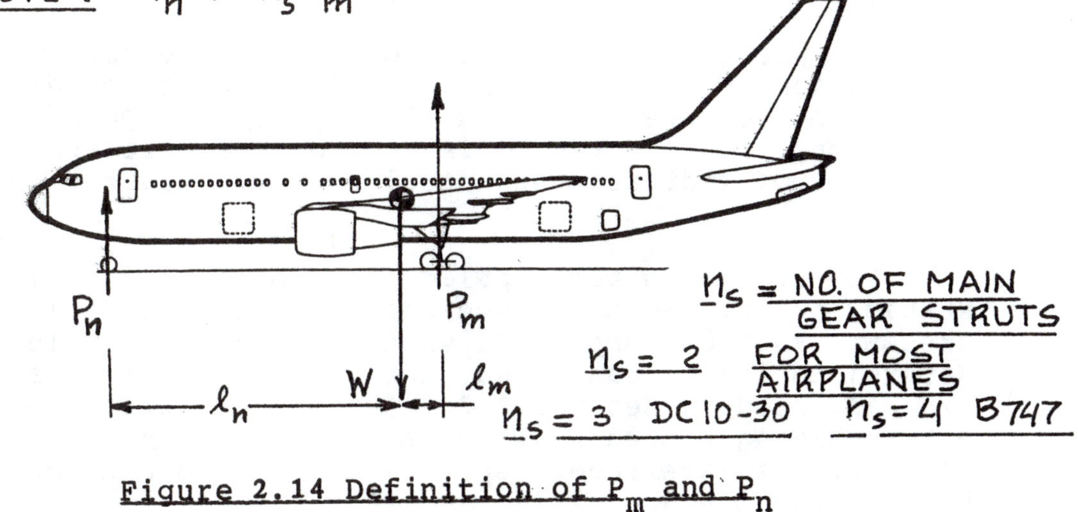

Figure 2.14 Definition of P_m and P_n

2.4 TIRES: TYPES, PERFORMANCE, SIZING AND DATA

The following information is presented in this section:

1. A discussion of tire types, tire construction and tire descriptions: see sub-section 2.4.1.
2. A discussion of tire performance: load, deflection and shock absorption capability: see sub-section 2.4.2.
3. A discussion of tire clearance requirements: see sub-section 2.4.3.
4. A method for determining the correct tire size for airplane applications: see sub-section 2.4.4.
5. Tabulated data on tire geometry, tire load carrying capability and tire applications: see sub-section 2.4.5.

2.4.1 Tire Types, Tire Construction, and Tire Descriptions

Figure 2.15 shows seven tire types which are frequently used in airplanes. A description of each type is given next to each tire.

Figure 2.16 shows how typical airplane tires are constructed.

Tire manufacturers rate tires in terms of:

1. Ply rating
2. Maximum allowable static loading
3. Recommended (unloaded) inflation pressure
4. Maximum allowable runway speed

The ply rating of a tire identifies the tire with its maximum recommended static load and corresponding inflation pressure when used in a specific type of operation. The ply rating is an index of tire strength and DOES NOT indicate the actual number of fabric core plies.

For any given tire application the maximum allowable static loading and the associated unloaded inflation pressure must be compatible with the allowable values determined from a runway surface viewpoint. The latter are discussed in Section 2.3. Sub-section 2.4.4 contains a tire sizing procedure which accounts for the tire/surface interface. Specific tire data are found in sub-section 2.4.5.

NEW DESIGN

New Design: This is a recent design. The outside tire dimensions are reflected in the type designation: D_oxW. All new tires will be designated with this system.

TYPE I

Type I: Smooth Contour. This type was designed for airplanes with non-retractable landing gears. Although this type is still available, its use in newly designed airplanes is discouraged because this tire type is considered obsolete.

Type II: High Pressure. This type, although still available is also considered obsolete. It was designed for airplanes with retractable gears. It has been replaced by Type VII which has considerably greater load carrying capacity.

TYPE III

Type III: Low Pressure. This type is comparable to Type I but has beads of smaller diameter. It also has larger volume and lower pressure. Any new sizes in this type will be listed under the 'New Design' designation.

Type VI: Low Profile (Inactive). This Type was designed for nosewheel applications only. It was designed to reduce wheel drop following complete deflation of the tire.

TYPE VII

Type VII: Extra High Pressure. This Type is almost universal on military and civil jets and turboprops. It has high load capacity and narrow width. Any new sizes in this type will be listed under the 'New Design' designation.

TYPE VIII

Type VIII: Low Profile High Pressure. This is a new design for very high take-off speeds. Any new sizes in this type will be listed under the 'New Design' designation.

Figure 2.15 Types of Airplane Tires

Aircraft Tire Construction

Aircraft tires, tubeless or tube type, provide a cushion of air that helps absorb the shocks and roughness of landings and takeoffs. They support the weight of the aircraft while on the ground and provide the necessary traction for braking and stopping aircraft on landing. Thus, aircraft tires must be carefully maintained to meet the rigorous demands of their basic job...to accept a variety of static and dynamic stresses in a wide range of operating conditions.

THE TREAD is a layer of rubber on the outer circumference of the tire, which serves as the wearing surface. With the sidewall, it helps protect the cord body from cuts, snags, bruises and moisture.

FABRIC TREAD REINFORCEMENT consists of plies added to reduce tire squirm and increase stability for high speed operation.

THE UNDERTREAD is a layer of special rubber which provides adhesion of the tread to the cord body, and enables the tire to be retreaded.

THE CORD BODY consists of layers (PLIES) of rubber-coated nylon cord. Since a layer of these cords (a ply) has all of its strength in only one direction, the cords of every succeeding ply run diagonally to each other to give balanced strength. The plies are folded around the wire beads, creating the PLY TURNUPS.

BEADS are layers of steel wire imbedded in rubber and then wrapped with fabric. They give a base around which the plies are anchored and provide a firm fit on the wheel.

THE SIDEWALL is a cover over the side of the cord body to protect the cords from injury and exposure.

CHAFER STRIPS protect the plies from damage when mounting or demounting the tire, and minimize the effects of chafing contact with the wheel.

THE LINER in tubeless tires is a layer of rubber specially compounded to resist diffusion of air. It is vulcanized to the inside of the tire, extending from bead to bead. In tube-type tires, a thin liner is provided to prevent tube chafing.

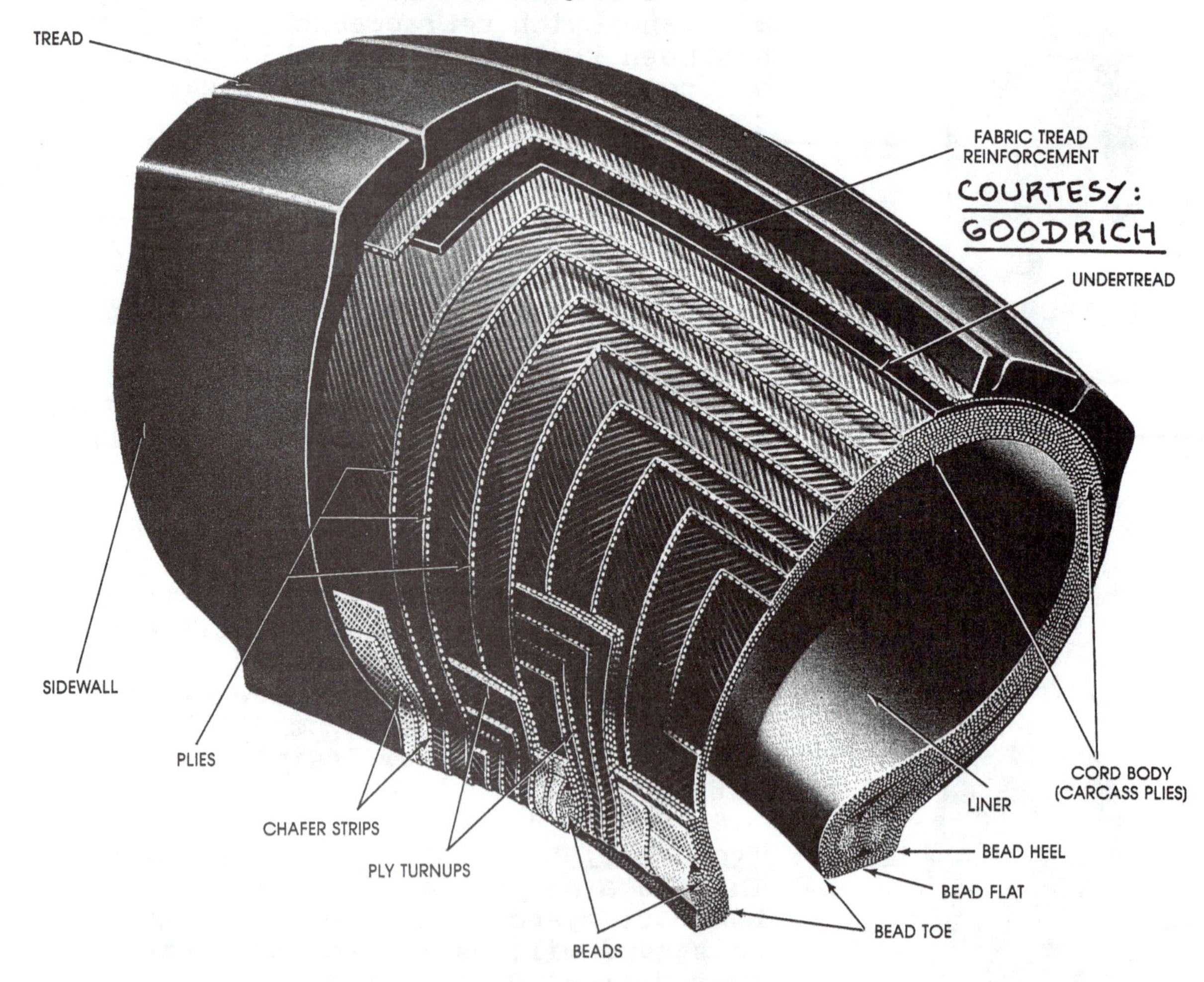

Figure 2.16 Conventional Airplane Tire Construction

Tires are described in terms of certain geometric parameters:

D_o or D_t, the tire outside diameter

W or b_t, the tire maximum width

D, the tire rim diameter

Figure 2.17 shows how these parameters are defined

With these geometric parameters, the following tire descriptions are being used:

Type I: D_o Type II: D_oxW Type III: W-D

Type VI: D_oxW-D Type VII or New Design: D_oxW

Type VIII or NS: D_oxW-D

There is a recent trend toward radial tires. See Figure 2.18 for how a radial tire is constructed. The advantages of radial tires over conventional tires are: up to 25 percent less weight and longer life. No systematic data on radial airplane tires were available when this text went to the printer.

2.4.2 Tire Performance: Load, Deflection and Shock Absorption Capability

Tires are subjected to rather severe static and dynamic loads during taxiing, during the take-off roll and during the landing roll. Figure 2.19 illustrates qualitatively what is typically demanded of an airplane tire.

Tires also participate significantly in the process of shock absorption following a touchdown. How much the tires participate, depends on the design of the shock absorbers. The amount of energy absorbed by the tires can be computed with the method discussed in Section 2.5.

Each tire is designed to operate at a so-called maximum allowable static load. The tire data tables in sub-section 2.4.4 indicate the maximum allowable static load for each tire. These loads must not be exceeded for the most critical weight/c.g. combination.

In selecting airplane tires, it is usually a good idea to keep future airplane growth capabilities in mind.

Once a given tire size is selected it is not always possible to accomodate larger tires within the geometric constraints of wheel wells. <u>It is recommended to allow for 25 percent growth in tire load</u> in selecting tires for a new airplane.

Nosewheel tires are designed for maximum allowable dynamic loads. These dynamic loads are obtained as follows:

$$\text{Dynamic load} = f_{dyn}(\text{static load}) \tag{2.3}$$

The factor f_{dyn} is defined as follows:

For tires of Types I and III: $f_{dyn} = 1.45$

For tires of Type II: $f_{dyn} = 1.25$

For tires of Types VI, VII, VIII and for New Design: $f_{dyn} = 1.50$

Allowable tire deflections are determined by the tire manufacturer. The allowable tire deflection, s_t may be computed from:

$$s_t = D_o/2 - (\text{loaded radius}) \tag{2.4}$$

Values for D_o and for the 'loaded radius' are given in tire tables such as presented in sub-section 2.4.5

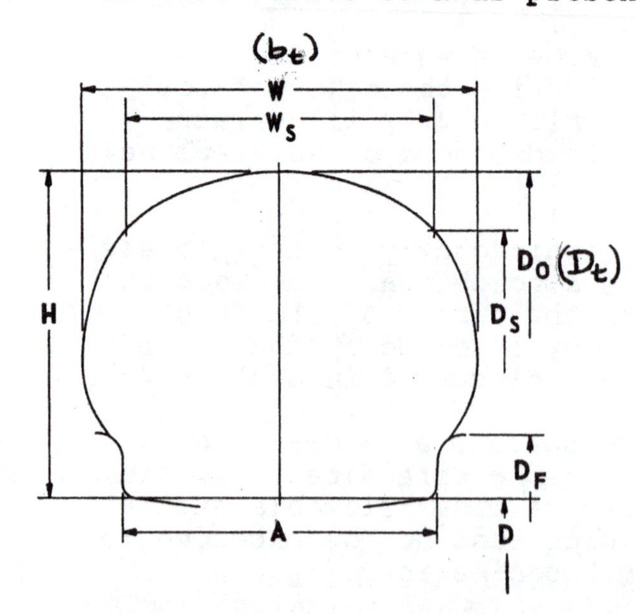

D = Bead Seat Diameter
D_F = Flange Diameter
D_O = Outside Diameter — Tire
D_S = Shoulder Diameter — Tire
W = Section Width — Tire
W_S = Shoulder Width — Tire
H = Section Height — Tire
W_S (max) = .85 W (max) for Type III Tires
W_S (max) = .88 W (max) for all other Types
D_S (max) = 1.64 H + D
$H = \frac{D_O - D}{2}$

Figure 2.17 Definition of Tire Geometry Parameters

Tread
Outer belt
Carcass
Chafer strips
Bead

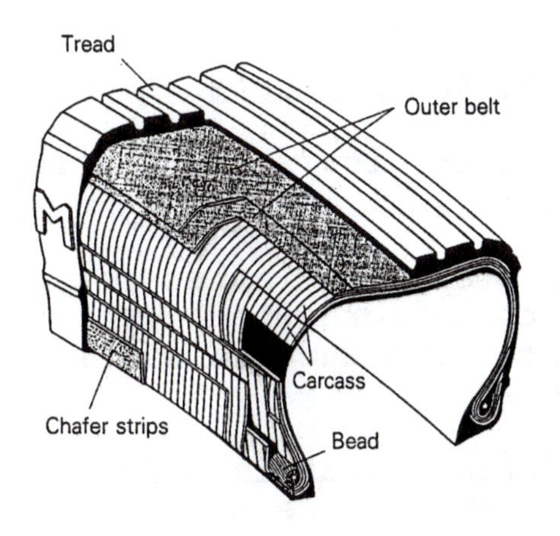

Figure 2.18 Radial Airplane Tire Construction

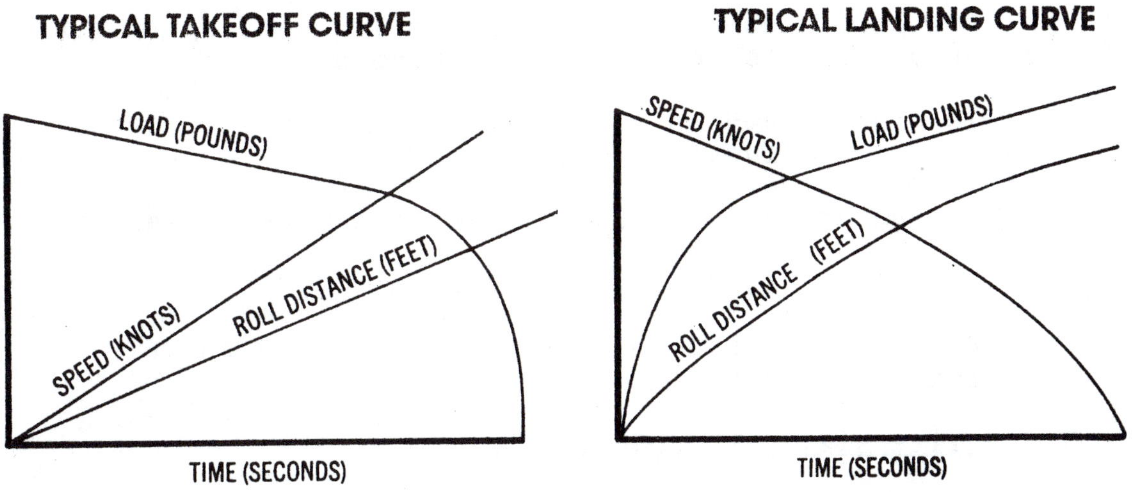

TYPICAL TAKEOFF CURVE
LOAD (POUNDS)
SPEED (KNOTS)
ROLL DISTANCE (FEET)
TIME (SECONDS)
TYPICAL LANDING CURVE
SPEED (KNOTS)
LOAD (POUNDS)
ROLL DISTANCE (FEET)
TIME (SECONDS)

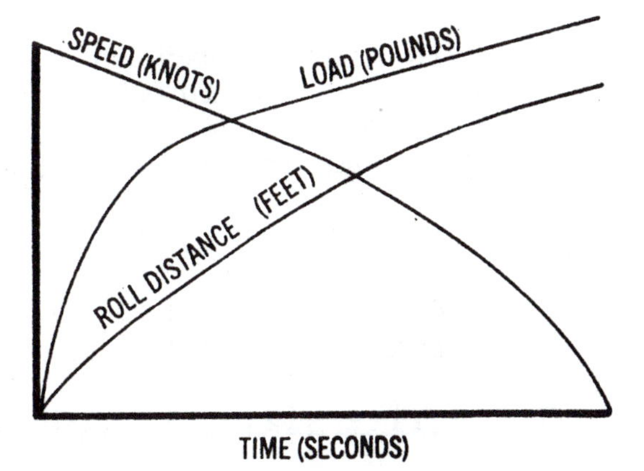

Figure 2.19 Typical Tire Performance Requirements During Take-off and Landing Rolls

2.4.3 Tire Clearance Requirements

The following tire clearance requirements must be observed:

1. Wheel well clearance (after retraction)
2. Tire-to-fork and/or tire-to-strut clearance
3. Tire-to-tire clearance in multiple wheel arrangements

Figure 2.20 defines what is meant by these clearance requirements. The physical reasons for these tire clearance requirements are:

a) Tires grow in size during their service life: 4 percent in width and 10 percent in diameter

b) Tires grow in size under the influence of centrifugal forces. This type of growth depends on the maximum tire operating speed on the ground

Figure 2.21 shows the required lateral (width) and radial clearances due to centrifugal forces as a function of the 'grown tire width'. The 'grown tire width' in Figure 2.21 may be taken as: 1.04W.

For preliminary design purposes it is acceptable to account for the following tire clearances:

In width: $0.04W$ + lateral clearance due to centrifugal forces + 1 inch

In radius: $0.1D_o$ + radial clearance due to centrifugal forces + 1 inch

For more precise methods to compute clearance requirements, see Ref.2.

2.4.4 Tire Sizing Procedure

The following procedure is recommended for airplane tire sizing:

1. Main gear tire(s): Determine the maximum static load on each main gear. This can be done with the help of Figure 2.14. Make sure that this load is computed for that weight/c.g. location which results in the maximum load per tire.

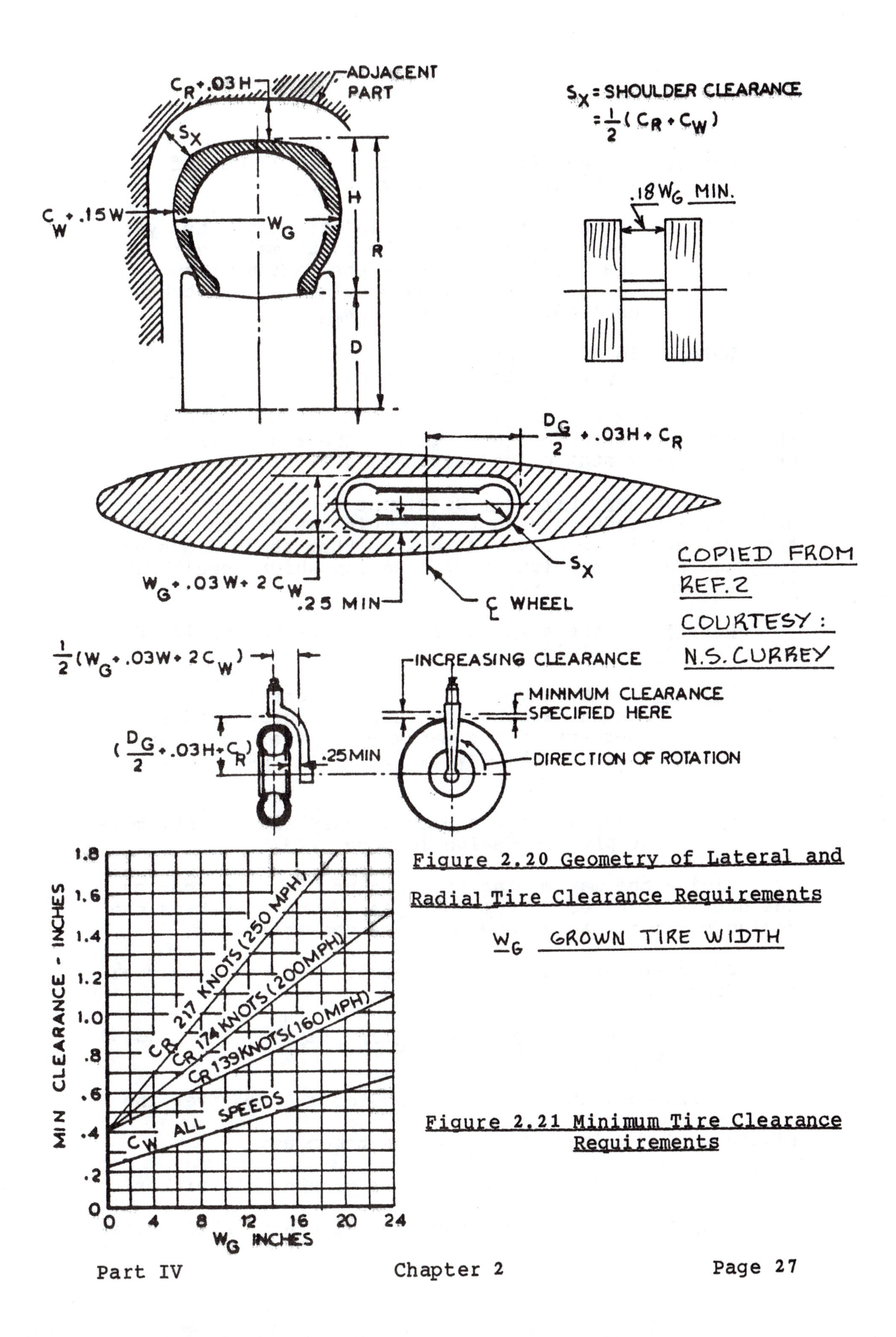

Figure 2.20 Geometry of Lateral and Radial Tire Clearance Requirements

W_G GROWN TIRE WIDTH

Figure 2.21 Minimum Tire Clearance Requirements

Note 1: If the airplane is to be FAR 25 certified, multiply this load by 1.07.

Note 2: If the airplane c.g. locations are not yet known, assume that the total main gear load is 90 percent of the maximum ramp weight. Maximum ramp weight may be taken as 1.005 to 1.01 times W_{TO}.

Note 3: To allow for growth in airplane weight, multiply the design load by 1.25.

Divide the maximum static load on each main gear by the number of tires per main gear. The result is the design maximum static load per main gear tire.

2. Nose gear tire(s): Determine the maximum static and dynamic loads on each tire. This can be done with the help of Figure 2.14. Make sure that these loads are computed for that weight/c.g. location which results in the maximum load per nose gear tire.

Note 1: If the airplane is to be FAR 25 certified, multiply these loads by 1.07.

Note 2: If the airplane c.g. locations are not yet known, assume that the nose gear load is 10 percent of the maximum ramp weight. Maximum ramp weight may be taken as 1.005 to 1.01 times W_{TO}.

Note 3: To allow for growth in airplane weight, multiply the design load by 1.25.

Divide the maximum static load on the nose gear by the number of tires on the nose gear. The result is the design maximum static load per nose gear tire.

3. Determine the maximum dynamic load per nose gear tire from:

$$P_{n_{dyn_t}} = W_{TO}\{l_m + (a_x/g)(h_{cg})\}/n_t(l_m + l_n) \qquad (2.5)$$

The definitions for l_m and for l_n are given in Figure 2.14. Values for a_x may be taken as:

a_x/g = 0.35 for dry concrete with simple brakes

a_x/g = 0.45 for dry concrete with anti-skid brakes

The design maximum static load may be obtained from the maximum dynamic load obtained with Eqn.(2.5) by dividing by the following factor:

For Type I and III tires: 1.45

For Type II tires: 1.25

For Type VI, VII and VIII tires: 1.50

For New Design tires: 1.50

This value for design maximum static load needs to be compared with the load computed under 2. The highest load value should be used.

4. Determine the maximum tire operating speed. This speed is the highest of the design take-off or landing speed of the airplane:

For landing: $V_{tire/max} = 1.2V_{s_L}$ (2.6)

For take-off: $V_{tire/max} = 1.1V_{s_{TO}}$ (2.7)

5. Using Tables 2.5 through 2.12 (Sub-section 2.4.5) list all tires which meet the load/speed conditions of the airplane.

6. Based on the ESWL calculation of Eqns (2.1) or (2.2) select the tires which meet the load/pressure criteria for surface compatibility.

7. For airplanes which operate from rough surfaces or from carriers the so-called 'crush-load' may be critical. This crush-load arises when the tire runs over a relatively sharp object such as sharp bump or a deck cable. A method for estimating the crush-load is given in Ref.2, p.6-14.

8. From the remaining candidate tires select a tire. This final selection can be made on the basis of the following criteria:

1. weight 2. minimum size

3. customer preference 4. wear and tear

Tire size and weight data are also listed in Tables 2.5-2.12 in sub-section 2.4.5.

Figures 2.22 present an example application of this tire sizing procedure.

2.4.5 Tire Data

For the most recent tire data the reader should consult tire catalogs published by tire manufacturers. In the USA airplane tires are most frequently selected from the following manufacturers:

Goodrich
Goodyear
Dunlop (United Kingdom)
Michelin (France)

In this text only a limited amount of tire data can be included. Tables 2.5 through 2.12 present tire data for Goodrich tires.

The reader should check his tire selection by comparison with the data in Tables 2.13 through 2.16. The latter tables contain information on which tire types were selected by a different manufacturers for their airplanes. For more information of this type Ref.6 should be consulted.

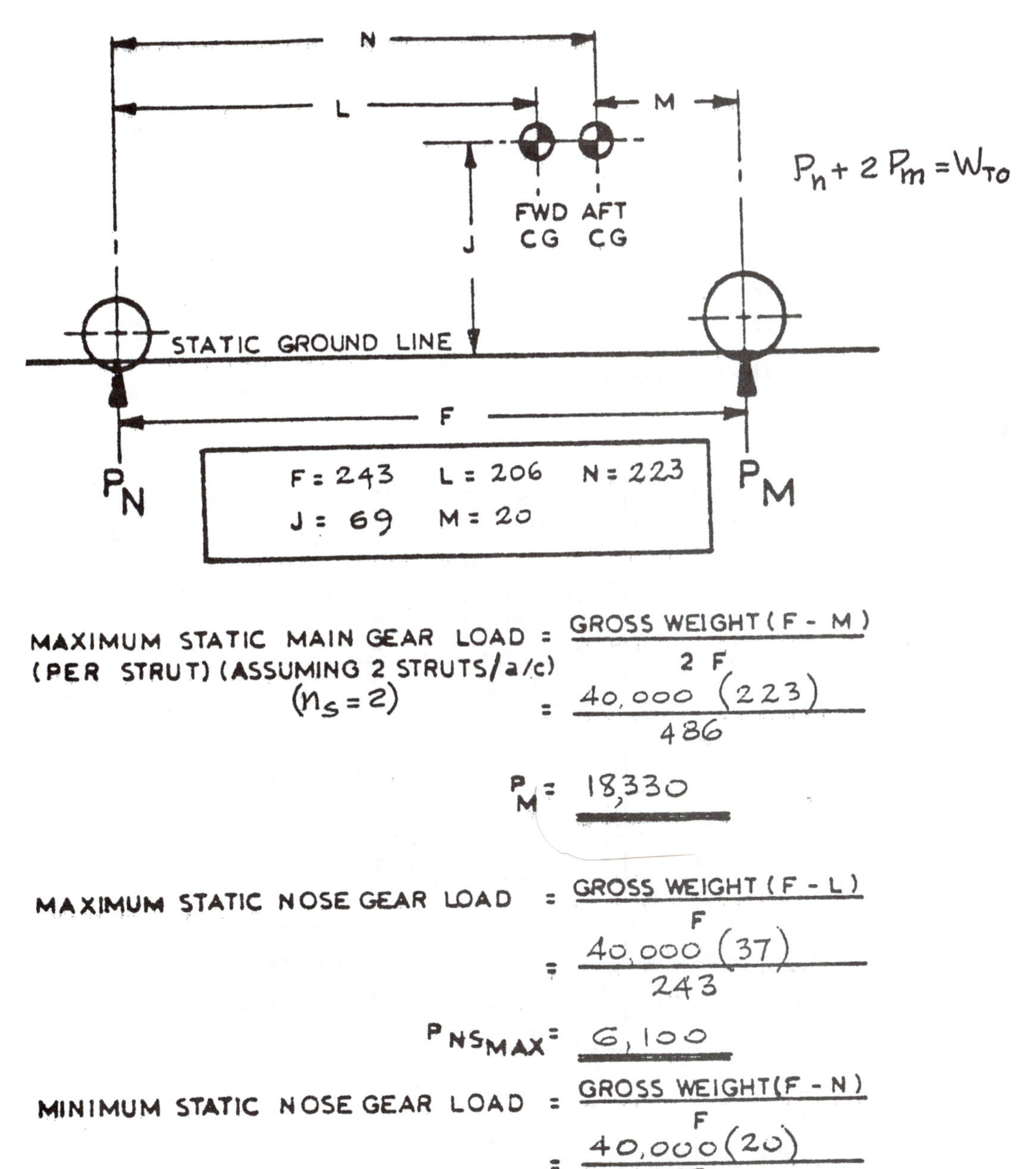

Figure 2.22 Tire Sizing Example

WITH (2) TIRES ON EACH MAIN GEAR, MAX STATIC TIRE LOAD = 9165 LB.
WITH (2) TIRES ON NOSE GEAR, MAX STATIC TIRE LOAD = 3050 LB.
MAX. DYN. NOSE TIRE LOAD = 4815 LB.

ALLOWING FOR 25% AIRPLANE GROWTH, USE THE FOLLOWING VALUES FOR BASIC SIZE SELECTION, & USE ABOVE VALUES FOR PLIES SELECTION:

MAIN GEAR TIRE STATIC LOAD = 11,450 LB.
NOSE GEAR TIRE STATIC LOAD = 3,810 LB.
NOSE GEAR TIRE DYNAMIC LOAD = 6,200 LB.

ASSUME MAX GROUND SPEED IS 180 MPH (156 KTS)

CANDIDATE TIRES MAIN GEAR

NO.	SIZE	PR	LOAD RATING STATIC LB.	LOAD RATING DYNAM. LB.	INFL PRESS PSI	SPEED RATING MPH (KTS)	TIRE O.D. INS	BEAD LEDGE DIA. INS	WIDTH INS	BUMP CAPAB INS	QUALIFIC N. STATUS
1	24 x 5.5	16	11,500	N.A.	355	200	24.15	14.0	5.70	1.6	MIL
2	25 x 6.0	16	12,000	N.A.	330	160	SPEED RATING INADEQUATE.				
3	22 x 6.6-10	20	12,000	N.A.	290	225	22.20	10.0	6.80	2.2	MIL
4	26 x 6.6	16	12,000	N.A.	270	200	25.75	14.0	6.65	1.9	MIL
5	30 x 6.6	14	12,950	N.A.	320	225	30.12	20.0	6.50	1.1	MIL
6	25 x 6.75	18	13,000	N.A.	300	275	25.50	14.00	6.85	1.6	MIL
7	29 x 7.7	16	13,800	N.A.	230	200	28.40	15.00	7.85	2.1	COMML.
8	26 x 8.0-14	16	12,700	N.A.	235	275	26.00	14.00	8.00	1.6	MIL.
9	32 x 11.50-15	12	11,200	N.A.	120	200	32.00	15.00	11.50	3.2	COMML.

SELECTION FACTORS
INFLATION PRESSURE ELIMINATES: 1, 2, 3, 4, 5, 6
WHEEL DIAMETER ELIMINATES: 5, 9 (D TOO LARGE)
DIMENSIONS ELIMINATES:
OTHER: (MINOR FACTOR: NOT QUAL. FOR MIL. USE) 7

SELECTED TIRES MAIN No. 8 26 x 8.0-14 16PR, OPER. AT 170 PSI
NOSE NOT DONE IN THIS EXAMPLE

COPIED FROM REF. 2
COURTESY: N.S. CURREY

Figure 2.22 (Cont'd) Tire Sizing Example

Table 2.5 Tire Data, Courtesy: B.F.Goodrich

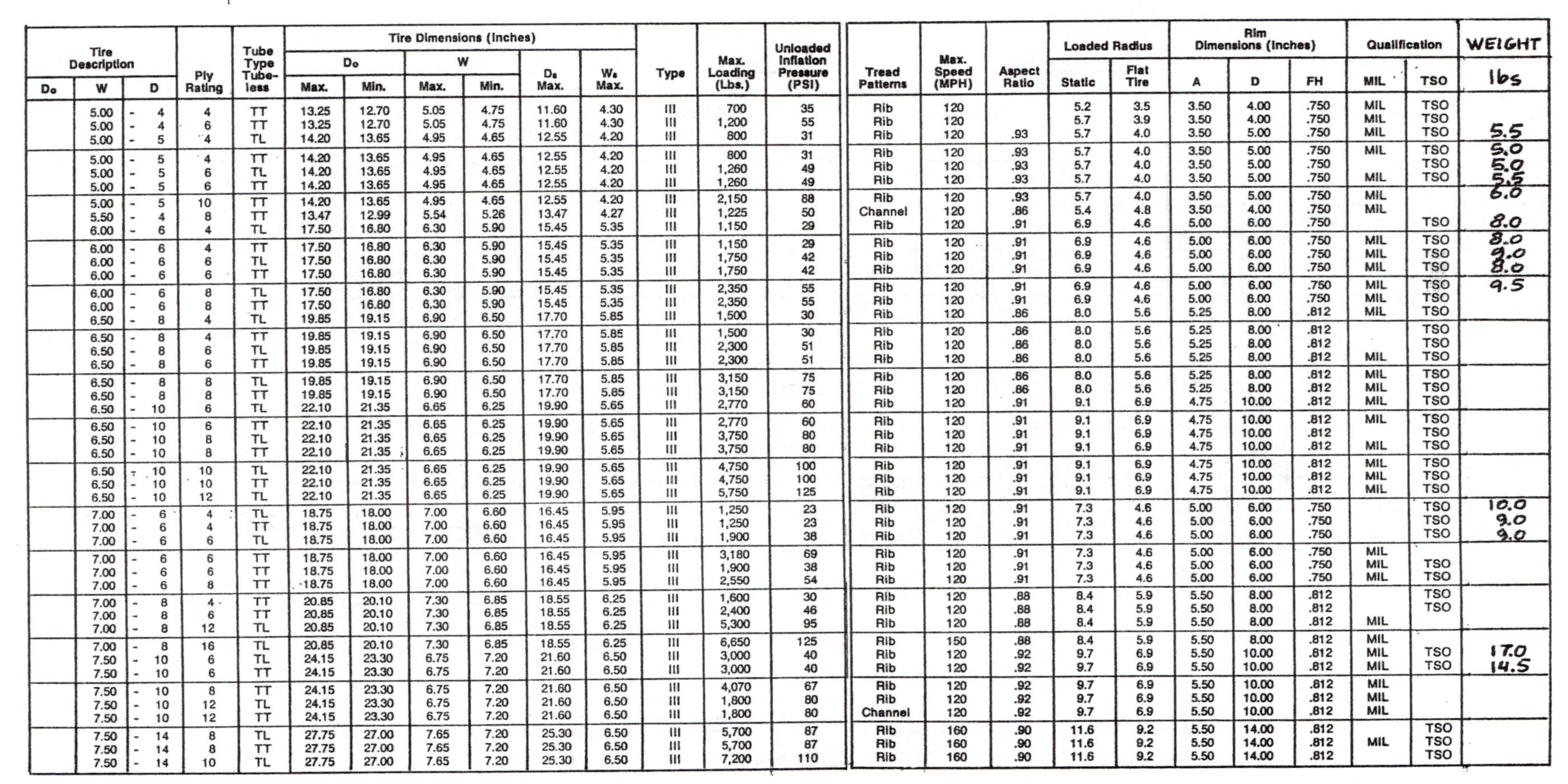

Tire Description			Ply Rating	Tube Type Tube-less	Tire Dimensions (Inches)						Type	Max. Loading (Lbs.)	Unloaded Inflation Pressure (PSI)	Tread Patterns	Max. Speed (MPH)	Aspect Ratio	Loaded Radius		Rim Dimensions (Inches)			Qualification		WEIGHT
					D_o		W		D_s	W_s														
D_o	W	D			Max.	Min.	Max.	Min.	Max.	Max.							Static	Flat Tire	A	D	FH	MIL	TSO	lbs
	5.00	- 4	4	TT	13.25	12.70	5.05	4.75	11.60	4.30	III	700	35	Rib	120		5.2	3.5	3.50	4.00	.750	MIL	TSO	
	5.00	- 4	6	TT	13.25	12.70	5.05	4.75	11.60	4.30	III	1,200	55	Rib	120		5.7	3.9	3.50	4.00	.750	MIL	TSO	
	5.00	- 5	4	TL	14.20	13.65	4.95	4.65	12.55	4.20	III	800	31	Rib	120	.93	5.7	4.0	3.50	5.00	.750	MIL	TSO	5.5
	5.00	- 5	4	TT	14.20	13.65	4.95	4.65	12.55	4.20	III	800	31	Rib	120	.93	5.7	4.0	3.50	5.00	.750	MIL	TSO	5.0
	5.00	- 5	6	TL	14.20	13.65	4.95	4.65	12.55	4.20	III	1,260	49	Rib	120	.93	5.7	4.0	3.50	5.00	.750		TSO	5.0
	5.00	- 5	6	TT	14.20	13.65	4.95	4.65	12.55	4.20	III	1,260	49	Rib	120	.93	5.7	4.0	3.50	5.00	.750	MIL	TSO	5.5
	5.00	- 5	10	TT	14.20	13.65	4.95	4.65	12.55	4.20	III	2,150	88	Rib	120	.93	5.7	4.0	3.50	5.00	.750	MIL		6.0
	5.50	- 4	8	TT	13.47	12.99	5.54	5.26	13.47	4.27	III	1,225	50	Channel	120	.86	5.4	4.8	3.50	4.00	.750	MIL		
	6.00	- 6	4	TL	17.50	16.80	6.30	5.90	15.45	5.35	III	1,150	29	Rib	120	.91	6.9	4.6	5.00	6.00	.750		TSO	8.0
	6.00	- 6	4	TT	17.50	16.80	6.30	5.90	15.45	5.35	III	1,150	29	Rib	120	.91	6.9	4.6	5.00	6.00	.750	MIL	TSO	8.0
	6.00	- 6	6	TL	17.50	16.80	6.30	5.90	15.45	5.35	III	1,750	42	Rib	120	.91	6.9	4.6	5.00	6.00	.750	MIL	TSO	9.0
	6.00	- 6	6	TT	17.50	16.80	6.30	5.90	15.45	5.35	III	1,750	42	Rib	120	.91	6.9	4.6	5.00	6.00	.750	MIL	TSO	8.0
	6.00	- 6	8	TL	17.50	16.80	6.30	5.90	15.45	5.35	III	2,350	55	Rib	120	.91	6.9	4.6	5.00	6.00	.750	MIL	TSO	9.5
	6.00	- 6	8	TT	17.50	16.80	6.30	5.90	15.45	5.35	III	2,350	55	Rib	120	.91	6.9	4.6	5.00	6.00	.750	MIL	TSO	
	6.50	- 8	4	TL	19.85	19.15	6.90	6.50	17.70	5.85	III	1,500	30	Rib	120	.86	8.0	5.6	5.25	8.00	.812	MIL	TSO	
	6.50	- 8	4	TT	19.85	19.15	6.90	6.50	17.70	5.85	III	1,500	30	Rib	120	.86	8.0	5.6	5.25	8.00	.812		TSO	
	6.50	- 8	6	TL	19.85	19.15	6.90	6.50	17.70	5.85	III	2,300	51	Rib	120	.86	8.0	5.6	5.25	8.00	.812		TSO	
	6.50	- 8	6	TT	19.85	19.15	6.90	6.50	17.70	5.85	III	2,300	51	Rib	120	.86	8.0	5.6	5.25	8.00	.812	MIL	TSO	
	6.50	- 8	8	TL	19.85	19.15	6.90	6.50	17.70	5.85	III	3,150	75	Rib	120	.86	8.0	5.6	5.25	8.00	.812	MIL	TSO	
	6.50	- 8	8	TT	19.85	19.15	6.90	6.50	17.70	5.85	III	3,150	75	Rib	120	.86	8.0	5.6	5.25	8.00	.812	MIL	TSO	
	6.50	- 10	6	TL	22.10	21.35	6.65	6.25	19.90	5.65	III	2,770	60	Rib	120	.91	9.1	6.9	4.75	10.00	.812	MIL	TSO	
	6.50	- 10	6	TT	22.10	21.35	6.65	6.25	19.90	5.65	III	2,770	60	Rib	120	.91	9.1	6.9	4.75	10.00	.812	MIL	TSO	
	6.50	- 10	8	TL	22.10	21.35	6.65	6.25	19.90	5.65	III	3,750	80	Rib	120	.91	9.1	6.9	4.75	10.00	.812		TSO	
	6.50	- 10	8	TT	22.10	21.35	6.65	6.25	19.90	5.65	III	3,750	80	Rib	120	.91	9.1	6.9	4.75	10.00	.812	MIL	TSO	
	6.50	- 10	10	TL	22.10	21.35	6.65	6.25	19.90	5.65	III	4,750	100	Rib	120	.91	9.1	6.9	4.75	10.00	.812	MIL	TSO	
	6.50	- 10	10	TT	22.10	21.35	6.65	6.25	19.90	5.65	III	4,750	100	Rib	120	.91	9.1	6.9	4.75	10.00	.812	MIL	TSO	
	6.50	- 10	12	TL	22.10	21.35	6.65	6.25	19.90	5.65	III	5,750	125	Rib	120	.91	9.1	6.9	4.75	10.00	.812	MIL	TSO	
	7.00	- 6	4	TL	18.75	18.00	7.00	6.60	16.45	5.95	III	1,250	23	Rib	120	.91	7.3	4.6	5.00	6.00	.750		TSO	10.0
	7.00	- 6	4	TT	18.75	18.00	7.00	6.60	16.45	5.95	III	1,250	23	Rib	120	.91	7.3	4.6	5.00	6.00	.750		TSO	9.0
	7.00	- 6	6	TL	18.75	18.00	7.00	6.60	16.45	5.95	III	1,900	38	Rib	120	.91	7.3	4.6	5.00	6.00	.750		TSO	9.0
	7.00	- 6	6	TT	18.75	18.00	7.00	6.60	16.45	5.95	III	3,180	69	Rib	120	.91	7.3	4.6	5.00	6.00	.750	MIL		
	7.00	- 6	6	TT	18.75	18.00	7.00	6.60	16.45	5.95	III	1,900	38	Rib	120	.91	7.3	4.6	5.00	6.00	.750	MIL	TSO	
	7.00	- 6	8	TT	18.75	18.00	7.00	6.60	16.45	5.95	III	2,550	54	Rib	120	.91	7.3	4.6	5.00	6.00	.750	MIL	TSO	
	7.00	- 8	4	TT	20.85	20.10	7.30	6.85	18.55	6.25	III	1,600	30	Rib	120	.88	8.4	5.9	5.50	8.00	.812		TSO	
	7.00	- 8	6	TT	20.85	20.10	7.30	6.85	18.55	6.25	III	2,400	46	Rib	120	.88	8.4	5.9	5.50	8.00	.812		TSO	
	7.00	- 8	12	TL	20.85	20.10	7.30	6.85	18.55	6.25	III	5,300	95	Rib	120	.88	8.4	5.9	5.50	8.00	.812	MIL		
	7.00	- 8	16	TL	20.85	20.10	7.30	6.85	18.55	6.25	III	6,650	125	Rib	150	.88	8.4	5.9	5.50	8.00	.812	MIL		
	7.50	- 10	6	TL	24.15	23.30	6.75	7.20	21.60	6.50	III	3,000	40	Rib	120	.92	9.7	6.9	5.50	10.00	.812	MIL	TSO	17.0
	7.50	- 10	6	TT	24.15	23.30	6.75	7.20	21.60	6.50	III	3,000	40	Rib	120	.92	9.7	6.9	5.50	10.00	.812	MIL	TSO	14.5
	7.50	- 10	8	TT	24.15	23.30	6.75	7.20	21.60	6.50	III	4,070	67	Rib	120	.92	9.7	6.9	5.50	10.00	.812	MIL		
	7.50	- 10	12	TL	24.15	23.30	6.75	7.20	21.60	6.50	III	1,800	80	Rib	120	.92	9.7	6.9	5.50	10.00	.812	MIL		
	7.50	- 10	12	TT	24.15	23.30	6.75	7.20	21.60	6.50	III	1,800	80	Channel	120	.92	9.7	6.9	5.50	10.00	.812	MIL		
	7.50	- 14	8	TL	27.75	27.00	7.65	7.20	25.30	6.50	III	5,700	87	Rib	160	.90	11.6	9.2	5.50	14.00	.812		TSO	
	7.50	- 14	8	TT	27.75	27.00	7.65	7.20	25.30	6.50	III	5,700	87	Rib	160	.90	11.6	9.2	5.50	14.00	.812	MIL	TSO	
	7.50	- 14	10	TL	27.75	27.00	7.65	7.20	25.30	6.50	III	7,200	110	Rib	160	.90	11.6	9.2	5.50	14.00	.812		TSO	

Table 2.6 Tire Data, Courtesy: B.F.Goodrich

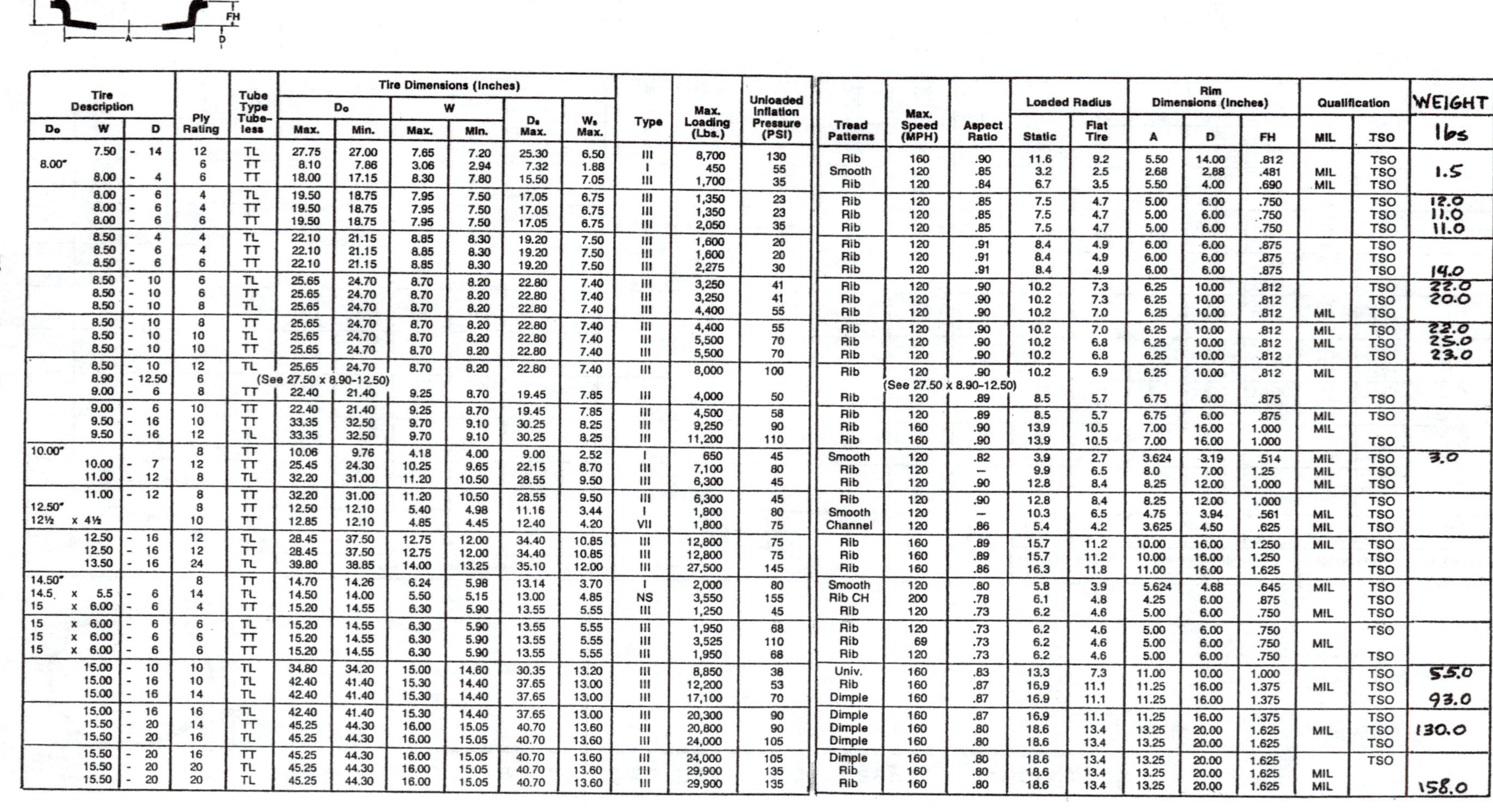

Tire Description			Ply Rating	Tube Type Tube-less	Tire Dimensions (Inches)						Type	Max. Loading (Lbs.)	Unloaded Inflation Pressure (PSI)	Tread Patterns	Max. Speed (MPH)	Aspect Ratio	Loaded Radius		Rim Dimensions (Inches)			Qualification		WEIGHT
					Do		W																	
Do	W	D			Max.	Min.	Max.	Min.	Ds Max.	Ws Max.							Static	Flat Tire	A	D	FH	MIL	TSO	lbs
	7.50	14	12	TL	27.75	27.00	7.65	7.20	25.30	6.50	III	8,700	130	Rib	160	.90	11.6	9.2	5.50	14.00	.812		TSO	
8.00"			6	TT	8.10	7.86	3.06	2.94	7.32	1.88	I	450	55	Smooth	120	.85	3.2	2.5	2.68	2.88	.481	MIL	TSO	1.5
	8.00	4	6	TT	18.00	17.15	8.30	7.80	15.50	7.05	III	1,700	35	Rib	120	.84	6.7	3.5	5.50	4.00	.690	MIL	TSO	
	8.00	6	4	TL	19.50	18.75	7.95	7.50	17.05	6.75	III	1,350	23	Rib	120	.85	7.5	4.7	5.00	6.00	.750		TSO	12.0
	8.00	6	4	TT	19.50	18.75	7.95	7.50	17.05	6.75	III	1,350	23	Rib	120	.85	7.5	4.7	5.00	6.00	.750		TSO	11.0
	8.00	6	6	TT	19.50	18.75	7.95	7.50	17.05	6.75	III	2,050	35	Rib	120	.85	7.5	4.7	5.00	6.00	.750		TSO	11.0
	8.50	4	4	TL	22.10	21.15	8.85	8.30	19.20	7.50	III	1,600	20	Rib	120	.91	8.4	4.9	6.00	6.00	.875		TSO	
	8.50	6	4	TT	22.10	21.15	8.85	8.30	19.20	7.50	III	1,600	20	Rib	120	.91	8.4	4.9	6.00	6.00	.875		TSO	
	8.50	6	6	TT	22.10	21.15	8.85	8.30	19.20	7.50	III	2,275	30	Rib	120	.91	8.4	4.9	6.00	6.00	.875		TSO	14.0
	8.50	10	6	TL	25.65	24.70	8.70	8.20	22.80	7.40	III	3,250	41	Rib	120	.90	10.2	7.3	6.25	10.00	.812		TSO	22.0
	8.50	10	6	TT	25.65	24.70	8.70	8.20	22.80	7.40	III	3,250	41	Rib	120	.90	10.2	7.3	6.25	10.00	.812		TSO	20.0
	8.50	10	8	TL	25.65	24.70	8.70	8.20	22.80	7.40	III	4,400	55	Rib	120	.90	10.2	7.0	6.25	10.00	.812	MIL	TSO	
	8.50	10	8	TT	25.65	24.70	8.70	8.20	22.80	7.40	III	4,400	55	Rib	120	.90	10.2	7.0	6.25	10.00	.812	MIL	TSO	22.0
	8.50	10	10	TL	25.65	24.70	8.70	8.20	22.80	7.40	III	5,500	70	Rib	120	.90	10.2	6.8	6.25	10.00	.812	MIL	TSO	25.0
	8.50	10	10	TT	25.65	24.70	8.70	8.20	22.80	7.40	III	5,500	70	Rib	120	.90	10.2	6.8	6.25	10.00	.812		TSO	23.0
	8.50	10	12	TL	25.65	24.70	8.70	8.20	22.80	7.40	III	8,000	100	Rib	120	.90	10.2	6.9	6.25	10.00	.812	MIL		
	8.90	12.50	6		(See 27.50 x 8.90-12.50)										(See 27.50 x 8.90-12.50)									
	9.00	6	8	TT	22.40	21.40	9.25	8.70	19.45	7.85	III	4,000	50	Rib	120	.89	8.5	5.7	6.75	6.00	.875		TSO	
	9.00	6	10	TT	22.40	21.40	9.25	8.70	19.45	7.85	III	4,500	58	Rib	120	.89	8.5	5.7	6.75	6.00	.875	MIL	TSO	
	9.50	16	10	TT	33.35	32.50	9.70	9.10	30.25	8.25	III	9,250	90	Rib	160	.90	13.9	10.5	7.00	16.00	1.000	MIL		
	9.50	16	12	TL	33.35	32.50	9.70	9.10	30.25	8.25	III	11,200	110	Rib	160	.90	13.9	10.5	7.00	16.00	1.000		TSO	
10.00"			8	TT	10.06	9.76	4.18	4.00	9.00	2.52	I	650	45	Smooth	120	.82	3.9	2.7	3.624	3.19	.514	MIL	TSO	3.0
	10.00	7	12	TT	25.45	24.30	10.25	9.65	22.15	8.70	III	7,100	80	Rib	120	—	9.9	6.5	8.0	7.00	1.25	MIL	TSO	
	11.00	12	8	TL	32.20	31.00	11.20	10.50	28.55	9.50	III	6,300	45	Rib	120	.90	12.8	8.4	8.25	12.00	1.000	MIL	TSO	
	11.00	12	8	TT	32.20	31.00	11.20	10.50	28.55	9.50	III	6,300	45	Rib	120	.90	12.8	8.4	8.25	12.00	1.000		TSO	
12.50"			8	TT	12.50	12.10	5.40	4.98	11.16	3.44	I	1,800	80	Smooth	120	—	10.3	6.5	4.75	3.94	.561	MIL	TSO	
12½	x 4½		10	TT	12.85	12.10	4.85	4.45	12.40	4.20	VII	1,800	75	Channel	120	.86	5.4	4.2	3.625	4.50	.625	MIL	TSO	
	12.50	16	12	TL	28.45	37.50	12.75	12.00	34.40	10.85	III	12,800	75	Rib	160	.89	15.7	11.2	10.00	16.00	1.250	MIL	TSO	
	12.50	16	12	TT	28.45	37.50	12.75	12.00	34.40	10.85	III	12,800	75	Rib	160	.89	15.7	11.2	10.00	16.00	1.250		TSO	
	13.50	16	24	TL	39.80	38.85	14.00	13.25	35.10	12.00	III	27,500	145	Rib	160	.86	16.3	11.8	11.00	16.00	1.625		TSO	
14.50"			8	TT	14.70	14.26	6.24	5.98	13.14	3.70	I	2,000	80	Smooth	120	.80	5.8	3.9	5.624	4.68	.645	MIL	TSO	
14.5	x 5.5	6	14	TL	14.50	14.00	5.50	5.15	13.00	4.85	NS	3,550	155	Rib CH	200	.78	6.1	4.8	4.25	6.00	.875		TSO	
15	x 6.00	6	4	TT	15.20	14.55	6.30	5.90	13.55	5.55	III	1,250	45	Rib	120	.73	6.2	4.6	5.00	6.00	.750	MIL	TSO	
15	x 6.00	6	6	TL	15.20	14.55	6.30	5.90	13.55	5.55	III	1,950	68	Rib	120	.73	6.2	4.6	5.00	6.00	.750		TSO	
15	x 6.00	6	6	TT	15.20	14.55	6.30	5.90	13.55	5.55	III	3,525	110	Rib	69	.73	6.2	4.6	5.00	6.00	.750	MIL		
15	x 6.00	6	6	TT	15.20	14.55	6.30	5.90	13.55	5.55	III	1,950	68	Rib	120	.73	6.2	4.6	5.00	6.00	.750		TSO	
	15.00	10	10	TL	34.80	34.20	15.00	14.60	30.35	13.20	III	8,850	38	Univ.	160	.83	13.3	7.3	11.00	10.00	1.000		TSO	55.0
	15.00	16	10	TL	42.40	41.40	15.30	14.40	37.65	13.00	III	12,200	53	Rib	160	.87	16.9	11.1	11.25	16.00	1.375	MIL	TSO	
	15.00	16	14	TL	42.40	41.40	15.30	14.40	37.65	13.00	III	17,100	70	Dimple	160	.87	16.9	11.1	11.25	16.00	1.375		TSO	93.0
	15.00	16	16	TL	42.40	41.40	15.30	14.40	37.65	13.00	III	20,300	90	Dimple	160	.87	16.9	11.1	11.25	16.00	1.375		TSO	
	15.50	20	14	TT	45.25	44.30	16.00	15.05	40.70	13.60	III	20,800	90	Dimple	160	.80	18.6	13.4	13.25	20.00	1.625	MIL	TSO	130.0
	15.50	20	16	TL	45.25	44.30	16.00	15.05	40.70	13.60	III	24,000	105	Dimple	160	.80	18.6	13.4	13.25	20.00	1.625		TSO	
	15.50	20	16	TT	45.25	44.30	16.00	15.05	40.70	13.60	III	24,000	105	Dimple	160	.80	18.6	13.4	13.25	20.00	1.625		TSO	
	15.50	20	20	TL	45.25	44.30	16.00	15.05	40.70	13.60	III	29,900	135	Rib	160	.80	18.6	13.4	13.25	20.00	1.625	MIL		
	15.50	20	20	TL	45.25	44.30	16.00	15.05	40.70	13.60	III	29,900	135	Rib	160	.80	18.6	13.4	13.25	20.00	1.625	MIL		158.0

Table 2.7 Tire Data, Courtesy: B.F.Goodrich

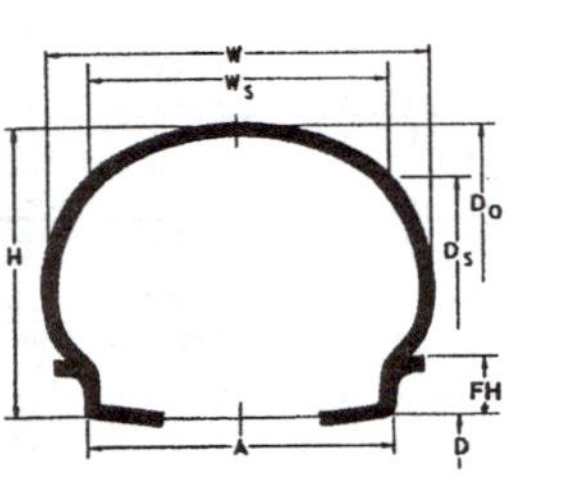

Tire Description Do	W	D	Ply Rating	Tube Type Tube-less	Do Max.	Do Min.	W Max.	W Min.	Ds Max.	Ws Max.	Type	Max. Loading (Lbs.)	Unloaded Inflation Pressure (PSI)	Tread Patterns	Max. Speed (MPH)	Aspect Ratio	Loaded Radius Static	Loaded Radius Flat Tire	Rim A	Rim D	Rim FH	MIL	TSO	WEIGHT lbs
	15.50	- 20	20	TT	45.25	44.30	16.00	15.05	40.70	13.60	III	29,900	135	Rib	160	.80	18.6	13.4	13.25	20.00	1.625	MIL		158.0
	15.50	- 20	20	TT	45.25	44.30	16.00	15.05	40.70	13.60	III	29,900	135	Dimple	160	.80	18.6	13.4	13.25	20.00	1.625		TSO	
16	x 4.4		4	TL	16.00	15.50	4.45	4.15	14.55	3.90	VII	1,100	55	Rib	200	.90	6.9	5.7	3.50	8.00	.812	MIL	TSO	7.5
16	x 4.4		6	TL	16.00	15.50	4.45	4.15	14.55	3.90	VII	1,700	85		160	.90	6.9	5.7	3.50	8.00	.812		TSO	5.5
16	x 4.4		6	TT	16.00	15.50	4.45	4.15	14.55	3.90	VII	1,700	85	Rib	160	.90	6.9	5.7	3.50	8.00	.812	MIL	TSO	
16	x 4.4		8	TL	16.00	15.50	4.45	4.15	14.55	3.90	VII	2,300	120	Rib	160	.90	6.9	5.7	3.50	8.00	.812	MIL		8.0
17.00"			10	TT	17.08	16.56	7.20	6.92	15.26	4.52	I	2,300	55	Smooth	120	.81	6.1	4.4	6.562	5.44	.733	MIL	TSO	
	17.00	- 16	10	TT	45.05	43.70	17.40	16.35	39.80	14.80	III	13,500	48	Rib	160	.84	17.7	11.2	13.25	16.00	1.875	MIL	TSO	
	17.00	- 16	12	TT	45.05	43.70	17.40	16.35	39.80	14.80	III	16,000	60	Rib	160	.84	17.7	11.2	13.25	16.00	1.875	MIL	TSO	125.0
	17.00	- 20	22	TL	48.75	47.70	17.25	16.40	43.60	14.65	III	34,500	130	Rib	160	.84	19.7	13.6	13.25	20.00	1.750	MIL	TSO	
	17.00	- 20	22	TT	48.75	47.70	17.25	16.40	43.60	14.65	III	34,500	130	Rib	160	.84	19.7	13.6	13.25	20.00	1.750	MIL	TSO	178.0
17.5	x 5.75	- 8	12	TL	17.5	16.95	5.75	5.40	15.80	5.10	NS	5,000	180	Rib	210	.83	7.4	4.3	4.25	8.00	.875		TSO	
17.5	x 6.25	- 6	10	TL	17.5	16.85	6.25	5.90	15.45	5.50	NS	3,750	90	Rib	120	.92	6.9	6.3	5.00	6.00	.750		TSO	
18	x 4.25	- 10	6	TL	18.25	17.75	4.70	4.45	16.75	4.15	NS	2,300	100	Rib CH	210	.87	8.0	6.6	3.625	10.00	.600		TSO	
18	x 4.4		6	TL	17.90	17.40	4.45	4.15	14.55	3.79	VII	2,100	100	Rib	200	.89	7.9	6.8	3.50	10.00	.812	MIL	TSO	8.0
18	x 4.4		6	TL	17.90	17.40	4.45	4.15	14.55	3.79	VII	2,100	100	Rib Du CH	200	.89	7.9	6.8	3.50	10.00	.812		TSO	
18	x 4.4		8	TL	17.90	17.40	4.45	4.15	14.55	3.79	VII	2,850	150	Rib CH	210	.89	7.9	6.8	3.50	10.00	.812		TSO	
18	x 4.4		10	TL	17.90	17.40	4.45	4.15	14.55	3.79	VII	3,550	185	Rib CH	200	.89	7.9	6.8	3.50	10.00	.812		TSO	11.5
18	x 4.4		10	TL	17.90	17.40	4.45	4.15	14.55	3.79	VII	3,550	185	Rib Du CH	200	.89	7.9	6.8	3.50	10.00	.812		TSO	
18	x 4.4		12	TL	17.90	17.40	4.45	4.15	14.55	3.79	VII	4,350	225	Rib CH	300	.89	7.9	6.8	3.50	10.00	.812	MIL		11.5
18	x 4.4		12	TL	17.90	17.40	4.45	4.15	14.55	3.79	VII	4,350	225	Rib CH	200	.89	7.9	6.8	3.50	10.00	.812		TSO	
18	x 4.4		12	TT	17.90	17.40	4.45	4.15	14.55	3.79	VII	4,350	225	Rib	250	.89	7.9	6.8	3.50	10.00	.812	MIL		10.4
18	x 5.5		8	TL	17.90	17.30	5.70	5.35	16.20	5.00	VII	3,050	105	Rib	160	.87	7.5	6.2	4.25	8.00	.875	MIL		11.0
18	x 5.5		8	TL	17.90	17.30	5.70	5.35	16.20	5.00	VII	3,050	105	Rib	180	.87	7.5	6.2	4.25	8.00	.875	MIL	TSO	13.5
18	x 5.5		8	TT	17.90	17.30	5.70	5.35	16.20	5.00	VII	3,050	110	Rib	180	.87	7.5	6.2	4.25	8.00	.875	MIL		
18	x 5.5		10	TL	17.90	17.30	5.70	5.35	16.20	5.00	VII	4,000	140	Rib	210	.87	7.5	6.2	4.25	8.00	.875		TSO	
18	x 5.5		12	TL	17.90	17.30	5.70	5.35	16.20	5.00	VII	5,050	170	Rib	160	.87	7.5	6.2	4.25	8.00	.875	MIL		14.0
18	x 5.5		12	TT	17.90	17.30	5.70	5.35	16.20	5.00	VII	5,050	170	Rib	160	.87	7.5	6.2	4.25	8.00	.875	MIL		13.0
18	x 5.5		14	TL	17.90	17.30	5.70	5.35	16.20	5.00	VII	6,200	215	Rib	275	.87	7.5	6.2	4.25	8.00	.875	MIL		
18	x 5.5		14	TL	17.90	17.30	5.70	5.35	16.20	5.00	VII	6,200	200	Rib	160	.87	7.5	6.2	4.25	8.00	.875	MIL		15.0
18	x 5.7	- 8	14	TL	17.8	17.25	5.60	5.25	16.20	5.00	NS	6,200	215	Rib	250	.87	7.46	6.1	4.25	8.00	.875	MIL		
	19.00	- 23	16	TT	55.10	53.15	19.38	18.25	49.30	16.50	III	29,000	85	Rib	160	.83	22.6	15.6	14.75	23.00	2.000		TSO	174.0
19.5	x 6.75	- 8	10	TL	19.50	18.90	6.75	6.35	17.45	5.95	NS	4,270	110	Rib	160	.85	8.1	5.5	5.25	8.00	.812		TSO	
19.5	x 6.75	- 8	10	TT	19.50	18.90	6.75	6.35	17.45	5.95	NS	4,270	110	Rib	160	.85	8.1	5.5	5.25	8.00	.812		TSO	
20	x 4.4		10	TL	20.00	19.50	4.45	4.15	19.45	3.95	VII	4,250	190	Rib	160	.90	8.9	8.0	3.50	12.00	.812	MIL		
20	x 4.4		10	TT	20.00	19.50	4.45	4.15	19.45	3.95	VII	4,250	190	Rib	160	.90	8.9	8.0	3.50	12.00	.812	MIL		13.5
20	x 4.4		12	TL	20.00	19.50	4.45	4.15	19.45	3.95	VII	5,150	225	Rib	225	.90	8.9	8.0	3.50	12.00	.812	MIL		
20	x 4.4		14	TL	20.00	19.50	4.45	4.15	19.45	3.95	VII	6,000	265	Rib	235	.90	8.9	8.0	3.50	12.00	.812	MIL		
20	x 5.5		12	TL	20.15	19.55	5.70	5.35	19.30	4.75	VII	6,150	180	Rib	160	.89	8.6	7.1	4.25	10.00	.875	MIL		17.0
20	x 5.5		14	TL	20.15	19.55	5.70	5.35	19.30	4.75	VII	7,200	230	Rib	200	.89	8.6	7.1	4.25	10.00	.875	MIL		
20	x 5.5		16	TL	20.15	19.55	5.70	5.35	19.30	4.75	VII	8,750	270	Rib	185	.89	8.6	7.1	4.25	10.00	.875	MIL		20.5
	20.00	- 20	20	TL	56.00	54.30	20.10	19.20	49.50	17.10	III	46,500	125	Rib	200	.89	22.0	13.9	15.50	20.00	2.000	MIL		
	20.00	- 20	22	TL	56.00	54.30	20.10	19.20	49.50	17.10	III	38,500	95	Rib	200	.89	22.0	13.9	15.50	20.00	2.000	MIL		202.0
	20.00	- 20	26	TL	56.00	54.30	20.10	19.20	49.50	17.10	III	46,500	125	Rib	200	.89	22.0	13.9	15.50	20.00	2.000	MIL		260.0
21	x 7.25	- 10	8	TL	21.25	20.60	7.20	6.80	19.25	6.35	NS	4,000	95	Rib CH	200	.78	9.1	7.1	5.50	10.00	1.000		TSO	

Table 2.8 Tire Data, Courtesy: B.F.Goodrich

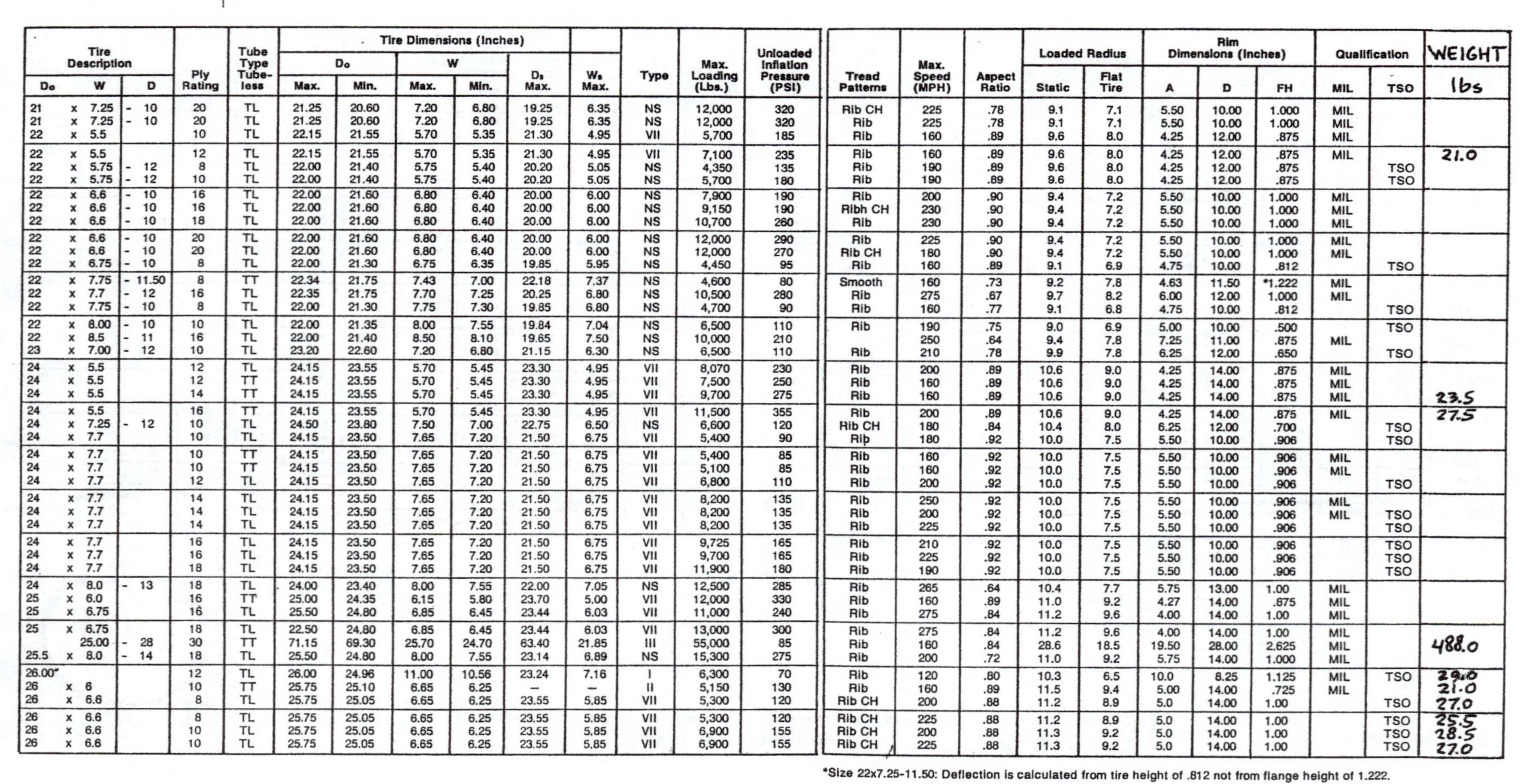

Tire Description D₀	Tire Description W	Tire Description D	Ply Rating	Tube Type Tube-less	Tire Dimensions (Inches) D₀ Max.	D₀ Min.	W Max.	W Min.	Dₛ Max.	Wₛ Max.	Type	Max. Loading (Lbs.)	Unloaded Inflation Pressure (PSI)	Tread Patterns	Max. Speed (MPH)	Aspect Ratio	Loaded Radius Static	Loaded Radius Flat Tire	Rim Dimensions (Inches) A	Rim D	Rim FH	Qualification MIL	Qualification TSO	WEIGHT lbs
21	x 7.25	- 10	20	TL	21.25	20.60	7.20	6.80	19.25	6.35	NS	12,000	320	Rib CH	225	.78	9.1	7.1	5.50	10.00	1.000	MIL		
21	x 7.25	- 10	20	TL	21.25	20.60	7.20	6.80	19.25	6.35	NS	12,000	320	Rib	225	.78	9.1	7.1	5.50	10.00	1.000	MIL		
22	x 5.5		10	TL	22.15	21.55	5.70	5.35	21.30	4.95	VII	5,700	185	Rib	160	.89	9.6	8.0	4.25	12.00	.875	MIL		
22	x 5.5		12	TL	22.15	21.55	5.70	5.35	21.30	4.95	VII	7,100	235	Rib	160	.89	9.6	8.0	4.25	12.00	.875	MIL		21.0
22	x 5.75	- 12	8	TL	22.00	21.40	5.75	5.40	20.20	5.05	NS	4,350	135	Rib	190	.89	9.6	8.0	4.25	12.00	.875		TSO	
22	x 5.75	- 12	10	TL	22.00	21.40	5.75	5.40	20.20	5.05	NS	5,700	180	Rib	190	.89	9.6	8.0	4.25	12.00	.875		TSO	
22	x 6.6	- 10	16	TL	22.00	21.60	6.80	6.40	20.00	6.00	NS	7,900	190	Rib	200	.90	9.4	7.2	5.50	10.00	1.000	MIL		
22	x 6.6	- 10	16	TL	22.00	21.60	6.80	6.40	20.00	6.00	NS	9,150	190	Ribh CH	230	.90	9.4	7.2	5.50	10.00	1.000	MIL		
22	x 6.6	- 10	18	TL	22.00	21.60	6.80	6.40	20.00	6.00	NS	10,700	260	Rib	230	.90	9.4	7.2	5.50	10.00	1.000	MIL		
22	x 6.6	- 10	20	TL	22.00	21.60	6.80	6.40	20.00	6.00	NS	12,000	290	Rib	225	.90	9.4	7.2	5.50	10.00	1.000	MIL		
22	x 6.6	- 10	20	TL	22.00	21.60	6.80	6.40	20.00	6.00	NS	12,000	270	Rib CH	180	.90	9.4	7.2	5.50	10.00	1.000	MIL		
22	x 6.75	- 10	8	TL	22.00	21.30	6.75	6.35	19.85	5.95	NS	4,450	95	Rib	160	.89	9.1	6.9	4.75	10.00	.812		TSO	
22	x 7.75	- 11.50	8	TT	22.34	21.75	7.43	7.00	22.18	7.37	NS	4,600	80	Smooth	160	.73	9.2	7.8	4.63	11.50	*1.222	MIL		
22	x 7.7	- 12	16	TL	22.35	21.75	7.70	7.25	20.25	6.80	NS	10,500	280	Rib	275	.67	9.7	8.2	6.00	12.00	1.000	MIL		
22	x 7.75	- 10	8	TL	22.00	21.30	7.75	7.30	19.85	6.80	NS	4,700	90	Rib	160	.77	9.1	6.8	4.75	10.00	.812		TSO	
22	x 8.00	- 10	10	TL	22.00	21.35	8.00	7.55	19.84	7.04	NS	6,500	110	Rib	190	.75	9.0	6.9	5.00	10.00	.500		TSO	
22	x 8.5	- 11	16	TL	22.00	21.40	8.50	8.10	19.65	7.50	NS	10,000	210		250	.64	9.4	7.8	7.25	11.00	.875	MIL		
23	x 7.00	- 12	10	TL	23.20	22.60	7.20	6.80	21.15	6.30	NS	6,500	110	Rib	210	.78	9.9	7.8	6.25	12.00	.650		TSO	
24	x 5.5		12	TL	24.15	23.55	5.70	5.45	23.30	4.95	VII	8,070	230	Rib	200	.89	10.6	9.0	4.25	14.00	.875	MIL		
24	x 5.5		12	TT	24.15	23.55	5.70	5.45	23.30	4.95	VII	7,500	250	Rib	160	.89	10.6	9.0	4.25	14.00	.875	MIL		
24	x 5.5		14	TT	24.15	23.55	5.70	5.45	23.30	4.95	VII	9,700	275	Rib	160	.89	10.6	9.0	4.25	14.00	.875	MIL		23.5
24	x 5.5		16	TT	24.15	23.55	5.70	5.45	23.30	4.95	VII	11,500	355	Rib	200	.89	10.6	9.0	4.25	14.00	.875	MIL		27.5
24	x 7.25	- 12	10	TL	24.50	23.80	7.50	7.00	22.75	6.50	NS	6,600	120	Rib CH	180	.84	10.4	8.0	6.25	12.00	.700		TSO	
24	x 7.7		10	TL	24.15	23.50	7.65	7.20	21.50	6.75	VII	5,400	90	Rib	180	.92	10.0	7.5	5.50	10.00	.906		TSO	
24	x 7.7		10	TT	24.15	23.50	7.65	7.20	21.50	6.75	VII	5,400	85	Rib	160	.92	10.0	7.5	5.50	10.00	.906	MIL		
24	x 7.7		10	TT	24.15	23.50	7.65	7.20	21.50	6.75	VII	5,100	85	Rib	160	.92	10.0	7.5	5.50	10.00	.906	MIL		
24	x 7.7		12	TL	24.15	23.50	7.65	7.20	21.50	6.75	VII	6,800	110	Rib	200	.92	10.0	7.5	5.50	10.00	.906		TSO	
24	x 7.7		14	TL	24.15	23.50	7.65	7.20	21.50	6.75	VII	8,200	135	Rib	250	.92	10.0	7.5	5.50	10.00	.906	MIL		
24	x 7.7		14	TL	24.15	23.50	7.65	7.20	21.50	6.75	VII	8,200	135	Rib	200	.92	10.0	7.5	5.50	10.00	.906	MIL	TSO	
24	x 7.7		14	TL	24.15	23.50	7.65	7.20	21.50	6.75	VII	8,200	135	Rib	225	.92	10.0	7.5	5.50	10.00	.906		TSO	
24	x 7.7		16	TL	24.15	23.50	7.65	7.20	21.50	6.75	VII	9,725	165	Rib	210	.92	10.0	7.5	5.50	10.00	.906		TSO	
24	x 7.7		16	TL	24.15	23.50	7.65	7.20	21.50	6.75	VII	9,700	165	Rib	225	.92	10.0	7.5	5.50	10.00	.906		TSO	
24	x 7.7		18	TL	24.15	23.50	7.65	7.20	21.50	6.75	VII	11,900	180	Rib	190	.92	10.0	7.5	5.50	10.00	.906		TSO	
24	x 8.0	- 13	18	TL	24.00	23.40	8.00	7.55	22.00	7.05	NS	12,500	285	Rib	265	.64	10.4	7.7	5.75	13.00	1.00	MIL		
25	x 6.0		16	TT	25.00	24.35	6.15	5.80	23.70	5.00	VII	12,000	330	Rib	160	.89	11.0	9.2	4.27	14.00	.875	MIL		
25	x 6.75		16	TL	25.50	24.80	6.85	6.45	23.44	6.03	VII	11,000	240	Rib	275	.84	11.2	9.6	4.00	14.00	1.00	MIL		
25	x 6.75		18	TL	22.50	24.80	6.85	6.45	23.44	6.03	VII	13,000	300	Rib	275	.84	11.2	9.6	4.00	14.00	1.00	MIL		
	25.00	- 28	30	TT	71.15	69.30	25.70	24.70	63.40	21.85	III	55,000	85	Rib	160	.84	28.6	18.5	19.50	28.00	2.625	MIL		488.0
25.5	x 8.0	- 14	18	TL	25.50	24.80	8.00	7.55	23.14	6.89	NS	15,300	275	Rib	200	.72	11.0	9.2	5.75	14.00	1.000	MIL		
26.00"			12	TL	26.00	24.96	11.00	10.56	23.24	7.16	I	6,300	70	Rib	120	.80	10.3	6.5	10.0	8.25	1.125	MIL	TSO	29.0
26	x 6		10	TT	25.75	25.10	6.65	6.25	—	—	II	5,150	130	Rib	160	.89	11.5	9.4	5.00	14.00	.725	MIL		21.0
26	x 6.6		8	TL	25.75	25.05	6.65	6.25	23.55	5.85	VII	5,300	120	Rib CH	200	.88	11.2	8.9	5.0	14.00	1.00		TSO	27.0
26	x 6.6		8	TL	25.75	25.05	6.65	6.25	23.55	5.85	VII	5,300	120	Rib CH	225	.88	11.2	8.9	5.0	14.00	1.00		TSO	25.5
26	x 6.6		10	TL	25.75	25.05	6.65	6.25	23.55	5.85	VII	6,900	155	Rib CH	200	.88	11.3	9.2	5.0	14.00	1.00		TSO	28.5
26	x 6.6		10	TL	25.75	25.05	6.65	6.25	23.55	5.85	VII	6,900	155	Rib CH	225	.88	11.3	9.2	5.0	14.00	1.00		TSO	27.0

*Size 22x7.25-11.50: Deflection is calculated from tire height of .812 not from flange height of 1.222.

Table 2.9 Tire Data, Courtesy: B.F.Goodrich

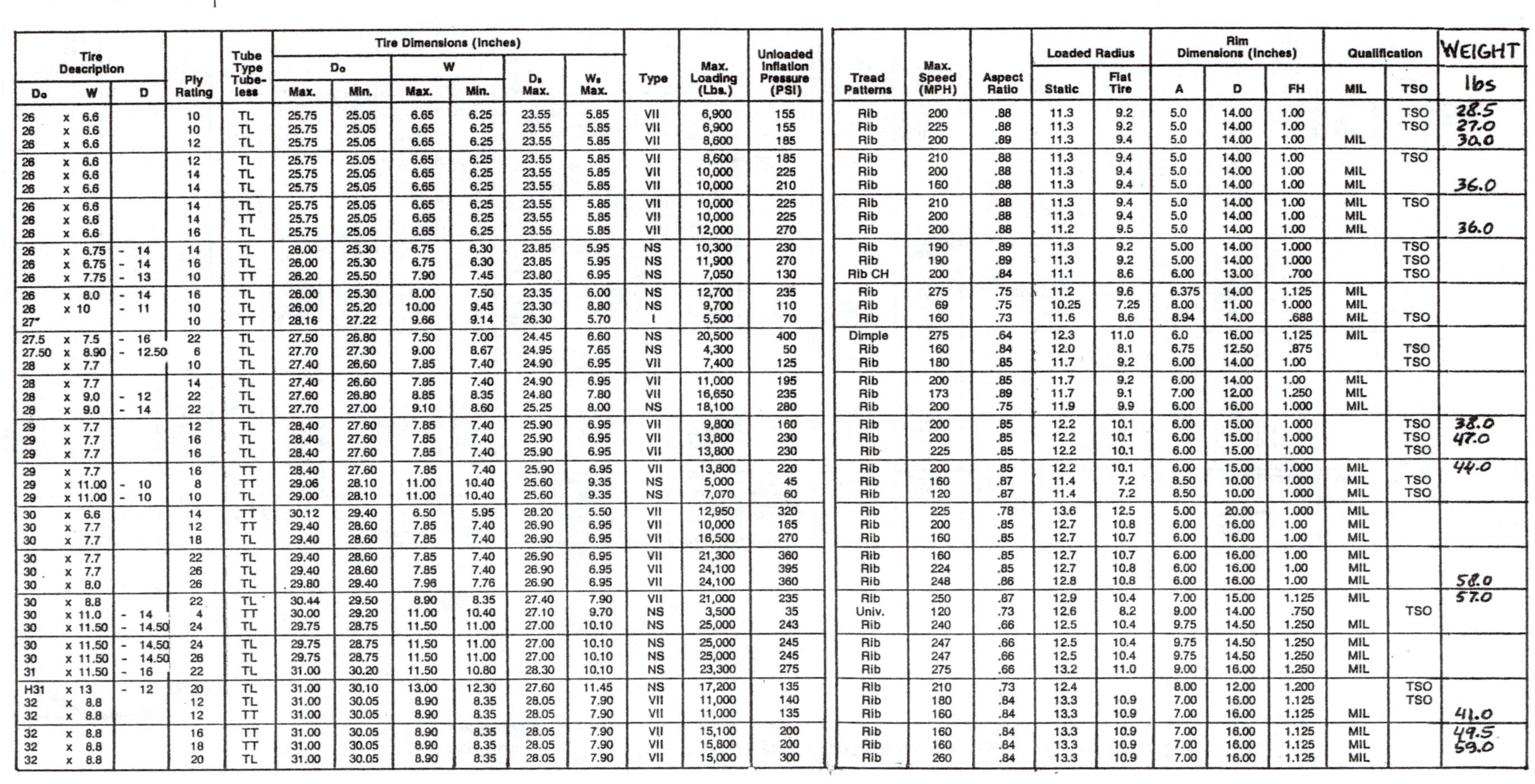

Tire Description D₀	W	D	Ply Rating	Tube Type Tube-less	D₀ Max.	D₀ Min.	W Max.	W Min.	D_s Max.	W_s Max.	Type	Max. Loading (Lbs.)	Unloaded Inflation Pressure (PSI)	Tread Patterns	Max. Speed (MPH)	Aspect Ratio	Loaded Radius Static	Loaded Radius Flat Tire	Rim A	Rim D	Rim FH	MIL	TSO	WEIGHT lbs
26	x 6.6		10	TL	25.75	25.05	6.65	6.25	23.55	5.85	VII	6,900	155	Rib	200	.88	11.3	9.2	5.0	14.00	1.00		TSO	28.5
26	x 6.6		10	TL	25.75	25.05	6.65	6.25	23.55	5.85	VII	6,900	155	Rib	225	.88	11.3	9.2	5.0	14.00	1.00		TSO	27.0
26	x 6.6		12	TL	25.75	25.05	6.65	6.25	23.55	5.85	VII	8,600	185	Rib	200	.89	11.3	9.4	5.0	14.00	1.00	MIL		30.0
26	x 6.6		12	TL	25.75	25.05	6.65	6.25	23.55	5.85	VII	8,600	185	Rib	210	.88	11.3	9.4	5.0	14.00	1.00		TSO	
26	x 6.6		14	TL	25.75	25.05	6.65	6.25	23.55	5.85	VII	10,000	225	Rib	200	.88	11.3	9.4	5.0	14.00	1.00	MIL		
26	x 6.6		14	TL	25.75	25.05	6.65	6.25	23.55	5.85	VII	10,000	210	Rib	160	.88	11.3	9.4	5.0	14.00	1.00	MIL		36.0
26	x 6.6		14	TL	25.75	25.05	6.65	6.25	23.55	5.85	VII	10,000	225	Rib	210	.88	11.3	9.4	5.0	14.00	1.00	MIL	TSO	
26	x 6.6		14	TT	25.75	25.05	6.65	6.25	23.55	5.85	VII	10,000	225	Rib	200	.88	11.3	9.4	5.0	14.00	1.00	MIL		
26	x 6.6		16	TL	25.75	25.05	6.65	6.25	23.55	5.85	VII	12,000	270	Rib	200	.88	11.2	9.5	5.0	14.00	1.00	MIL		36.0
26	x 6.75	- 14	14	TL	26.00	25.30	6.75	6.30	23.85	5.95	NS	10,300	230	Rib	190	.89	11.3	9.2	5.00	14.00	1.000		TSO	
26	x 6.75	- 14	16	TL	26.00	25.30	6.75	6.30	23.85	5.95	NS	11,900	270	Rib	190	.89	11.3	9.2	5.00	14.00	1.000		TSO	
26	x 7.75	- 13	10	TT	26.20	25.50	7.90	7.45	23.80	6.95	NS	7,050	130	Rib CH	200	.84	11.1	8.6	6.00	13.00	.700		TSO	
26	x 8.0	- 14	16	TL	26.00	25.30	8.00	7.50	23.35	6.00	NS	12,700	235	Rib	275	.75	11.2	9.6	6.375	14.00	1.125	MIL		
26	x 10	- 11	10	TL	26.00	25.20	10.00	9.45	23.30	8.80	NS	9,700	110	Rib	69	.75	10.25	7.25	8.00	11.00	1.000	MIL		
27"			10	TT	28.16	27.22	9.66	9.14	26.30	5.70	I	5,500	70	Rib	160	.73	11.6	8.6	8.94	14.00	.688	MIL	TSO	
27.5	x 7.5	- 16	22	TL	27.50	26.80	7.50	7.00	24.45	6.60	NS	20,500	400	Dimple	275	.64	12.3	11.0	6.0	16.00	1.125	MIL		
27.50	x 8.90	- 12.50	6	TL	27.70	27.30	9.00	8.67	24.95	7.65	NS	4,300	50	Rib	160	.84	12.0	8.1	6.75	12.50	.875		TSO	
28	x 7.7		10	TL	27.40	26.60	7.85	7.40	24.90	6.95	VII	7,400	125	Rib	180	.85	11.7	9.2	6.00	14.00	1.00		TSO	
28	x 7.7		14	TL	27.40	26.60	7.85	7.40	24.90	6.95	VII	11,000	195	Rib	200	.85	11.7	9.2	6.00	14.00	1.00	MIL		
28	x 9.0	- 12	22	TL	27.60	26.80	8.85	8.35	24.80	7.80	VII	16,650	235	Rib	173	.89	11.7	9.1	7.00	12.00	1.250	MIL		
28	x 9.0	- 14	22	TL	27.70	27.00	9.10	8.60	25.25	8.00	NS	18,100	280	Rib	200	.75	11.9	9.9	6.00	16.00	1.000	MIL		
29	x 7.7		12	TL	28.40	27.60	7.85	7.40	25.90	6.95	VII	9,800	160	Rib	200	.85	12.2	10.1	6.00	15.00	1.000		TSO	38.0
29	x 7.7		16	TL	28.40	27.60	7.85	7.40	25.90	6.95	VII	13,800	230	Rib	200	.85	12.2	10.1	6.00	15.00	1.000		TSO	47.0
29	x 7.7		16	TL	28.40	27.60	7.85	7.40	25.90	6.95	VII	13,800	230	Rib	225	.85	12.2	10.1	6.00	15.00	1.000		TSO	
29	x 7.7		16	TT	28.40	27.60	7.85	7.40	25.90	6.95	VII	13,800	220	Rib	200	.85	12.2	10.1	6.00	15.00	1.000	MIL		44.0
29	x 11.00	- 10	8	TT	29.06	28.10	11.00	10.40	25.60	9.35	NS	5,000	45	Rib	160	.87	11.4	7.2	8.50	10.00	1.000	MIL	TSO	
29	x 11.00	- 10	10	TL	29.00	28.10	11.00	10.40	25.60	9.35	NS	7,070	60	Rib	120	.87	11.4	7.2	8.50	10.00	1.000	MIL	TSO	
30	x 6.6		14	TT	30.12	29.40	6.50	5.95	28.20	5.50	VII	12,950	320	Rib	225	.78	13.6	12.5	5.00	20.00	1.000	MIL		
30	x 7.7		12	TT	29.40	28.60	7.85	7.40	26.90	6.95	VII	10,000	165	Rib	200	.85	12.7	10.8	6.00	16.00	1.00	MIL		
30	x 7.7		18	TL	29.40	28.60	7.85	7.40	26.90	6.95	VII	16,500	270	Rib	160	.85	12.7	10.7	6.00	16.00	1.00	MIL		
30	x 7.7		22	TL	29.40	28.60	7.85	7.40	26.90	6.95	VII	21,300	360	Rib	160	.85	12.7	10.7	6.00	16.00	1.00	MIL		
30	x 7.7		26	TL	29.40	28.60	7.85	7.40	26.90	6.95	VII	24,100	395	Rib	224	.85	12.7	10.8	6.00	16.00	1.00	MIL		
30	x 8.0		26	TL	29.80	29.40	7.96	7.76	26.90	6.95	VII	24,100	360	Rib	248	.86	12.8	10.8	6.00	16.00	1.00	MIL		58.0
30	x 8.8		22	TL	30.44	29.50	8.90	8.35	27.40	7.90	VII	21,000	235	Rib	250	.87	12.9	10.4	7.00	15.00	1.125	MIL		57.0
30	x 11.0	- 14	4	TT	30.00	29.20	11.00	10.40	27.10	9.70	NS	3,500	35	Univ.	120	.73	12.6	8.2	9.00	14.00	.750		TSO	
30	x 11.50	- 14.50	24	TL	29.75	28.75	11.50	11.00	27.00	10.10	NS	25,000	243	Rib	240	.66	12.5	10.4	9.75	14.50	1.250	MIL		
30	x 11.50	- 14.50	24	TL	29.75	28.75	11.50	11.00	27.00	10.10	NS	25,000	245	Rib	247	.66	12.5	10.4	9.75	14.50	1.250	MIL		
30	x 11.50	- 14.50	26	TL	29.75	28.75	11.50	11.00	27.00	10.10	NS	25,000	245	Rib	247	.66	12.5	10.4	9.75	14.50	1.250	MIL		
31	x 11.50	- 16	22	TL	31.00	30.20	11.50	10.80	28.30	10.10	NS	23,300	275	Rib	275	.66	13.2	11.0	9.00	16.00	1.250	MIL		
H31	x 13	- 12	20	TL	31.00	30.10	13.00	12.30	27.60	11.45	NS	17,200	135	Rib	210	.73	12.4		8.00	12.00	1.200		TSO	
32	x 8.8		12	TL	31.00	30.05	8.90	8.35	28.05	7.90	VII	11,000	140	Rib	180	.84	13.3	10.9	7.00	16.00	1.125		TSO	
32	x 8.8		12	TT	31.00	30.05	8.90	8.35	28.05	7.90	VII	11,000	135	Rib	160	.84	13.3	10.9	7.00	16.00	1.125	MIL		41.0
32	x 8.8		16	TT	31.00	30.05	8.90	8.35	28.05	7.90	VII	15,100	200	Rib	160	.84	13.3	10.9	7.00	16.00	1.125	MIL		49.5
32	x 8.8		18	TT	31.00	30.05	8.90	8.35	28.05	7.90	VII	15,800	200	Rib	160	.84	13.3	10.9	7.00	16.00	1.125	MIL		59.0
32	x 8.8		20	TL	31.00	30.05	8.90	8.35	28.05	7.90	VII	15,000	300	Rib	260	.84	13.3	10.9	7.00	16.00	1.125	MIL		

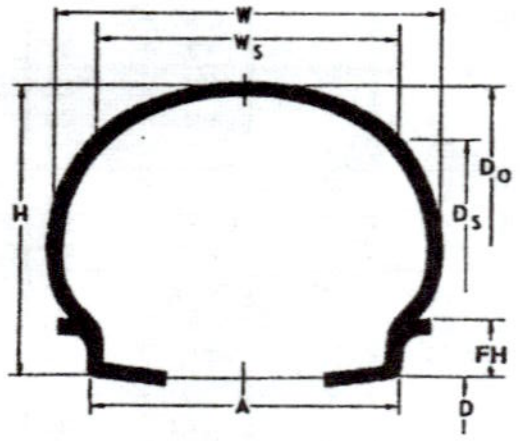

Table 2.10 Tire Data, Courtesy: B.F.Goodrich

Tire Description			Ply Rating	Tube Type Tube-less	Tire Dimensions (Inches)						Type	Max. Loading (Lbs.)	Unloaded Inflation Pressure (PSI)	Tread Patterns	Max. Speed (MPH)	Aspect Ratio	Loaded Radius		Rim Dimensions (Inches)			Qualification		WEIGHT
					D_o		W																	
D_o	W	D			Max.	Min.	Max.	Min.	D_s Max.	W_s Max.							Static	Flat Tire	A	D	FH	MIL	TSO	lbs
32	x 8.8		24	TL	31.00	30.05	8.90	8.35	28.05	7.90	VII	23,300	335	Rib	275	.84	13.3	10.9	7.00	16.00	1.125	MIL		
32	x 11.50	- 15	12	TL	32.00	31.10	11.50	10.80	29.00	10.50	NS	11,200	120	Rib CH	210	.74	13.5	10.3	9.00	15.00	1.250		TSO	67.0
32	x 11.50	- 15	12	TL	32.00	31.10	11.50	10.80	29.00	10.50	NS	11,200	120	Rib CH	225	.74	13.5	10.3	9.00	15.00	1.250		TSO	65.0
33"			10	TT	33.06	32.06	11.30	10.84	31.30	6.60	I	8,000	70	Rib	160	.73	13.8	10.3	10.78	16.50	.813	MIL	TSO	
34	x 9.25	- 16	16	TL	34.00	33.15	9.25	8.75	30.75	8.15	NS	15,500	155	Rib	200	.98	14.3	10.4	7.00	16.00	1.125		TSO	55.5
34	x 9.9		12	TL	33.40	32.45	10.20	9.55	30.10	8.80	VII	11,200	115	Rib	200	.86	14.2	10.7	8.00	16.00	1.250		TSO	53.0
34	x 9.9		14	TL	33.40	32.45	10.20	9.55	30.10	8.80	VII	14,000	150	Rib	160	.86	14.2	10.7	8.00	16.00	1.250	MIL		55.0
34	x 9.9		14	TT	33.40	32.45	10.20	9.55	30.10	8.80	VII	14,000	150	Rib	160	.86	14.2	10.7	8.00	16.00	1.250	MIL		
34	x 11		18	TL	33.40	32.60	11.30	10.60	29.90	9.95	VII	16,100	145	Rib	200	.87	13.9	10.0	9.00	14.00	1.500		TSO	75.0
34	x 11		20	TL	33.40	32.60	11.30	10.60	29.90	9.95	VII	18,300	165	Rib	200	.87	13.9	10.0	9.00	14.00	1.500		TSO	77.0
34	x 11		22	TL	33.40	32.60	11.30	10.60	29.90	9.95	VII	20,500	185	Rib	200	.87	13.9	10.0	9.00	14.00	1.500		TSO	75.5
34	x 11		22	TL	33.40	32.60	11.30	10.60	29.90	9.95	VII	20,500	185	Rib	225	.87	13.9	10.0	9.00	14.00	1.500		TSO	76.0
34	x 12	- 12	12	TL	34.00	33.10	12.00	11.35	30.00	10.55	NS	10,400	70	Rib	160	.92	13.4	8.3	9.00	12.00	1.125		TSO	68.0
34	x 12	- 12	14	TL	34.00	33.10	12.00	11.35	30.00	10.55	NS	11,300	85	Rib	180	.92	13.4	8.3	9.00	12.00	1.125		TSO	64.5
34.5	x 9.75	- 18	22	TL	34.50	33.76	9.75	9.15	31.55	8.60	NS	23,400	260	Rib	260	.85	14.9	11.6	7.50	18.00	1.125	MIL		
34.5	x 9.75	- 18	26	TL	34.50	33.76	9.75	9.15	31.55	8.60	NS	30,100	340	Rib	259	.85	14.9	11.6	7.50	18.00	1.125	MIL		
35	x 9.00	- 17	16	TL	34.80	33.95	9.40	8.90	31.60	8.25	NS	16,300	170	Rib	200	.95	14.8	11.6	7.25	17.00	1.100		TSO	
35	x 11.50	- 16	22	TL	35.00	31.80	11.50	10.90	31.80	10.10	NS	23,000	210	Rib	225	.83	14.8	9.7	9.00	16.00	1.375	MIL		
36"			12	TT	36.86	35.40	13.08	12.56	34.84	7.02	I	10,500	70	Rib	160	.72	15.3	11.1	12.46	17.75	.875	MIL	TSO	
36	x 11		20	TL	35.10	34.00	11.50	10.80	31.65	10.10	VII	21,000	185	Rib	225	.83	14.7	11.3	9.00	16.00	1.375		TSO	78.0
36	x 11		22	TL	35.10	34.00	11.50	10.80	31.65	10.10	VII	23,300	200	Rib	200	.83	14.7	11.3	9.00	16.00	1.375	MIL		84.5
36	x 11		22	TL	35.10	34.00	11.50	10.80	31.65	10.10	VII	23,300	200	Rib	225	.83	14.7	11.3	9.00	16.00	1.375		TSO	
36	x 11		24	TL	35.10	34.00	11.50	10.80	31.65	10.10	VII	26,000	235	Rib	160	.83	14.7	11.3	9.00	16.00	1.375	MIL		90.5
36	x 11		24	TL	35.10	34.00	11.50	10.80	31.65	10.10	VII	26,000	235	Rib	250	.83	14.7	11.3	9.00	16.00	1.375	MIL		85.0
36	x 11		28	TL	35.10	34.00	11.50	10.80	31.65	10.10	VII	31,500	290	Rib	200	.83	14.7	11.3	9.00	16.00	1.375	MIL		97.0
37	x 11.50	- 16	28	TL	37.00	36.10	11.50	10.90	33.20	10.10	NS	31,200	245	Rib	180	.91	15.1	10.4	9.00	16.00	1.375	MIL		90.0
37	x 13	- 16	20	TL	37.00	36.10	13.00	12.30	33.20	11.45	NS	22,200	165	Rib	210	.81	15.4	10.8	9.00	16.00	1.375		TSO	
37	x 13	- 16	26	TL	37.00	36.10	13.00	12.30	33.20	11.45	NS	29,300	220	Rib	225	.81	15.4	10.8	9.00	16.00	1.375		TSO	
37	x 14	- 14	24	TL	37.00	36.05	14.00	13.30	32.85	12.30	NS	25,000	160	Rib	225	.83	15.1	10.4	11.00	14.00	1.500		TSO	90.0
H37	x 14.0	- 15	22	TL	37.00	36.10	14.00	13.30	33.05	12.30	NS	24,100	165	Rib	225	.79	15.0		9.00	15.00	1.300		TSO	
38	x 11		14	TL	37.10	36.00	11.50	10.80	33.65	10.10	VII	15,400	130	Rib	225	.83	15.7	12.3	9.00	18.00	1.375	MIL		80.5
39	x 13		14	TL	38.25	37.30	13.00	12.25	34.25	11.45	VII	15,000	100	Rib	180	.86	15.8	11.0	10.00	16.00	*1.375		TSO	
39	x 13		14	TL	38.25	37.30	13.00	12.25	34.25	11.45	VII	15,000	100	Rib	200	.86	15.8	11.0	10.00	16.00	*1.375		TSO	87.0
39	x 13		14	TT	38.25	37.30	13.00	12.25	34.25	11.45	VII	15,000	100	Rib	180	.86	15.8	11.0	10.00	16.00	*1.375		TSO	
39	x 13		16	TL	38.25	37.30	13.00	12.25	34.25	11.45	VII	17,200	115	Rib	210	.86	15.8	11.0	10.00	16.00	*1.375		TSO	
39	x 13		16	TL	38.25	37.30	13.00	12.25	34.25	11.45	VII	17,200	115	Rib	225	.86	15.8	11.0	10.00	16.00	*1.375	MIL	TSO	94.0
39	x 13		20	TL	38.25	37.30	13.00	12.25	34.25	11.45	VII	22,300	150	Rib	200	.86	15.8	11.0	10.00	16.00	1.375		TSO	107.0
39	x 13		22	TL	38.25	37.30	13.00	12.25	34.25	11.45	VII	24,600	165	Rib	200	.86	15.8	11.0	10.00	16.00	1.375		TSO	114.0
39	x 13		22	TL	38.25	37.30	13.00	12.25	34.25	11.45	VII	24,600	165	Rib	225	.86	15.8	11.0	10.00	16.00	1.375		TSO	109.0
40	x 12		14	TT	39.40	38.04	12.35	11.70	35.50	10.90	VII	14,500	95	Rib	160	.87	16.6	12.3	10.00	18.00	1.500	MIL		78.0
40	x 12		16	TL	39.40	38.40	12.35	11.70	35.50	10.90	VII	18,500	130	Rib	180	.87	16.6	12.3	10.00	18.00	1.500		TSO	95.0
40	x 12		18	TL	39.40	38.40	12.35	11.70	35.50	10.90	VII	21,000	150	Rib	180	.87	16.6	12.3	10.00	18.00	1.500		TSO	102.0
40	x 12		18	TL	39.40	38.40	12.35	11.70	35.50	10.90	VII	21,000	150	Rib	200	.87	16.6	12.3	10.00	18.00	1.500		TSO	
40	x 12		22	TL	39.40	38.40	12.35	11.70	35.50	10.90	VII	26,700	190	Rib	180	.87	16.6	12.3	10.00	18.00	1.500		TSO	107.0
40	x 14		20	TL	39.80	38.95	14.00	13.25	35.10	12.00	VII	22,300	200	Rib	200	.86	16.5	11.6	11.00	16.00	1.625		TSO	

*Rim with FH of 1.25 is optional for 39x13, 14 and 16 PR tires.

Table 2.11 Tire Data, Courtesy: B.F.Goodrich

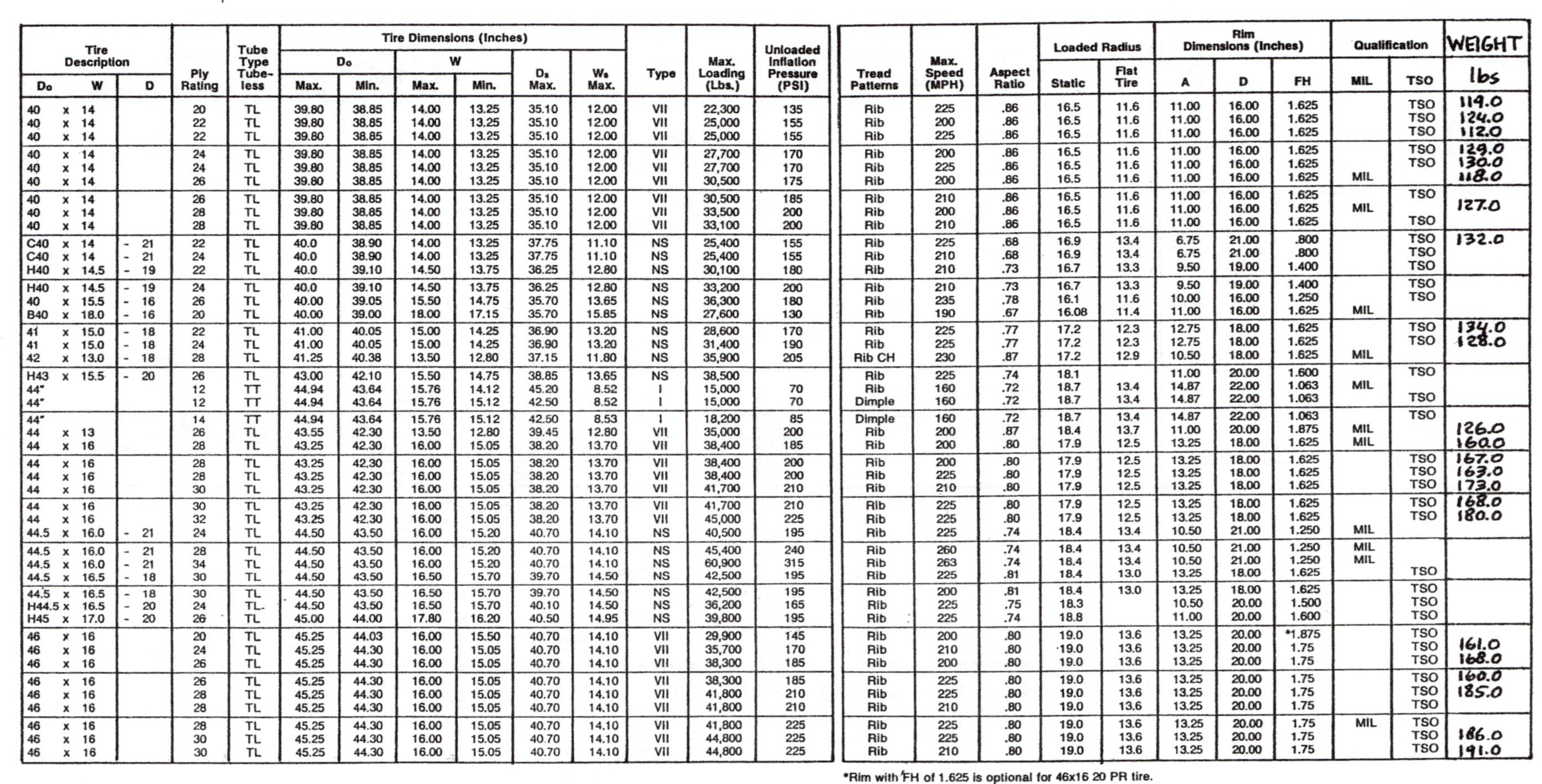

Tire Description			Ply Rating	Tube Type Tube-less	Tire Dimensions (Inches)						Type	Max. Loading (Lbs.)	Unloaded Inflation Pressure (PSI)	Tread Patterns	Max. Speed (MPH)	Aspect Ratio	Loaded Radius		Rim Dimensions (Inches)			Qualification		WEIGHT
					Do		W		Ds Max.	Ws Max.							Static	Flat Tire	A	D	FH	MIL	TSO	lbs
Do	W	D			Max.	Min.	Max.	Min.																
40	x 14		20	TL	39.80	38.85	14.00	13.25	35.10	12.00	VII	22,300	135	Rib	225	.86	16.5	11.6	11.00	16.00	1.625		TSO	119.0
40	x 14		22	TL	39.80	38.85	14.00	13.25	35.10	12.00	VII	25,000	155	Rib	200	.86	16.5	11.6	11.00	16.00	1.625		TSO	124.0
40	x 14		22	TL	39.80	38.85	14.00	13.25	35.10	12.00	VII	25,000	155	Rib	225	.86	16.5	11.6	11.00	16.00	1.625		TSO	112.0
40	x 14		24	TL	39.80	38.85	14.00	13.25	35.10	12.00	VII	27,700	170	Rib	200	.86	16.5	11.6	11.00	16.00	1.625		TSO	129.0
40	x 14		24	TL	39.80	38.85	14.00	13.25	35.10	12.00	VII	27,700	170	Rib	225	.86	16.5	11.6	11.00	16.00	1.625		TSO	130.0
40	x 14		26	TL	39.80	38.85	14.00	13.25	35.10	12.00	VII	30,500	175	Rib	200	.86	16.5	11.6	11.00	16.00	1.625	MIL		118.0
40	x 14		26	TL	39.80	38.85	14.00	13.25	35.10	12.00	VII	30,500	185	Rib	210	.86	16.5	11.6	11.00	16.00	1.625		TSO	
40	x 14		28	TL	39.80	38.85	14.00	13.25	35.10	12.00	VII	33,500	200	Rib	200	.86	16.5	11.6	11.00	16.00	1.625	MIL		127.0
40	x 14		28	TL	39.80	38.85	14.00	13.25	35.10	12.00	VII	33,100	200	Rib	210	.86	16.5	11.6	11.00	16.00	1.625		TSO	
C40	x 14	- 21	22	TL	40.0	38.90	14.00	13.25	37.75	11.10	NS	25,400	155	Rib	225	.68	16.9	13.4	6.75	21.00	.800		TSO	132.0
C40	x 14	- 21	24	TL	40.0	38.90	14.00	13.25	37.75	11.10	NS	25,400	155	Rib	210	.68	16.9	13.4	6.75	21.00	.800		TSO	
H40	x 14.5	- 19	22	TL	40.0	39.10	14.50	13.75	36.25	12.80	NS	30,100	180	Rib	210	.73	16.7	13.3	9.50	19.00	1.400		TSO	
H40	x 14.5	- 19	24	TL	40.0	39.10	14.50	13.75	36.25	12.80	NS	33,200	200	Rib	210	.73	16.7	13.3	9.50	19.00	1.400		TSO	
40	x 15.5	- 16	26	TL	40.00	39.05	15.50	14.75	35.70	13.65	NS	36,300	180	Rib	235	.78	16.1	11.6	10.00	16.00	1.250		TSO	
B40	x 18.0	- 16	20	TL	40.00	39.00	18.00	17.15	35.70	15.85	NS	27,600	130	Rib	190	.67	16.08	11.4	11.00	16.00	1.625	MIL		
41	x 15.0	- 18	22	TL	41.00	40.05	15.00	14.25	36.90	13.20	NS	28,600	170	Rib	225	.77	17.2	12.3	12.75	18.00	1.625		TSO	134.0
41	x 15.0	- 18	24	TL	41.00	40.05	15.00	14.25	36.90	13.20	NS	31,400	190	Rib	225	.77	17.2	12.3	12.75	18.00	1.625		TSO	128.0
42	x 13.0	- 18	28	TL	41.25	40.38	13.50	12.80	37.15	11.80	NS	35,900	205	Rib CH	230	.87	17.2	12.9	10.50	18.00	1.625	MIL		
H43	x 15.5	- 20	26	TL	43.00	42.10	15.50	14.75	38.85	13.65	NS	38,500		Rib	225	.74	18.1		11.00	20.00	1.600		TSO	
44"			12	TT	44.94	43.64	15.76	14.12	45.20	8.52	I	15,000	70	Rib	160	.72	18.7	13.4	14.87	22.00	1.063	MIL		
44"			12	TT	44.94	43.64	15.76	15.12	42.50	8.52	I	15,000	70	Dimple	160	.72	18.7	13.4	14.87	22.00	1.063		TSO	
44"			14	TT	44.94	43.64	15.76	15.12	42.50	8.53	I	18,200	85	Dimple	160	.72	18.7	13.4	14.87	22.00	1.063		TSO	
44	x 13		26	TL	43.55	42.30	13.50	12.80	39.45	12.80	VII	35,000	200	Rib	200	.87	18.4	13.7	11.00	20.00	1.875	MIL		126.0
44	x 16		28	TL	43.25	42.30	16.00	15.05	38.20	13.70	VII	38,400	185	Rib	200	.80	17.9	12.5	13.25	18.00	1.625	MIL		160.0
44	x 16		28	TL	43.25	42.30	16.00	15.05	38.20	13.70	VII	38,400	200	Rib	200	.80	17.9	12.5	13.25	18.00	1.625		TSO	167.0
44	x 16		28	TL	43.25	42.30	16.00	15.05	38.20	13.70	VII	38,400	200	Rib	225	.80	17.9	12.5	13.25	18.00	1.625		TSO	163.0
44	x 16		30	TL	43.25	42.30	16.00	15.05	38.20	13.70	VII	41,700	210	Rib	210	.80	17.9	12.5	13.25	18.00	1.625		TSO	173.0
44	x 16		30	TL	43.25	42.30	16.00	15.05	38.20	13.70	VII	41,700	210	Rib	225	.80	17.9	12.5	13.25	18.00	1.625		TSO	168.0
44	x 16		32	TL	43.25	42.30	16.00	15.05	38.20	13.70	VII	45,000	225	Rib	225	.80	17.9	12.5	13.25	18.00	1.625		TSO	180.0
44.5	x 16.0	- 21	24	TL	44.50	43.50	16.00	15.20	40.70	14.10	NS	40,500	195	Rib	225	.74	18.4	13.4	10.50	21.00	1.250	MIL		
44.5	x 16.0	- 21	28	TL	44.50	43.50	16.00	15.20	40.70	14.10	NS	45,400	240	Rib	260	.74	18.4	13.4	10.50	21.00	1.250	MIL		
44.5	x 16.0	- 21	34	TL	44.50	43.50	16.00	15.20	40.70	14.10	NS	60,900	315	Rib	263	.74	18.4	13.4	10.50	21.00	1.250	MIL		
44.5	x 16.5	- 18	30	TL	44.50	43.50	16.50	15.70	39.70	14.50	NS	42,500	195	Rib	225	.81	18.4	13.0	13.25	18.00	1.625		TSO	
44.5	x 16.5	- 18	30	TL	44.50	43.50	16.50	15.70	39.70	14.50	NS	42,500	195	Rib	200	.81	18.4	13.0	13.25	18.00	1.625		TSO	
H44.5	x 16.5	- 20	24	TL	44.50	43.50	16.50	15.70	40.10	14.50	NS	36,200	165	Rib	225	.75	18.3		10.50	20.00	1.500		TSO	
H45	x 17.0	- 20	26	TL	45.00	44.00	17.80	16.20	40.50	14.95	NS	39,800	195	Rib	225	.74	18.8		11.00	20.00	1.600		TSO	
46	x 16		20	TL	45.25	44.03	16.00	15.50	40.70	14.10	VII	29,900	145	Rib	200	.80	19.0	13.6	13.25	20.00	*1.875		TSO	
46	x 16		24	TL	45.25	44.30	16.00	15.05	40.70	14.10	VII	35,700	170	Rib	210	.80	19.0	13.6	13.25	20.00	1.75		TSO	161.0
46	x 16		26	TL	45.25	44.30	16.00	15.05	40.70	14.10	VII	38,300	185	Rib	200	.80	19.0	13.6	13.25	20.00	1.75		TSO	168.0
46	x 16		26	TL	45.25	44.30	16.00	15.05	40.70	14.10	VII	38,300	185	Rib	225	.80	19.0	13.6	13.25	20.00	1.75		TSO	160.0
46	x 16		28	TL	45.25	44.30	16.00	15.05	40.70	14.10	VII	41,800	210	Rib	225	.80	19.0	13.6	13.25	20.00	1.75		TSO	185.0
46	x 16		28	TL	45.25	44.30	16.00	15.05	40.70	14.10	VII	41,800	210	Rib	210	.80	19.0	13.6	13.25	20.00	1.75		TSO	
46	x 16		28	TL	45.25	44.30	16.00	15.05	40.70	14.10	VII	41,800	225	Rib	225	.80	19.0	13.6	13.25	20.00	1.75	MIL	TSO	
46	x 16		30	TL	45.25	44.30	16.00	15.05	40.70	14.10	VII	44,800	225	Rib	225	.80	19.0	13.6	13.25	20.00	1.75		TSO	186.0
46	x 16		30	TL	45.25	44.30	16.00	15.05	40.70	14.10	VII	44,800	225	Rib	210	.80	19.0	13.6	13.25	20.00	1.75		TSO	191.0

*Rim with FH of 1.625 is optional for 46x16 20 PR tire.

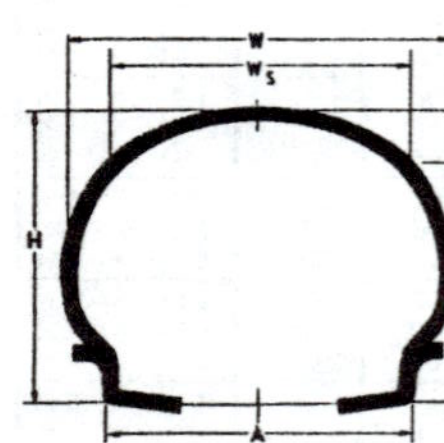

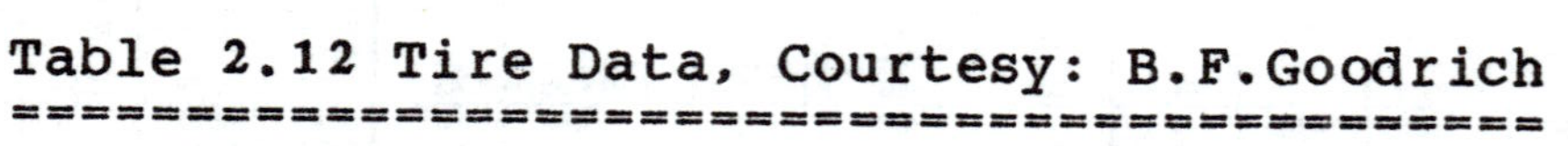

Table 2.12 Tire Data, Courtesy: B.F.Goodrich

Tire Description			Ply Rating	Tube Type Tube-less	Tire Dimensions (Inches)						Type	Max. Loading (Lbs.)	Unloaded Inflation Pressure (PSI)	Tread Patterns	Max. Speed (MPH)	Aspect Ratio	Loaded Radius		Rim Dimensions (Inches)			Qualification		WEIGHT
					D_o		W																	
D_o	W	D			Max.	Min.	Max.	Min.	D_s Max.	W_s Max.							Static	Flat Tire	A	D	FH	MIL	TSO	lbs
46	x 16		30	TL	45.25	44.30	16.00	15.05	40.70	14.10	VII	44,800	225	Rib	225	.80	19.0	13.6	13.25	20.00	1.750		TSO	186.0
B46	x 16.0	- 23.5	26	TL	46.00	45.10	16.00	15.20	41.95	14.10	NS	45,400	220	Rib	225	.71	19.4	14.75	10.50	23.50	1.250	MIL		
H46	x 18.0	- 20	26	TL	46.00	45.00	18.00	17.15	41.30	15.85	NS	40,700	185	Rib	225	.73	19.1		11.00	20.00	1.600		TSO	
47"			14	TT	44.98	47.02	17.00	16.32	44.24	10.66	I	17,500	70	Rib	160	.72	20.1	14.2	16.25	23.50	1.125	MIL		108.0
47	x 18	- 18	26	TL	46.90	46.00	18.00	17.25	41.60	16.34	NS	38,100	150	Rib	230	.81	19.2	12.5	14.75	18.00	1.625	MIL		150.0
47	x 18	- 18	30	TL	46.90	46.00	18.00	17.25	41.60	16.34	NS	43,700	175	Rib	225	.81	19.2	12.5	14.75	18.00	1.625	MIL		170.0
47	x 18	- 18	36	TL	46.90	46.00	18.00	17.25	41.60	16.34	NS	54,000	215	Rib	250	.81	19.2	12.5	14.75	18.00	1.625	MIL		
49	x 17		26	TL	48.75	47.70	17.25	16.40	43.00	14.50	VII	39,600	170	Rib	200	.84	20.2	14.0	13.25	20.00	1.750	MIL		
49	x 17		26	TL	48.75	47.70	17.25	16.40	43.00	14.50	VII	39,600	170	Rib	225	.84	20.2	14.0	13.25	20.00	1.750	MIL		204.0
49	x 17		28	TL	48.75	47.70	17.25	16.40	43.00	14.50	VII	43,200	180	Rib	210	.84	20.2	14.0	13.25	20.00	1.750		TSO	212.0
49	x 17		28	TL	48.75	47.70	17.25	16.40	43.00	14.50	VII	43,200	180	Rib	225	.84	20.2	14.0	13.25	20.00	1.750		TSO	206.0
49	x 17		30	TL	48.75	47.70	17.25	16.40	43.00	14.50	VII	50,400	195	Rib	225	.84	20.2	14.0	13.25	20.00	1.750		TSO	
49	x 17		30	TL	48.75	47.70	17.25	16.40	43.00	14.50	VII	50,400	210	Rib	225	.84	20.2	14.0	13.25	20.00	1.750		TSO	
49	x 17		30	TL	48.75	47.70	17.25	16.40	43.00	14.50	VII	46,700	195	Rib	210	.84	20.2	14.0	14.00	20.00	1.750		TSO	
49	x 17		30	TL	48.75	47.70	17.25	16.40	43.00	14.50	VII	46,700	195	Rib	210	.84	20.2	14.0	14.00	20.00	1.750		TSO	
49	x 17		30	TL	48.75	47.70	17.25	16.40	43.00	14.50	VII	46,700	195	Rib	225	.84	20.2	14.0	13.25	20.00	1.750		TSO	
49	x 17		32	TL	48.75	47.70	17.25	16.40	43.00	14.50	VII	50,400	210	Rib	225	.84	20.2	14.0	14.00	20.00	1.750		TSO	
49	x 19	- 20	32	TL	49.00	48.00	19.00	18.15	43.80	16.70	NS	51,900	195	Rib	235	.77	20.3	13.9	13.25	20.00	1.875		TSO	
50	x 20	- 20	24	TL	50.00	49.00	20.00	19.10	44.60	17.60	NS	38,200	135	Rib	200	.75	20.6	13.6	16.25	20.00	1.875		TSO	220.0
50	x 20	- 20	26	TL	50.00	49.00	20.00	19.10	44.60	17.60	NS	41,800	150	Rib	200	.75	20.6	13.6	16.25	20.00	1.875		TSO	238.0
50	x 20	- 20	30	TL	50.00	49.00	20.00	19.10	44.60	17.60	NS	49,400	175	Rib	210	.75	20.6	13.6	16.25	20.00	1.875		TSO	245.0
50	x 20	- 20	32	TL	50.00	49.00	20.00	19.10	44.60	17.60	NS	53,800	190	Rib	225	.75	20.6	13.6	16.25	20.00	1.875		TSO	240.0
50	x 20	- 20	34	TL	50.00	49.00	20.00	19.10	44.60	17.60	NS	57,000	205	Rib	225	.75	20.6	13.6	16.25	20.00	1.875		TSO	
50	x 21	- 20	30	TL	50.00	49.00	21.00	20.10	44.60	18.50	NS	49,000	160	Rib	210	.72	20.2	14.6	13.25	20.00	1.750		TSO	
50	x 21	- 20	30	TL	50.00	49.00	21.00	20.10	44.60	18.50	NS	49,000	160	Rib	225	.72	20.2	14.6	13.25	20.00	1.750		TSO	
52	x 20.5	- 20	34	TL	52.00	51.00	20.50	19.60	46.25	18.05	NS	57,800	185	Rib	235	.79	21.3	14.3	16.25	20.00	1.875		TSO	
52	x 20.5	- 20	36	TL	52.00	51.00	20.50	19.60	46.25	18.05	NS	62,500	200	Rib	225	.79	21.3	14.3	16.25	20.00	1.875		TSO	
52	x 20.5	- 23	26	TL	52.00	51.00	20.50	19.60	46.80	18.05	NS	55,000	165	Rib	235	.71	21.3	15.9	13.00	23.00	1.500		TSO	
52	x 20.5	- 23	28	TL	52.00	51.00	20.50	19.60	46.80	18.05	NS	59,400	180	Rib	235	.71	21.3	15.9	13.00	23.00	1.500		TSO	
56"			20	TT	56.62	55.44	19.92	19.12	53.44	11.44	I	35,000	100	Rib	160	.74	23.6	16.9	18.94	27.00	1.375	MIL		203.0
56"			20	TT	56.62	55.44	19.92	19.12	53.44	11.44	I	35,000	100	Rib	160	.74	23.6	16.9	18.94	27.00	1.375	MIL		187.0
56"			22	TL	56.62	55.44	19.92	19.12	53.44	11.44	I	37,500	110	Dimple	160	.74	23.6	16.9	18.94	27.00	1.375		TSO	238.0
56"			22	TT	56.62	55.44	19.92	19.12	53.44	11.44	I	37,500	110	Dimple	160	.74	23.6	16.9	18.94	27.00	1.375		TSO	226.0
56	x 16		24	TL	55.90	54.80	16.20	15.50	50.85	14.30	VII	45,000	178	Rib	200	.88	24.1	18.7	12.75	28.00	2.25	MIL		
56	x 16		32	TL	55.90	54.80	16.20	15.50	50.85	14.30	VII	60,000	250	Rib	250	.88	24.1	18.7	12.75	28.00	2.25	MIL		
56	x 16		38	TL	55.90	54.80	16.20	15.50	50.85	14.30	VII	76,000	315	Rib	250	.88	24.1	18.7	12.75	28.00	2.25	MIL		
56	x 20	- 20 -	24	TL	56.00	54.80	20.00	19.10	49.50	17.60	NS	38,500	110	Rib	200	.90	22.7	15.2	15.50	20.00	2.000		TSO	

Table 2.13 Tire Applications, Courtesy: B.F.Goodrich

Business, Personal, Utility

Aircraft	Model	Popular Name	MAIN GEAR Tire Size	MAIN GEAR Ply Rating	AUXILIARY GEAR Tire Size	AUXILIARY GEAR Ply Rating
Air Products Co.	F-1A	Aircoupe	6.00-6	4 TT	6.00-6	4 TT
Bede Aircraft Corp.	XBD-2		15x6.00-6	4 TT	5.00-5	4 TT
	BD-3, BD-5, BD-5J, BD-4		7.00-6	6 TT	5.00-5	6 TT
Beech Aircraft Corp.	18	Twin Beech	8.50-10	8 TL	8.50-10	8 TL
	19	Musketeer Sport	6.00-6	4 TT	6.00-6	4 TT
	23	Musketeer	6.00-6	4 TT	6.00-6	4 TT
	23-C	Sundowner	6.00-6	4 TT	6.00-6	4 TT
	24	Musketeer	6.00-6	4 TT	15x6.00-6	4 TT
	24-R, 25-R	Super Musketeer	6.00-6	4 TT	5.00-5	4 TT
	33, 33A	Debonair	6.00-6	6 TT	5.00-5	4 TT
	35, 35B, 36	Bonanza	7.00-6	6 TT	5.00-5	4 TT
	50	Twin Bonanza	8.50-10	6 TT	6.50-10	6 TT
	B55, C55, E55	Baron	6.50-8	8 TT	5.00-5	6 TT
	5670, 56TO, 58	Baron	6.50-8	6 TT	5.00-5	6 TT
	58TC, 58P	Baron	6.50-8	8 TT	5.00-5	6 TT
	60	Duke	19.5x6.75-8	10 TT	6.00-6	4 TT
	D18S, D18C, E18S, R18S, 9700, G18S	Twin Beech	11.00-12	8	14.50 (Tail)	6 TT
	H18	Twin Beech	8.50-10	10	14.50 (Tail)	6 TT
	H18 (opt. equip.)	Twin Beech	8.50-10	8/10	8.50-10 (NW)	8/10
	65, A65, 70, 80, A80, B80, 88	Queen Air	8.50-10	8 TL	6.50-10	6 TL
	90, A90, B90, C90	King Air	8.50-10	8 TL	6.50-10	6 TL
	E90	King Air	8.50-10	10 TL	6.50-10	6 TL
	99, 99A, A99, A99A, B99	Airliner	18x5.5	8 TL	6.50-10	6 TL
	99, 99A, A99, A99A, B99 (opt. equip.)	Airliner	6.50-10	6 TL	6.50-10	6 TL
	95	Travelair	7.00x6	6 TT	5.00-5	6 TT
	100, A100, B100	King Air	18x5.5	8 TL	6.50x10	6 TL
	100, A110, B100 (opt.)	King Air	6.50-10	6 TL	6.50x10	6 TL
	200	Super King Air	18x5.5	8 TL	22x6.75-10	8 TL
	200 (opt.)	Super King Air	22x6.75-10	8 TL	22x6.75-10	8 TL
Bellanca Inter Air	260A	Bellanca	6.00-6	6 TL	6.00-6	4 TT
		Viking	6.00-6	6 TL	6.00-6	4 TT
	17-30A	Viking	6.00-6	6/8	15x6.00-6	6/8 TT
	17-31A	Super Viking	6.00-6	6/8	15x6.00-6	6/8 TT
	17-31ATC	Turbo Viking	6.00-6	6/8	15x6.00-6	6/8 TT
	7ECA, 7GCAA	Citabria	6.00-6	4/6		
	7BCBC	Citabria	7.00-6	4/6 TT		
	7KCAB	Citabria	8.00-6	4 TT		
	8GC5BC	Scout	8.50-6	4/6		
Cessna	150, 152, A152		6.00-6	4 TT	5.00-5	4 TT
	172	Skyhawk	6.00-6	4 TT	5.00-5	4 TT
	R172	Hawk XP	6.00-6	6 TL	5.00-5	6 TT
	F172		6.00-6	4 TT	5.00-5	4 TT
	175	Sky Lark	6.00-6	4 TT	5.00-5	4 TT
	177	Cardinal	6.00-6	6 TT	5.00-5	4 TT
	177RG	Cardinal RG	15x6.00-6	6 TT	5.00-5	4 TT
	180	Skywagon	6.00-6	6 TL	8" SC	6 TT
	182, 182-N	Skylane	6.00-6	6 TL	5.00-5	6 TT
	R182	Skylane RG	15x6.00-6	6 TT	5.00-5	6 TT
	185	Skywagon	6.00-6	6 TT, TL	10" SC	8 TT
	A185	Skywagon	6.00-6	6 TT	8.00-3	4 TT
	A185 (opt.)	Skywagon	8.00-6	6 TT	8.00-3	4 TT
	188	AG Wagon	8.00-6	6 TT	8.00" SC	8 TT
	188 (opt.)	AG Wagon	6.50-10	6 TT	10.00"	8 TT
	A188	AG Wagon & AG Truck	8.00-6	6 TT	10.00-3.5	4 TT
	A188 (opt.)	AG Wagon & AG Truck	8.5-10	6 TT	10.00-3.5	4 TT
	206, P206	Super Skylane	6.00-6	6 TT	5.00-5	6 TT
	TP206	Super Skylane Turbo	6.00-6	6 TT	5.00-5	6 TT
	U206/	Super Skywagon	6.00-6/	6 TT	6.00-6	6 TT
	Opt.		8.00-6	6 TT	6.00-6	4 TT
	TU206/	Super Skywagon Turbo	6.00-6/	6 TT	6.00-6	6 TT
	Opt.		8.00-6	6 TT	6.00-6	4 TT
	U206E	Station Air	6.00-6	6 TT	5.00-5	6 TT
	TV206E	Station Air Turbo	6.00-6	6 TT	5.00-5	6 TT
	207	Skywagon	6.00-6	8 TT	5.00-5	6 TT
	207 (opt.)	Skywagon	8.00-6	6 TT	6.00-6	4 TT

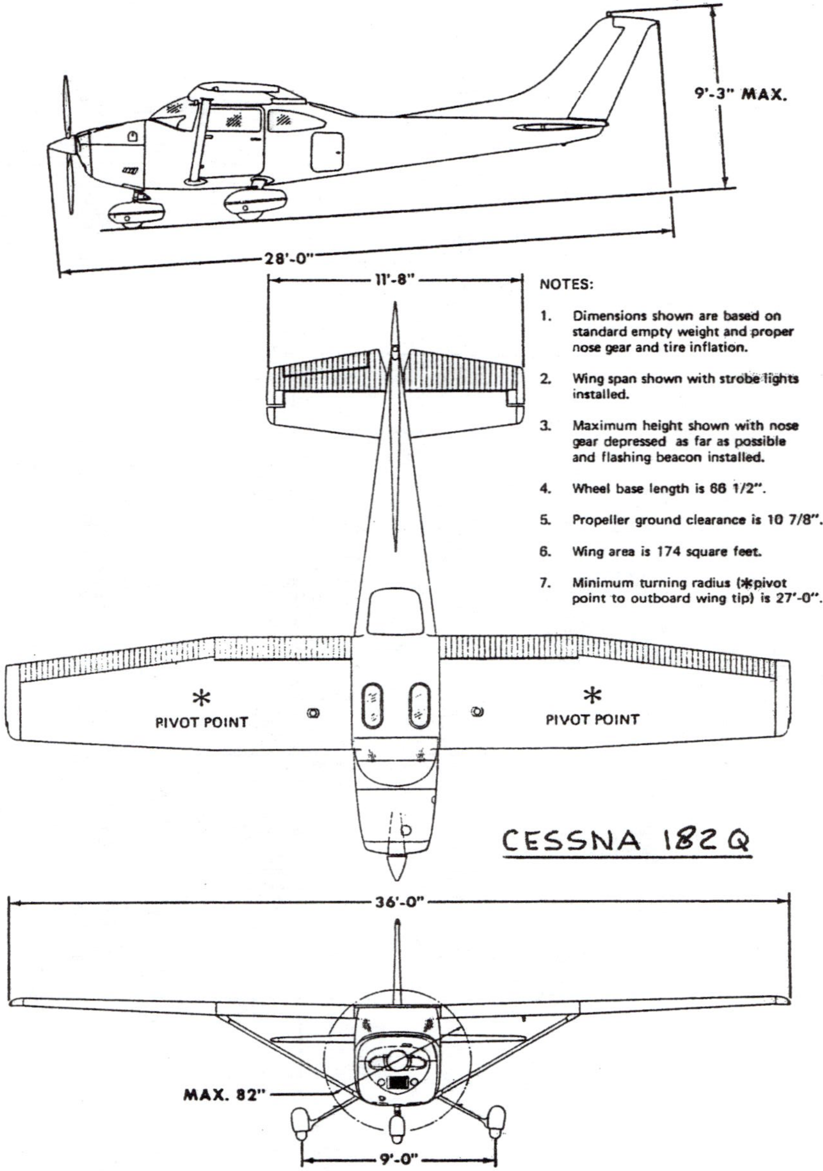

9'-3" MAX.

28'-0"

11'-8"

PIVOT POINT

PIVOT POINT

CESSNA 182Q

36'-0"

MAX. 82"

9'-0"

NOTES:

1. Dimensions shown are based on standard empty weight and proper nose gear and tire inflation.
2. Wing span shown with strobe lights installed.
3. Maximum height shown with nose gear depressed as far as possible and flashing beacon installed.
4. Wheel base length is 66 1/2".
5. Propeller ground clearance is 10 7/8".
6. Wing area is 174 square feet.
7. Minimum turning radius (*pivot point to outboard wing tip) is 27'-0".

Table 2.14 Tire Applications, Courtesy: B.F.Goodrich

Business, Personal, Utility (Continued)

Aircraft	Model	Popular Name	MAIN GEAR Tire Size	MAIN GEAR Ply Rating	AUXILIARY GEAR Tire Size	AUXILIARY GEAR Ply Rating
Cessna (Cont.)	T207/	Turbo Skywagon	6.00-6	8 TT	5.00-5	6 TT
	Opt.		6.00-6	6 TT	6.00-6	4 TT
	210, 210-2	Centurion	6.00-6	6 TT	5.00-5	6 TT
	T210, T210-2	Turbo Centurion	6.00-6	6 TT	5.00-5	6 TT
	P210	Pressurized Centurion	6.00-6	8 TL, TT	5.00-5	10 TL
	310, 310-2, 310 (Turbo)		6.50-10	6 TL, TT	6.00-6	6 TT
	320	Skynight	6.50-10	6 TL	6.00-6	6 TL
	336		6.00-6	6 TL, TT	15x6.00-6	4 TT
	337	Super Skymaster	6.00-6	6 TL, TT	6.00-6	6 TL
	337 (opt.), T337, P337	Super Skymaster	18x5.5	8 TT	15x6.00-6	4 TT
	340, 340-A		6.50-10	8 TT	6.00-6	6 TT
	401, 402, 402-B		6.50-10	8 TT	6.00-6	6 TT
	404, 441		22x7.75-10	8 TL	6.00-6	6 TT
	411, 414, 421		6.50-10	8 TT	6.00-6	6 TT
	500	Citation I	22x8.0-10	10 TL	18x4.4 CH-DU	10 TL
	550	Citation II	22x8.0-10	10 TL	18x4.4 CH-DU-LG	10 TL
Champion Aircraft	75RS	Citabria	6.00-6	4 TT	5.00-5	4 TT
	7EC	Traveler	6.00-6	4 TT	5.00-5	4 TT
	7FC	Tri-Traveler	6.00-6	4 TT	5.00-5	4 TT
	7GCB	Challenger	7.00-6	4 TT	5.00-5	4 TT
	402	Lancer	7.00-6	4 TT	5.00-5	4 TT
	7GCB-A	Challenger	8.00-6	4 TT	5.00-5	4 TT
Custer Channel Wing	CCW5		6.50-10	6 TL	6.00-6	4 TL, TT
Grumman Aircraft	AA1B, AA5	Trainer	6.00-6	4 TT	5.00-5	4 TT
	AA1B	TR-2	6.00-6	4 TT	5.00-5	4 TT
	158A	AG-CAT	8.50-10	6 TT	10.00"	8 TT
	159	Gulfstream I	7.50-14	12 TL	6.50-8	6 TL
	164	Gulfstream II	34x9.25-16	16 TL	21x7.25-10 CH	8 TL
	164	AG-CAT	8.50-6	6 TT	10"	8 TT
Hawker-Siddeley	HS125 Up to Series 700		23x7.00-12	10 TL	18x4.25-10 CH	6 TL
Helio Aircraft Corp.	250	Courier	6.50-8	6 TL	6.00-6	6 TL, TT
	295	Super Courier	8.00-6	6 TT	6.00-6	6 TT
	550-A	Stallion	7.50-10	6 TT	5.00-5	4 TT
	634	Twin Stallion	7.50-10	6 TT	5.00-5	4 TT
Howard Aero Mfg.		500	15.00-16	14/16 TT/TL	17" SC	10 TT
		Super Ventura	16.00-16	14 TT	17" SC	10 TT
Lake Aircraft Corp.	4	Buccaneer	6.00-6	4 TT	5.00-4	4 TT
	4, 4-200	(Amphibian)	6.00-6	4 TT	5.00-4	4 TT
Lear	23	Learjet	18x5.5	8 TL	18x4.4 CH-DU	10 TL
	24, 25	Learjet	18x5.5	10 TL	18x4.4 CH-DU	10 TL
	35, 36	Learjet	17.5x5.75-8	12 TL	18x4.4 CH-DU	10 TL
		Learstar	15.00-16	10 TT	17" SC	10 TT
Lockheed Aircraft Co.		Lodestar	15.00-16	10 TT	17" SC	10 TT
		Jetstar-6	26x6.6	14 TL	18x4.4 CH	10 TL
		Jetstar-8/ II	26x6.6	14 TL	18x4.4 CH	12 TL
	SA60	Azacarte-60	6.50-8	4 TT	6.00-6	4 TT
Maule	4-220C	Lunar Rocket	8.50-6	6 TT	–	–
	5-210-C		8.00-6	4 TT	–	–
			6.00-6	4 TT	–	–
Meyers	200B		6.50-8	6 TT	6.00-6	4 TT
Mitsubishi	MU-2		8.50-10	8 TL	5.00-5	6 TT
	MU-2		8.50-10	10 TL	5.00-5	6 TT
Mooney	Mark 20	Ranger	6.00-6	6 TT	5.00-5	4 TT
	Mark 21C	Super	6.00-6	6 TT	5.00-5	4 TT
	Mark 22	Mustang	6.00-6	6 TT	15x6.00-6	4 TT
	20E	Chaparral	6.00-6	6 TT	5.00-5	4 TT
	20F	Executive	6.00-6	6 TT	5.00-5	6 TT
	20J	201	6.00-6	6 TT	5.00-5	6 TT
Mystere (Dassault)	Fanjet Falcon 10		22x5.75-12	8/10 TL	18x5.75-8 CH	8 TL
	Fanjet Falcon 20 (thru F Models)		26x6.6	10 TL	14.5x5.5-6 CH	14 TL
	Fanjet Falcon 20G and 50		26x6.6	14 TL	14.5x5.5-6 CH	14 TL
Navion	Navion G.H.T.	Rangemaster	6.50-8	4 TT	6.00-6	4 TT
Piper	18-135	Cub	8.00-4	4 TT	6x2.00	Solid
	18A-150	Super Cub	8.00-4	4 TT	6x2.00	Solid
	22	Caribbean	6.00-6	4 TT	6.00-6	4 TT

Aircraft	Model	Popular Name	MAIN GEAR Tire Size	MAIN GEAR Ply Rating	AUXILIARY GEAR Tire Size	AUXILIARY GEAR Ply Rating
Piper (Cont.)	22-108	Colt	6.00-6	4 TT	6.00-6	4 TT
	22-160	Tri-Pacer	6.00-6	4 TT	6.00-6	4 TT
	23, 23-235, 23-160	Apache	7.00-6	6 TT	6.00-6	4 TT
	23-250	Aztec	7.00-6	8 TT	6.00-6	4 TT
	24-180	Comanche	6.00-6	4 TT	6.00-6	4 TT
	24-250, 24-260	Comanche	6.00-6	6 TT	6.00-6	4 TT
	24-400, 25-400	Comanche	6.00-6	6 TT	6.00-6	6 TT
	25-150	Pawnee	7.00-6	4 TT	8.00x3-4	4 TT
	25-135, 25-235, 25-260	Pawnee	8.00-6	4 TT	8.00x3-4	4 TT
	28, 28-180R, 28-200R	Arrow	6.00-6	4 TT	5.00-5	4 TT
	28-140, 28-235, 28-160, 28-180, 28-150	Cherokee	6.00-6	4 TT	6.00-6	4 TT
	28-151	Warrior	6.00-6	4 TT	5.00-5	4 TT
	30	Twin Comanche	6.00-6	6 TT	6.00-6	4 TT
	31, 31-350, 31-300, 31-310	Navajo	6.50-10	8 TT	6.00-6	6 TT
	31P-425	Navajo	6.50-10	8 TT	6.00-6	8 TT
	31T	Navajo	6.50-10	10 TT, TL	18x4.4	6 TT, TL
	34, 34-200	Seneca	6.00-6	8 TT	6.00-6	6 TT
	32, 32-300	Cherokee (6 ply)	6.00-6	6 TT	6.00-6	6 TT
(See Ted	32-260	Cherokee	6.00-6	8 TT	6.00-6	6 TT
Smith Aircraft)	36-285, 300, 375	Brave	8.50-10	6 TT	6x2.00	solid
	42	Cheyenne	6.50-10	12 TL	17.5x6.25-6	10 TT
Rockwell Int. (Aero Commander)	112, 112A	Commander	6.00-6	4/6 TT	5.00-5	4 TT
	200	Commander	7.00-6	6 TT	6.00-6	4 TT
	500A	Commander	8.50-10	6 TL	6.00-6	6 TL, TT
	680FP	Commander	8.50-10	6 TT	6.00-6	6 TL, TT
	(Turbo 11) 681	Commander	8.50-10	10 TT, TL	6.00-6	6 TL, TT
	500B	Commander	8.50-10	8 TL	6.00-6	6 TL, TT
	500S	Shrike Commander	8.50-10	8 TL, TT	6.00-6	6 TL, TT
	720, 560E, 560F, 680F	Aero Commander	8.50-10	8 TL	6.00-6	6 TL, TT
	680FL, 680FLP	Grand Commander	8.50-10	8 TL	6.00-6	6 TL, TT
	680T	Turbo Commander	8.50-10	8 TL	6.00-6	6 TL, TT
	685, 690, 690A, 690B	Commander	8.50-10	10 TL	6.00-6	6 TL, TT
	1121	Jet Commander	24x7.7	14 TL	16x4.4	4 TL
	5ZR	Thrvs H Commander	8.50-10	10 TL	12.5-4.5	4 TT
Rockwell Int. (North American)	40	Sabreliner (Low Pressure)	26x6.75-14	14 TL	18x4.4 CH	10 TL
		(Standard)	26x6.6	14 TL	18x4.4 CH	10 TL
	60	Sabreliner (Low Pressure)	26x6.75-14	14 TL	18x4.4 CH	10 TL
		(Standard)	26x6.6	14 TL	18x4.4 CH	10 TL
	70	Sabreliner	22x5.75-12	10 TL	18x4.4 CH	10 TL
	75-A	Sabreliner	22x5.75-12	10 TL	18x4.4 CH	10 TL
	265	Sabreliner	26x6.6	14 TL	18x4.4	6 TL
					18x4.4 CH	10 TL
Smith (L.B.) Aircraft	C465 Conver		17.00-20	20 TT	10.00-7	12 TT
	B26 Conver	Tempo II	17.00-20	16 TT		
Swearingen Aircraft		Metro	18x5.5	8/10 TL	16x4.4	6 TT
		Merlin	8.50-10	8 TL	6.50-10	6 TT
	226-R-TC, AT		19.5x6.75-8	10 TT, TL	16x4.4	6 TT
	3, 4		22x6.75-10	8 TL	16x4.4	6 TT
Taylorcraft, Inc.	F19		6.00-6	4 TT	6.00-2	
	20	Topper	7.00-6	4 TT	Tail Skid	
Ted Smith Aircraft (Piper)	600, 601	Aerostar	7.00-6	8 TT	6.00-6	6 TL, TT
	600, 601, 601P		6.50-8	8 TT	6.00-6	6 TT
	600, 601, 601P		19.5x6.75-8	8 TT	6.00-6	6 TT
Thurston		Teal	6.00-6	4 TT	8x3.00-4	
Ransland Aircraft	T-17		5.00-5	6 TT	5.00-5	4 TT
Trecker Aircraft	166		8.50-10	8 TT	6.00-6	6, 8 TT
	L-2, L-2	Gull Amphibian	8.50-10	6 TT		
Volaircraft	10	1900	6.00-6	4 TT	6.00-6	4 TT
			6.00-6	4 TT	6.00-6	4 TT
			15x6.00-6	4 TT		
Volpar		Tri Gear	8.50-10	10 TL	8.50-10	10 TL
		Turboliner	8.50-10	10 TL	8.50-10	10 TL
Waco Aircraft	250	Vega	6.00-6	6 TT	5.00-5	6 TT
	260	Meteor	6.00-6	6 TT	5.00-5	6 TT

Table 2.15 Tire Applications, Courtesy: B.F.Goodrich

Commercial Transport

Manufacturer	Model	MLG Size	MLG PR	NLG Size	NLG PR
Airbus	A300B	46x16	24 TL	40x14	20 TL
	A300B1-B2-B4	46x16	24/28 TL	40x14	20 TL
	A300B1-B2-B4 (Optional)	49x17	28 TL	40x14	22 TL
Boeing	707-120	46x16	24 TL	39x13	14 TL
	707-320B	46x16	26 TL	39x13	16 TL
	707-320C	46x16	28 TL	39x13	16 TL
	720	40x14	24 TL	34x9.9	12 TL
	720B	40x14	24 TL	39x13	14 TL
	727-100	49x17	26 TL	32x11.50-15 CH	12 TL
	727/200	49x17	28 TL	32x11.50-15 CH	12 TL
		50x20-20	30 TL	32x11.50-15 CH	12 TL
	727 ADV	50x21-20	28/30 TL	32x11.50-15 CH	12 TL
	737-100	40x14	22 TL	24x7.7	14 TL
	737-200	40x14	24 TL	24x7.7	14 TL
	737-200	C40x14-21	22, 24 TL	24x7.7	14 TL
		C40x18-17	20 TL	C24.5x8.5-12	12 TL
	747-100	46x16	28 TL	46x16	26 TL
	747-200	49x17	30 TL	49x17	30 TL
	747-200F	49x19-20	32 TL	49x19-20	32 TL
	747SP	46x16	26 TL	46x16	26 TL
	757	H40x14.5-19	22 TL	H31x13-12	20 PR TL
	767	H45x17-20	26 TL	H37x14-15	22 TL
	767	H46x18-20	26 TL	H37x14-15	22 TL
British Aircraft	BAC-111-200	40x12	16 TL	24x7.25-12 CH	10 TL
	BAC-111-400	40x12	18 TL	24x7.25-12 CH	10 TL
	500/510	40x12	20 TL	24x7.25-12 CH	10 TL
	475	40x12	16/18 TL	24x7.75-12	12 TL
Bristol Freighter		48x18.0-18	14		
Britannia	300	15.50-20			
		40x12	22 TL	32x8.8	12
Canadair	CL440D-2	40x12	22 TL	32x8.8	12 TL
	CL600	26x6.6	12 TL	18x4.4	8 TL
Caravelle	48C, 50C	35x9.00-17	14 TL	26x7.75-13 CH	10 TL
	52C, 56C	35x9.00-17	16 TL	26x7.75-13 CH	10 TL
	58C	35x9.00-17	18 TL	26x7.75-13 CH	10 TL
Convair	240	34x9.9	14 TT	26x6.6	12 TL
	340	12.50-16	12 TT	7.50-14	8 TT
	440	12.50-16	12 TT	7.50-14	8 TT
	540	12.50-16	12 TT	7.50-14	8 TT
	580/600/640	12.50-16	14 TT	7.50-14	8 TT
		39x13	14 TT	7.50-14	8 TT
	880	39x13	20 TL	29x7.7	12 TL
	880M	39x13	22 TL	29x7.7	12 TL
	990	41x15.0-18	22 TL	29x7.7	16 TL
	990HV	41x15.0-18	24 TL	29x7.7	16 TL
Comet		35x9.00-17	16	30x9.10-15	10
Curtis	C46	19.00-23	16 TT	10.00-7	12 TT
De Havilland	DHC-6 Twin Otter	11.00-12	8 TT	8.90-12.50	6 TT
	Flotation	15.00-12	10 TL	15.00-12	10 TL
	DHC-7	30x9.00-15	12 TL	8.90-12.50	6 TL
Fairchild Hiller	F-27/227	9.50-16	10-12 TL	8.50-10	10 TL
	F-28	39x13	14 TL	24x7.7	10 TL
Fokker	F-27	34x10.75-16	10 TL	8.50-10	10 TL
	F-28	39x13	14 TL	24x7.7	10 TL
	F-28 (Flotation)	40x14	14 TL	24.5x8.5-10 CH	10 TL
	F-227	9.50-16	12 TL	8.50-10	10 TL
Hawker-Siddeley	HS 748, HFB-330	32x8.8	12 TL	25.65x8.5-10	
	HS 748	32x10.75-14	12 TL	8.50-10	6 TT/TL
Lockheed	049	17.00-20	20 TT	33"	10 TT
	747	17.00-20	20 TT	33"	10 TT
	1049	17.00-20	22 TT	34x9.9	10 TT
	188	13.50-16	24 TL	7.50-14	10 TL
	188 (Alt.)	40x14	24 TL	28x7.7	10-14 TL
	L-1001-1	56x20.0-20	24 TL	39x13	14 TL
	L-1011-1	50x20.0-20	32 TL	36x11	20-22 TL
	L-1011-14/15	52x20.5-20	34/36 TL	37x13-16	20 TL
	L-1011-500	52x20.5-20	36 TL	37x13-16	26 TL
Martin	202/404	12.50-16	12 TL	9.50-16	10 TT
McDonnell-Douglas Co. (Douglas Div.)	DC-3 (Super)	17.00-16	12 TT	9.00-6	10 TT
	DC-4	15.50-20	14-16 TT	44" SC	12 TT
	DC-6, DC-6A	15.50-20	16 TT	44" SC	12 TT
	DC-6B	15.50-20	20 TL	44" SC	12 TT
	DC-7, DC-7B	17.00-20	20 TL	44" SC	14 TL
	DC-7C, DC-7C/F	17.00-20	22 TL	15.00-16	14 TL
	DC-8	44x16	26 TL	34x11	18 TL
	DC-8 (HV)	44x16	28 TL	34x11	20 TL
	DC-8 (50F)	44x16	28 TL	34x11	20-22 TL
	DC-8 (61, 62)	44x16	30 TL	34x11	22 TL
	DC-8 (62F, 62H)	44.5x16.5-18	30 TL	34x11	22 TL
	DC-8 (63)	44x16	32 TL	34x11	22 TL
	DC-8 (63, 63F)	44.5x16.5-18	30 TL	34x11	22 TL
	DC-8F	44x16	28 TL	34x11	22 TL
	DC-9 (10, 11, 12, 14, 15)	40x14	20 TL	26x6.6 (CH)	8 TL
	DC-9 (20)	40x14	22 TL	26x6.6 (CH)	10 TL
	DC-9 (30, 31)	40x14	24 TL	26x6.6 (CH)	10 TL
	DC-9 (30, 32)	42x15.0-16	20 TL	26x6.6 (CH)	10 TL
	DC-9 (30, 33, 41, 50)	41x15.0-18	22, 24 TL	26x6.6 (CH)	10 TL
	DC-10 (10)	50x20.0-20	32 TL	37x14-14	24 TL
	DC-10 (20, 30, 40)	52x20.5-23	26/28 TL	40x15.5-16	26 TL
Mercure	02	46x16	20	30x8.8	12
Nissau	YS-11	12.50-16	12 TL	24x7.7	10 TL
Nord	262	12.50-16	10 TT	6.00-6	8 TT
Potez	840	26x7.75-13	10 TL		
Swiss	Caravelle 48t, 50t	35x9.00-17	14 TL	26x7.75-13 CH	10 TL
	52t, 56t	35x9.00-17	16 TL	26x7.75-13 CH	10 TL
	58t	35x9.00-17	18 TL	26x7.75-13 CH	10 TL
	Concorde Preprod	47x15.75-22	26	31x10.75-14	20
	Prod	1195x400-22		785x275-14	
Shorts	SD3-30	34x10.75-16	10 TL	9.00-6	10 TL
Trident	1	34x9.50-18	12 TL	29x8.0-15	12 TL
	1E, 2E, 3E	36x10.0-18	16 TL	29x8.0-15	12 TL
Vickers	742/745	36x10.75-16.50	12 TT	24x7.25-12 (CH)	8 TT
	VC-10	50x18	26 TL	39x13	16 TL
	Vanguard	49x17	24 TL	33x9.75-16	10 TL

Table 2.16 Tire Applications, Courtesy: B.F.Goodrich

Military Airplanes

Manufacturer	DOD Designation	Popular Name	Old Designation	MAIN GEAR		AUXILIARY GEAR	
				Tire Size	Ply Rating	Tire Size	Ply Rating
Beech	T34B	Mentor	T34B	6.50-8	6 TT	5.00-5	6 TT
	U8F	Seminole	L23F	8.50-10	8 TT	6.50-10	6 TT
	T42A			6.50-8	6 TL	5.00-5	6 TL
	VC6A	King Air		8.50-10	8 TL	6.50-10	6 TL, TT
Boeing	B47E	Stratojet	B47E	56x16	36 TL	26x6.6	14 TT
	B52F, G, H	Stratofortress	B52, F, G, H	56x16	38 TL	32x8.8	12 TT
	C97D	Stratofreighter	C97D	56x16	32 TT	36" SC	12 TT
	C135B	Stratolifter	C135B	49x17	26 TL	38x11	14 TL
	VC137C		VC137C	46x16	28 TL	39x13	16 TL
	E3A	707		46x16	38 TL	39x16	16 TL
	E4A	747		49x17	30 TL	49x17	30 TL
	T43A	737		40x14	24 TL	24x7.7	14 TL
	YC14	STOL		B40x18.0-16	20 TL	B40x18.0-16	20 TL
Cessna	O-1E	Bird Dog	L19A	8.00-6	6 TL	8x3.0-4	4 TT
	T37B		T37B	20x4.4	10 TL	16x4.4	6 TT
	A37		T-37A	7.00-8	12 TL	6.00-6	6 TT
	T41A	Skyhawk	172	6.00-6	4 TL	5.00-5	4 TL
	U3B		U3B	6.50-10	6 TT	6.00-6	6 TT
	U17A	Skywagon		6.00-6	6 TT	10" SC	8 TT
	O2A, B			6.00-6	8 TT	15x6.00-6	4 TT
De Havilland	CV-2B	Caribou	AC-1A	11.00-12	8 TL	7.50-10	6 TL
	CV-7A	Buffalo	AC-2	15.00-10			
				15.00-12	10 TL	8.90-12.50	6 TL
	U-1A	Otter		11.00-12	6 TT	6.00-6	6 TT
	U-6A	Beaver	L-20	8.50-10	6 TT	5.50-4	6 TT
Fairchild-Hiller	C-119J	Flying Boxcar	C-119J	15.50-20	14 TT	9.50-16	10 TT
	C-123B	Provider	C-123B	17.00-20	22 TT, TL	11.00-12	8 TT
	C-123J	Provider	C-123J	17.00-20	22 TT, TL	9.50-16	10 TT
(Republic Division)	F-84F	Thunder Streak	F-84F	30x6.6	14 TT	20x4.4	12 TL,TT
	F-105	Thunder Chief	F-105	36x11	24 TL	24x7.7	14 TL
	A-10A	—	—	36x11	22 TL	24x7.7	10 TL
General Dynamics	B-58A	Hustler	B-58A	22x7.7-12	16 TL	22x7.7-12	16 TL
	F102A	Delta Dagger	F102A	30x8.8	22 TT	24x5.5	12 TT
	F106A	Delta Dart	F106A	30x8.8	22 TL	18x4.4	12 TT
	C131A	Samaritan	C131A	34x9.9	14 TL	26x6.6	14 TL
	C131B, C, D, E, F	Samaritan	C131B, C, D, E, F	12.50-16	12 TT	7.50-14	8 TT
	RB-57F			44x13	26 TT	24x5.5	12 TT
	T29D		T29D	34x9.9	14 TT	26x6.6	14 TL
	F-16			25.5x8.0-14	18 TL	18x5.5	14 TL
	F111A			47x18-18	30 TL	22x6.6-10	16 TL
	FB111A			47x18-18	36 TL	22x6.6-10	20 TL
	FB-111B			47x18-18	36 TL	21x7.25-10	20 CH, TL
Grumman	A6A	Intruder	A2F-1	36x11	24 TL	20x5.5	12 TL
	E16B	Intruder		36x11	24 TL	20x5.5	16 TL
	F11A	Tiger	F11F-1	26x6.6	16 TL	18x5.5	8 TL
	OV-1C	Mohawk	AO-1C	8.50-10	12 TL	6.50-8	6 TL
	S-2D	Tracker	S2F-3	34x9.9	14 TT	18x5.5	12 TT
	F-14A	Tomcat		37x11.50-16	28 TL	22x6.6-10	20 TL
	E-1B	Tracer	WF-2	34x9.9	14 TT	18x5.5	12 TT
	E-2A	Hawkeye	W2F-1	36x11	24 TL	20x5.5	12 TL
	C-1A	Trader	TF-1	34x9.9	14 TT	18x5.5	12 TT
	C-2A		C-2A	36x11	24 TL	20x5.5	12 TL
	HU-16E	Albatross	UF-2G	40x12	14 TT	26x6	10 TT
Helio	U10A	Courier	L28A	6.50-8	6 TT	10" SC	8 TT
Ling-Temco-Vought (Vought Aeronautics Div.)	A7A, B, E	Corsair II		28x9.0-12	22 TL	22x5.5	12 TL
	A7D	Corsair II		28x9.0-14	22 TL	22x5.5	10 TL
	F8H, J	Crusader	F8E	26x6.6	16 TL	20x5.5	14 TL
	XC142A		XC-142A	11.00-12.	8 TL	8.50-10	10 TL
Lockheed (Early Version)	F104A, B	Starfighter		25x6.75	18 TL	18x5.5	14 TT
	F104C, D, J, DJ	Starfighter		25x6.75	18 TL	18x5.5	14 TT
	F104G	Starfighter		25x6.75	18 TL	18x5.5	14 TT
	F104G	Starfighter	F104FG	26x8.0-14	16 TL	18x5.5	14 TL
	YF12A			27.5x7.5-16	22 TL	25x6.75	16 TL
	SR71			27.5x7.5-16	22 TL	25x6.75	16 TL
	P2H	Neptune	P2V-7	56" SC	20 TL	34x9.9	14 TL
	P3A	Orion	P3V	40x14	26 TL	28x7.7	14 TL
	P3B, C	Orion	P3V-1	40x14	28 TL	28x7.7	14 TL
	C121G	Super Constellation	C121G	17.00-20	22 TL	33" SC	10 TT
	C130A, B, D, E	Hercules	—	20.00-20	22 TT	12.50-16	12 TT
	S3A	Viking	—	30x11.5-14.5	24 TL	22x6.75-10	18 TL
	HC130H	Hercules	C130	20.00-20	26 TL	12.50-16	12 TT
	C140A	Jetstar	C140A	26x6.6	14 TL	18x4.4 CH	10 TL
	C141	Starlifter	C141	44x16	28 TL	36x11	22 TL
	T33	Shooting Star	T33	26x6.6	14 TL	22x7.25-11.50	8 TL
	C5A	Galaxy		49x17	26 TL	49x17	26 TL
	AH56A	Cheyenne		29x11.00-10	10 TL	5.00-4	8 TL
Martin McDonnell-Douglas Co. (Douglas Division)	P5B	Marlin	P5M-2	15.50-20	14 TT	10.00-7	12 TT
	A1	Skyraider	AD-7	32x8.8	16 TT	12½x4½	14 TT
	A3	Skywarrior	A3D-2	44x13	26 TL	32x8.8	18 TT
	A4	Skyhawk	A4D-5	24x5.5	16 TT	18x5.5	12 TL
	A-26A	Counter Invader		15.50-20	14 TT	36" SC	12 TT
	B66	Destroyer	B66D	49x17	26 TL	36x11	22 TT
	F6A	Skyray	F4D-1	26x6.6	16 TL	22x5.5	12 TT
	C47	Skytrain	C47E	17.00-16	12 TT	9.00-6	10 TT
	C54	Skymaster	C54	15.50-20	14 TT	44" SC	12 TL
	C9A	Nightingale		40x14	24 TL	26.4-6-CH	10 TL
	C117		R4D8	17.00-16	12 TT	9.00-6	10 TT
	VC118	Liftmaster	C118A	15.50-20	20 TL	44" SC	12 TL
	C124	Globemaster	C124C	25.00-28	30 TT	15.50-20	14 TL
	C133	Cargomaster	C133B	20.00-20	22 TT	15.00-16	10 TL
McDonnell-Douglas Co. (McDonnell)	F-101A, C	Voodoo	F-101A, C	32.x8.8	24 TL	18x5.5	14 TL
	F-101B	Voodoo	F-101B	31x11.50-16	22 TL	18x5.5	14 TL
	F4B	Phantom II	F4H	30x8.0	26 TL	18x5.7	14 TL
	F4C	Phantom II		30x11.50-14.5	24 TL	18x5.5	14 TL
	F4J	Phantom II		30x11.50-14.5	24 TL	18x5.7	14 TL
	F-15	Eagle	101	34.5x9.75-18	22 TL	22x6.6-10	16 TL
	F-15		103	34.5x9.75-18	26 TL	22x6.6-10	20 TL
	F-18A			30x11.5-14.5	26 TL	22x6.6-10	20 TL
Rockwell Int.	F86	Sabre	F86A	26x6.6	14 TL	22x7.25-11.50	8 TL
	F100	Super Sabre	F100	30x8.8	22 TL	18x4.4	12 TL
	RA5C	Vigilante	A3J-3	36x11	28 TL	26x6.6	16 TL
	OV10A	Bronco	Coin	29x11.00-10	10 TL	7.50-10	12 TL
	U4A	Aero Commander	U4B	8.50-10	6 TT	6.00-16	6 TT
	T-2B	Buckeye	T2J	24x5.5	12 TL	20x4.4	10 TL
	T-28D	Trojan		24x7.7	10 TT	24x7.7	10 TT
	T-39	Sabreliner	T39A	26x6.6	14 TL	18x4.4	6 TL
	X-15			Skids		18x4.4	12 TL
	XB70A	Valkyrie		40x17.50-18	40 TL	14x4.5-8	8 TL
	Shuttle			44.5x16.0-21	28 TL	32x8.0	28 TL
Northrop	F51A, B	Freedom Fighter		22x8.5-11	16 TL	18x6.5-8	12 TL
	F5E			24x8.0-13	16 TL	18x6.5-8	12 TL
	T38	Talon	T38A	20x4.4	12 TL	18x4.4	6 TL
	F19L			28x9.5-15	20 TL	18x6.5-8	16 TL
Piper	U7A		L-21	8.00-4	4 TT	TAILSKID	
	U11A	Aztec	UO-1	7.00-6	8 TT	6.00-6	4 TT

2.5 STRUT-WHEEL INTERFACE, STRUTS AND SHOCK ABSORBERS

The landing gear must absorb the shocks of landing as well as taxiing. Two elements for shock absorption are incorporated in most landing gears: the tire(s) and the shock absorber(s). This section will discuss various ways in which shocks can be absorbed through the struts and through the tires. A rapid method for shock absorber sizing is also presented. A method for tire sizing was presented in Section 2.4. This section is organized as follows:

2.5.1 Strut-wheel interface
2.5.2 Devices used for shock absorption
2.5.3 Shock absorption capability of tires and shock absorbers: sizing of struts

2.5.1 Strut-Wheel Interface

Landing gear wheels are attached to some type of strut. Figure 2.23 illustrates two important strut-wheel interface parameters: the 'rake' and the 'trail'. These parameters are important to the static and dynamic stability behavior of the wheel relative to the strut.

The 'rake' is defined as the angle between the wheel swivel axis and a line vertical to the runway surface. Figure 2.23a defines positive and negative rake. The wheel swivel axis is identified as X-X.

The 'trail' is defined as the distance between the runway-wheel contact point and the point where the wheel swivel axis intersects the ground. Positive and negative trail are defined in Figures 2.23c and 2.23d.

The wheel rotation axis is the line perpendicular to the paper through point P.

In Figure 2.23a the wheel swivel axis, X-X passes below the wheel rotation axis P. This arrangement is statically stable because any wheel swivel about X-X would tend to 'lift' the airplane.

In Figure 2.23b the wheel swivel axis passes above the wheel rotation axis P. This arrangement is statically unstable because any wheel swivel about X-X would tend to 'lower' the airplane: the wheel therefore has a tendency to 'flop over'!

The arrangement shown in Figure 2.23c (positive trail) is called dynamically stable: when the wheel has

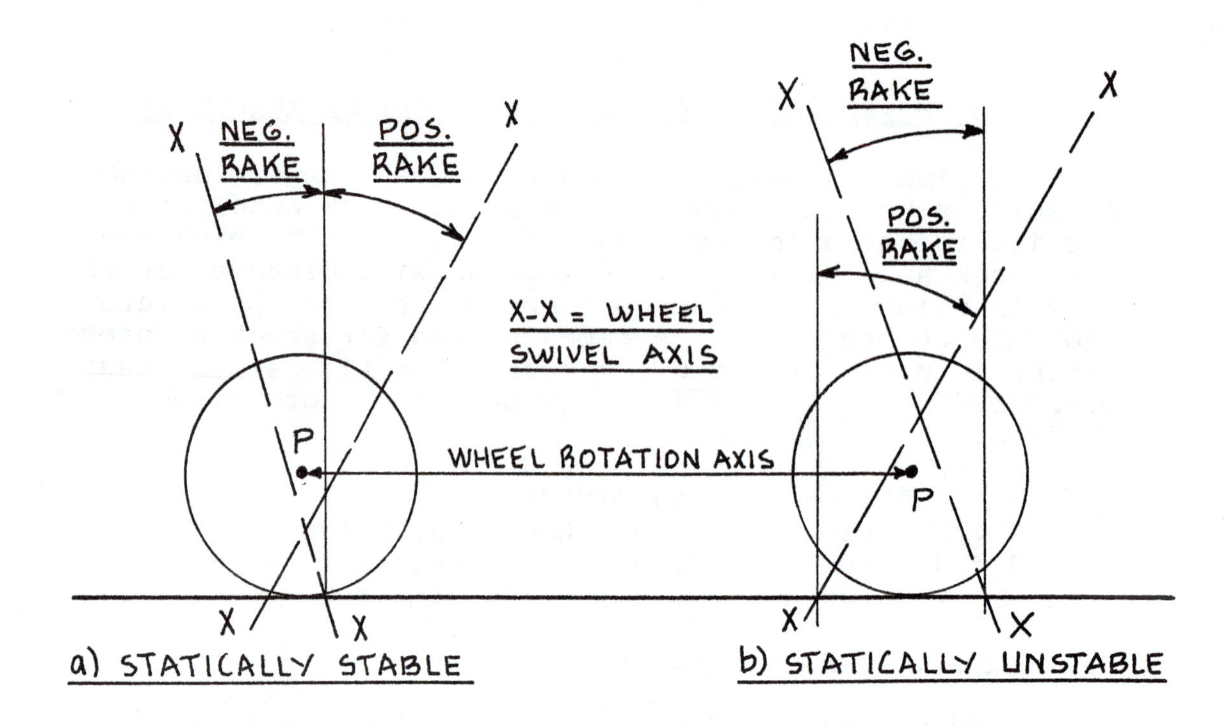

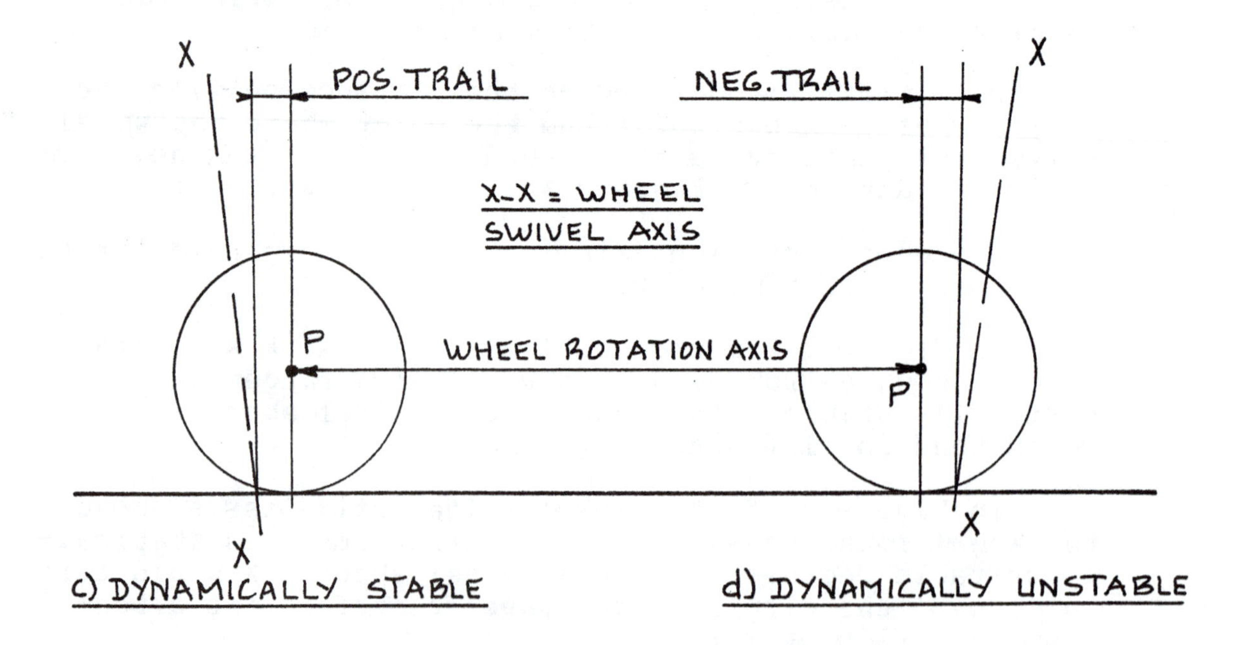

Figure 2.23 Definition of Rake and Trail

swiveled about X-X, the runway-to-tire friction would tend to rotate the wheel back to its original position.

Figure 2.23d (negative trail) depicts a dynamically unstable situation: when the wheel has swiveled about X-X, the runway-to-tire friction force would tend to rotate the wheel away from its original position.

In most airplane applications either stable or neutrally stable strut-wheel arrangements are favored. Unstable combinations do occur, but rarely so.

Figure 2.24 shows a number of strut-wheel combinations used in nose gear, tail gear and main gear designs.

Another form of dynamic instability is 'shimmy'. When a wheel shimmies, it oscillates about the wheel swivel axis. This is not only annoying, but can result in structural failure(s).

The physical causes for shimmy can be any combination of the following factors:

1. Overall torsional stiffness of the gear is insufficient. Torsional stiffness is defined about the swivel axis. Torsional movement is restricted by the 'scissors' and by the combined 'stiffness of the gear attachment to the structure, including any side- and(or) drag-bracing. Fig.2.7 shows a scissor (=torque-link) application.

2. Inadequate trail. Positive trail reduces shimmy.

3. Improper wheel mass balancing about the wheel rotation axis P.

To reduce and/or to damp shimmy, a so-called shimmy damper is often installed. A shimmy damper 'acts' like a shock absorber, but in a 'rotary' fashion.

2.5.2 Devices Used For Shock Absorption

The following devices are available for absorbing energy associated with landing and taxi loads:

*Tires
*Air springs
*Oleo-pneumatic struts
*Shock chords and/or rubbers
*Cantilever springs
*Liquid springs

Figure 2.25 shows examples of each device for shock absorption as well as typical load-deflection relation-

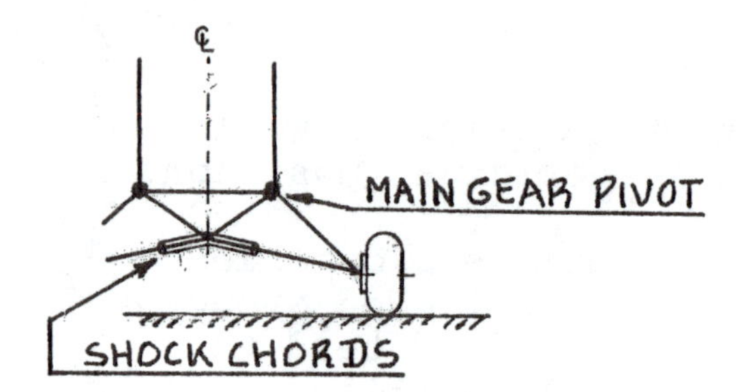

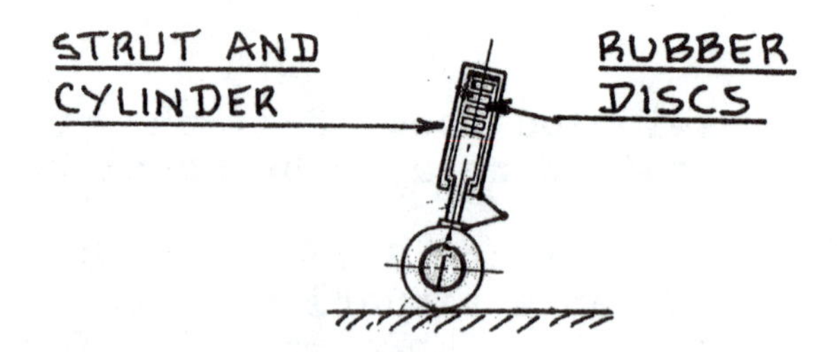

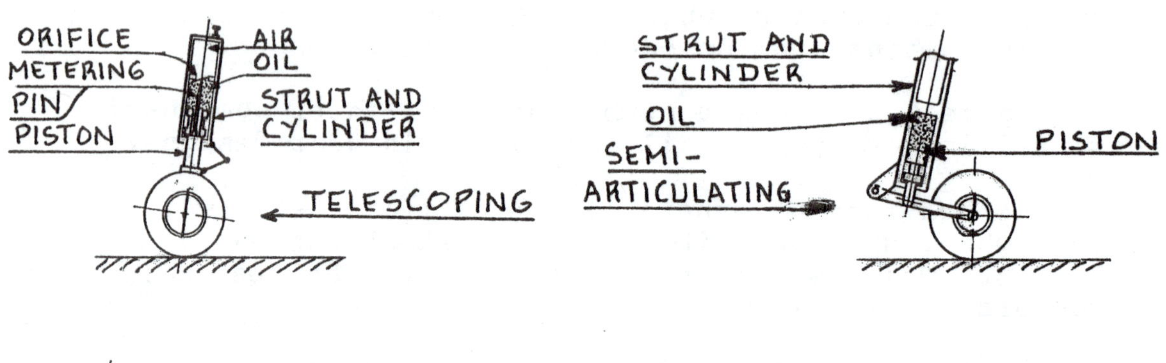

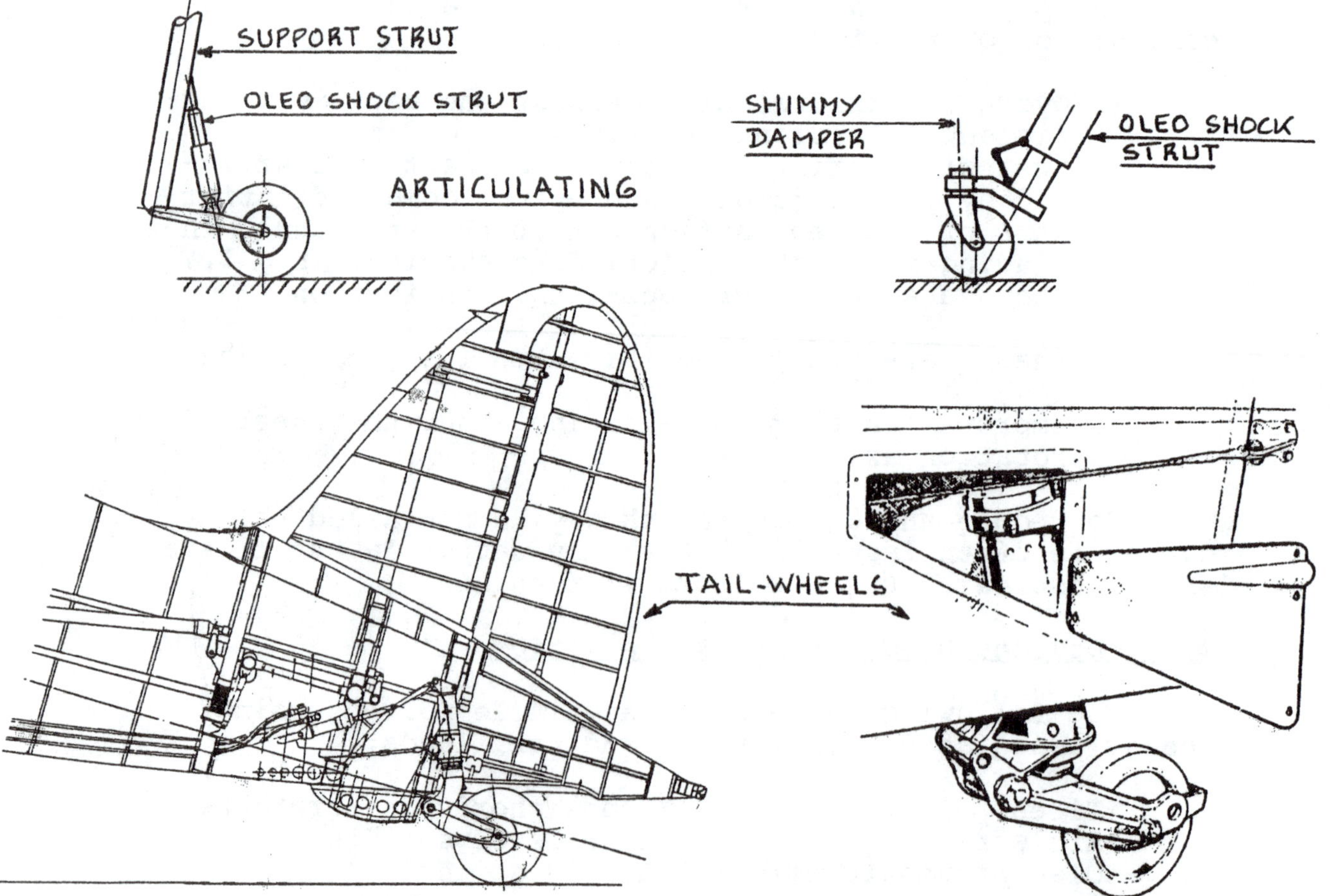

Figure 2.24 Examples of Strut-Wheel Combinations

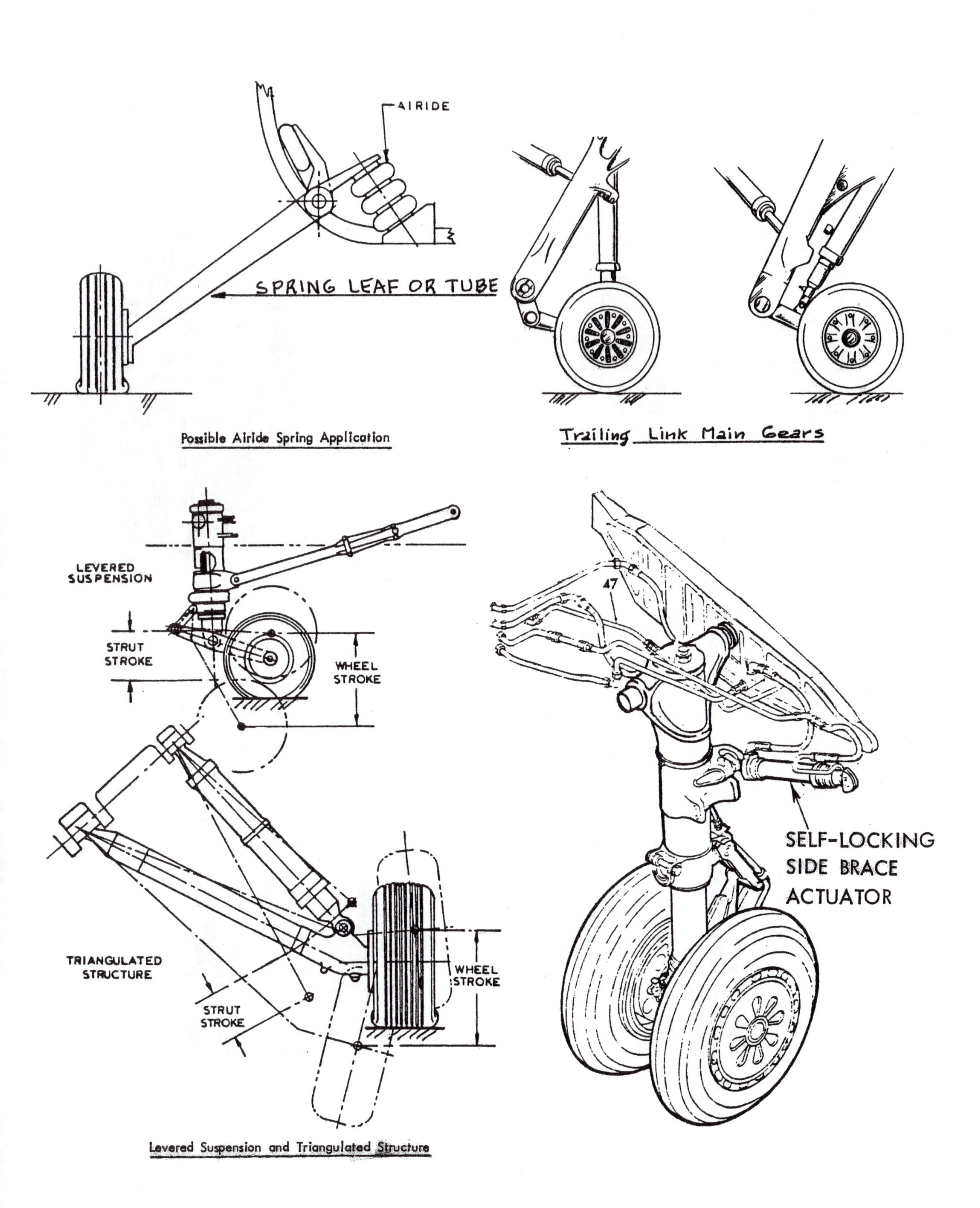

Figure 2.24 (Cont'd) Examples of Strut-Wheel Combinations

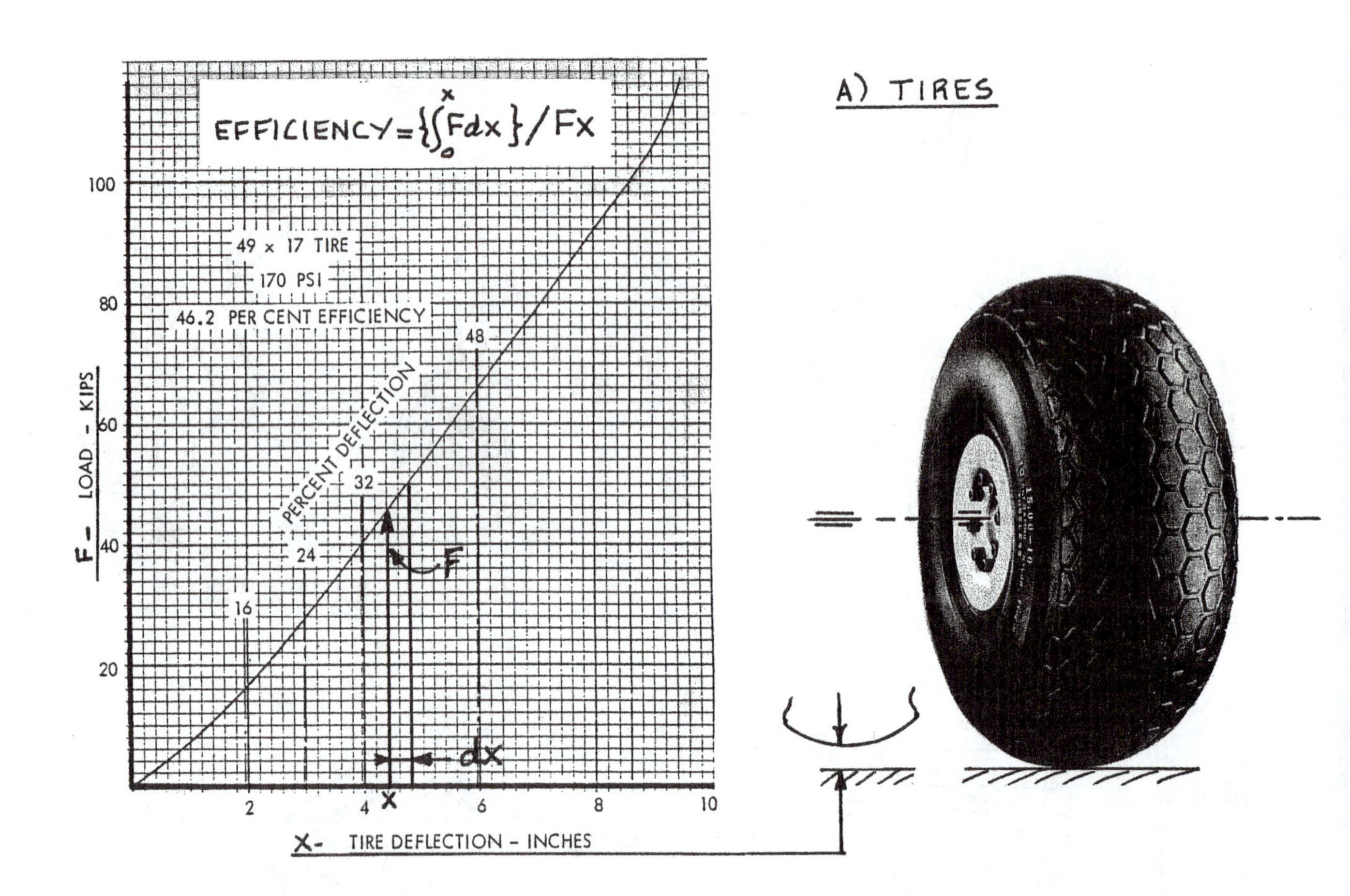

B) LIQUID SPRINGS

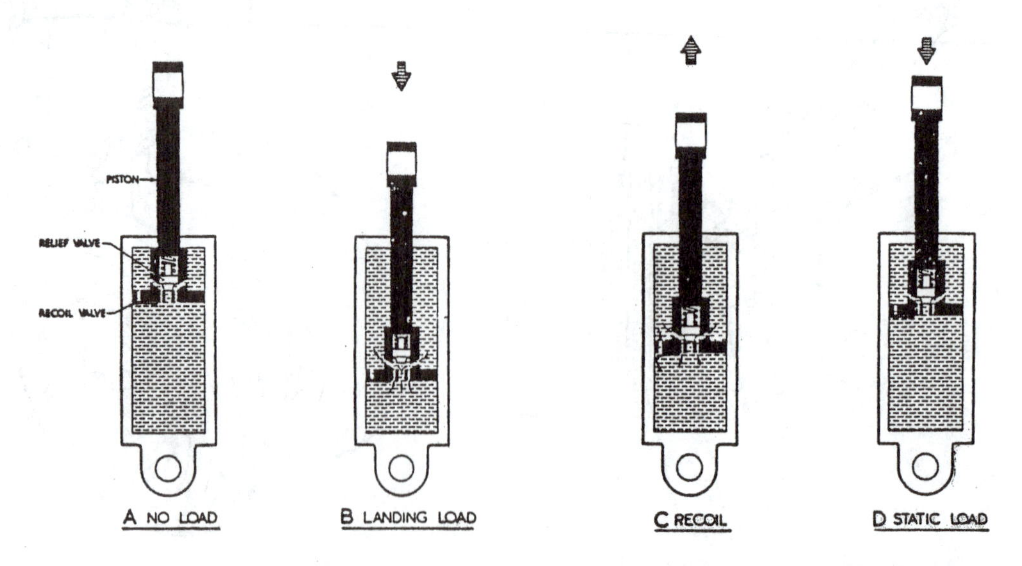

Figure 2.25 Examples of Shock Absorbing Devices

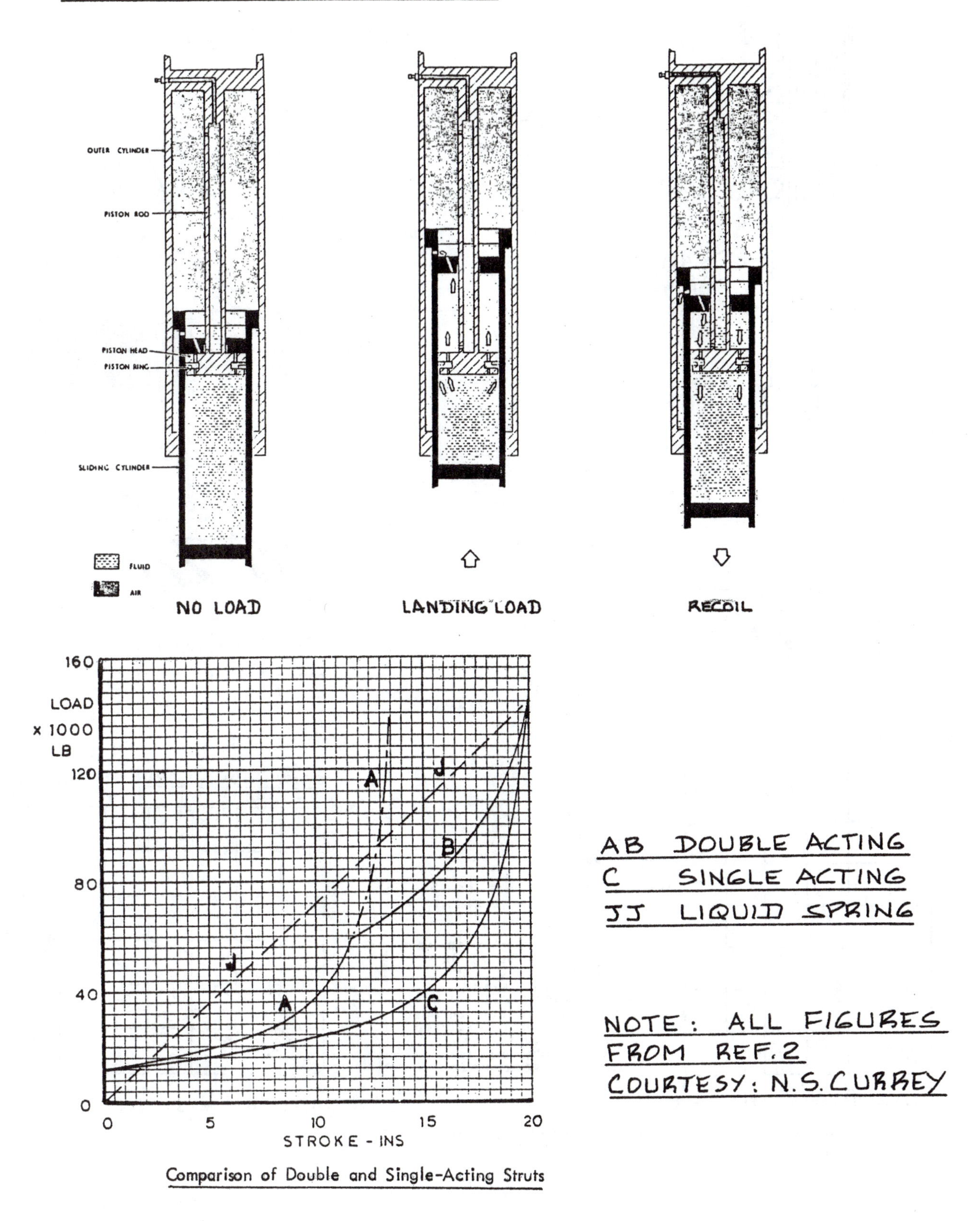

Comparison of Double and Single-Acting Struts

Figure 2.25 (Cont'd) Examples of Shock Absorbing Devices

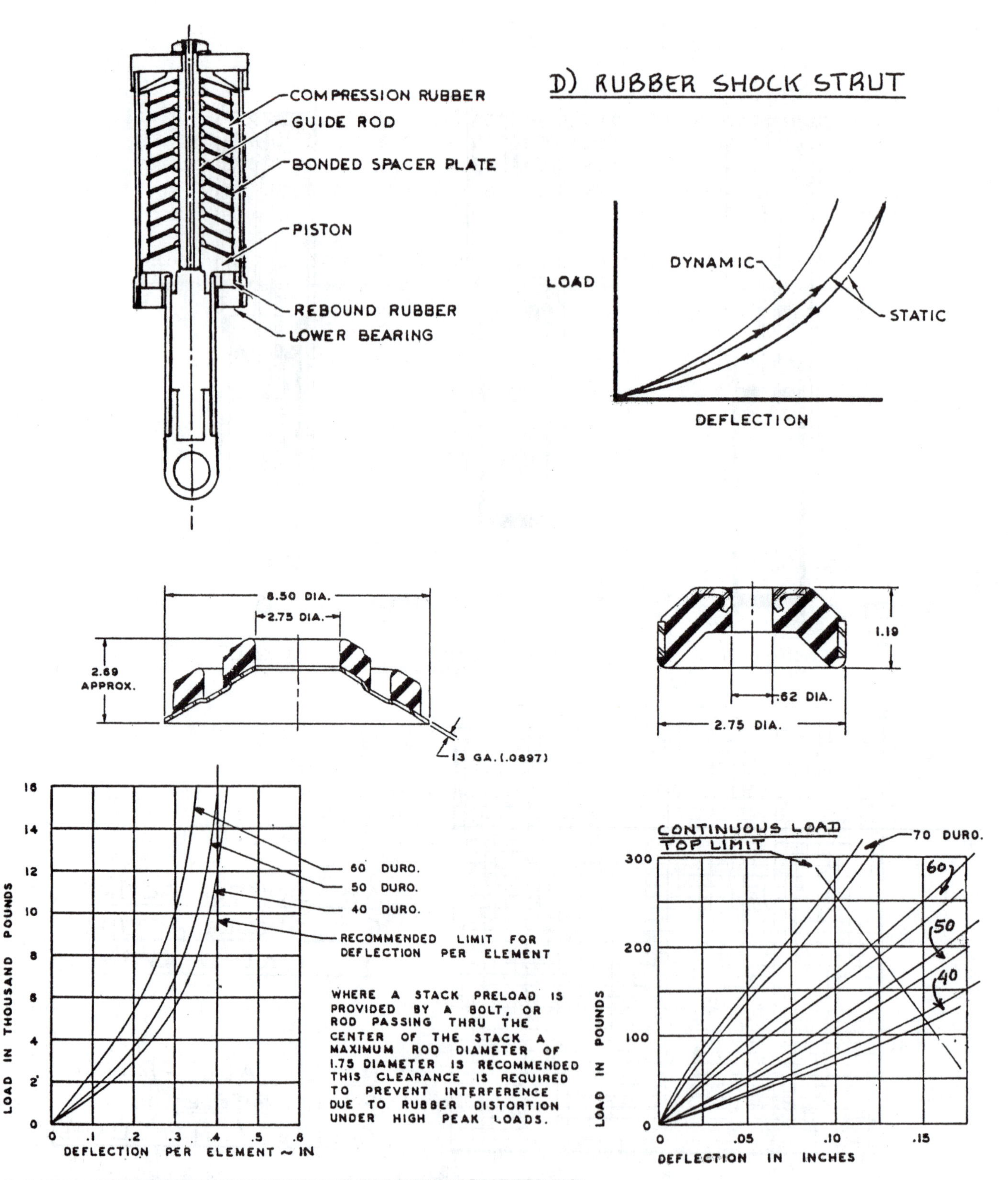

Figure 2.25 (Cont'd) Examples of Shock Absorbing Devices

ships for each. The energy absorbed by each type follows from the area under its load-deflection curve. The ratio of that area to the rectangular area around the load-deflection curve is called the shock absorption efficiency. Table 2.17 presents typical values for tire and shock absorber efficiencies.

Figure 2.24 showed a number of ways in which these shock absorbing elements can be integrated into a landing gear/strut arrangement.

2.5.3 Shock Absorption Capability of Tires and Shock Absorbers: Sizing of Struts

When an airplane touches down, the maximum kinetic energy which needs to be absorbed is:

$$E_t = 0.5(W_L)(w_t)^2/g \tag{2.8}$$

The landing weight, W_L follows from Table 3.3 in Part I. The design vertical touchdown rate, w_t follows from Section 2.1, page 4.

The energy of Eqn.(2.8) needs to be aborbed by the landing gear. How this energy is distributed over the landing gear components is discussed in the following.

For the main landing gear it is assumed that the entire touch-down kinetic energy is absorbed by the main landing gear. This is a conservative assumption. The participants in this shock absorption process are: the tires and the shock absorbers.

The following equation is used:

$$E_t = n_s P_m N_g (\eta_t s_t + \eta_s s_s) \tag{2.9}$$

In Eqn.(2.9) it is assumed that:

$$W_L = n_s P_m \tag{2.10}$$

The various quantities in Eqns(2.9) are defined as:

n_s is the number of main gear stuts (or legs), assumed to be equal to the number of shock absorbers. Note: $n_s = 2$ for most main gears.

P_m is the maximum static load per main gear strut.

N_g is the landing gear load factor: ratio of maximum

Table 2.17 Energy Absorption Efficiency of Tires and Shock Absorbers

Element:	Energy Absorption Efficiency:
Tires:	η_t = 0.47
Shock absorbers:	
air springs	η_s = 0.60 t0 0.65
metal springs with oil damping	= 0.70
liquid springs	= 0.75 to 0.85
oleo-pneumatic	= 0.80
cantilever spring	= 0.50

Table 2.18 Suggested Landing Gear Load Factors

Certification Base:	Landing Gear Load Factor, N_g:
FAR 23	N_g = 3.0
FAR 25	N_g = 1.5 to 2.0
Fighters and Trainers	N_g = 3.0 - 8.0: See Fig.2.26 for more details
Military transports	N_g = 1.5 - 2.0

load per leg to the maximum static load per leg.

η_t is the tire energy absorption efficiency.

η_s is the energy absorption efficiency of the shock absorber.

s_t is the maximum allowable tire deflection as determined from Eqn.(2.4).

s_s is the stroke of the shock absorber.

Eqn.(2.9) may be used to compute the required shock absorber length:

$$s_s = [\{0.5(W_L/g)(w_t)^2/(n_s P_m N_g)\} - \eta_t s_t]/\eta_s \qquad (2.11)$$

It is suggested to add one inch to this length:

$$s_{s_{design}} = s_s + 1/12 \qquad (2.12)$$

Table 2.18 shows the values for landing gear load factors which may be used in preliminary design. How these landing gear load factors are related to design touchdown rate and to shock absorber stroke is shown in Figure 2.26 for some example airplanes.

The diameter of the shock absorber (strut) may be estimated from:

$$d_s = 0.041 + 0.0025(P_m)^{1/2} \qquad (2.13)$$

Note that Eqn.(2.9) tacitly assumes that the main gear reaction load is transferred directly into the shock absorber. This condition is not satisfied for gears where the reaction load is not 'in line' with the shock absorber. A landing gear where the reaction load is aligned with the shock absorber is given in Fig.(2.7). Examples of landing gears where this alignment is absent, are given in Fig.(2.24). For the latter type gears, the required value for shock absorber stroke (and thus strut length) must be determined for the particular landing gear geometry at hand. No general rules can be given.

For main gears where the design calls for a simple cantilever (leaf or tube) spring, Ref.8 contains an example sizing calculation. Figure 2.27 shows an example application, seen mostly in light airplanes.

For the nose gear, the shock absorber length may be computed from Eqn.(2.11), BUT WITH ALTERATIONS:

1. Replace W_L with P_n (See Figure 2.14).
2. The load P_m must be replaced by the maximum dynamic nose wheel load which is equal to the maximum dynamic nose wheel tire load from Eqn.(2.5) multiplied by the number of tires. For the nose gear: $n_s = 1.0$
3. The tire deflection, s_t in this case is the nose gear tire deflection.

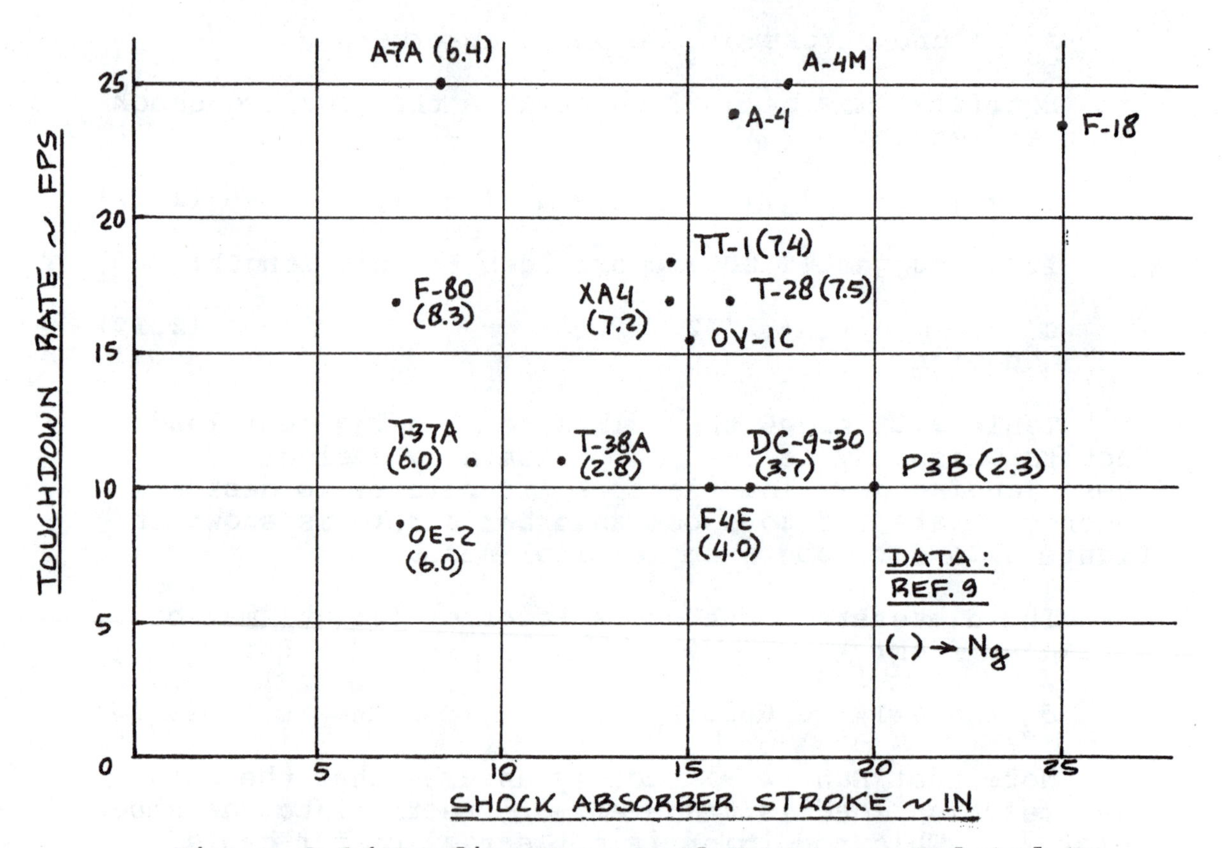

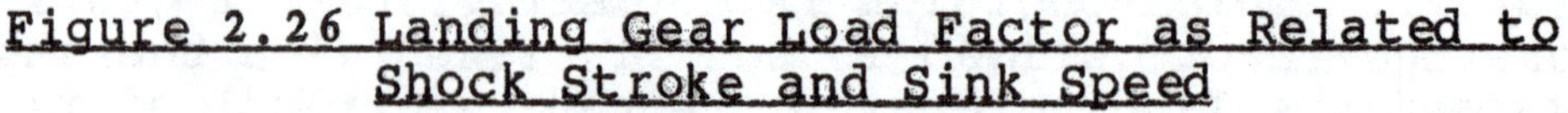

Figure 2.26 Landing Gear Load Factor as Related to Shock Stroke and Sink Speed

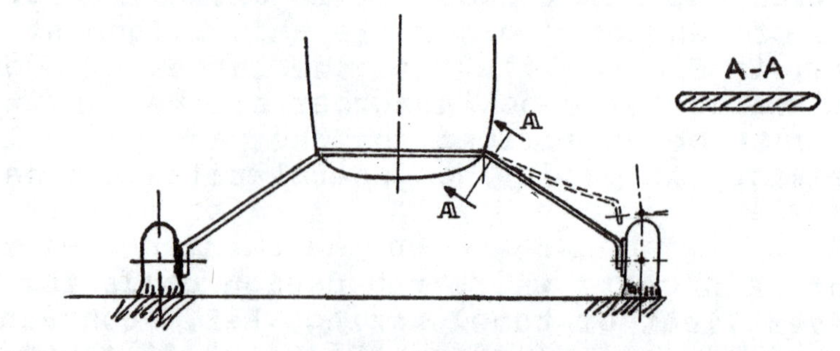

Figure 2.27 Example of Cantilever Spring Leaf Main Gear

2.6 BRAKES AND BRAKING CAPABILITY

2.6.1 Braking and Brakes

The purpose of brakes is to:

1. help stop an airplane
2. help steer an airplane by differential braking action
3. hold the airplane when parked
4. hold the airplane while running up the engines
5. control speed while taxiing.

Since virtually all modern airplanes use disc type brakes, only these will be discussed.

Figure 2.28 shows a cross section through a tire/wheel/brake assembly: wheels are made of two halves.

Figure 2.29 shows the wheel construction for a B767. Note the heat shields used to prevent overheating of the tire from the inside. A cross section of the B767 brakes which fit into the wheel cavity is shown in Figure 2.30.

Brakes turn kinetic energy (due to forward motion of the airplane) into heat energy through friction. This friction generated heat is dissipated to the immediate environment of the brake: wheel, tire and surrounding air. Figure 2.31 indicates how the braking heat flows into the wheel and into the tire: these elements act as heat sinks. The capacity of wheel and tire to absorb heat is limited. This must be accounted for in the design of wheel and tire. The methods used in heat sink sizing for brakes and wheels are beyond the scope of this text: Ref.2 deals with such methods.

Ultimately, it is the rolling friction (while applying braking action) between the tire(s) and the runway surface which slows the airplane. Fig.2.32 shows how the rolling friction coefficient is related to the so-called slip ratio. The slip ratio of a wheel is defined as:

$$\text{Slip ratio} = \{1 - (\text{Wheel RPM})_{\text{brakes on}} / (\text{Wheel RPM})_{\text{brakes off}}\} \qquad (2.14)$$

As shown in Figure 2.32, at zero slip ratio (free rolling wheel without braking) the friction coefficient is 0.02 to 0.05 depending on the surface characteristics. At a slip ratio of 1.0, the brakes have 'locked' the wheel and the friction coefficient is about 0.4, corresponding to a skidding condition (This will wear out a

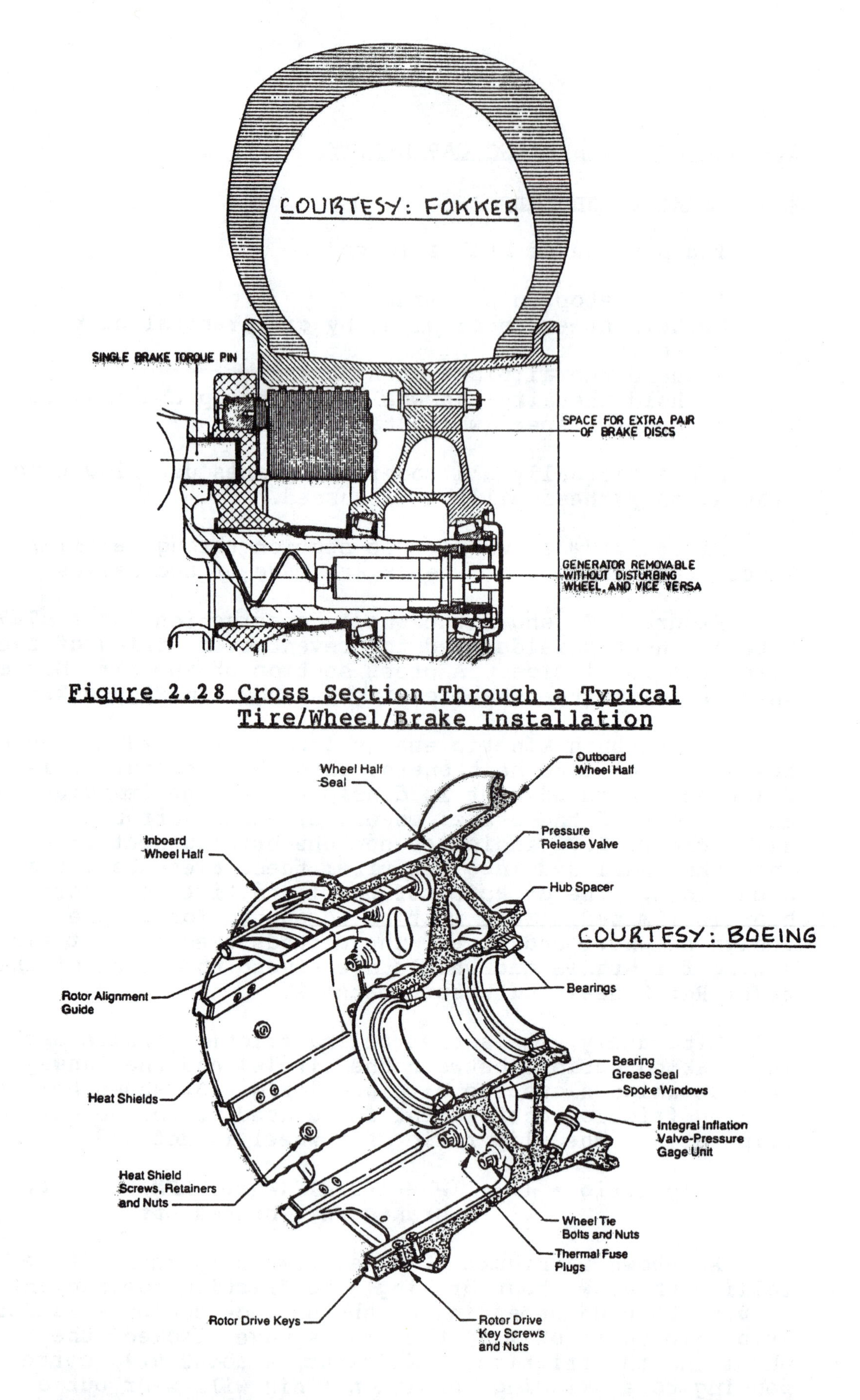

Figure 2.28 Cross Section Through a Typical Tire/Wheel/Brake Installation

Figure 2.29 Wheel Construction for a Boeing 767

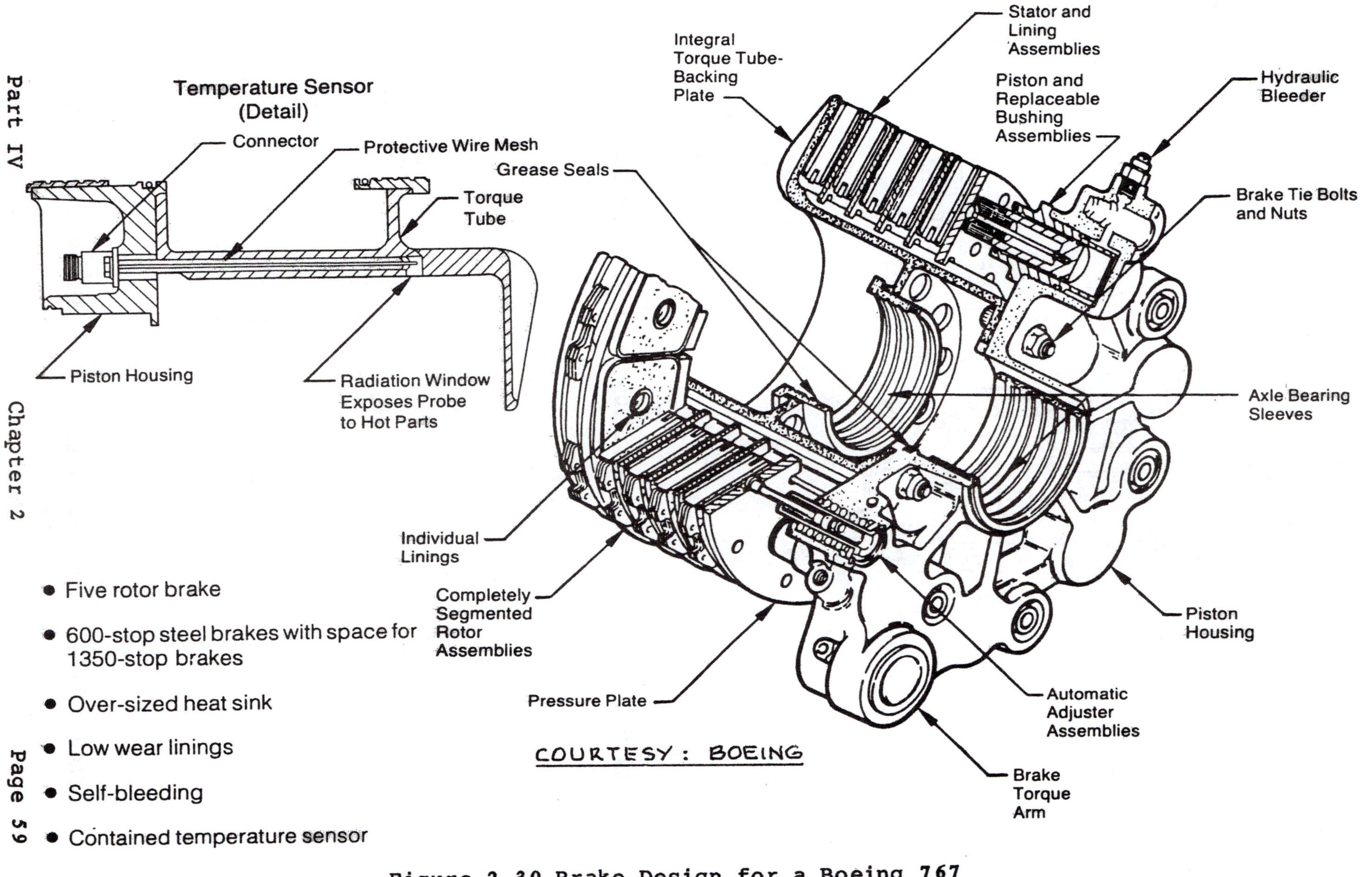

Figure 2.30 Brake Design for a Boeing 767

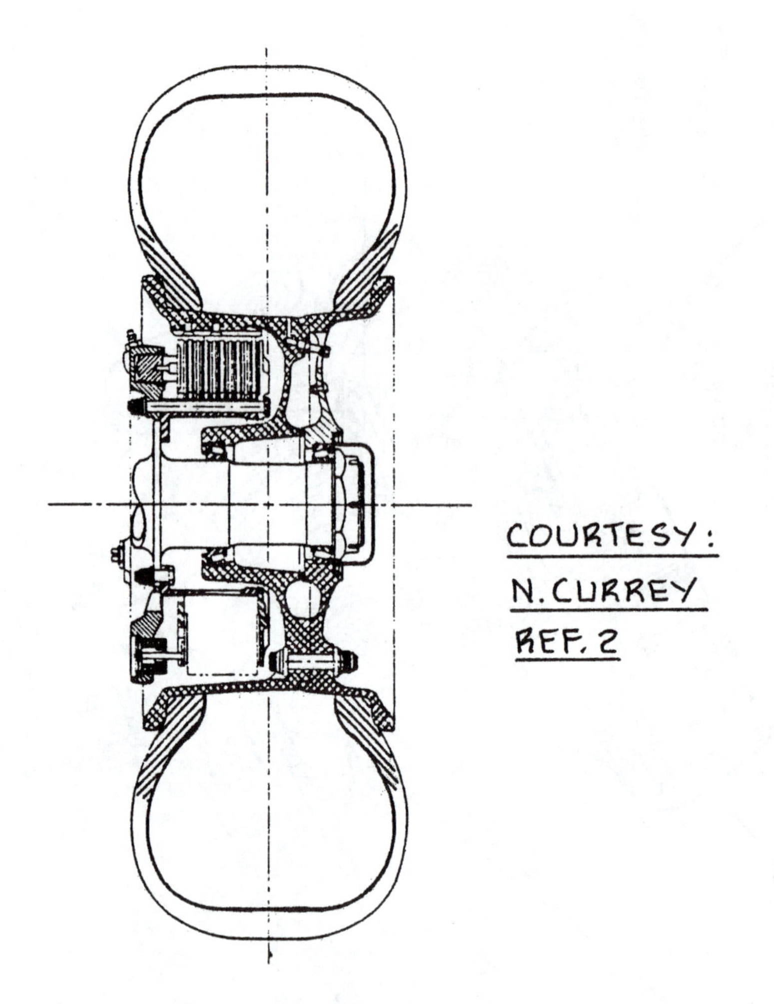

Figure 2.31 Example of Heat Flow into Wheel and Tire

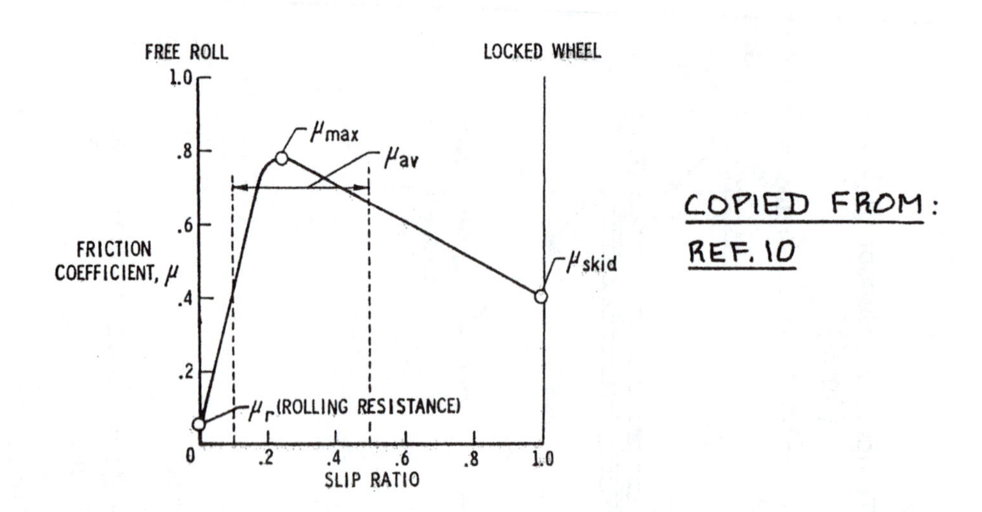

Figure 2.32 Effect of Slip Ratio on Ground Friction Coefficient

tire and cause tire blow-out in less than 100 ft!). If an anti-skid system is used to control wheel RPM during braking, the average value of friction coefficient which can be attained is about 0.70.

In practical situations, accounting for the fact that tires will be a bit worn and that brakes do not operate at their best efficiency, the following deceleration values can be obtained during roll-out:

Conventional brakes:	0.35g on a dry surface
Carbon brakes:	0.40g on a dry surface
Anti-skid brakes:	0.45g on a dry surface
Anti-skid carbon brakes:	0.50g on a dry surface

Note that carbon brakes offer a significant improvement. Carbon brakes are also about 40 percent lighter than conventional brakes. Their cost is about twice the cost of conventional brakes.

On other than dry surfaces the friction coefficient while braking deteriorates significantly. Fig.2.33 shows the effect of surface condition on available braked friction coefficient values. Note that the braked friction coefficient also depends on the ground speed!

For jet transports, the average friction coefficients depicted in Figure 2.34 should be used in preliminary stop distance calculations. How to perform such calculations is shown in Part VII.

When a sufficient amount of water is present on a runway surface, a condition known as hydroplaning can arise. Fig.2.35 shows that the tire is actually 'lifted' off the runway and rides on a cushion of water: the friction coefficient values now are extremely low: approximately 0.05. When hydroplaning conditions prevail, the only way to slow down the airplane is with reverse thrust.

For a detailed discussion of available braked friction coefficients under a wide variety of conditions, Ref.10 should be consulted.

2.6.2 Brake Actuation

Brakes are normally actuated with the help of a hydraulic system. Figure 2.36 shows the main gear brake actuation system used in a Piper PA-38 Tomahawk. Figure 2.27 depicts the main gear brake actuation system used in a B767.

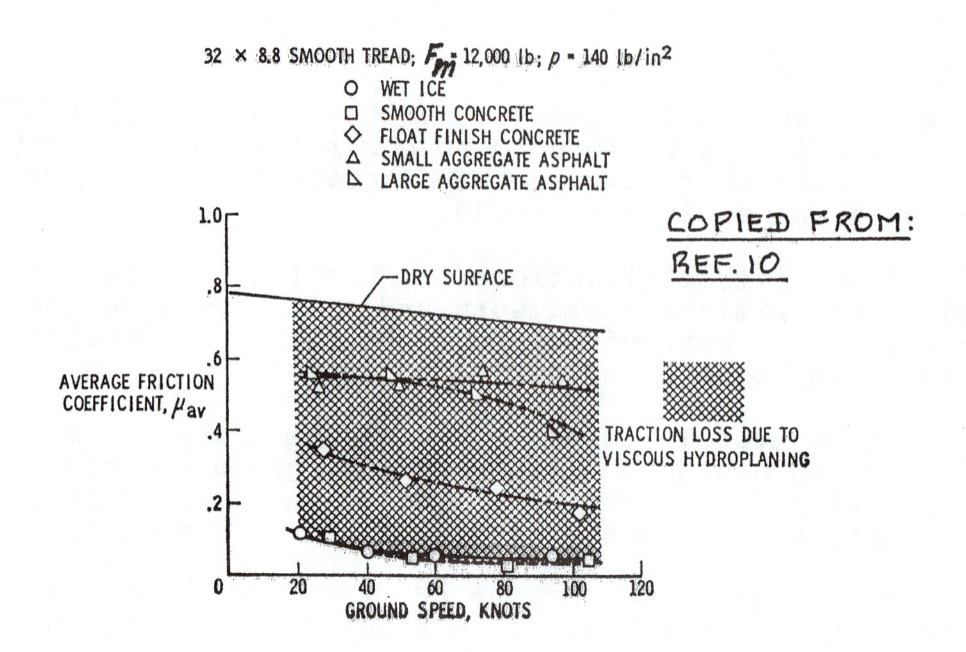

Figure 2.33 Effect of Surface Condition on Ground Friction Coefficient

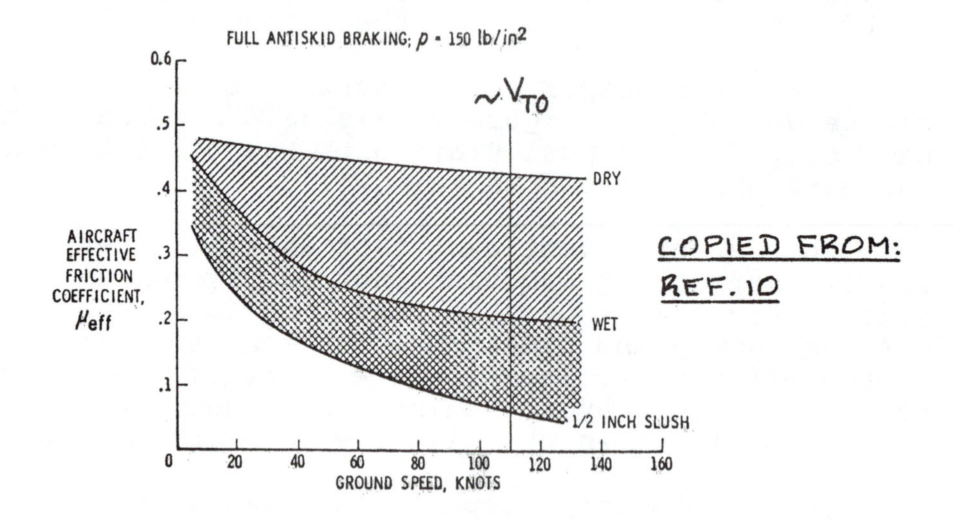

Figure 2.34 Typical Ground Friction Coefficients Encountered by Jet Transports

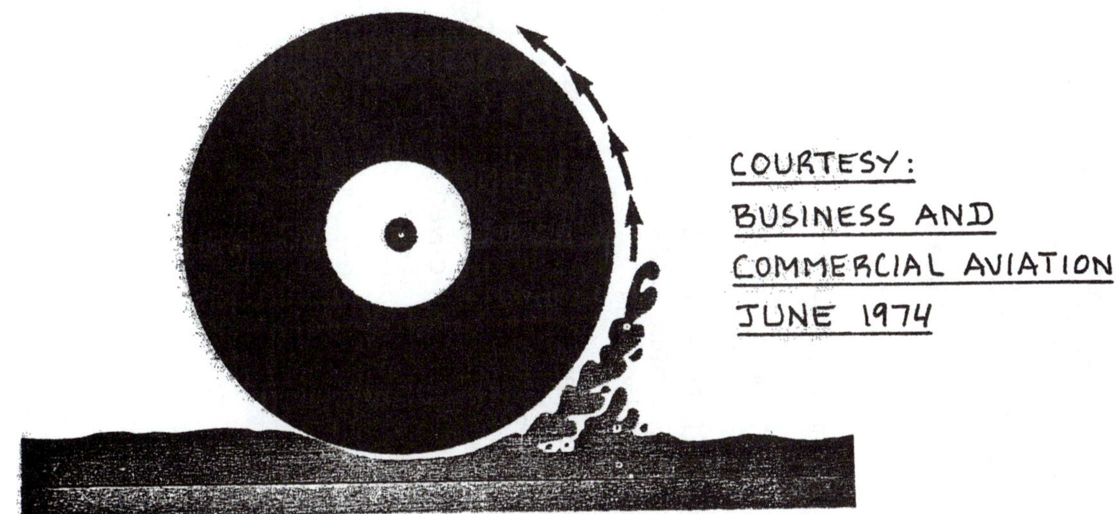

In standing water a wedge will build at the forward rolling point on the tire, lift it from the runway and impart vectors that cause the wheel to begin rotating backwards. NASA has demonstrated this phenomenon on aircraft as large as a DC-8.

Figure 2.35 Hydroplaning Tire

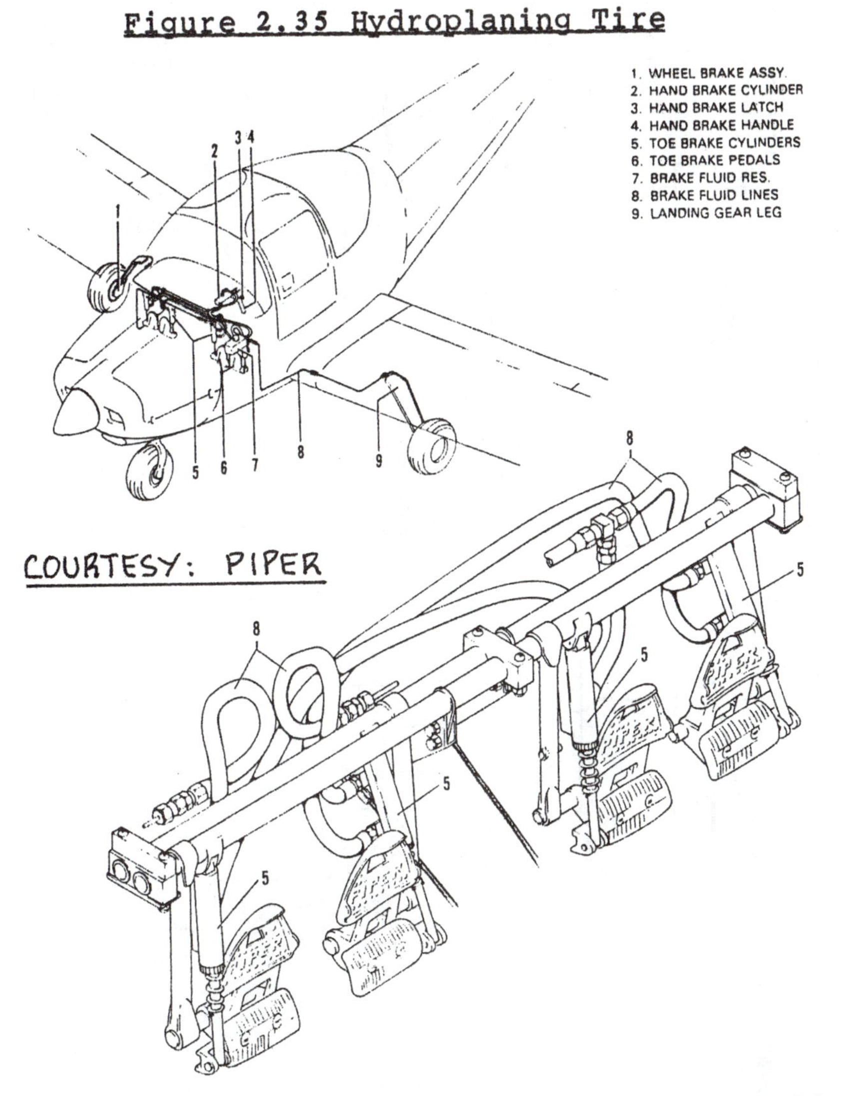

Figure 2.36 Brake Installation Piper PA-38-112

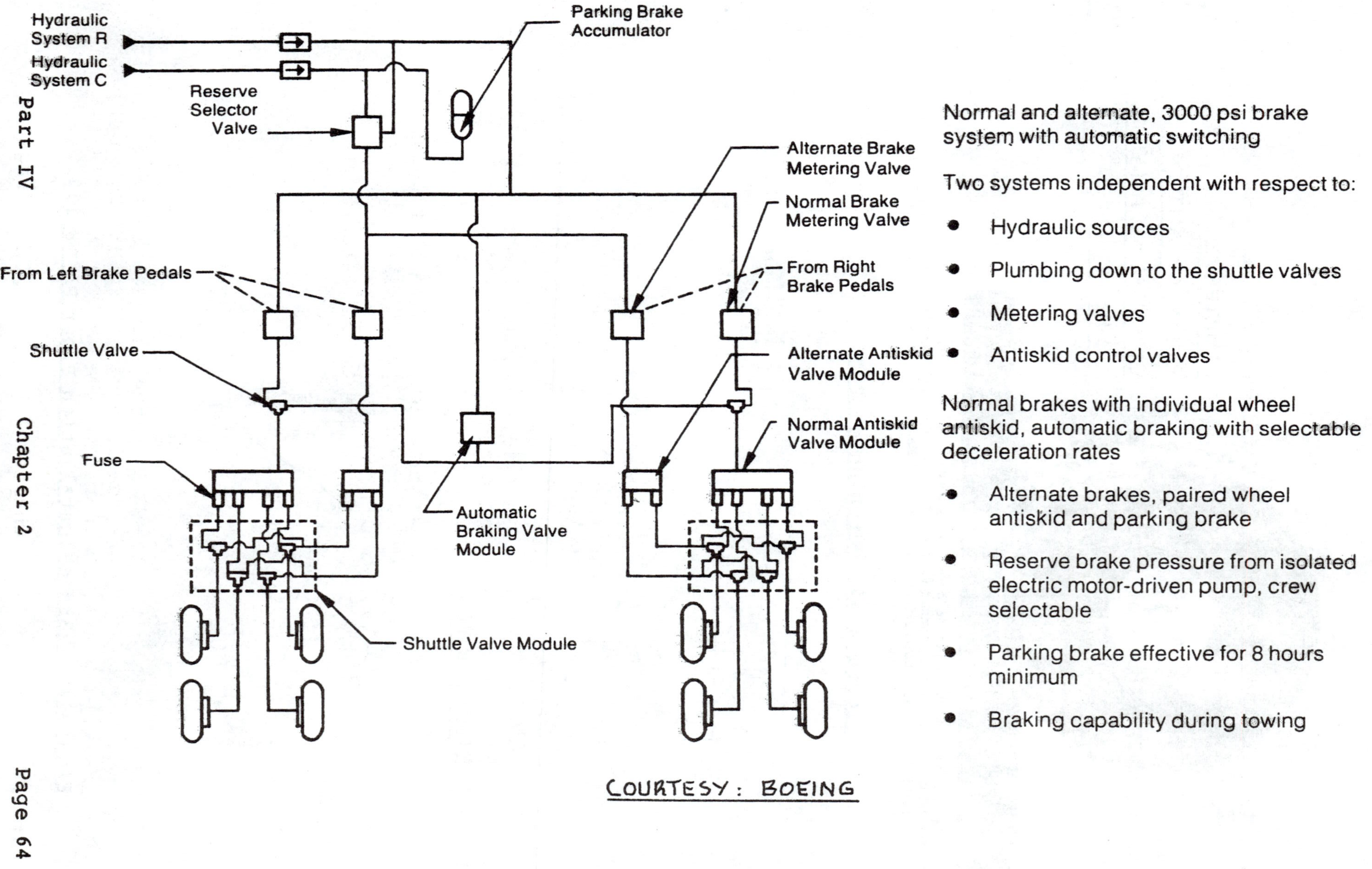

Figure 2.37 Main Gear Brake System Boeing 767

2.7 DESIGN CONSIDERATIONS FOR LANDING GEARS OF CARRIER BASED AIRPLANES

Because of the requirements to catapult and arrest carrier based airplanes, special provisions must be made in their landing gear design. The purpose of this section is to discuss the most important aspects of landing gear design for carrier based airplanes. The material is organized as follows:

2.7.1 Description of flight deck features
2.7.2 Description of a catapult system
2.7.3 Catapulting procedures and required landing gear provisions
2.7.4 Description of arresting gear system
2.7.5 Arresting procedures and required landing gear provisions

2.7.1 Description of Flight Deck Features

Figure 2.38 shows the flight deck layout of the aircraft carrier Enterprise (CVN 65). Note the four catapults and the four deck pendants of the arresting gear. Also note the four elevators. All naval aircraft must fit within the geometry of the elevators: it must be possible to move aircraft from the flight deck to the hangar deck for repairs and for special outfittings.

It is essential to fit as many airplanes as possible on the flight deck and on the hangar deck below. A carrier airplane must also fit on an elevator to transport it from the flight deck to the hangar deck. This requires naval airplanes to be sufficiently small and/or to have folding surfaces. Examples of wing folding are given in Chapter 4 of Part III.

To determine how many airplanes fit on the flight deck during launching and recovery operations, so-called 'deck-spotting' studies are performed. Figure 2.39 shows a typical spotting study result for a small escort carrier (CVE 2/53: never built).

2.7.2 Description of a Catapult System

For detailed in formation on catapulting systems Ref.11 should be consulted. Figure 2.40 shows an example of how a catapult system works.

The airplane is attached to the shuttle and to a so-called holdback unit. The launching force is applied to the airplane via the shuttle. The shuttle is attached

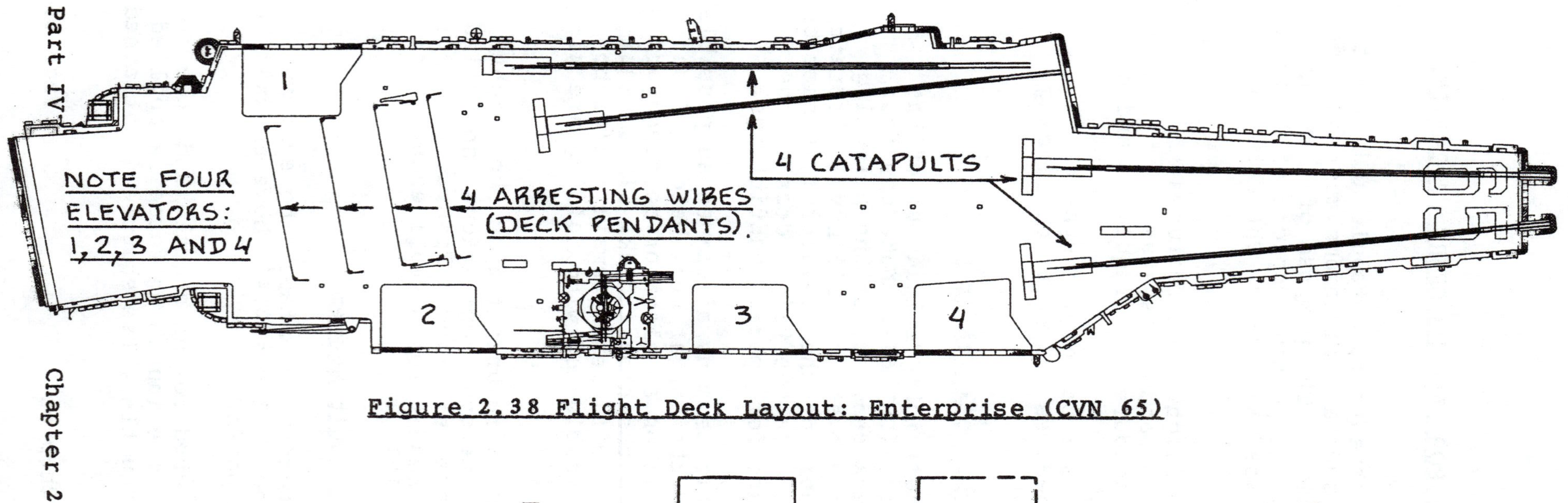

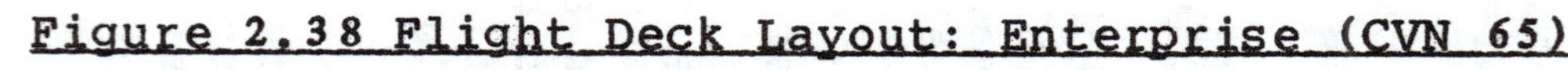

Figure 2.38 Flight Deck Layout: Enterprise (CVN 65)

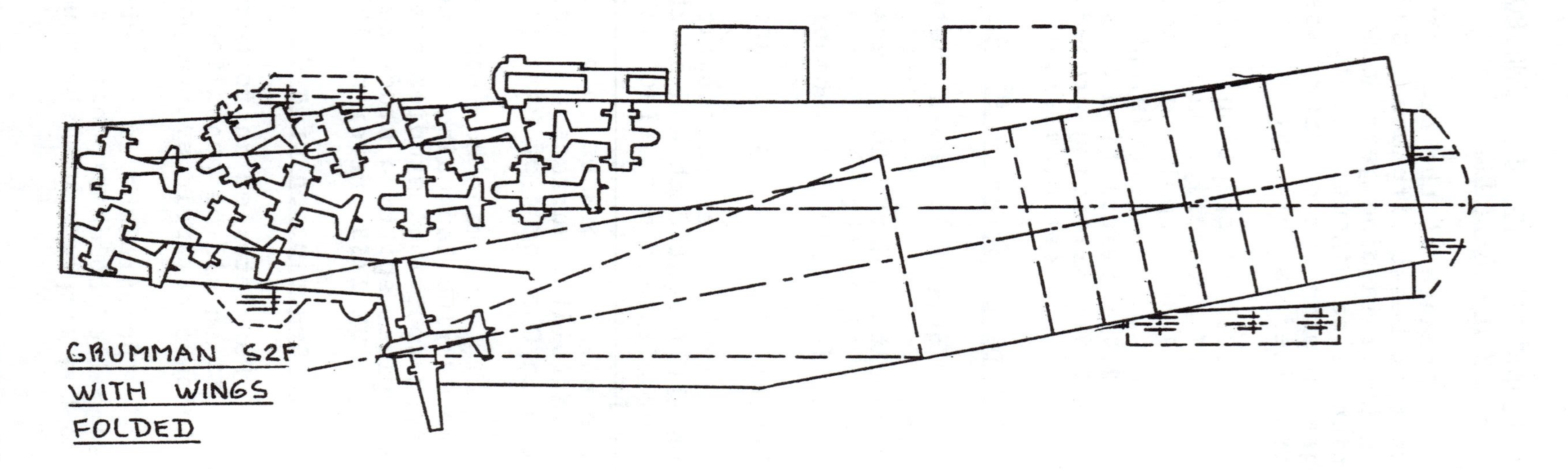

Figure 2.39 Example of a Flight Deck Spotting Study

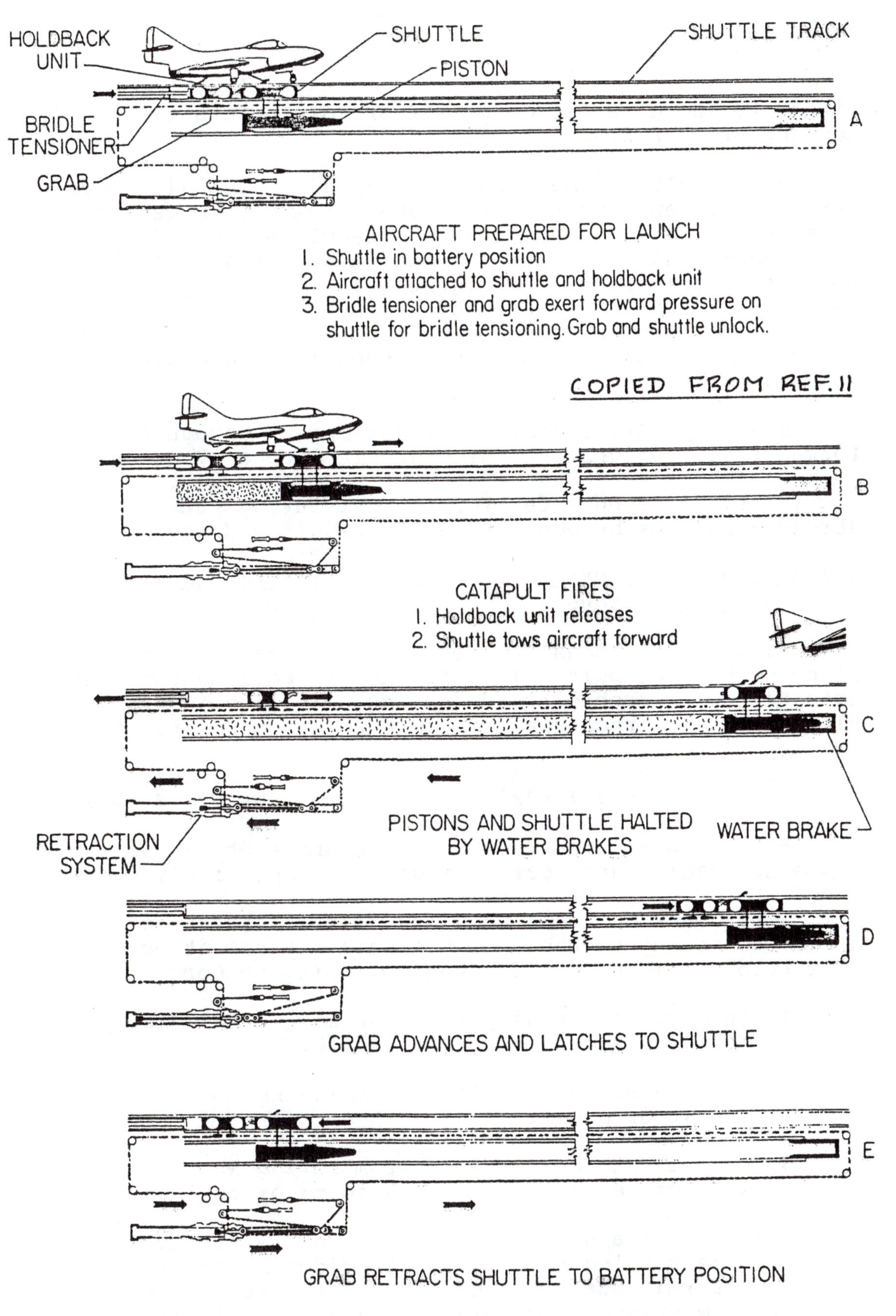

Figure 2.40 Typical Catapult Operation

either to the nosegear or to one or more points with a pendant or a bridle.

A catapult can be thought of as a device which accelerates a given mass (the airplane) from zero speed relative to the carrier deck to the catapult end-speed. The aerodynamic characteristics of the airplane are not important in this process. Figure 2.41 shows a typical catapult performance relationship. Any point within the limits represents a possible mass/end-speed combination. The steam pressure used during catapulting determines the actual performance obtained.

The airplane 'sizing' implications of catapult performance limitations are discussed in Chapter 3 of Pt I.

Note: To account for full thrust during the catapult stroke 3 kts may be added to the catapult end speed.

2.7.3 Catapulting Procedures and Required Landing Gear Provisions

Figure 2.42 shows the general arrangement of an airplane on the catapult. The airplane is positioned on the catapult and attached to the catapult tow fitting. The element which attaches an airplane to the tow fitting is:

1. a bridle   2. a pendant   3. a launch bar
(See Fig.2.42a)   (See Fig.2.42b)

Figures 2.43 and 2.44 show an example of a bridle and a nosegear launch bar arrangement respectively.

A holdback device (Fig.2.42) connects the airplane holdback fitting to the deck holdback attachment point. This holdback is needed for the following reasons:

1. to prevent the airplane from moving forward due to pitching of the carrier deck.

2. to prevent the airplane from moving under full thrust or power.

3. to prevent the airplane from moving when the bridle (or pendant) is tensioned prior to launch.

When the catapult is fired the shuttle attempts to move the tow fitting forward. The holdback device continues to restrain the airplane until a pre-set force level is reached whereupon the holdback releases the airplane.

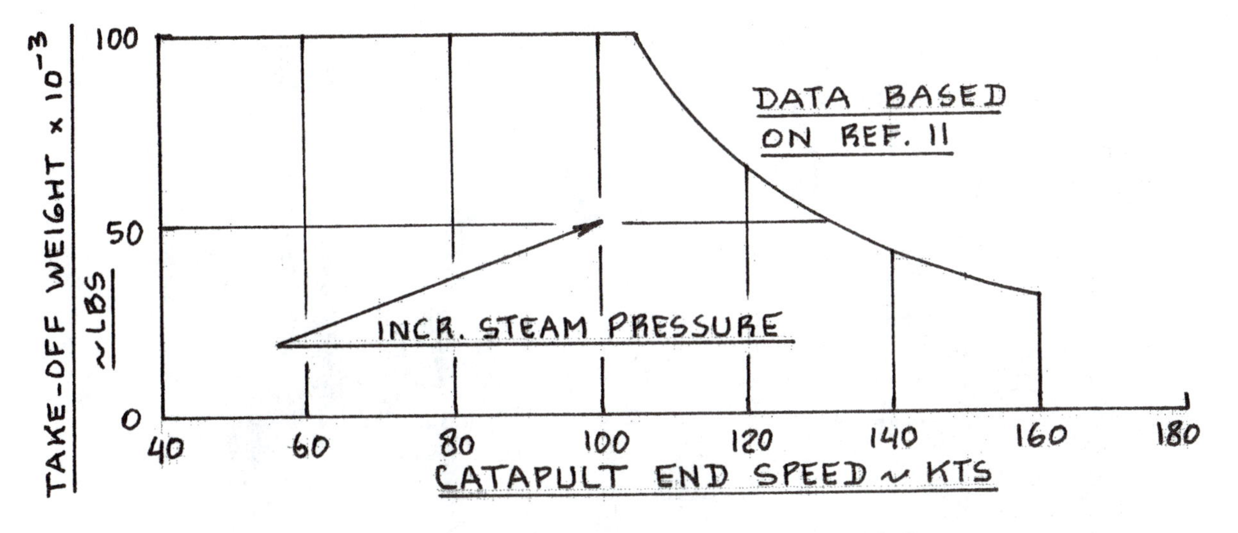

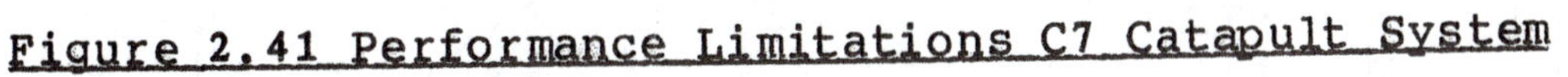

Figure 2.41 Performance Limitations C7 Catapult System

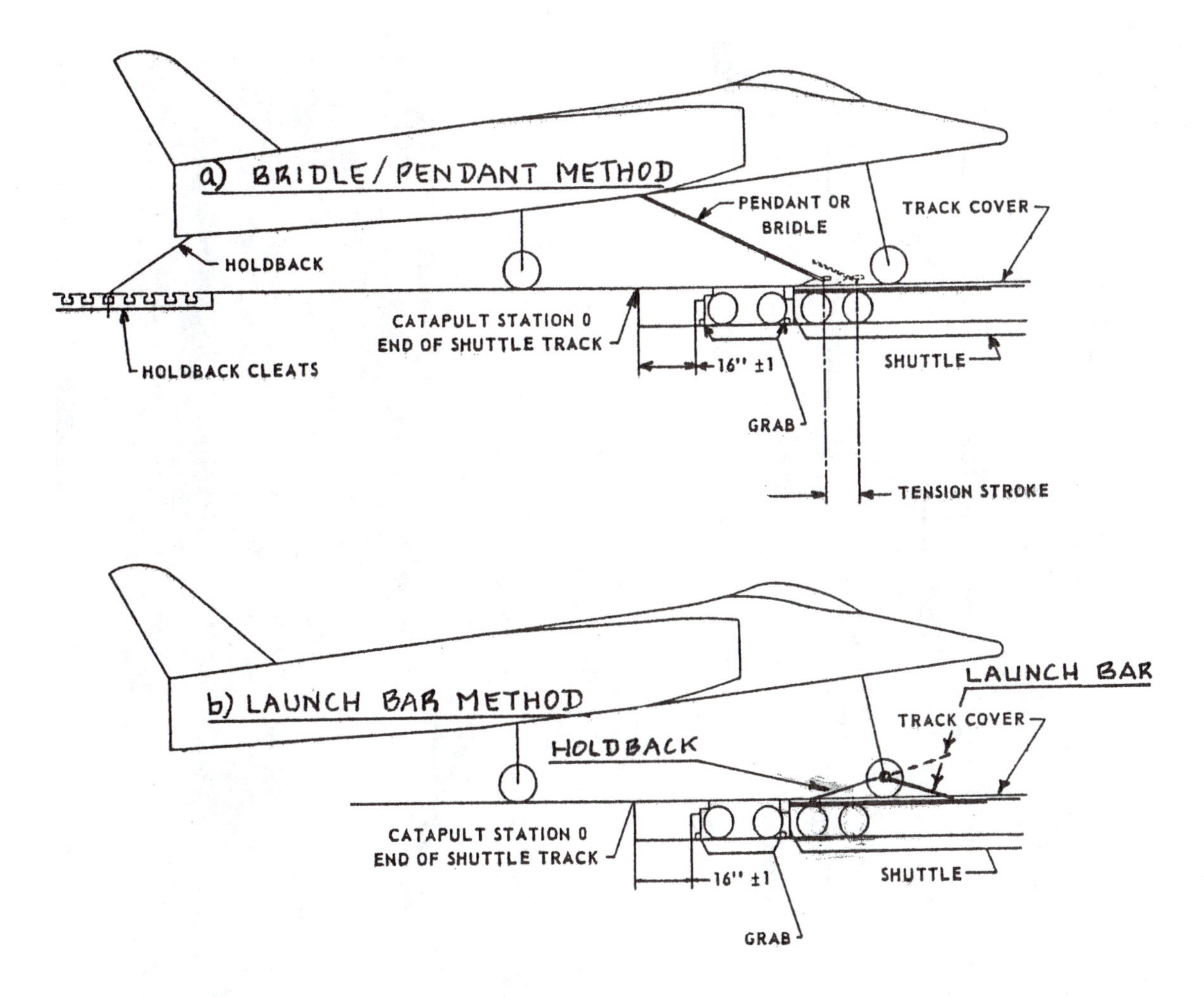

Figure 2.42 Airplane/Catapult Arrangements

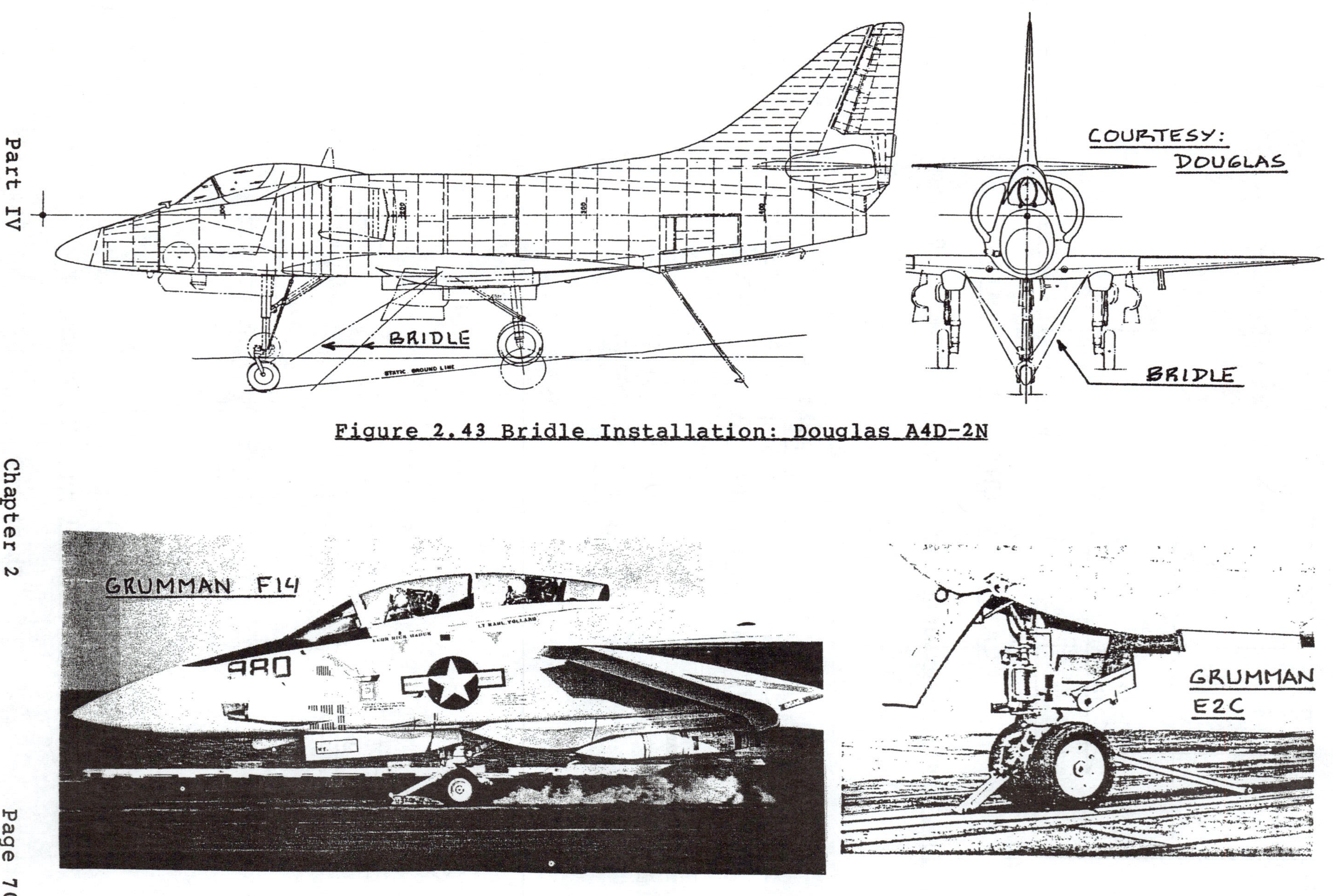

Figure 2.43 Bridle Installation: Douglas A4D-2N

Figure 2.44 Nosegear Launch Bar Installations

At the end of the catapult stroke the airplane automatically disengages from the catapult tow fitting. The airplane then continues under its own thrust or power.

When using the bridle/pendant method, the airplane is attached to the catapult tow fitting with a V-shaped bridle (requires two attachment points on the airplane) or with a pendant (requires one attachment point on the airplane). This method requires the deck crew to manually attach the airplane to the catapult tow fitting and to the holdback cleats: See Figure 2.42a.

When using the nose-gear launch-bar method, the airplane is coupled to the catapult tow fitting with a nose-gear mounted launch-bar. This launch-bar retracts with the nose gear into the airplane after launch. The hold-back device is also attached to the nose gear.

Note: All new carrier based airplanes must be designed for the nosegear launch method.

The forces which act on an airplane on the catapult are large: structural provisions must be made to transmit these forces into the airframe. In the case of the F14, the design load imposed by the shuttle on the nosegear is 120,000 lbs! Couple this with the touchdown rates of p.4 and it is clear why naval airplanes have much higher landing gear weight fractions than other types of airplanes.

2.7.4 Description of Arresting Gear System

The arresting gear consists of an arresting engine, reeved with a purchase cable coupled to a deck pendant which is the arresting cable on the flight deck. A typical arresting gear arrangement is shown in Figure 2.45.

An arresting hook, located in the rear of the airplane (See Figs 2.43 and 2.46) engages the deck pendant to stop the airplane. On modern carriers there are four arresting cables. Older carriers have six such cables.

The kinetic energy of the airplane is absorbed in the arresting engine which (like the catapult) is located below the flight deck. Figure 2.47 shows typical performance capabilities for various types of arresting gear: the allowable engaging speed of the arresting hook is a function of the airplane weight.

The airplane 'sizing' implications of arresting gear performance limitations are discussed in Ch.3 of Part I.

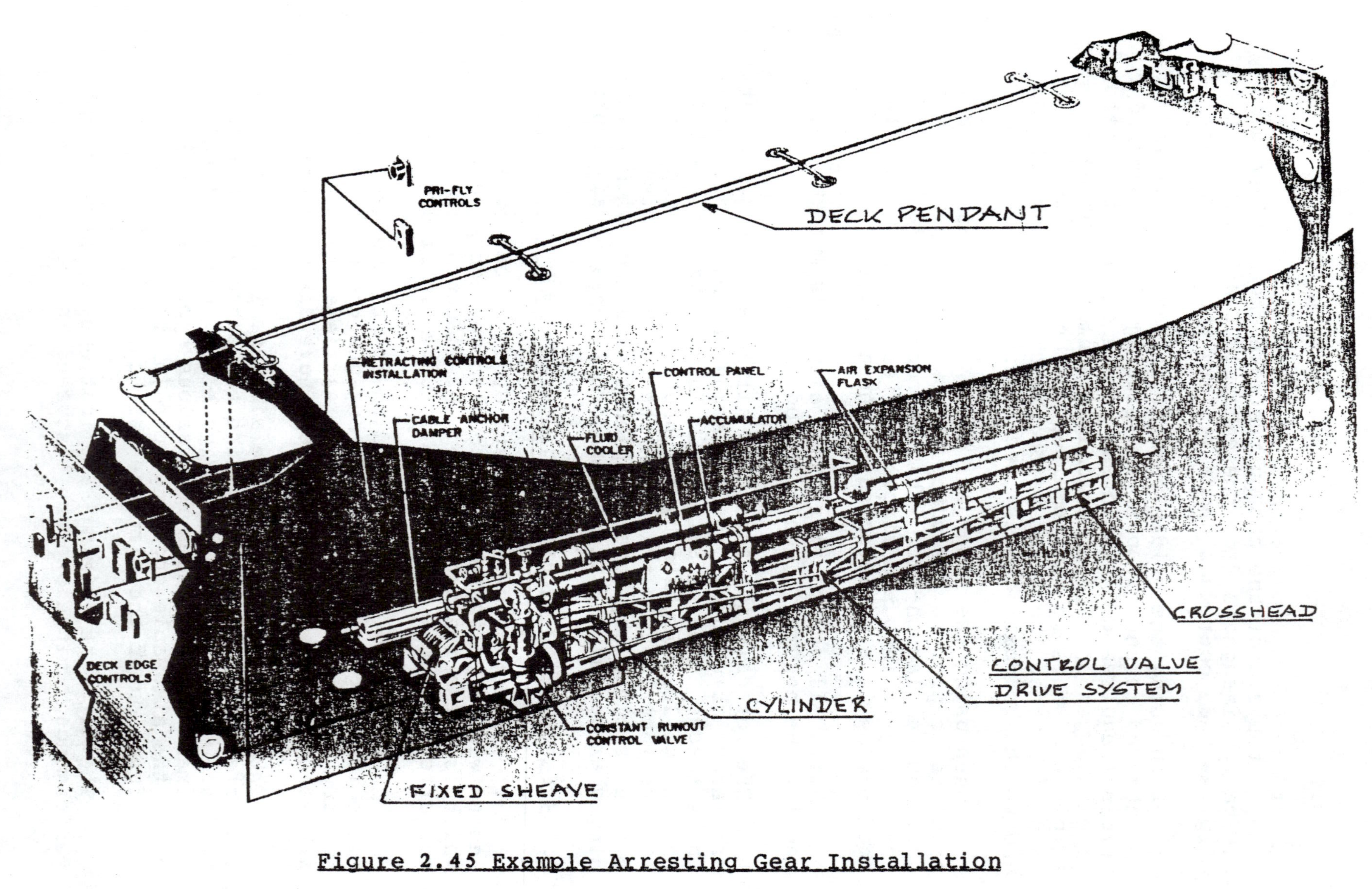

Figure 2.45 Example Arresting Gear Installation

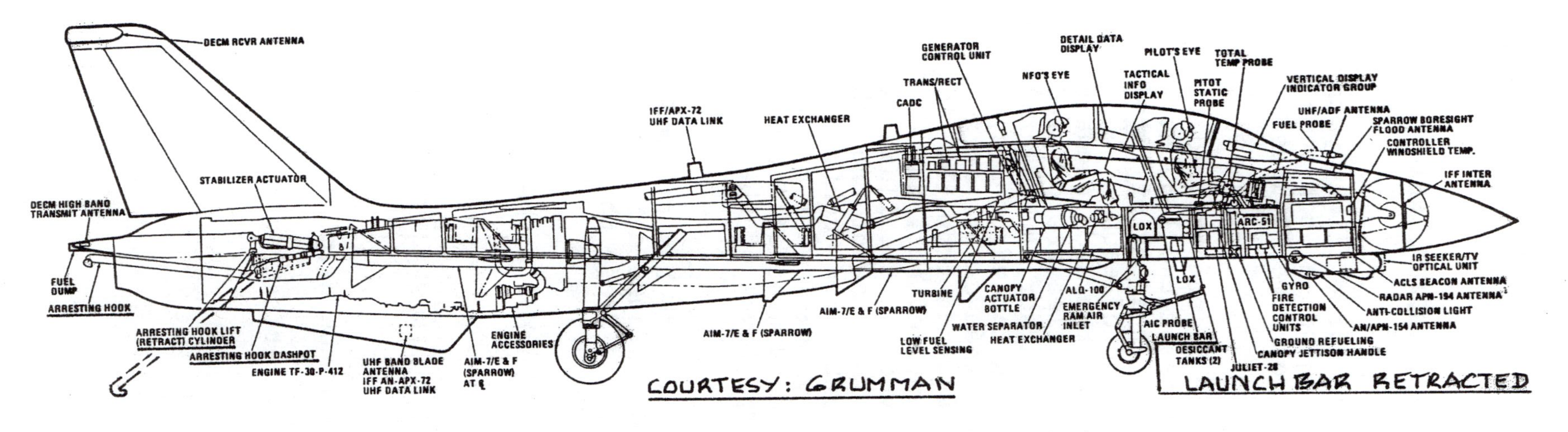

Figure 2.46 Arresting Hook Installation: Grumman F14

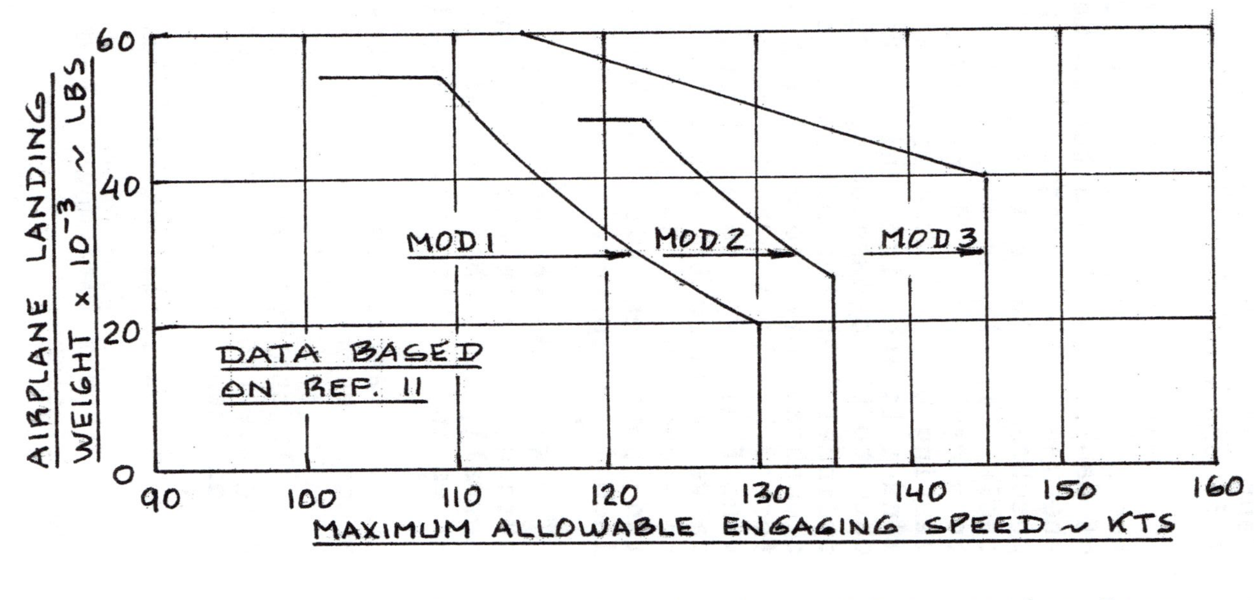

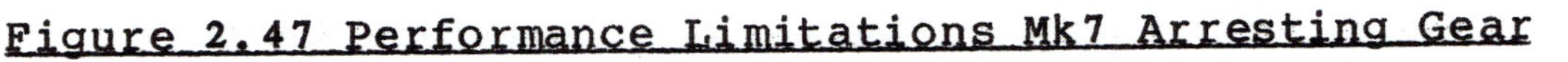
Figure 2.47 Performance Limitations Mk7 Arresting Gear

2.7.5 Arresting Procedures and Required Landing Gear and Arresting Hook Provisions

Carrier based airplanes fly a 3-4 degree flight path toward the deck. The arresting hook engages a deck pendant resting on wire supports on the deck: See Figures 2.38 and 2.45.

The forward motion of the airplane is transferred to the purchase cable system which consists of two lengths of cable reeved on a set of movable and on a set of fixed sheaves on the arresting engine. One end of each purchase cable is attached to one end of the deck pendant cable. The other ends of the purchase cables are coupled to a cable anchor damper or to a fixed anchor at the arresting engine. The fixed sheaves are attached to a cylinder filled with fluid. The movable sheaves are attached to the crosshead of a ram which extends from the cylinder. As the purchase cables are pulled out, the crosshead moves toward the fixed sheaves and the fluid is forced by the ram from the cylinder. The flow of the moving fluid is metered through a constant runout control valve to an accumulator. The control valve is set prior to engagement to determine the required degree of metering consistent with the kinetic energy level of the incoming airplane.

When the forward motion of the airplane has stopped, strain energy in the deck pendant and purchase cable will normally pull the airplane back enough to cause the arresting hook to automatically disengage.

The arresting hook of the airplane needs to be mounted such that:

1. the deceleration loads corresponding to 2.0g to 3.5g are transferred into the structure without significant deformations.

2. no large moments are imposed on the airplane.

Figures 2.43 and 2.46 show typical arresting hook installations.

2.8 REVIEW OF LANDING GEAR LAYOUT GEOMETRY

The purpose of this section is to review the overall landing gear layout geometry. The material is organized as follows:

2.8.1 Overall landing gear disposition

2.8.2 Critical landing gear dimensions:
tires, struts, drag links and side braces

2.8.3 Landing gear layout checklist

2.8.1 Review of Overall Landing Gear Disposition

Figures 2.48 and 2.49 summarize the pertinent ground clearance and tip-over requirements (criteria) already mentioned in Chapter 9 of Part II. If these criteria are not met, the proposed landing gear disposition MUST be changed!

Figure 2.50 shows typical FOD (=foreign object damage) angles which should be observed. It may be possible in some cases to 'relocate' spray patterns by the use of deflector devices. Figure 2.51 shows two possibilities: a chine tire and a nosegear splash guard (sometimes called deflector shield). If an airplane must operate from 'gravel' covered runways it is essential that tests be carried out to demonstrate that any nosegear deflector shields really work!

Tests under actual slush conditions must be conducted to verify their proper operation. Figure 2.52 shows a typical spray pattern thrown up by the nose gear when travelling through water on the runway. Note that the sprays do not enter into the engine inlets.

2.8.2 Critical Landing Gear Dimensions: Tires, Struts, Drag Links and Side Braces

Figure 2.53 and 2.54 summarize the dimensions associated with the tire/wheel/shock-absorber system. Sections 2.3 through 2.5 address the sizing procedures for these landing gear components.

With the help of these dimensions and the landing gear disposition geometry of sub-section 2.8.1 it is now possible to identify the required attachment points to the airframe. The structural integration of the landing gear into the airframe is of major importance: if major additional structure is needed for the landing gear

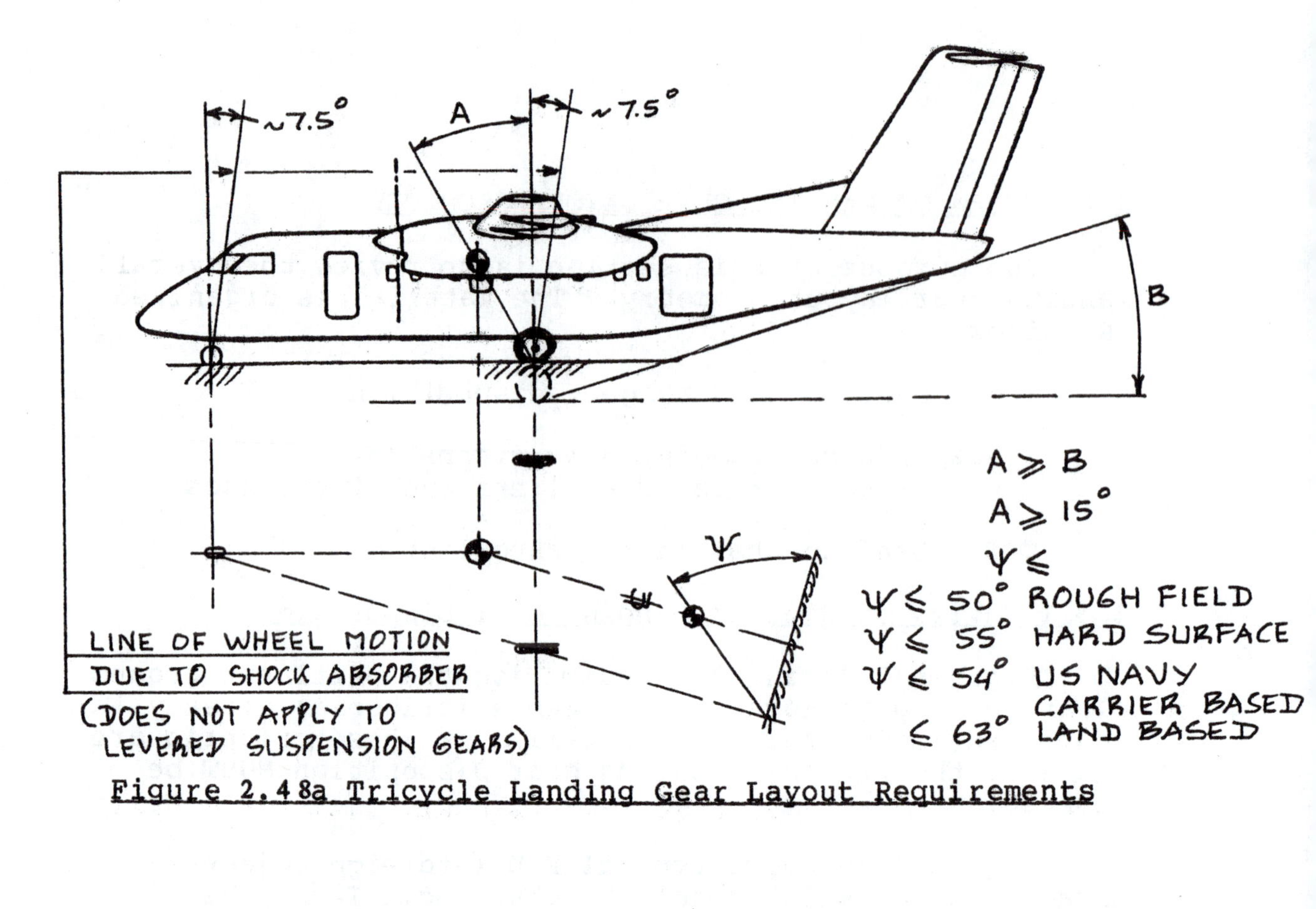

Figure 2.48a Tricycle Landing Gear Layout Requirements

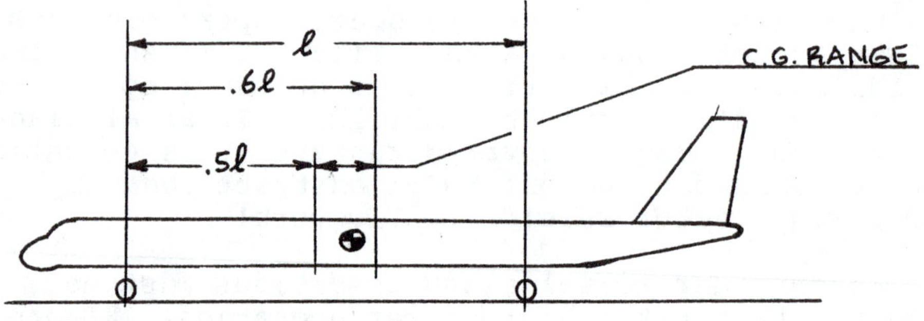

Figure 2.48b Tandem Landing Gear Layout Requirements

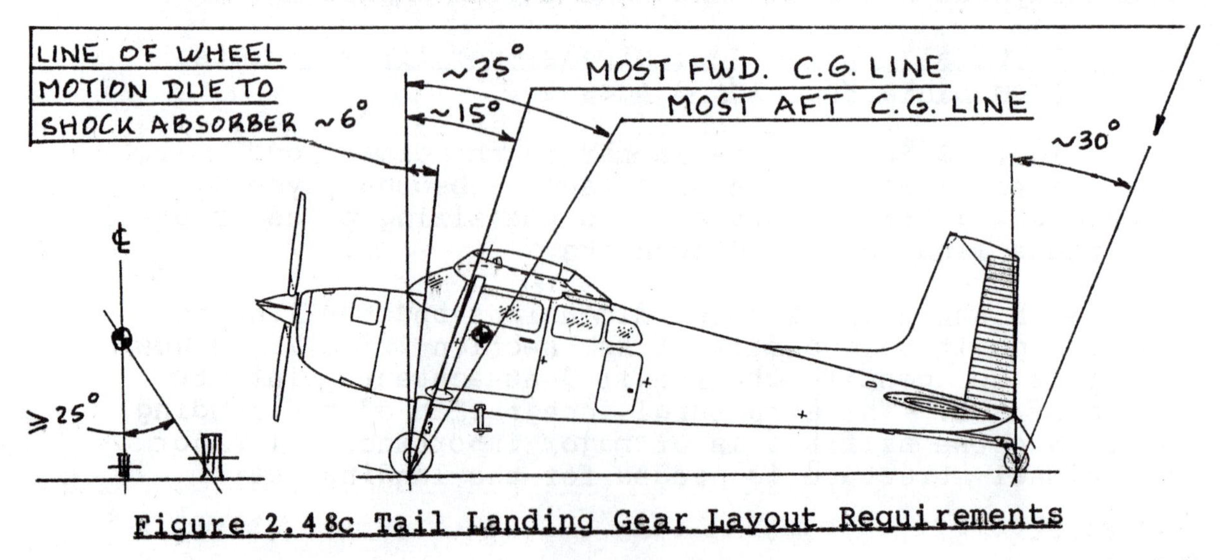

Figure 2.48c Tail Landing Gear Layout Requirements

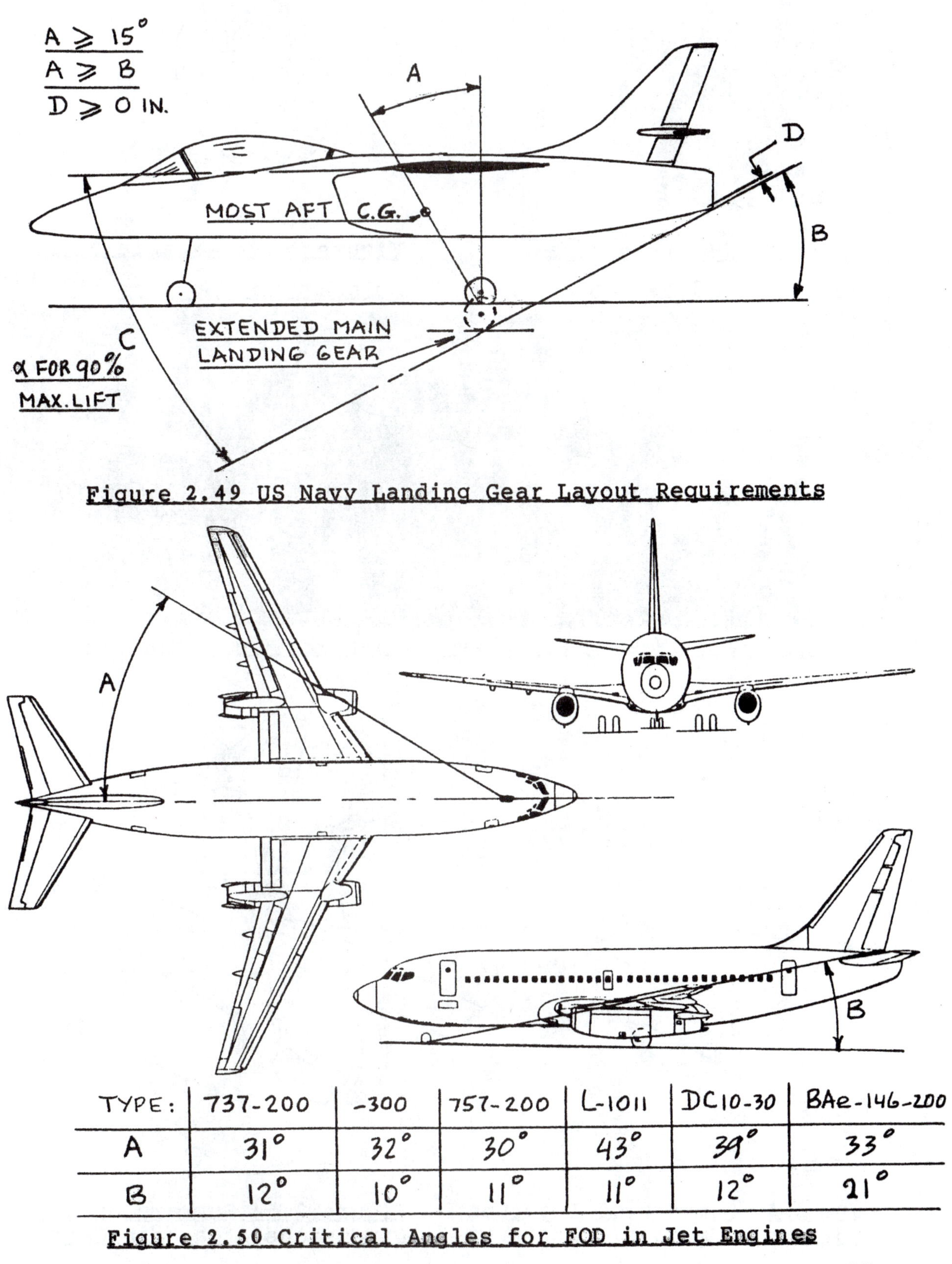

Figure 2.49 US Navy Landing Gear Layout Requirements

TYPE:	737-200	-300	757-200	L-1011	DC10-30	BAe-146-200
A	31°	32°	30°	43°	39°	33°
B	12°	10°	11°	11°	12°	21°

Figure 2.50 Critical Angles for FOD in Jet Engines

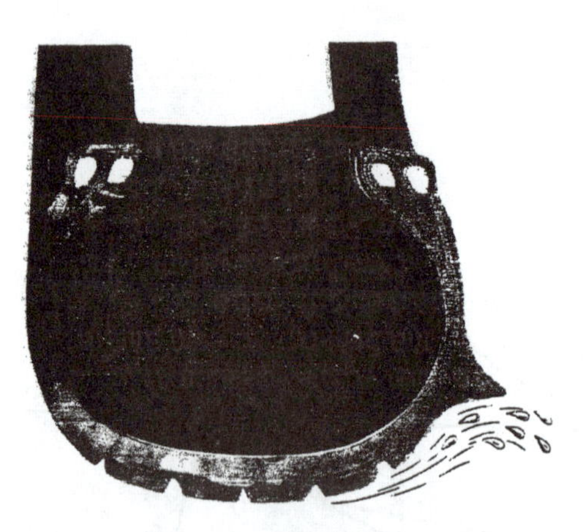

Figure 2.51a Example of a Chine Tire Application

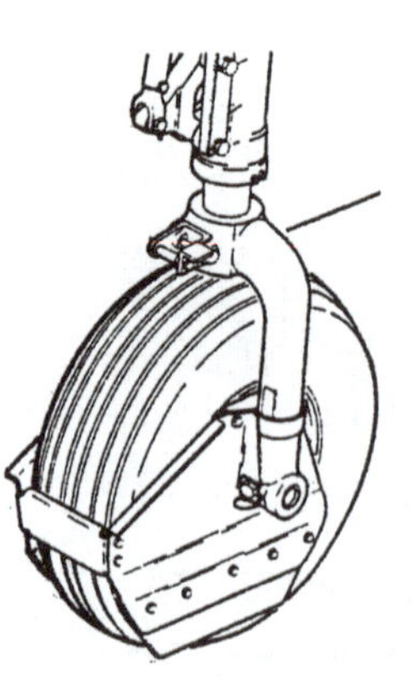

Figure 2.51b Example of a Splash Guard Application

Figure 2.52 Nose Gear Spray Pattern for the Boeing 767

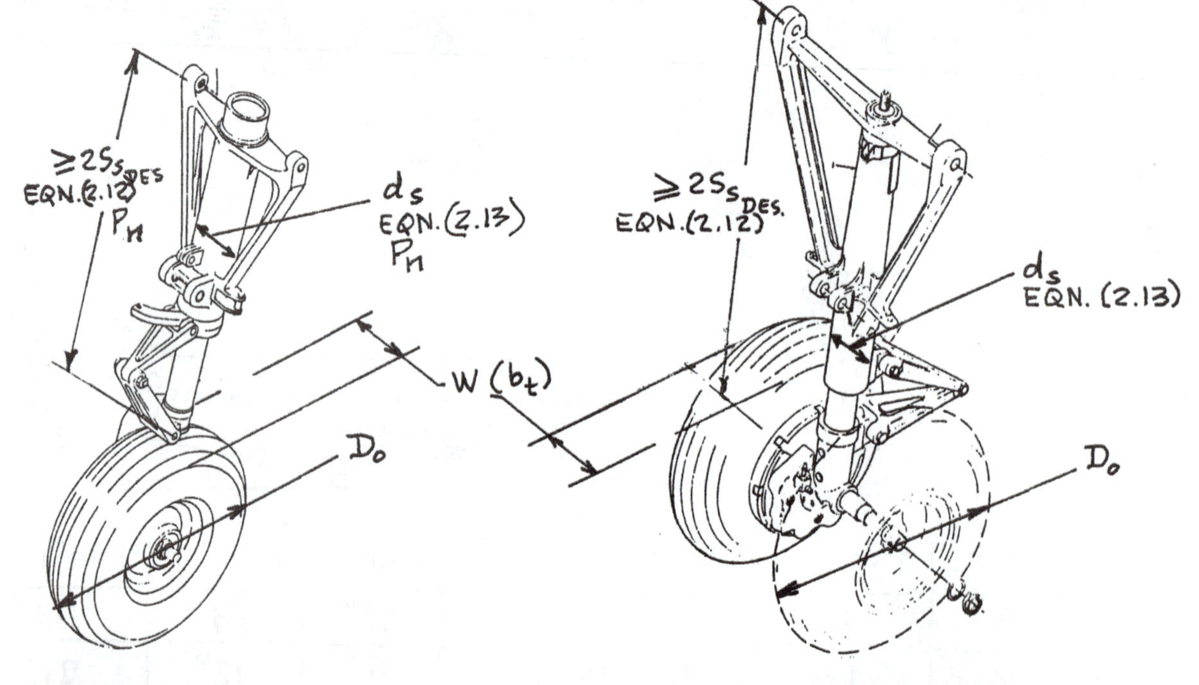

Figure 2.53 Summary of Important Dimensions of a Nose Gear

Figure 2.54 Summary of Important Dimensions of a Main Gear

attachments, the entire gear design should be reviewed.

Note: The reader should use the checklist of sub-section 2.8.3 to determine whether or not the proposed gear layout is satisfactory.

2.8.3 Landing Gear Layout Checklist

The reader is reminded again that a valid solution to a landing gear design problem cannot be claimed until ALL items in the following checklist have been addressed:

1. ALL geometric clearance and tip-over criteria are satisfied: See Chapter 9 in Part II and sub-section 2.8.1 in this part.

2. The proper tire size, shock absorber stroke and strut diameter have been determined: See Sections 2.3, 2.4 and 2.5 respectively.

3. The need for drag braces and side braces have been satisfied: See Section 2.1.

4. The structural attachment points for the gear should NOT require the introduction of significant additional structure.

5. Any 'spray' caused by the tires (particularly the nose gear) on wet or slushy runways must not enter the engine inlets. The spray (=FOD) angle criteria of Figure 2.50 must be met.

6. The gear can retract without interfering with other components of the airplane: the tire clearance requirements discussed in Section 2.4 must be met.

7. The retraction kinematics (to be discussed in Section 2.10) is feasible and does not require excessive actuator forces to retract or to lower.

<u>Important Comment:</u>

In Part II, Chapter 2, Step 9, the point is made that whether or not a 'valid' solution to the landing gear problem is obtained may well determine the viability of the entire airplane design.

2.9 STEERING, TURNRADII AND GROUND OPERATION

The purpose of this section is to show how steering during ground operations is accomplished. The material is organized as follows:

2.9.1 Steering systems

2.9.2 Turnradii and ground operation

2.9.1 Steering Systems

To operate airplanes on the ground it must be possible to 'steer' them. The steering feature used on most airplanes is a steerable nosegear or a steerable tailwheel. The steering capability in most airplanes is augmented by the use of differential braking on the main gear and in some cases by the use of differential thrust.

In very light airplanes, steering can be accomplished by a direct mechanical link from the rudder pedals. However, many light airplanes are equipped with a hydraulic type steering system. Section 2.11 contains a number of examples of light airplane nose gear installations with the steering system identified.

In the case of transport airplanes, nosegear steering requires a considerable force. This force is normally generated with the help of a hydraulic system. The hydraulic system itself receives its 'signal' to move from the rudder pedals and/or from a cockpit mounted 'steering tiller wheel'. Figure 2.55 shows an example of a nosegear steering system.

2.9.2 Turnradii and Ground Operation

It is essential that the airplane can be maneuvered on the ground without requiring excessive real estate. This is of particular importance to transport airplanes which need to maneuver in crowded terminal areas. The ground maneuvering capability of an airplane is normally expressed in terms of its minimum turn radius. This minimum turn radius is dependent on the location of the landing gears and on the nosewheel steering ability incorporated in the nose gear. Whther or not an airplane can turn about one main gear also influences its turnradius. Figures 2.56 and 2.57 illustrate the maneuvering capability of two typical jet transports.

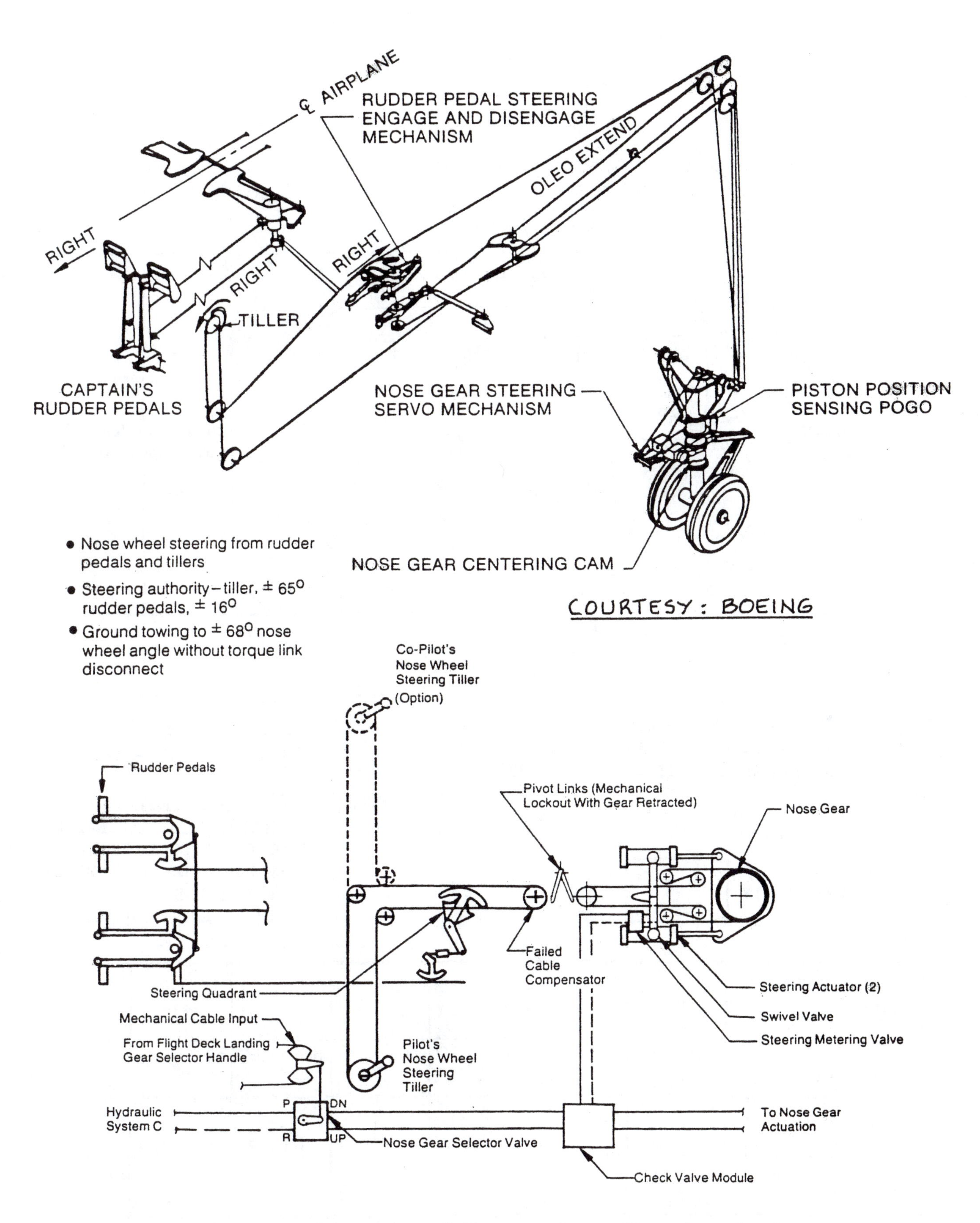

Figure 2.55 Boeing 767 Nose Gear Steering System

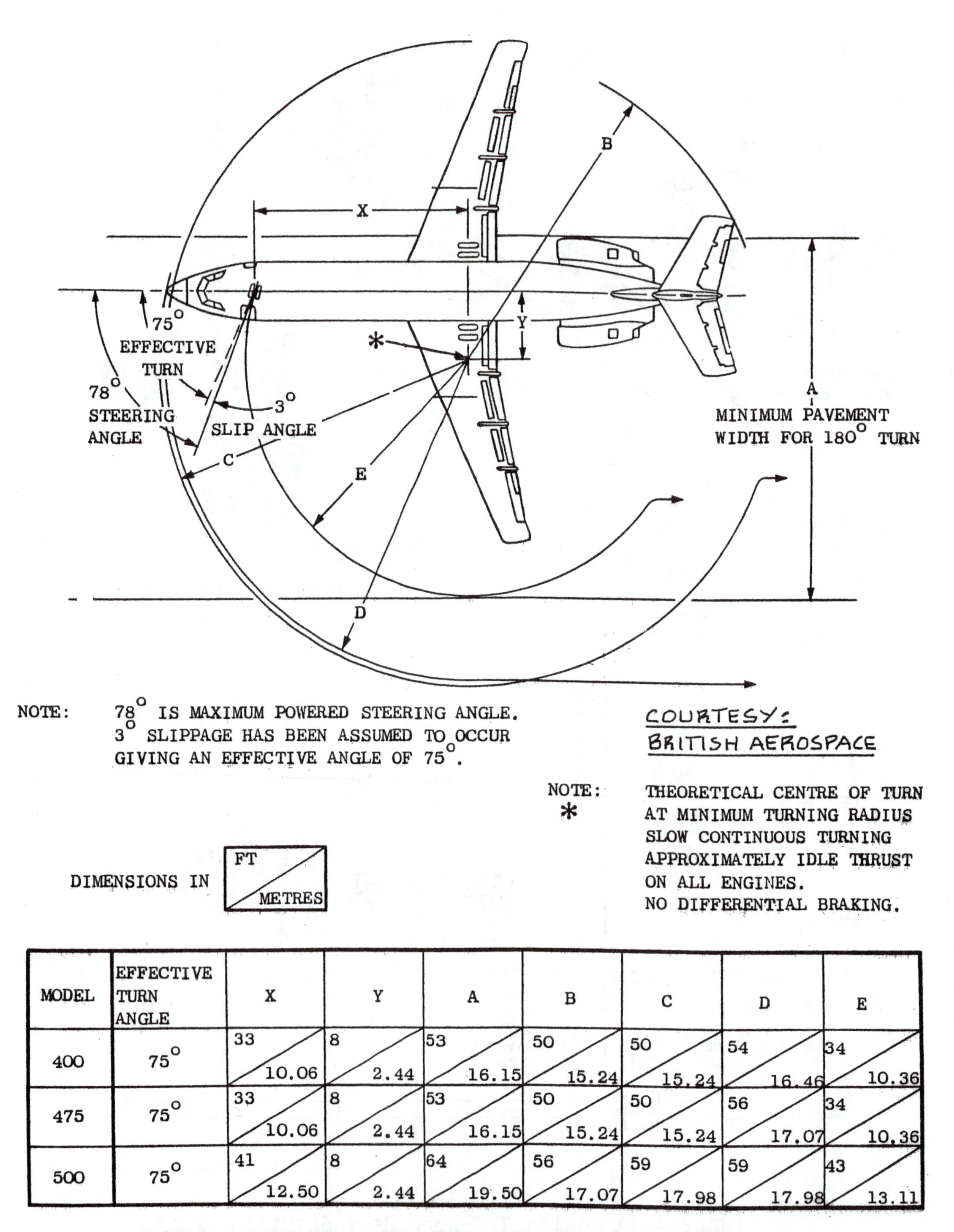

MODEL	EFFECTIVE TURN ANGLE	X	Y	A	B	C	D	E
400	75°	33 / 10.06	8 / 2.44	53 / 16.15	50 / 15.24	50 / 15.24	54 / 16.46	34 / 10.36
475	75°	33 / 10.06	8 / 2.44	53 / 16.15	50 / 15.24	50 / 15.24	56 / 17.07	34 / 10.36
500	75°	41 / 12.50	8 / 2.44	64 / 19.50	56 / 17.07	59 / 17.98	59 / 17.98	43 / 13.11

Figure 2.56 Ground Turning Capability of a BAC 111

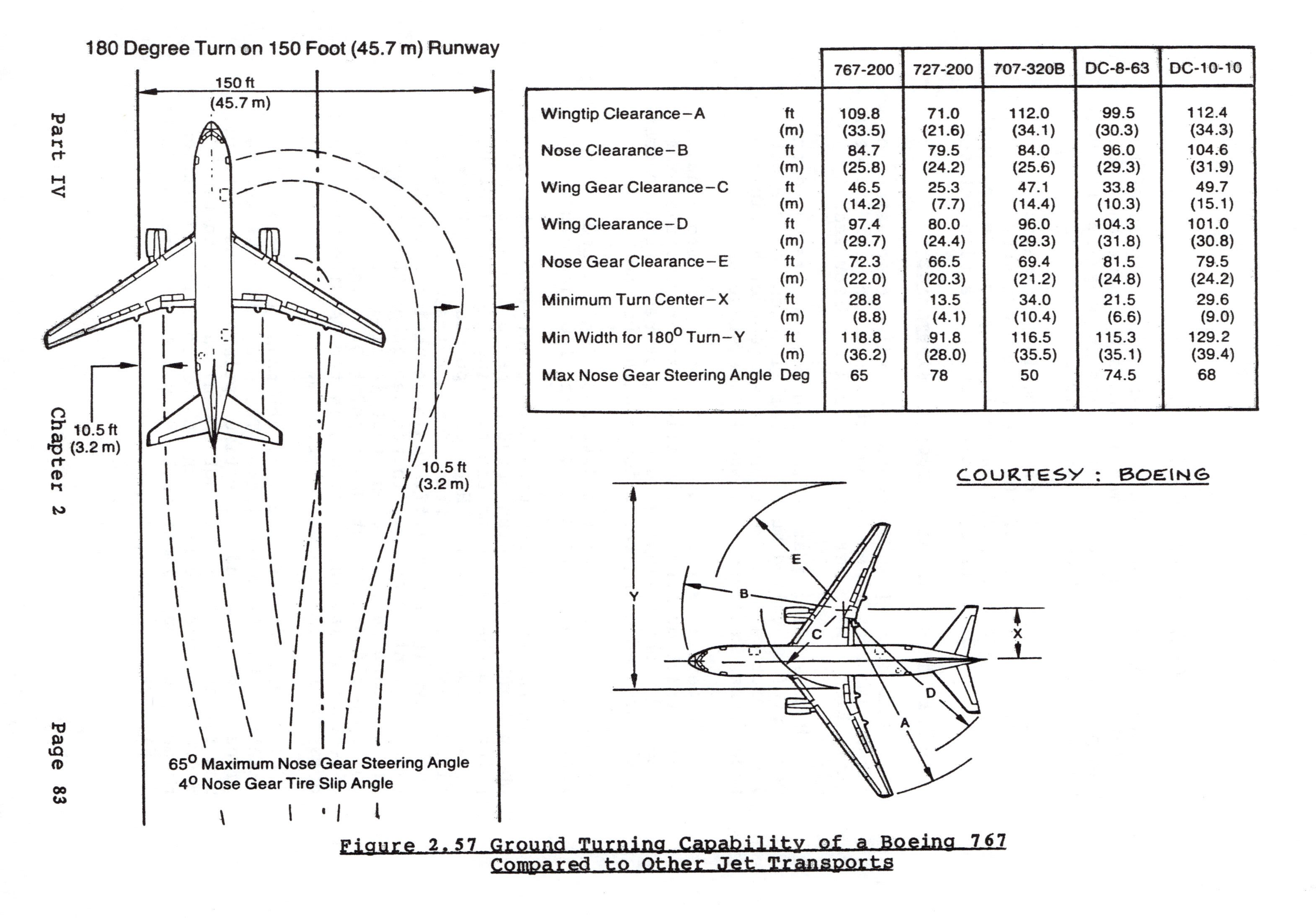

		767-200	727-200	707-320B	DC-8-63	DC-10-10
Wingtip Clearance – A	ft	109.8	71.0	112.0	99.5	112.4
	(m)	(33.5)	(21.6)	(34.1)	(30.3)	(34.3)
Nose Clearance – B	ft	84.7	79.5	84.0	96.0	104.6
	(m)	(25.8)	(24.2)	(25.6)	(29.3)	(31.9)
Wing Gear Clearance – C	ft	46.5	25.3	47.1	33.8	49.7
	(m)	(14.2)	(7.7)	(14.4)	(10.3)	(15.1)
Wing Clearance – D	ft	97.4	80.0	96.0	104.3	101.0
	(m)	(29.7)	(24.4)	(29.3)	(31.8)	(30.8)
Nose Gear Clearance – E	ft	72.3	66.5	69.4	81.5	79.5
	(m)	(22.0)	(20.3)	(21.2)	(24.8)	(24.2)
Minimum Turn Center – X	ft	28.8	13.5	34.0	21.5	29.6
	(m)	(8.8)	(4.1)	(10.4)	(6.6)	(9.0)
Min Width for 180° Turn – Y	ft	118.8	91.8	116.5	115.3	129.2
	(m)	(36.2)	(28.0)	(35.5)	(35.1)	(39.4)
Max Nose Gear Steering Angle	Deg	65	78	50	74.5	68

Figure 2.57 Ground Turning Capability of a Boeing 767 Compared to Other Jet Transports

2.10 RETRACTION KINEMATICS

In this section some fundamental principles of the operation of retraction mechanisms are discussed. Actual examples of landing gear retraction methods are provided. The material is presented as follows:

2.10.1 Fundamentals of Retraction Kinematics

References 1 and 2 contain detailed discussions of a large number of kinematic possibilities for landing gear retraction. A number of these will be discussed.

Most landing gear retraction schemes are based on a version of so-called multiple bar linkages. For a detailed theoretical treatment on the subject of multiple bar mechanisms, References 12 and 13 are suggested.

To determine the feasibility of a proposed landing gear retraction scheme, so-called 'stick diagrams' are drawn. In the following, a number of retraction schemes are discussed using the stick diagram method.

Figures 2.58 through 2.60 show landing gear retraction schemes with a so-called floating link.

In Figure 2.58 the main gear strut (AC) rotates about point A. The drag strut BE has a pin joint at point D so that BD forms the floating link and link DE rotates about point E. Note that points A and E are fixed points in the airframe. The retraction cylinder (not shown in Figure 2.58) could be attached anywhere between points D and E. In the design of this type of linkage care must be taken, that landing loads do not collapse the drag strut BE at point D. Since the main direction of landing loads are in the CA direction, this does not represent a major problem.

In Figures 2.59 and 2.60 the main gear strut BD itself is the floating link. The drag strut CE is one piece. Points A and E are fixed points in the airframe. The retraction cylinder (not shown) can be attached anywhere between points A and B. Note that the gear can be made to go up almost vertically with a system as shown in Fig.2.59.

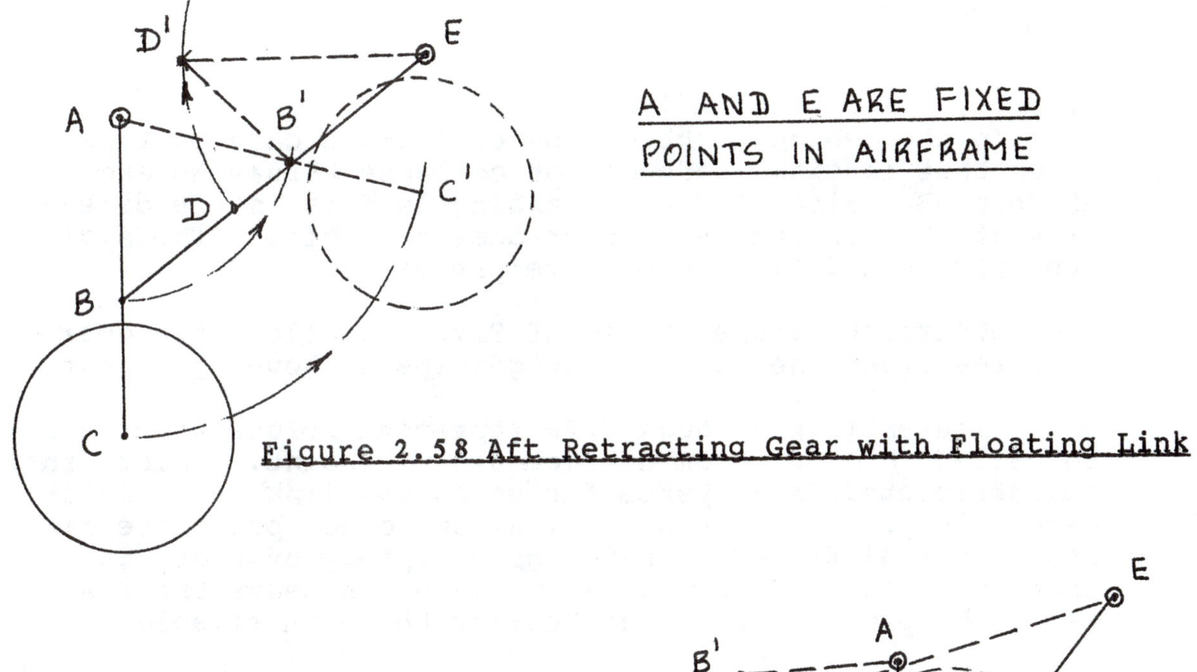

Figure 2.58 Aft Retracting Gear with Floating Link

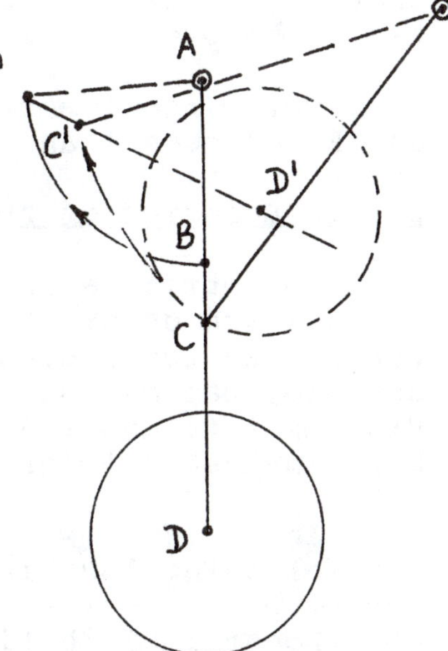

A AND E ARE FIXED
POINTS IN AIRFRAME

Figure 2.59 Upward Retracting Gear with Floating Link

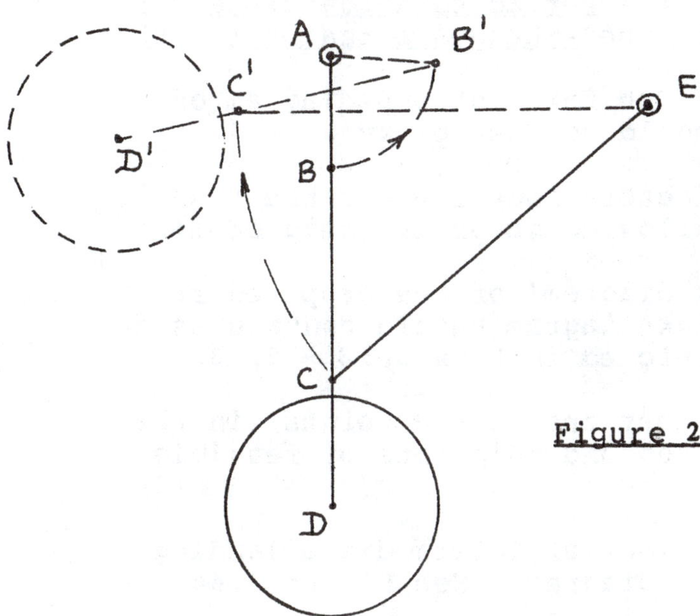

A AND E ARE FIXED
POINTS IN AIRFRAME

Figure 2.60 Forward Retracting Gear with Floating Link

In the design of this type of linkage care must be taken that landing loads do not collapse struts AB and BD at point B. Since the main landing load is in the direction of DB this can be a major design problem. The system of Figure 2.58 is therefore preferred.

Observe that the system of Fig.2.60 allows the gear to 'free drop': aerodynamic drag helps to lower the gear.

Figures 2.61 through 2.63 represent solutions with a so-called 'slot'. From a kinematic viewpoint, a slot can be interpreted as a system for which the link AB is infinitely long. The slot as well as one other point are fixed in the airframe. A major disadvantage of slots is that mud, snow and ice can accumulate and cause the system to be prone to wear and tear or to be unreliable.

For more kinematic options in landing gear retraction, References 1 and 2 should be consulted.

2.10.2 Location of the Retraction Actuator

To retract a landing gear, so-called retraction actuators are used. Retraction actuators are normally of the hydraulic or electro-mechanical type. The retraction actuator can be thought of as a link of variable length. Figures 2.64 and 2.65 show examples of retraction actuators in the 'up' and 'down' position.

In deciding where to locate the retraction actuator, the following location criteria must be kept in mind:

Location Criterion 1: The retracted actuator length can't be less than one half the extended actuator length.

Location Criterion 2: The force-stroke diagram of the retraction actuator should not be 'peaky'.

To determine a satisfactory location for the retraction cylinder the following steps are suggested:

Step 1: Draw a 'stick diagram' of the proposed retraction scheme. This stick diagram should contain as a minimum the information depicted in Figs 2.58 - 2.63.

Step 2: Make certain that the 'fixed points' in the proposed stick diagram correspond to points of feasible structure on the airplane.

Step 3: Construct a number of intermediate landing gear positions in the stick diagram. See Figure 2.66a.

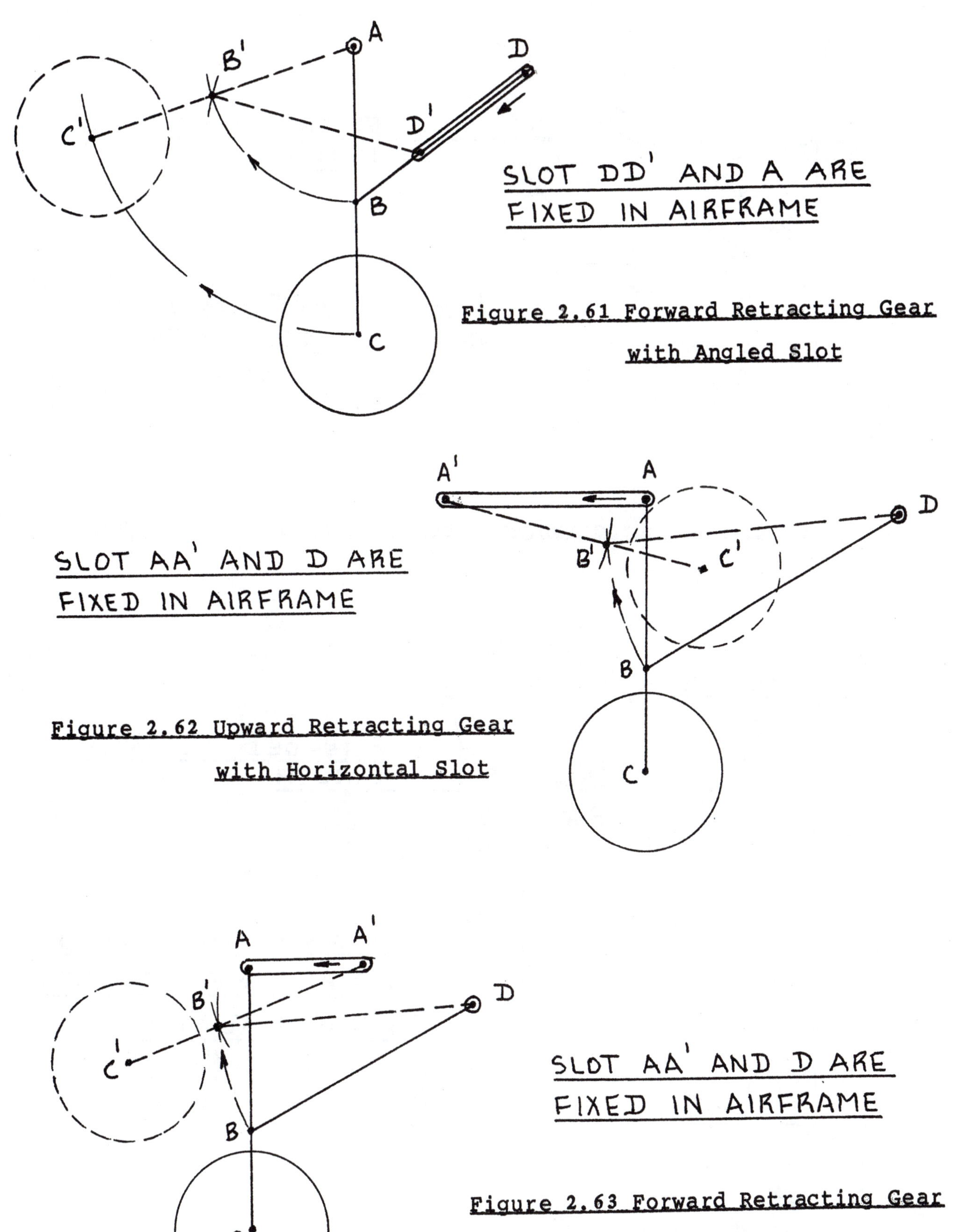

Figure 2.61 Forward Retracting Gear with Angled Slot

Figure 2.62 Upward Retracting Gear with Horizontal Slot

Figure 2.63 Forward Retracting Gear with Horizontal Slot

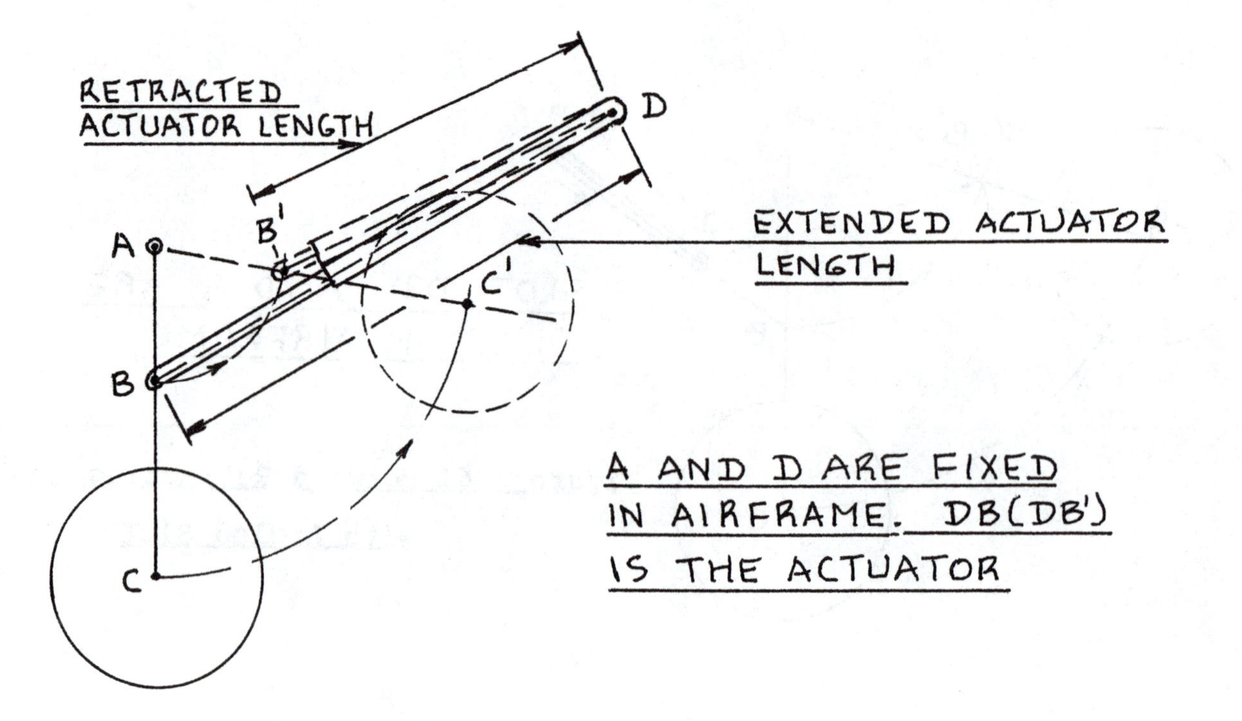

Figure 2.64 Retraction Actuator as Part of the Drag Brace

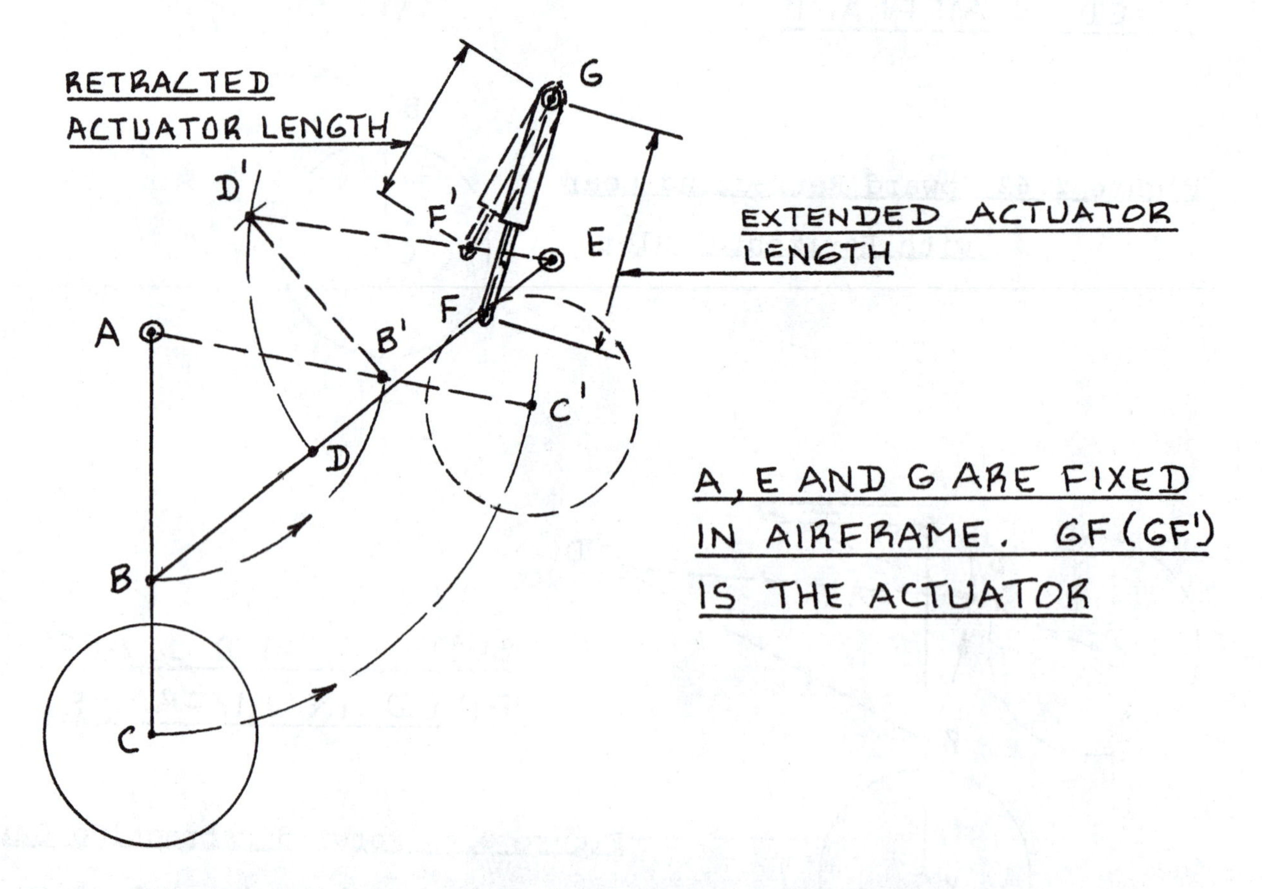

Figure 2.65 Retraction Actuator with Folding Main Strut

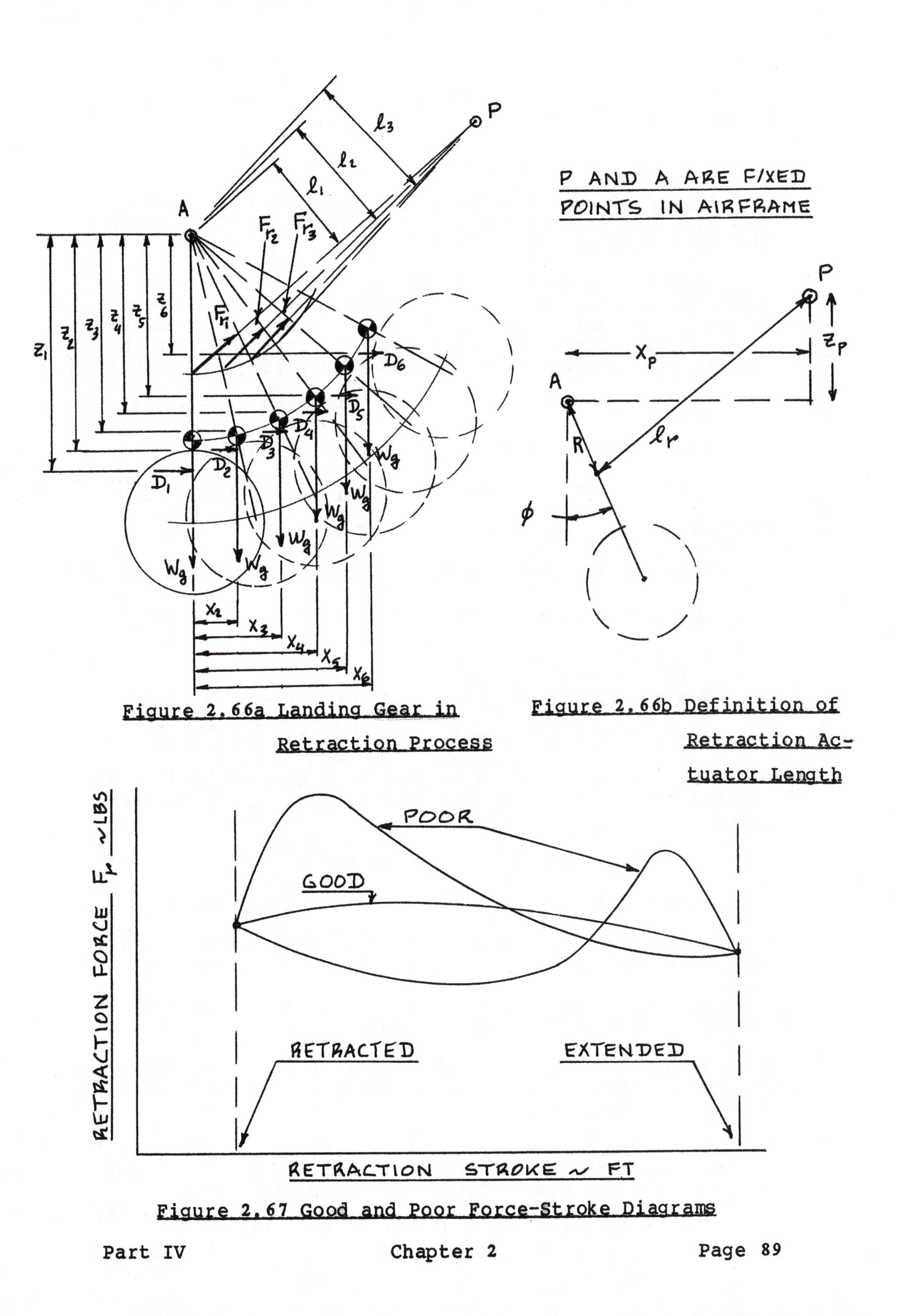

Figure 2.66a Landing Gear in Retraction Process

Figure 2.66b Definition of Retraction Actuator Length

Figure 2.67 Good and Poor Force-Stroke Diagrams

Step 4: Select a location for the retraction actuator and make certain that location criterion 1 is satisfied before moving on to Step 5.

Step 5: Indicate in the stick diagram the forces which act on the landing gear during retraction. These forces are normally the weight of the gear and the aerodynamic drag force. Figure 2.66a shows these forces: W_g for the gear weight and D_i for the gear drag.

Notes: 1. The gear weight, W_g may be determined from the Class II weight estimating methods of Part V.

2. The gear drag, D_i may be determined from the gear drag estimating methods of Part VI.

Figure 2.66a shows these forces properly drawn in.

Step 5: Compute the retraction force, F_{r_i} from:

$$F_{r_i} = (W_g x_i - D_i z_i)/l_i \qquad (2.15)$$

Moment arms x_i, z_i and l_i are defined in Fig.2.66a.

The retraction force F_{r_i} must be plottes versus the retraction stroke (change in length of the retraction cylinder). The retraction cylinder length, l_r may be computed from:

$$l_r = \{(z_p + R\cos\phi)^2 + (x_p - R\sin\phi)^2\}^{1/2} \qquad (2.16)$$

Figure 2.66b defines the quatities R, z_p and x_p.

The reader must realize that the form taken by Eqns(2.15) and (2.16) will be different for each gear layout.

Step 6: To satisfy location criterion 2, the plot of retraction force, F_r versus retraction length, l_r must not contain any major fluctuations in force (must not be 'peaky'). Figure 2.67 shows an example of a good and a poor force-stroke diagram.

Note: The retraction force/stroke diagram does not contain any information about the time needed to retract or to extend the gear. Table 2.19 summarizes the allowable 'retract' and 'extend' times for landing gears.

Table 2.19 Landing Gear Operating Times

Type of retraction and extension	Temperature	Max. time to extend and lock gear	Max. time to retract and lock gear
Power operated	above $-20^{o}F$	15 sec.	10 sec.
	from $-65^{o}F$ to $-20^{o}F$	30 sec.	10 sec.

Note: The gear must be retracted BEFORE the airplane reaches 75 percent of the gear placard speed at maximum forward acceleration

Type of retraction and extension	Temperature	Max. time to extend and lock gear	Max. time to retract and lock gear
Manually operated	above $-20^{o}F$	15 sec.	30 sec.
	from $-65^{o}F$ to $-20^{o}F$	30 sec.	60 sec.

Note: The power required to operate the system must NOT exceed 3000 ftlbs/min. and the force required on the operating handle shall NOT exceed 50 lbs.

Note: For multi-engine airplanes these requirements must be met with the critical engine inoperative.

These requirements apply to military and civilian airplanes

Examples of extension and retraction times are:

Airplane	Extend	Retract	Airplane	Extend	Retract
A-10	6-9	6-9	F-5	6	6
			F-100	6-8	6-8
B-52	10-12	8-10	F-105	5-9	4-8
B-66	8	10	F-111	26	18
C-5	20	20	T-37	8	10
C-123	6	9	T-38	6	6
C-130	19	19			
C-135	10	10			

These times together with the force/stroke diagram determine the power required for gear retraction.

2.10.3 Special Problems in Gear Retraction

The examples of retraction kinematics illustrated sofar were extremely simple. It is not always possible to use such simple methods. In many landing gear applications it is necessary to use additional kinematic features to make retraction possible. Examples are:

1. Wheel rotates relative to the strut while retracting.
2. Main strut must shorten while retracting.
3. Bogie rotates relative to the strut while retracting.
4. Tandem gears sometimes must retract synchronously.

Figure 2.68 shows examples of a retraction scheme where the wheel rotates 90 degrees while retracting. This scheme is useful when the 'vertical' volume for receiving the tire is limited. Note that an additional link is required to accomplish this.

Figure 2.69 shows an example of a main strut which must 'shorten' while retracting. This scheme is used when the 'horizontal' volume for receiving the gear is limited. Additional links are needed to accomplish this.

Figure 2.70 shows examples of a 'folding' bogie in the case of a four-wheel landing gear. In jet transports this is often used to prevent having to reroute major structural components. Note that an additional actuator is needed in cases a) and b) but not in case c).

Figure 2.71 is an illustration of a main landing gear consisting of two units in tandem. These units retract in a synchronous manner.

2.10.4 Examples of Landing Gear Retraction Methods

Figures 2.72 through 2.84 are presented to illustrate a variety of solutions to retraction problems which are in use. Note the so-called 'tilted' pivot used in retracting landing gears into fuselages: Figures 2.72b, 2.80 and 2.81 are typical examples. This method is most often used in fighters.

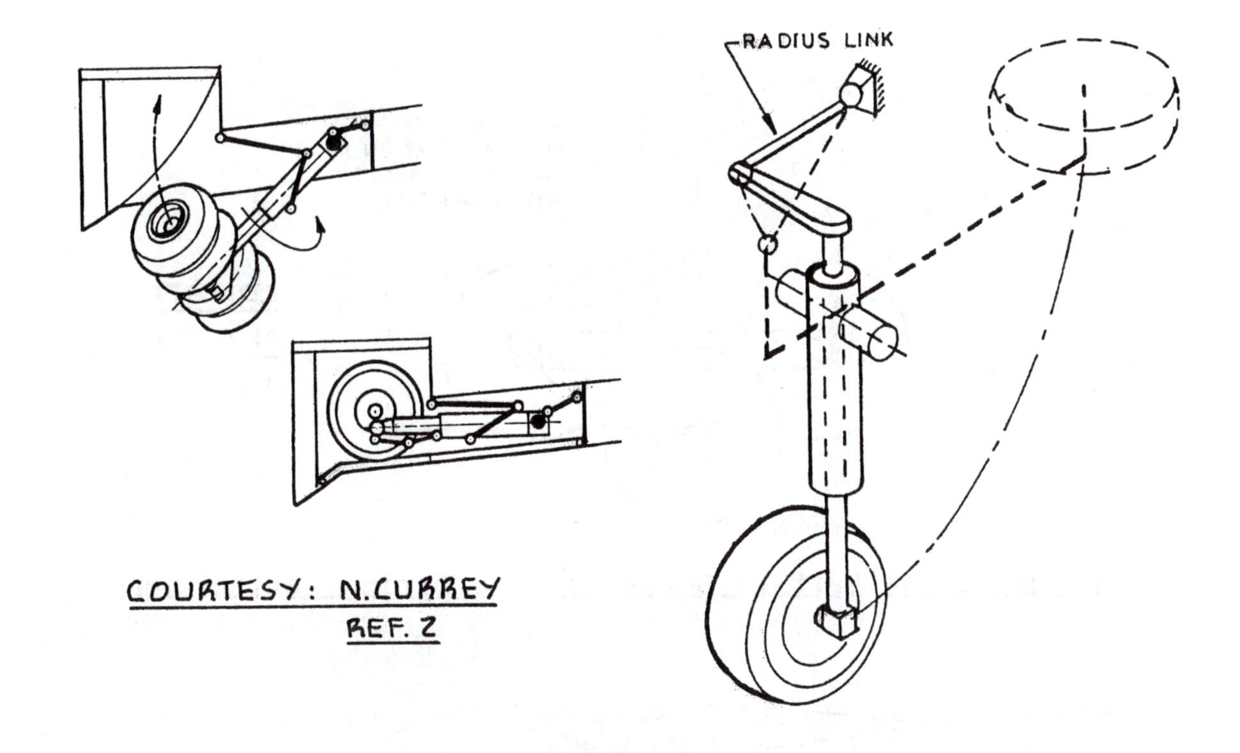

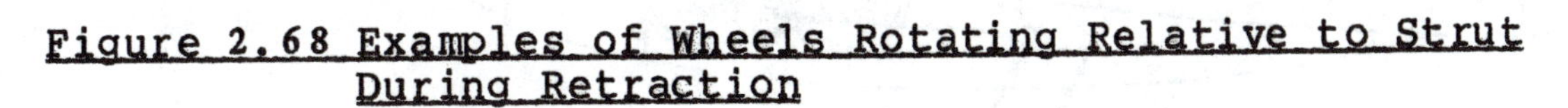

Figure 2.68 Examples of Wheels Rotating Relative to Strut During Retraction

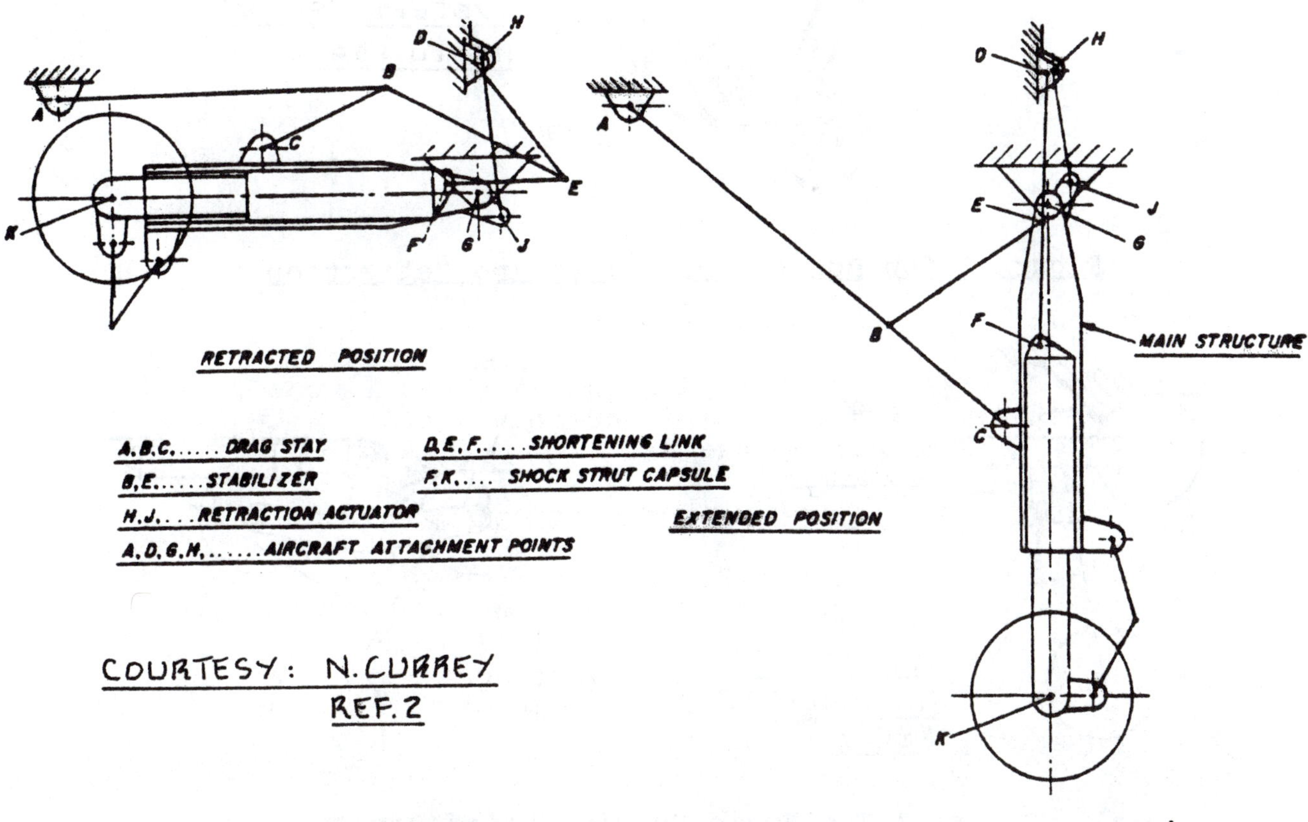

Figure 2.69 Example of Strut Shortening During Retraction

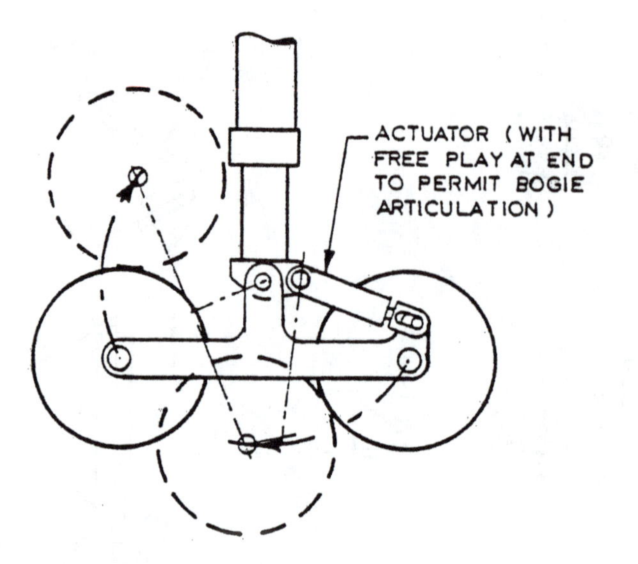

Figure 2.70a Partial Bogie Folding During retraction

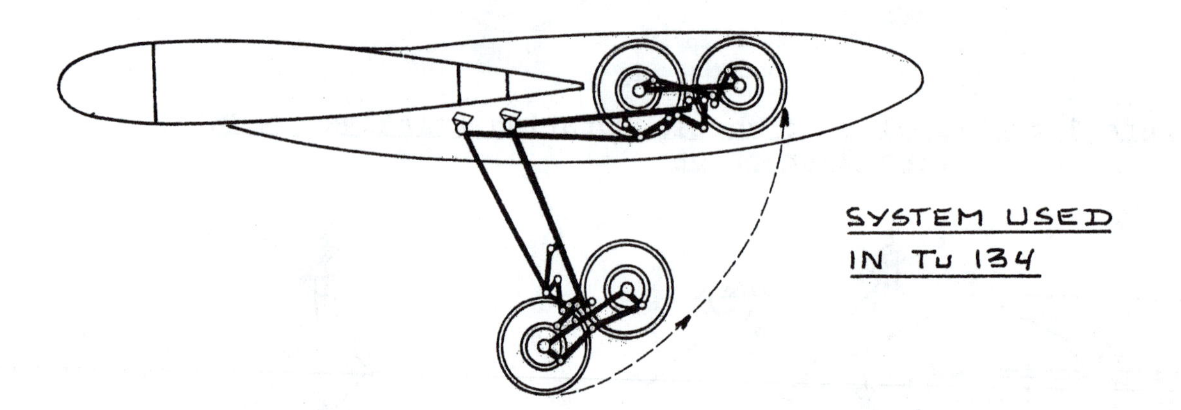

Figure 2.70b Bogie Reversal During Retraction

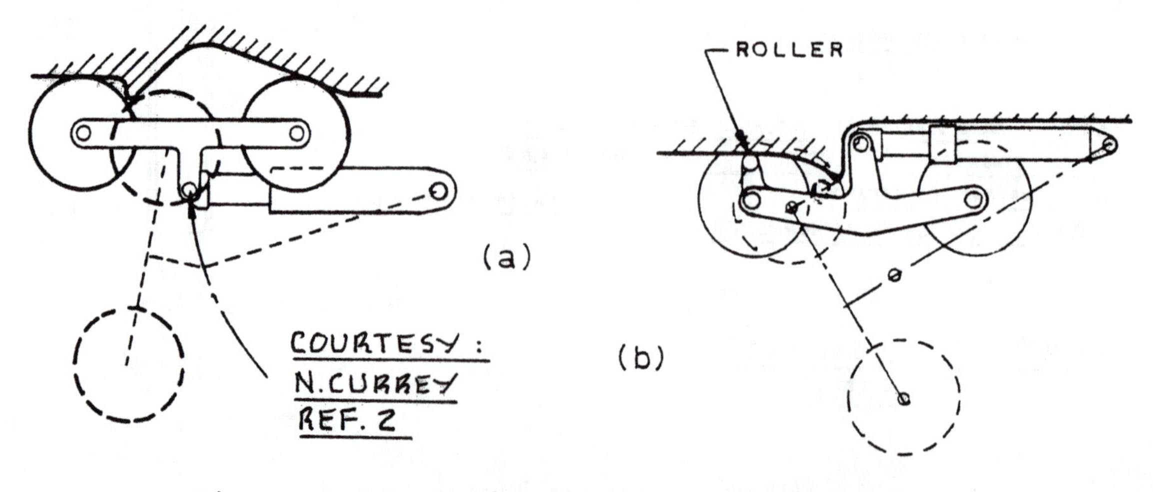

Figure 2.70c Bogie Folding Using Ramps

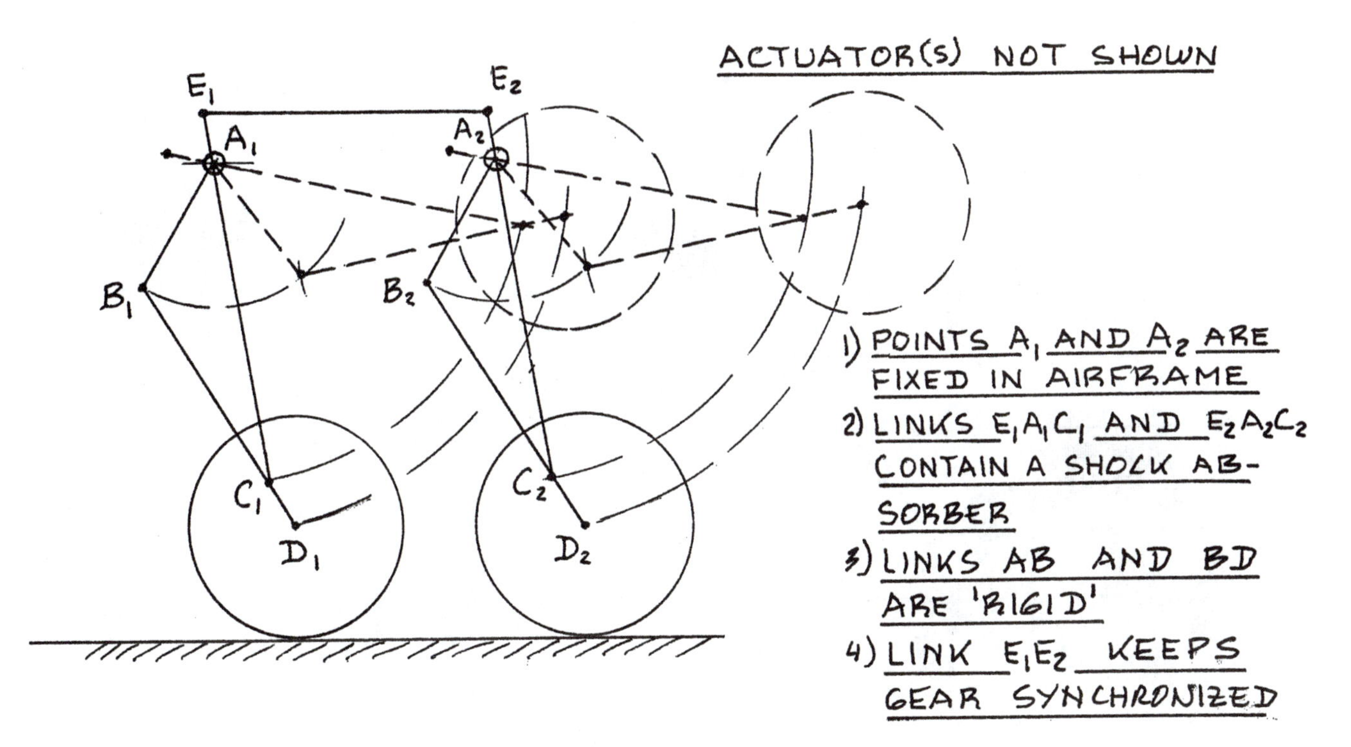

Figure 2.71 Tandem Gear Retraction Example

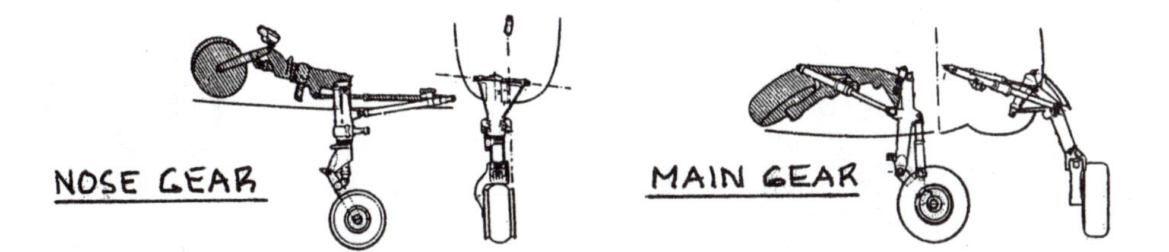

Figure 2.72 Gear Retraction: DBD Alphajet Trainer

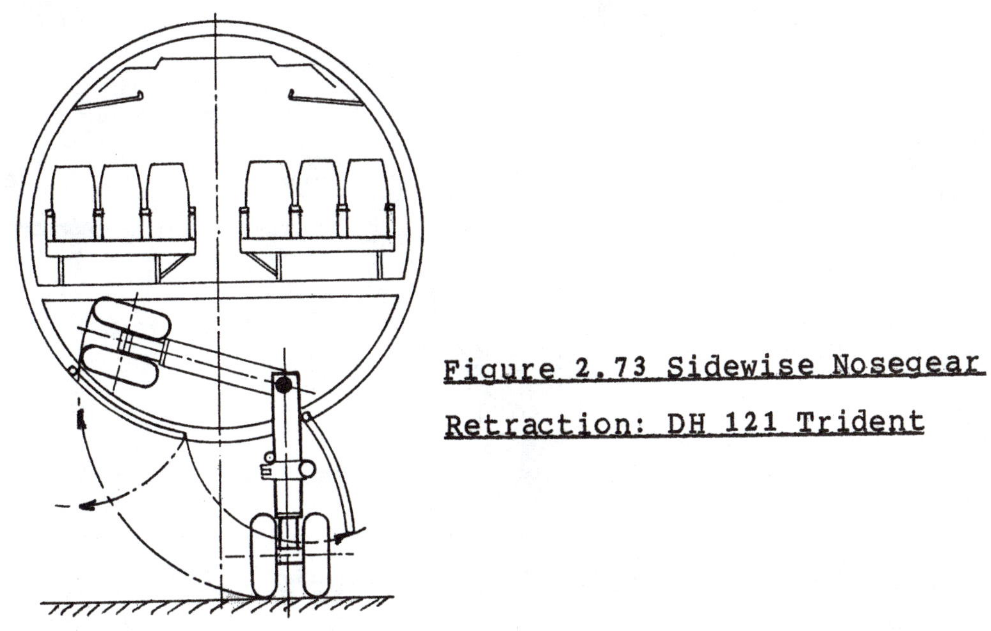

Figure 2.73 Sidewise Nosegear Retraction: DH 121 Trident

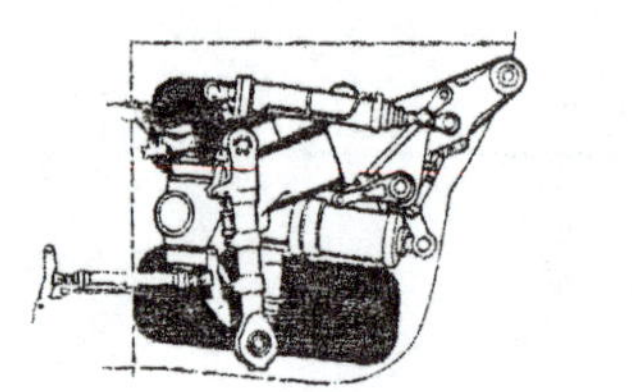

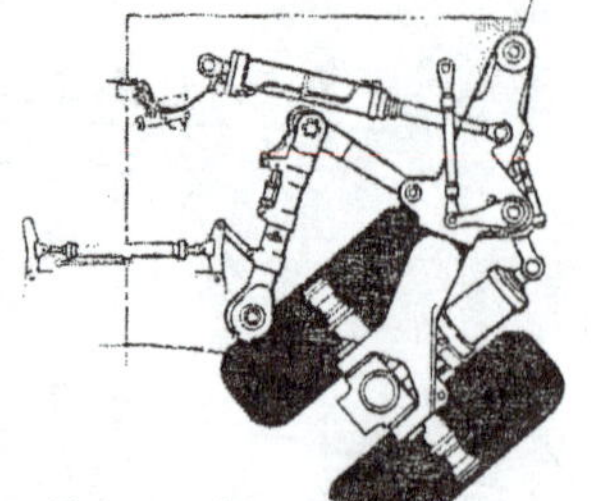

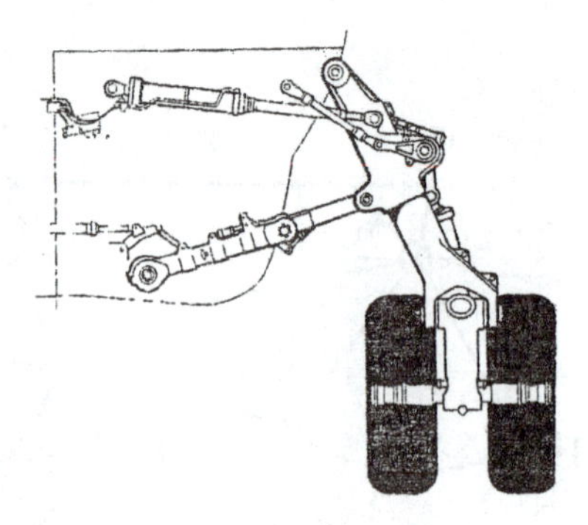

Landing gear extension sequence (front view, left leg). A single hydraulic jack effects extension/retraction of each leg; there is thus no jack sequencing. When viewed from rear, locking/bracing geometry and ease of oleo removal are clear.

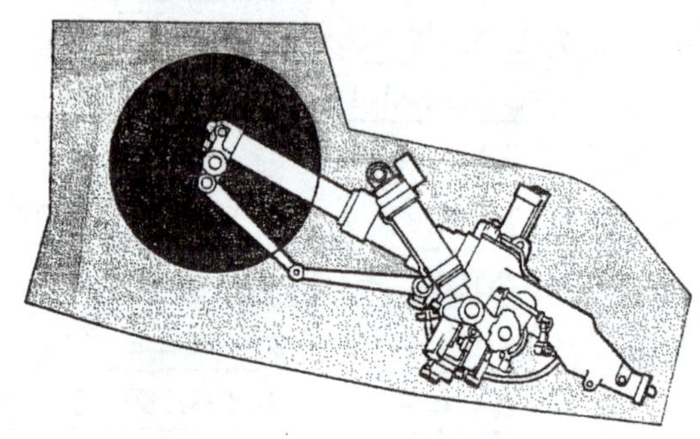

COURTESY:
BRITISH AEROSPACE

As with main gear, nose gear actuation is by a single jack with mechanically linked doors. Photo shows how closed doors keep nosewheel bay clean.

Figure 2.74 Gear Retraction: BAe 146

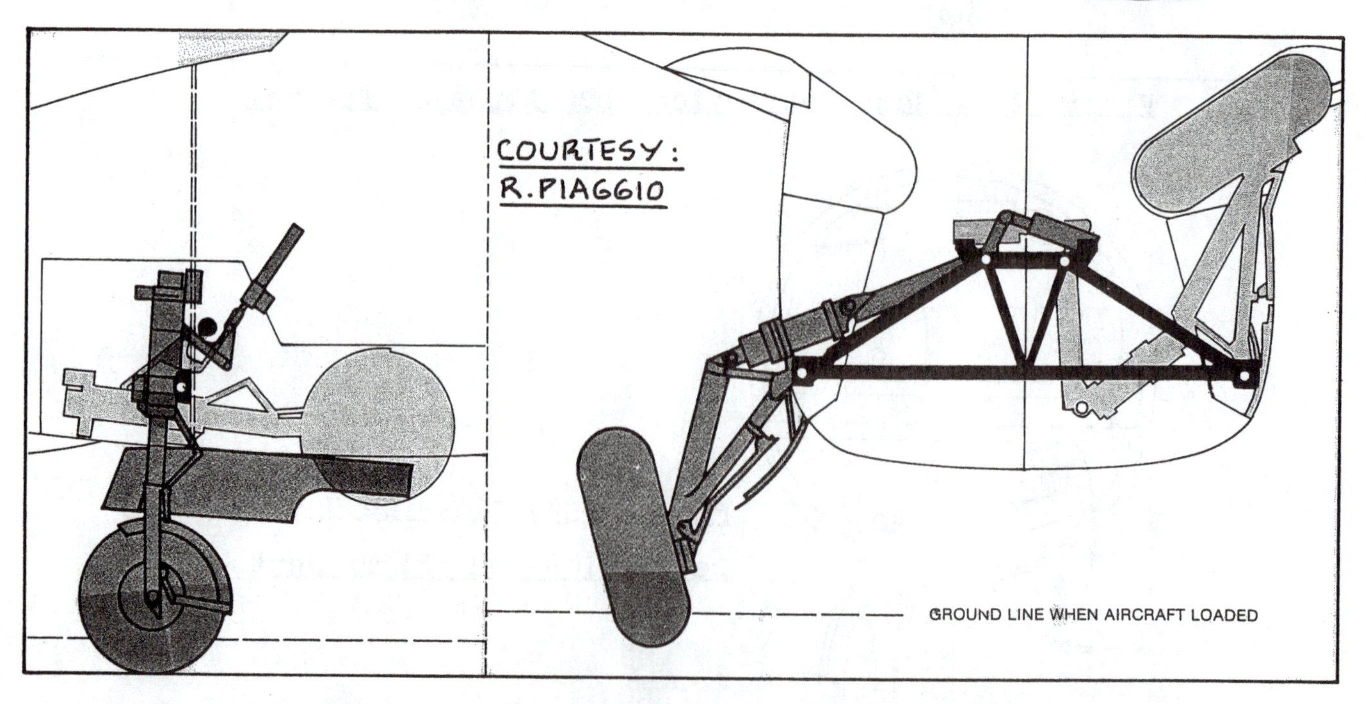

Figure 2.75 Gear Retraction: Piaggio P.166-DL3

Figure 2.76 Gear Retraction: Mig-23 Flogger

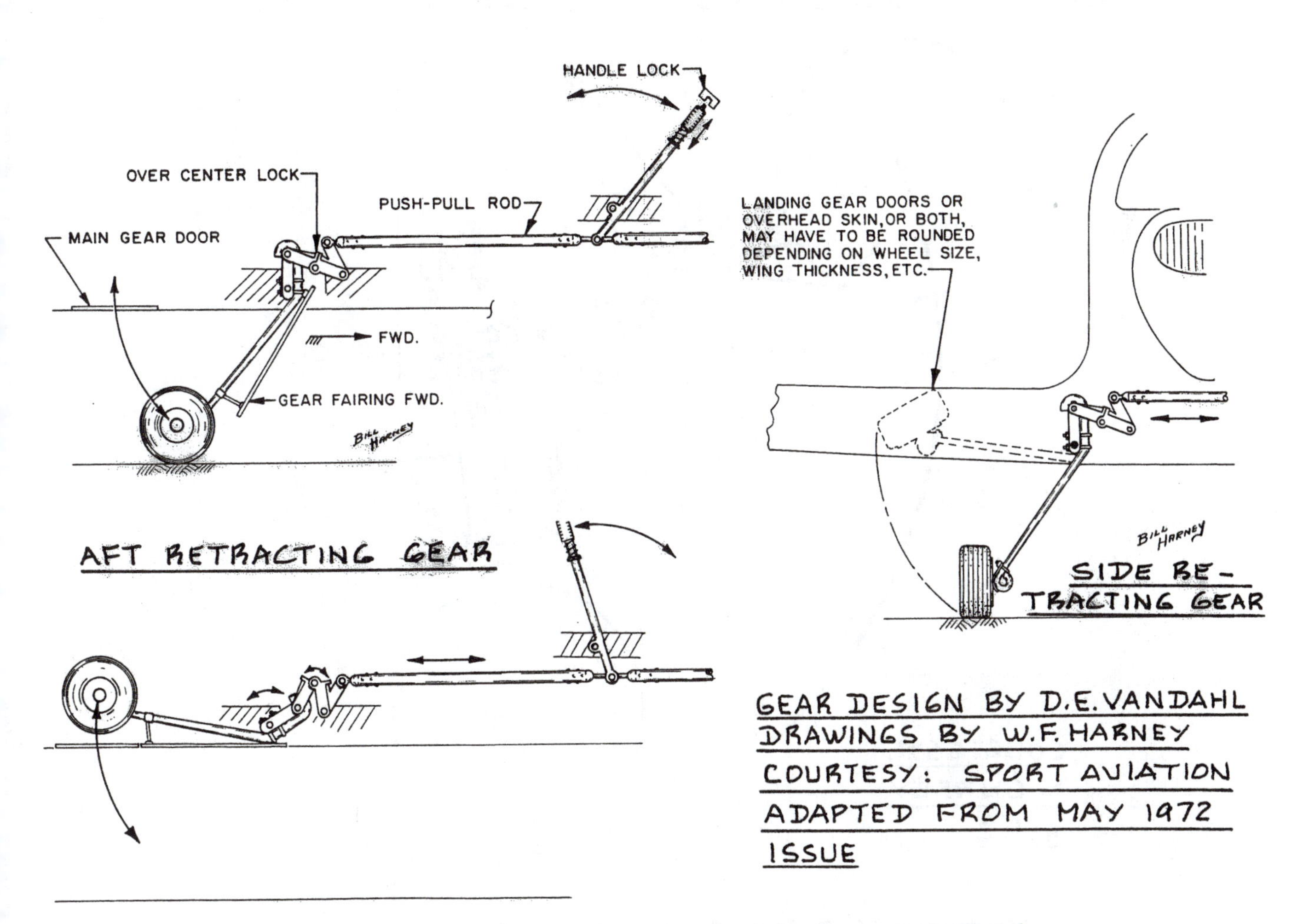

Figure 2.77 Example of Manual Gear Retraction

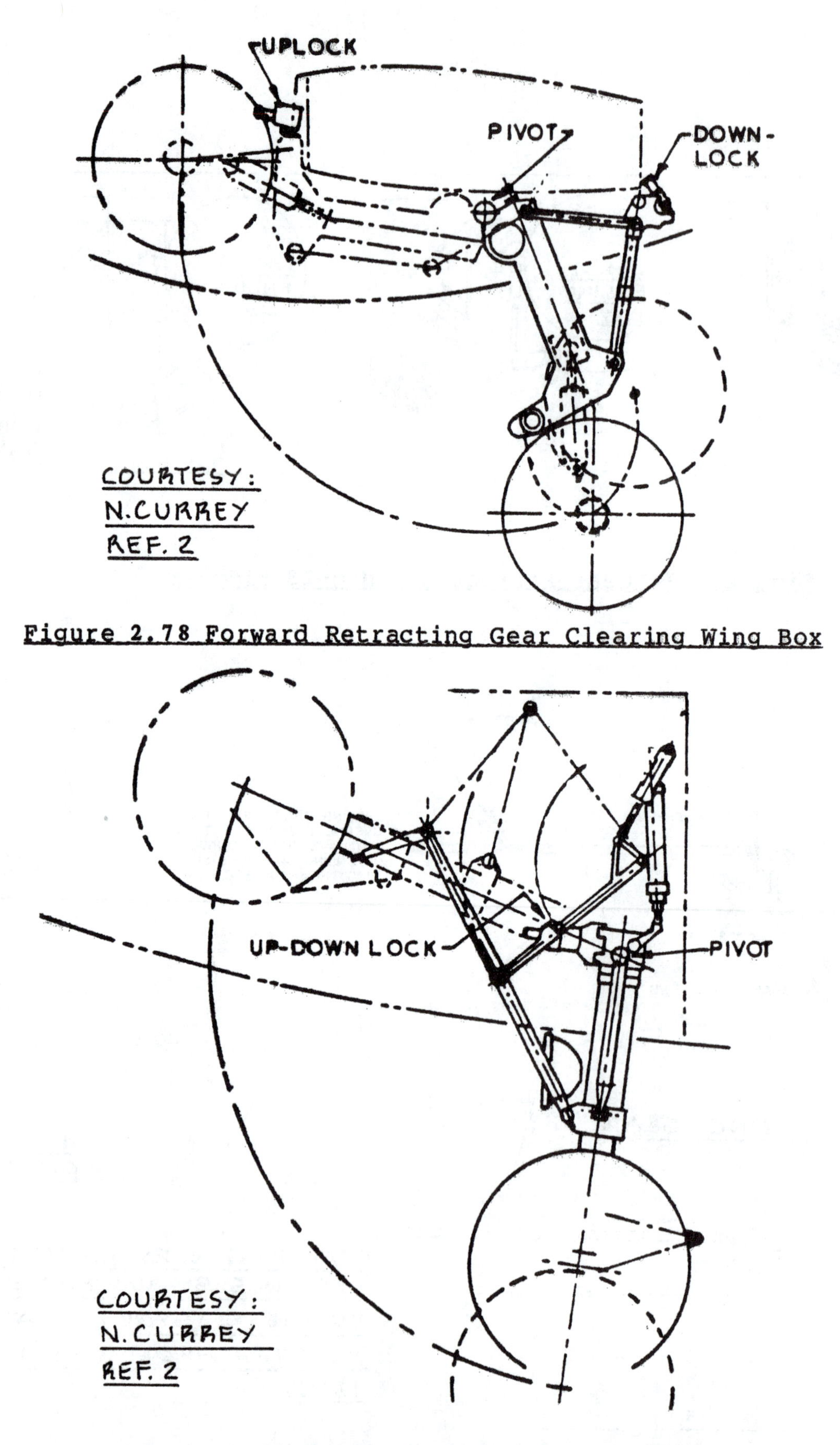

Figure 2.78 Forward Retracting Gear Clearing Wing Box

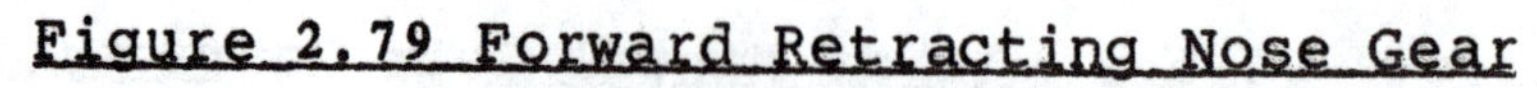

Figure 2.79 Forward Retracting Nose Gear

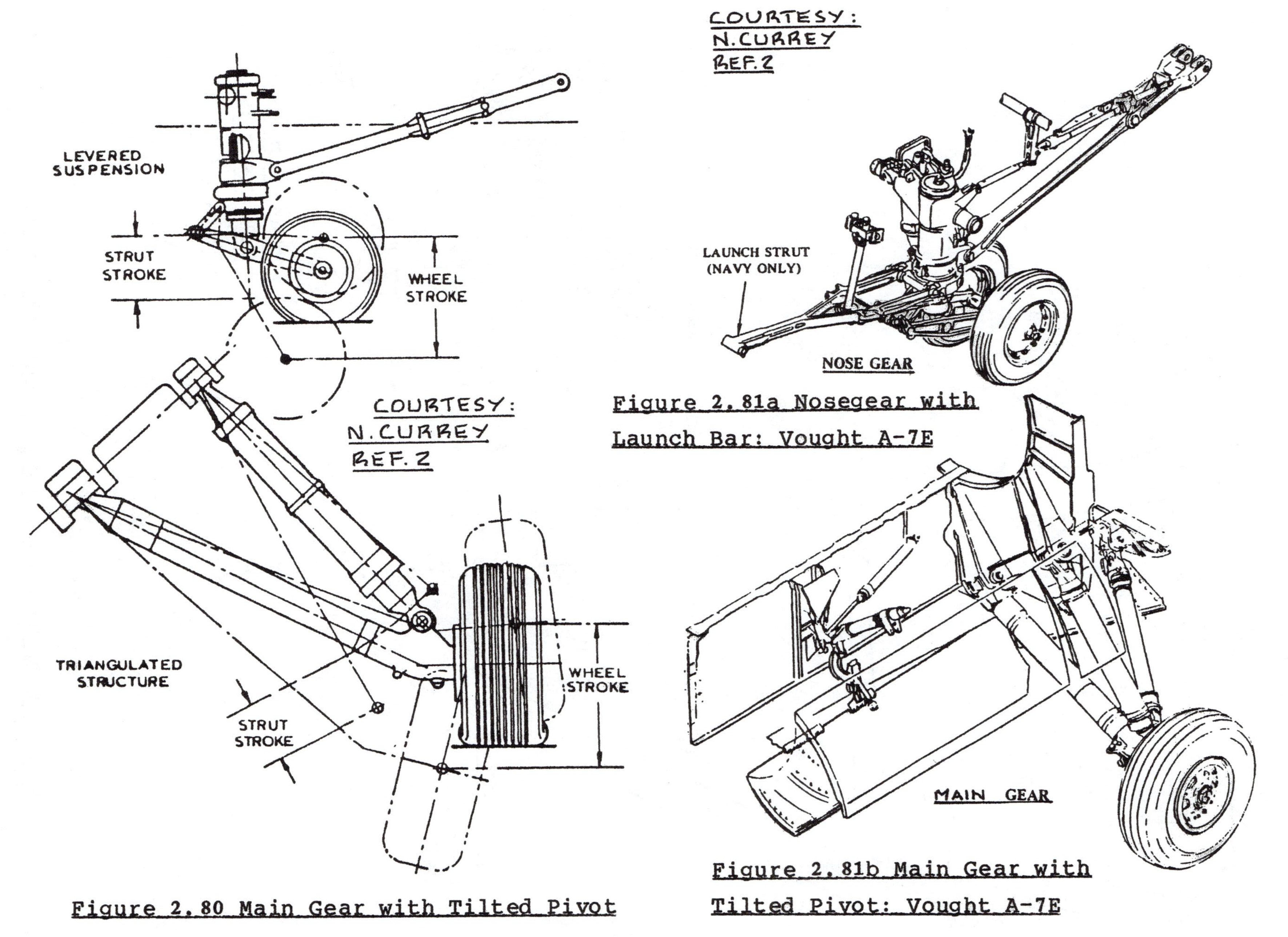

Figure 2.80 Main Gear with Tilted Pivot

Figure 2.81a Nosegear with Launch Bar: Vought A-7E

Figure 2.81b Main Gear with Tilted Pivot: Vought A-7E

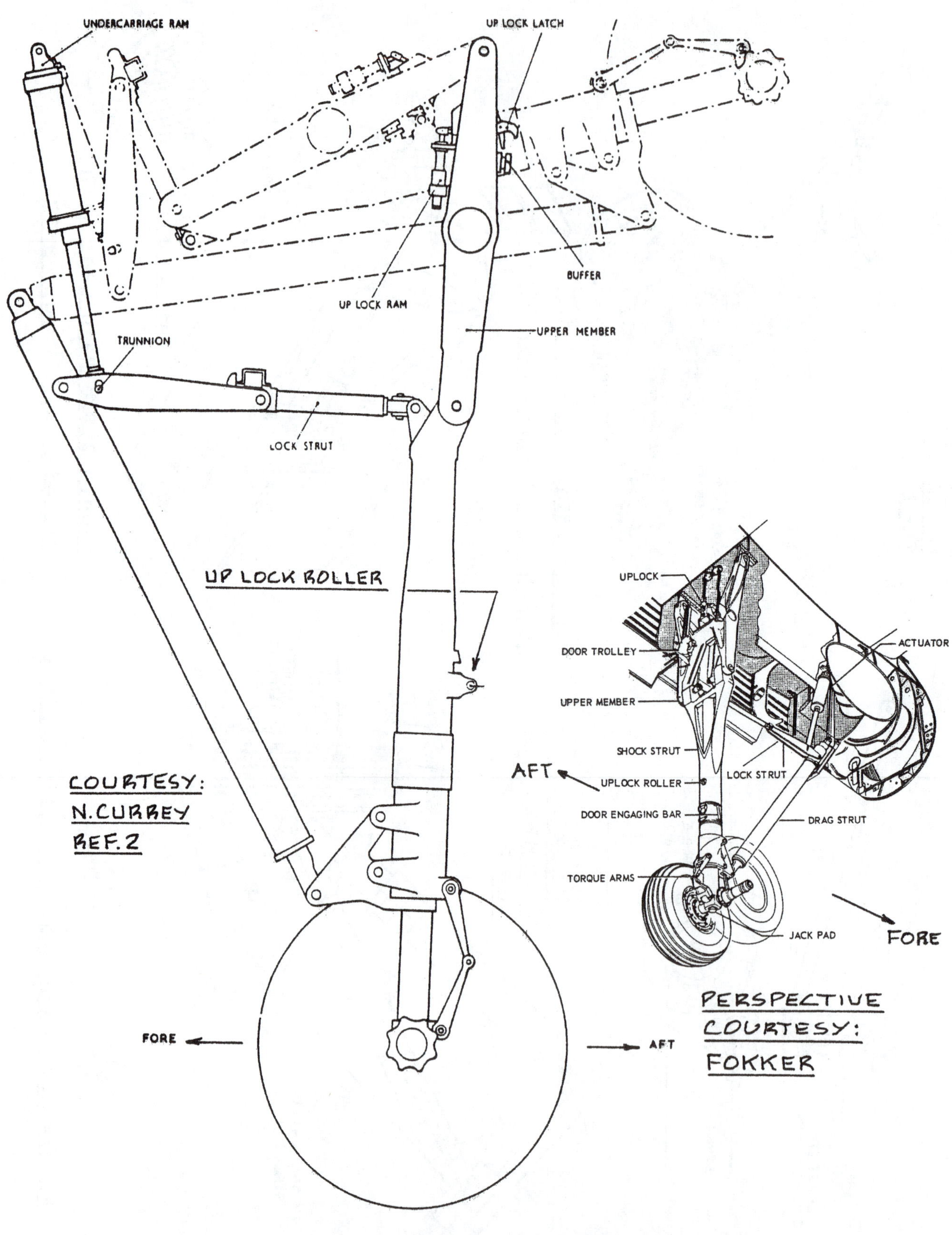

Figure 2.82 Main Gear Retraction: Fokker F-27

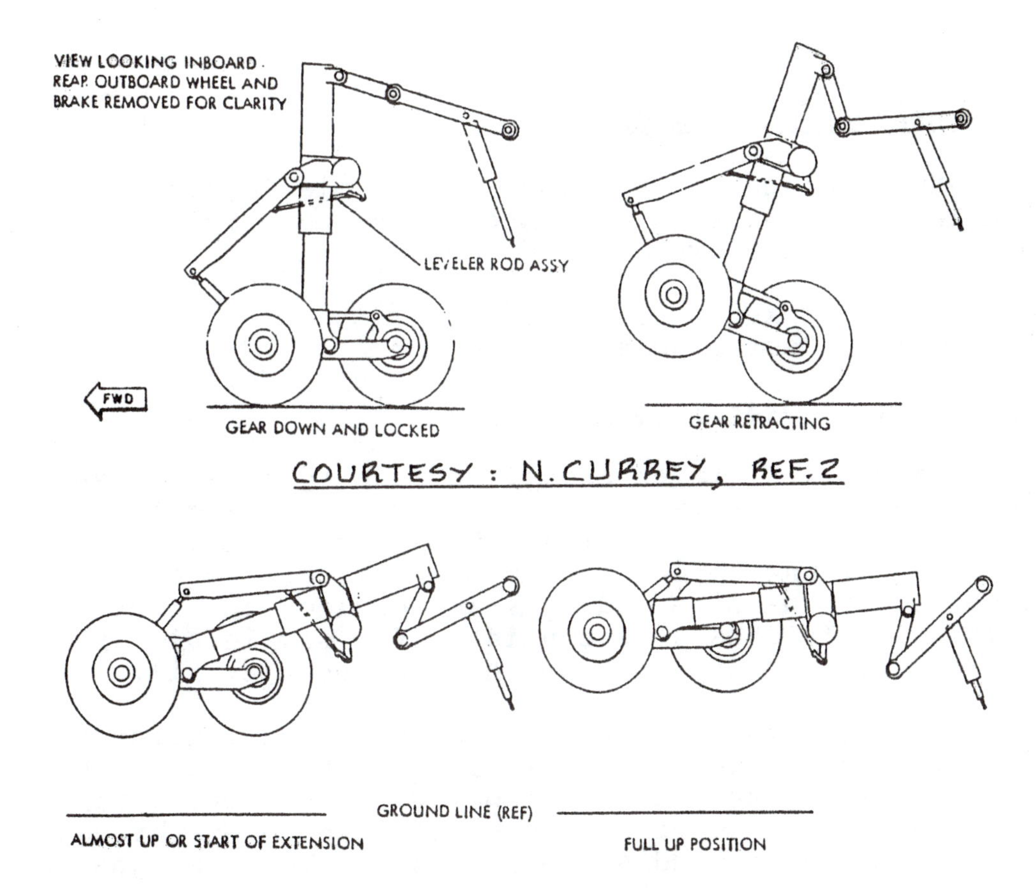

Figure 2.83 Main Gear Retraction: Lockheed C-141

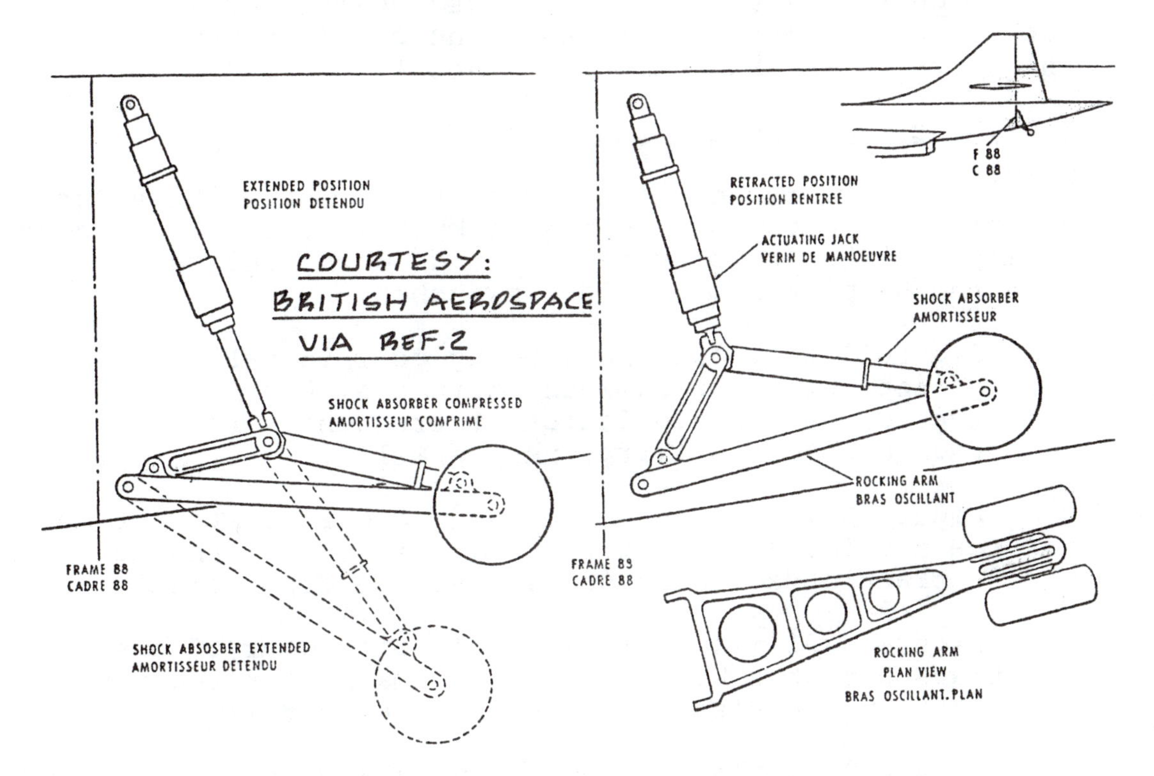

Figure 2.84 Tail Bumper Retraction: Concorde

2.11 EXAMPLE LANDING GEAR LAYOUTS

In this section a number of example layouts of landing gears will be presented. The material is presented as follows:

2.11.1 Fixed Gear Layouts
2.11.2 Retractable Gear Layouts

2.11.1 Fixed Gear Layouts

Figure 2.85 is an example of a simple fixed gear installation in a two-place, single engine light airplane: the Piper PA-38-112 Tomahawk. The main gear strut is of the spring-leaf type. All shocks are absorbed by the tire and by strut deflection. The nose gear which is steerable consists of a simple oleo strut attached to the engine truss and to the firewall.

Figure 2.86 shows the fixed gear installation for a typical four-place, single engine light airplane: the Cessna 172. Note that the tubular strut (spring-tube instead of spring-leaf) is one unit extending through the fuselage and attached at four points. The nose gear strut is attached to the firewall at two points.

Figure 2.87 also shows a fixed gear installation, but for a much larger airplane, the DHC Twin Otter. Note the rubber disk shock absorber installation.

2.11.2 Retractable Gear Layouts

Any retractable landing gear requires positive 'up' and 'down' indications in the cockpit. Figure 2.88 shows a simple method to provide 'up' and 'down' indications with the help of simple micro-switches.

Figure 2.89 shows the landing gear installation of a twin engine turboprop commuter airplane: the SF-340. This type of gear installation if frequently found in twin engined business airplanes as well.

Figure 2.90 presents the landing gear installation for the amphibious Canadair CL215. Note that the main gear retraction actuator doubles as the drag strut!

Figure 2.91 illustrates the landing gear of a small jet transport: the Fokker F28. The main gear retracts into the wing-fuselage area behind the rear wing spar.

Figure 2.92 shows the gear installation for the McDD

DC10, a heavy jet transport. The main gear retracts into the wing-fuselage junction behind the rear wing spar.

For very large transports, more than two main gears may be needed. Figure 2.93 shows the 'four-poster' main gear design for the B747.

Figures 2.94 show a detailed layout for the Boeing 767 landing gear including the retraction system.

Fighters present many landing gear design problems due to severe space limitations. Figures 2.76, 2.80, 2.81 and 2.95 illustrate several fighter landing gear arrangements.

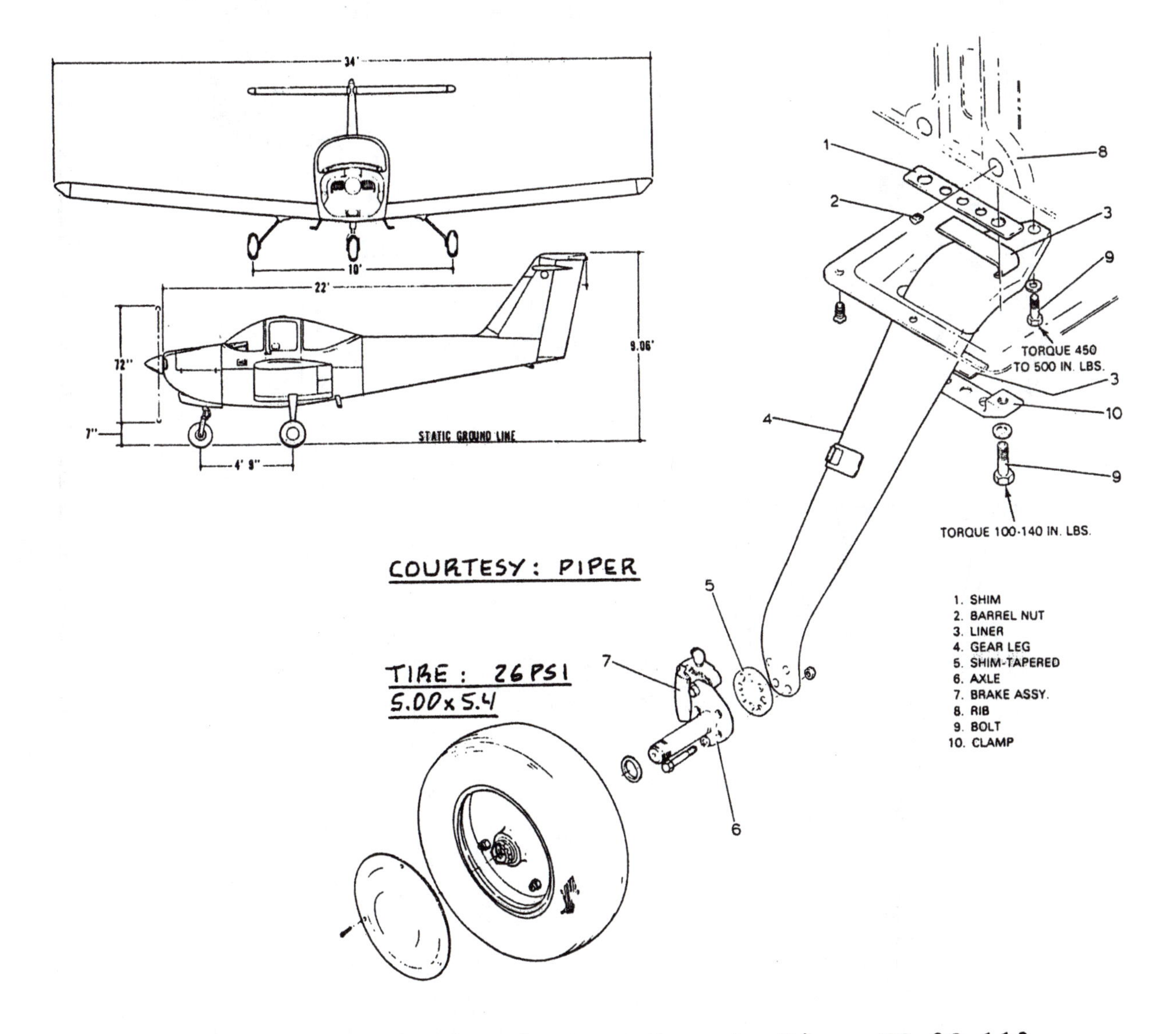

Figure 2.85a Fixed Main Gear Layout: Piper PA-38-112

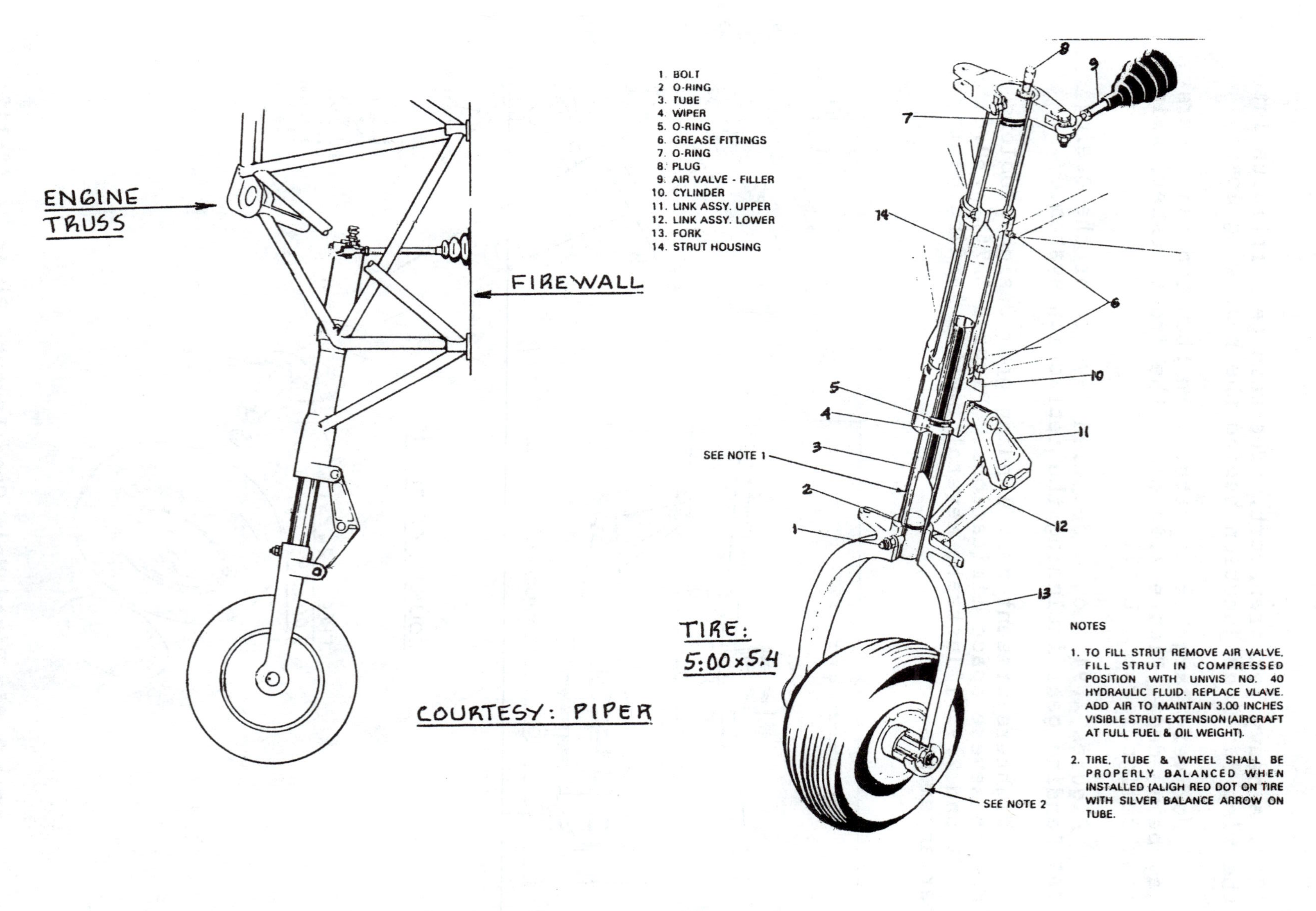

Figure 2.85b Fixed Nosegear Layout: Piper PA-38-112

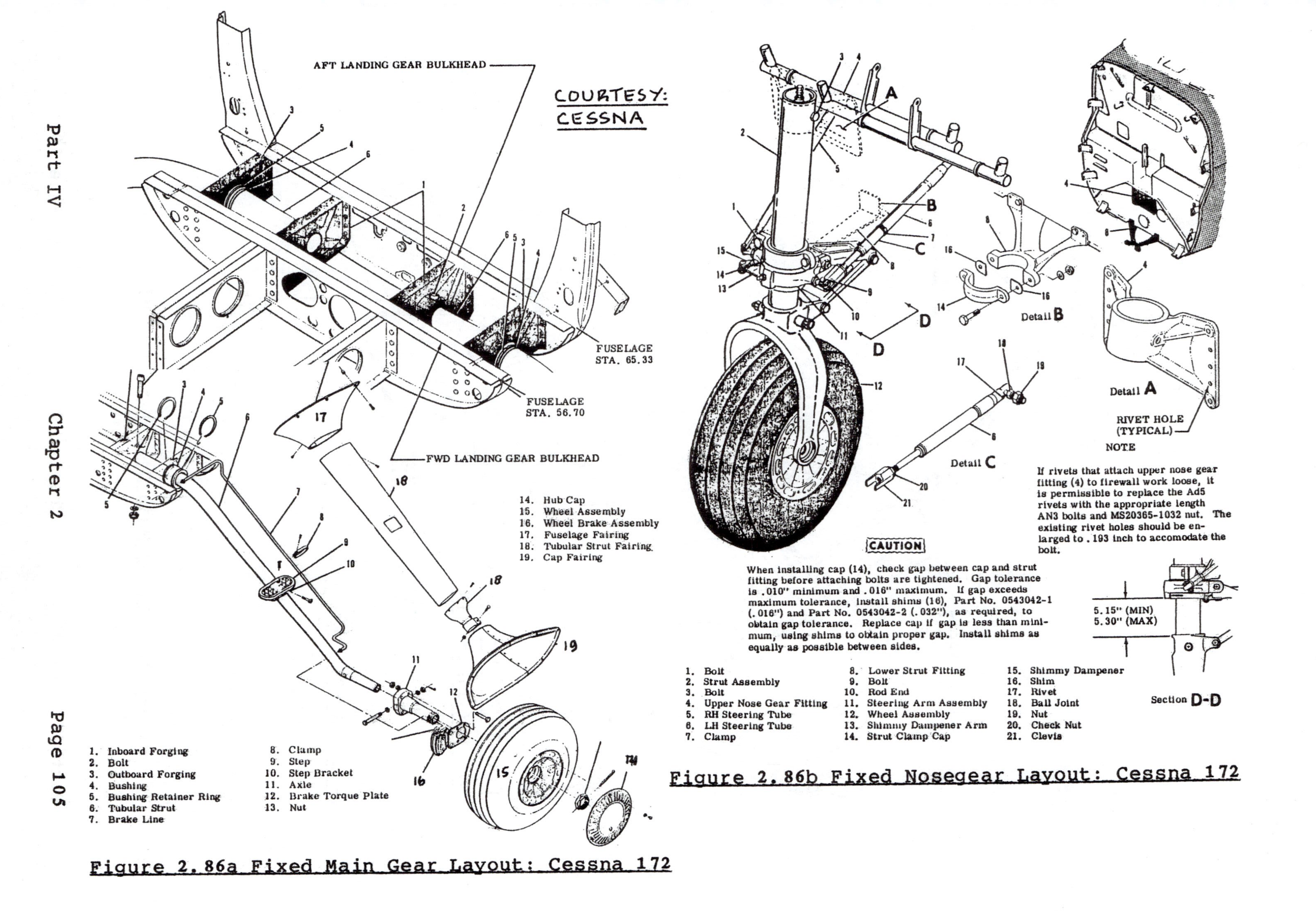

Figure 2.86a Fixed Main Gear Layout: Cessna 172

Figure 2.86b Fixed Nosegear Layout: Cessna 172

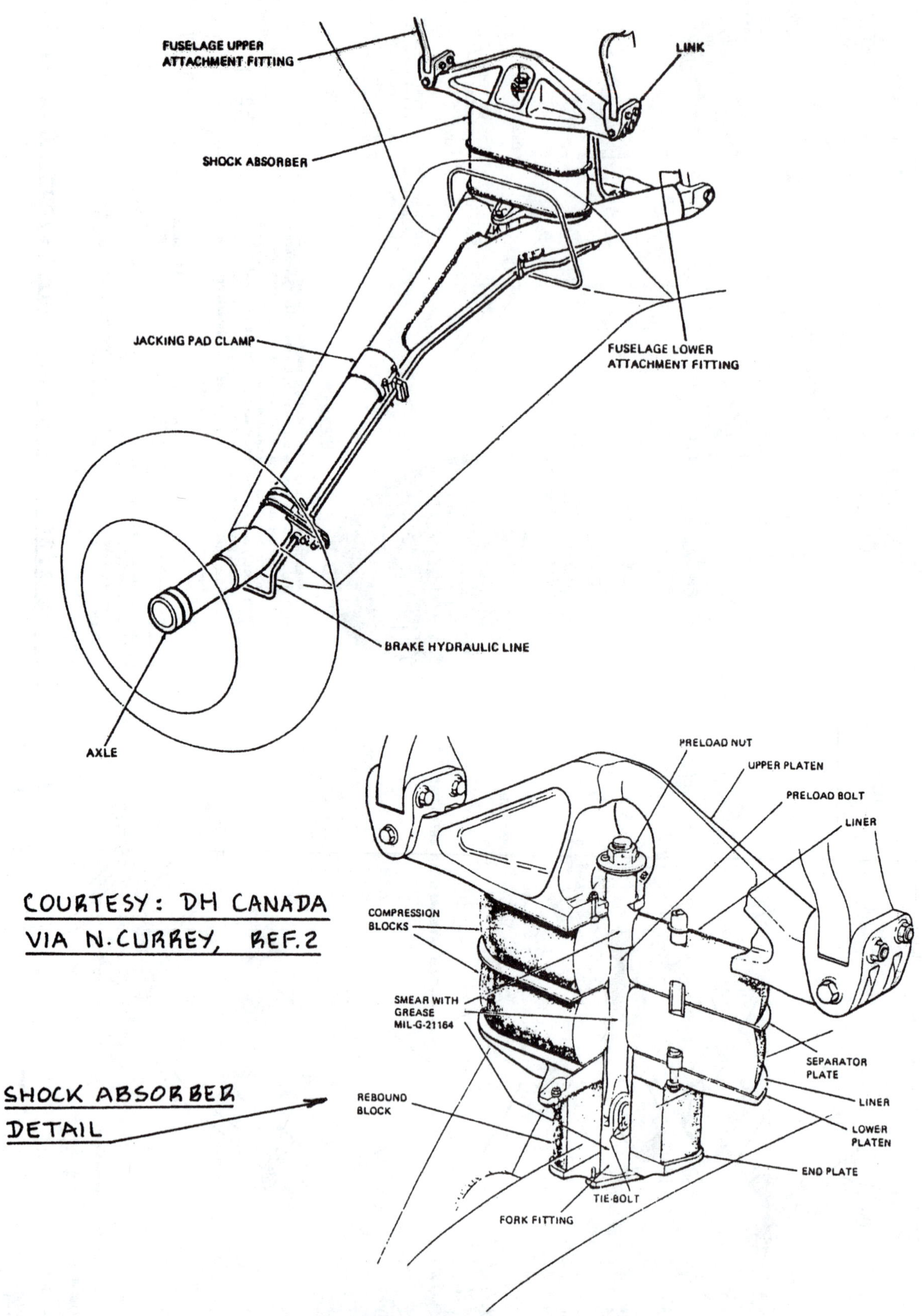

Figure 2.87 Fixed Main Gear Layout: DHC-Twin Otter

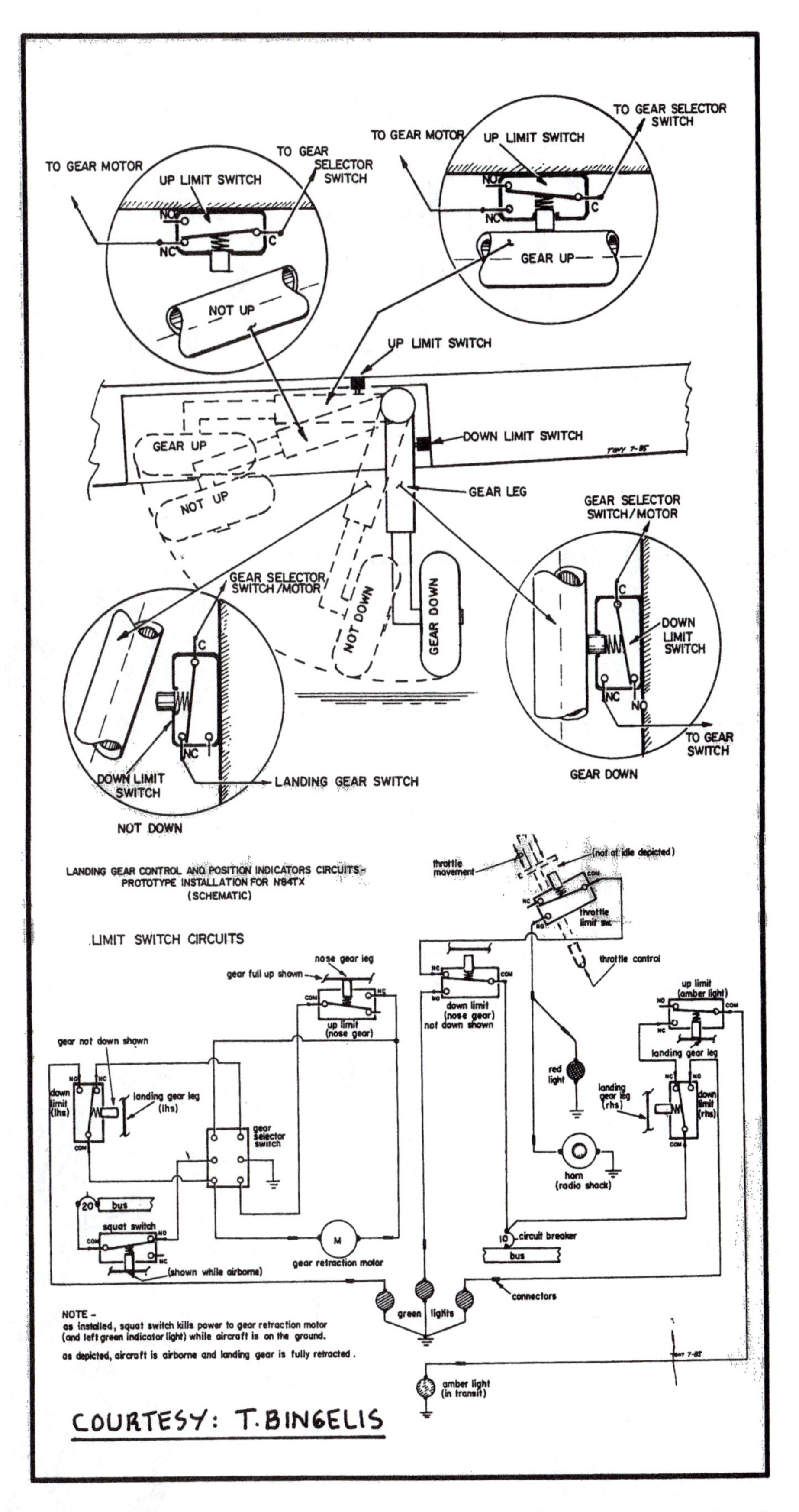

Figure 2.88 Up and Down Limit Switch Installation

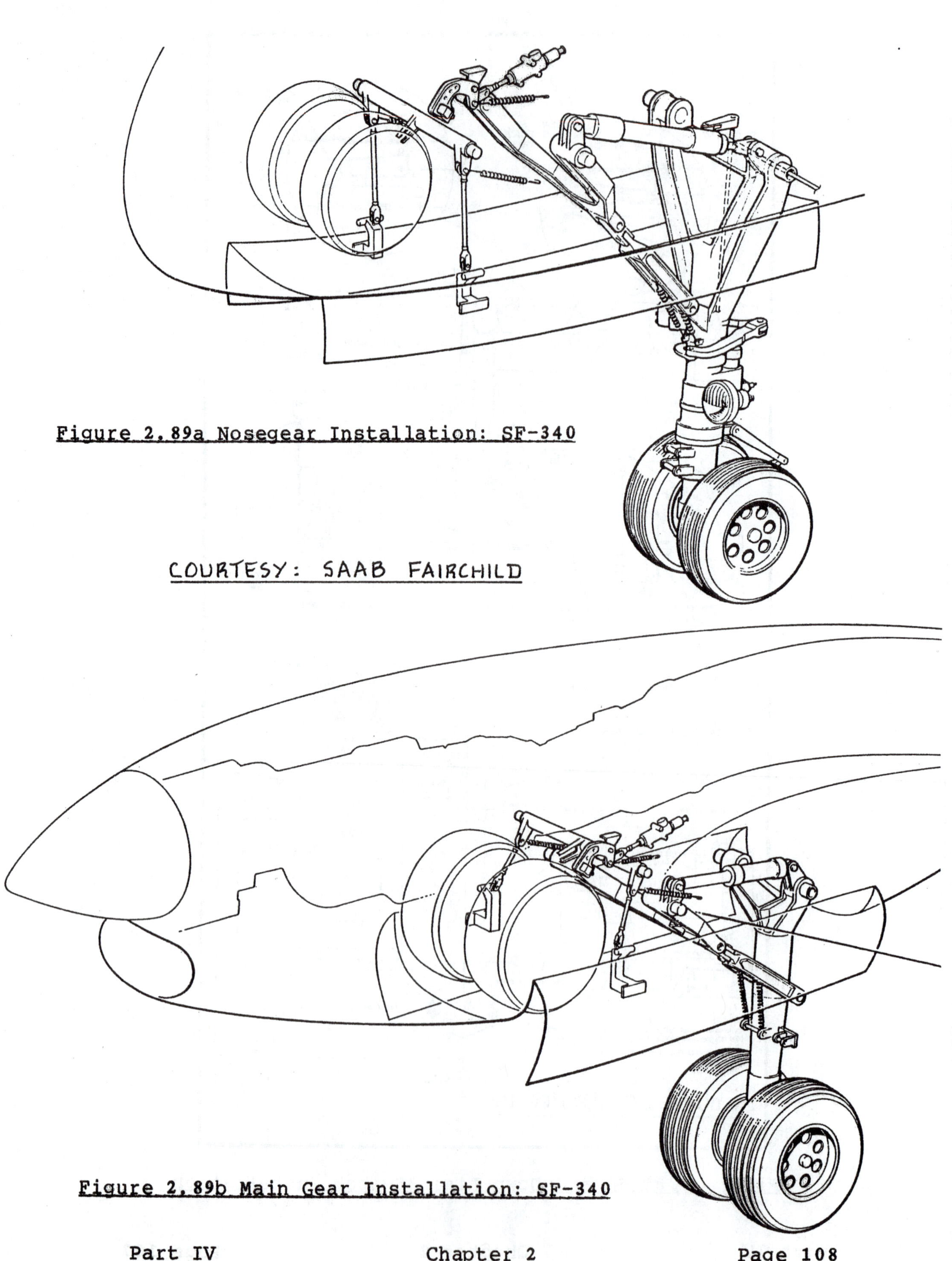

Figure 2.89a Nosegear Installation: SF-340

COURTESY: SAAB FAIRCHILD

Figure 2.89b Main Gear Installation: SF-340

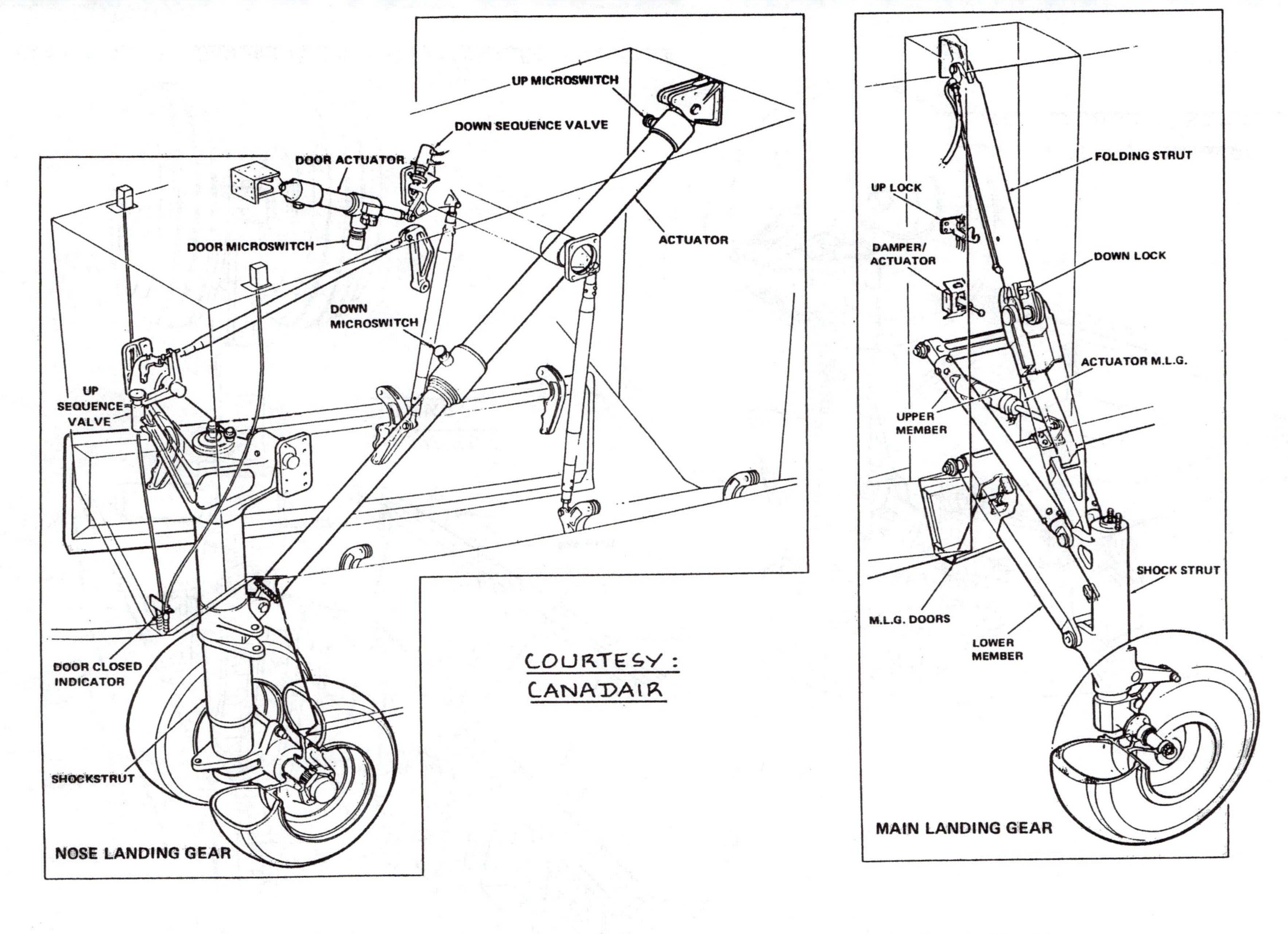

Figure 2.90 Landing Gear Installation: Canadair CL-215

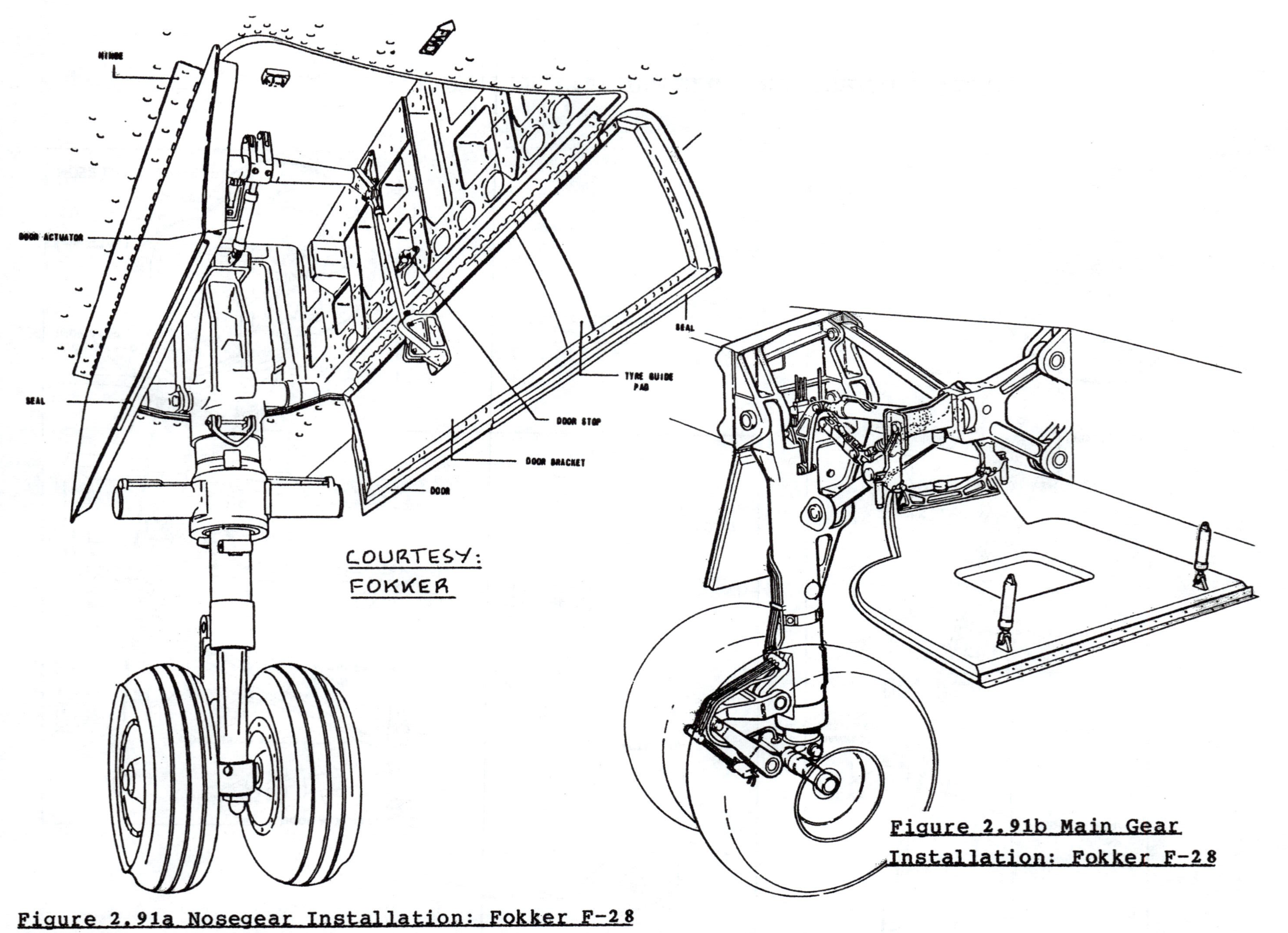

Figure 2.91a Nosegear Installation: Fokker F-28

Figure 2.91b Main Gear Installation: Fokker F-28

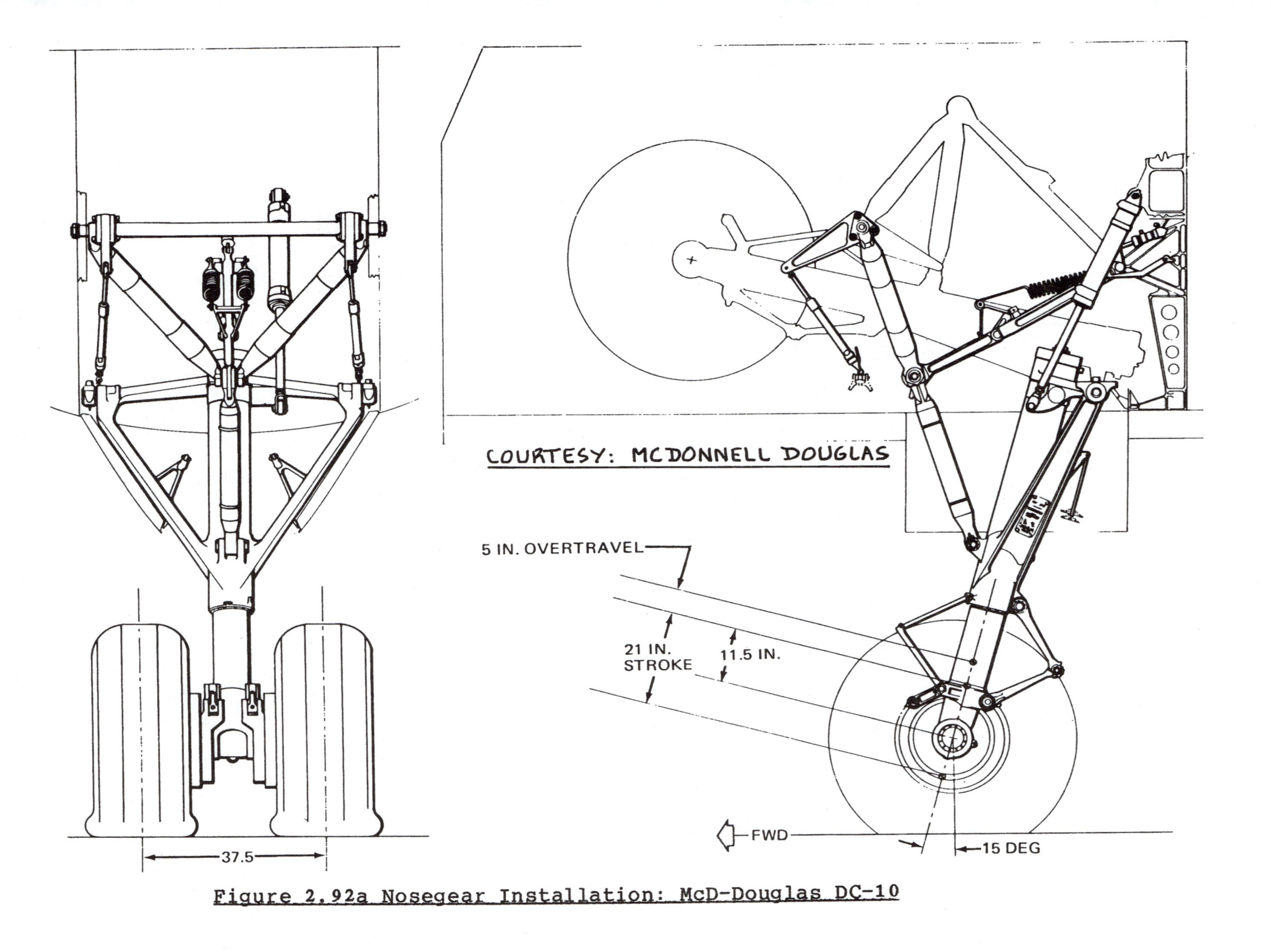

Figure 2.92a Nosegear Installation: McD-Douglas DC-10

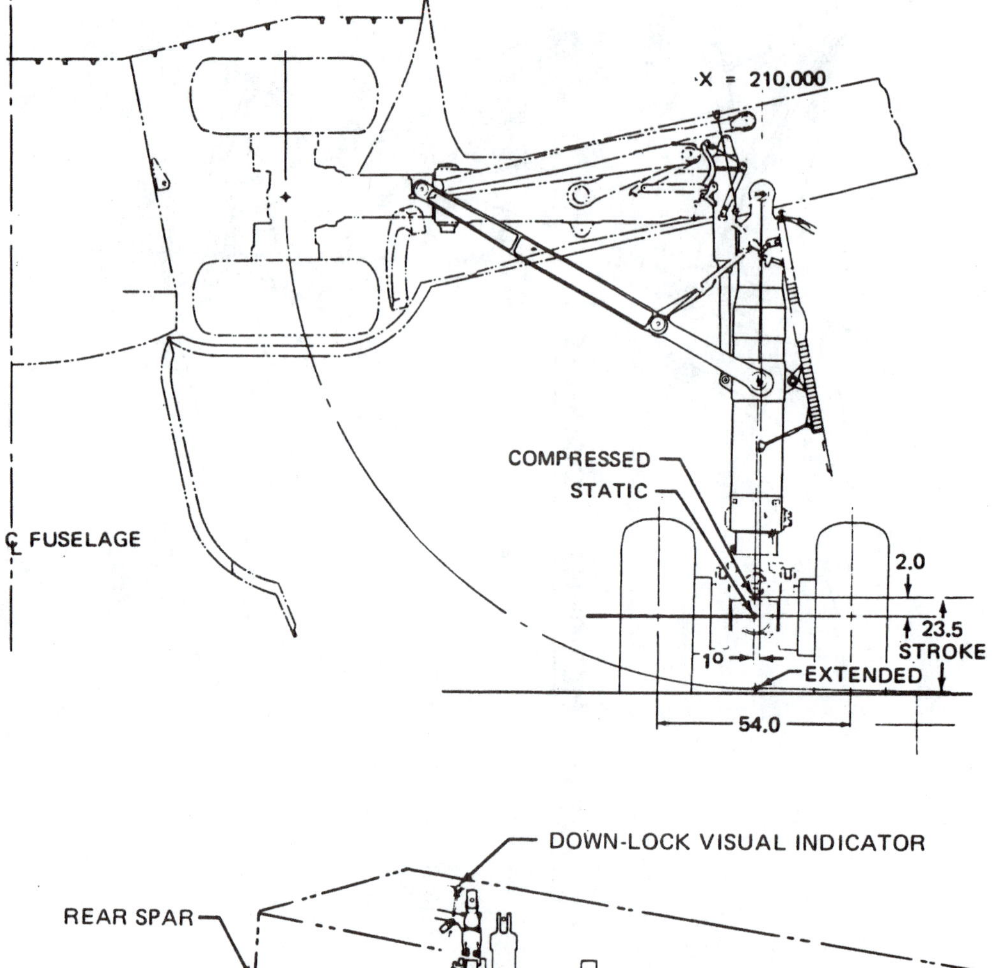

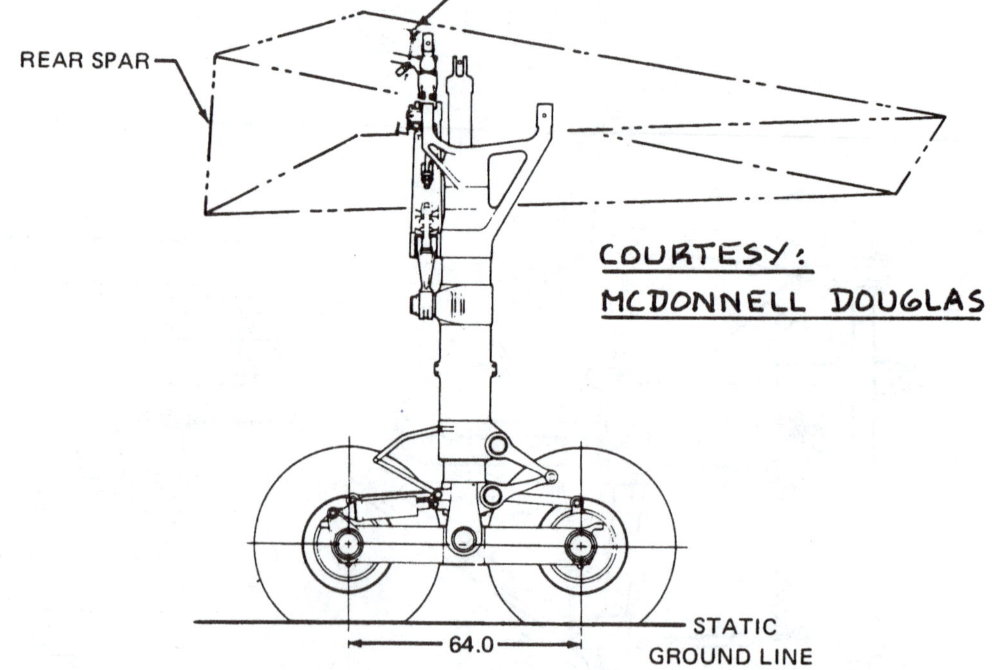

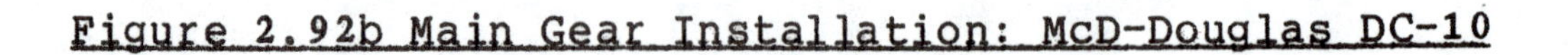
Figure 2.92b Main Gear Installation: McD-Douglas DC-10

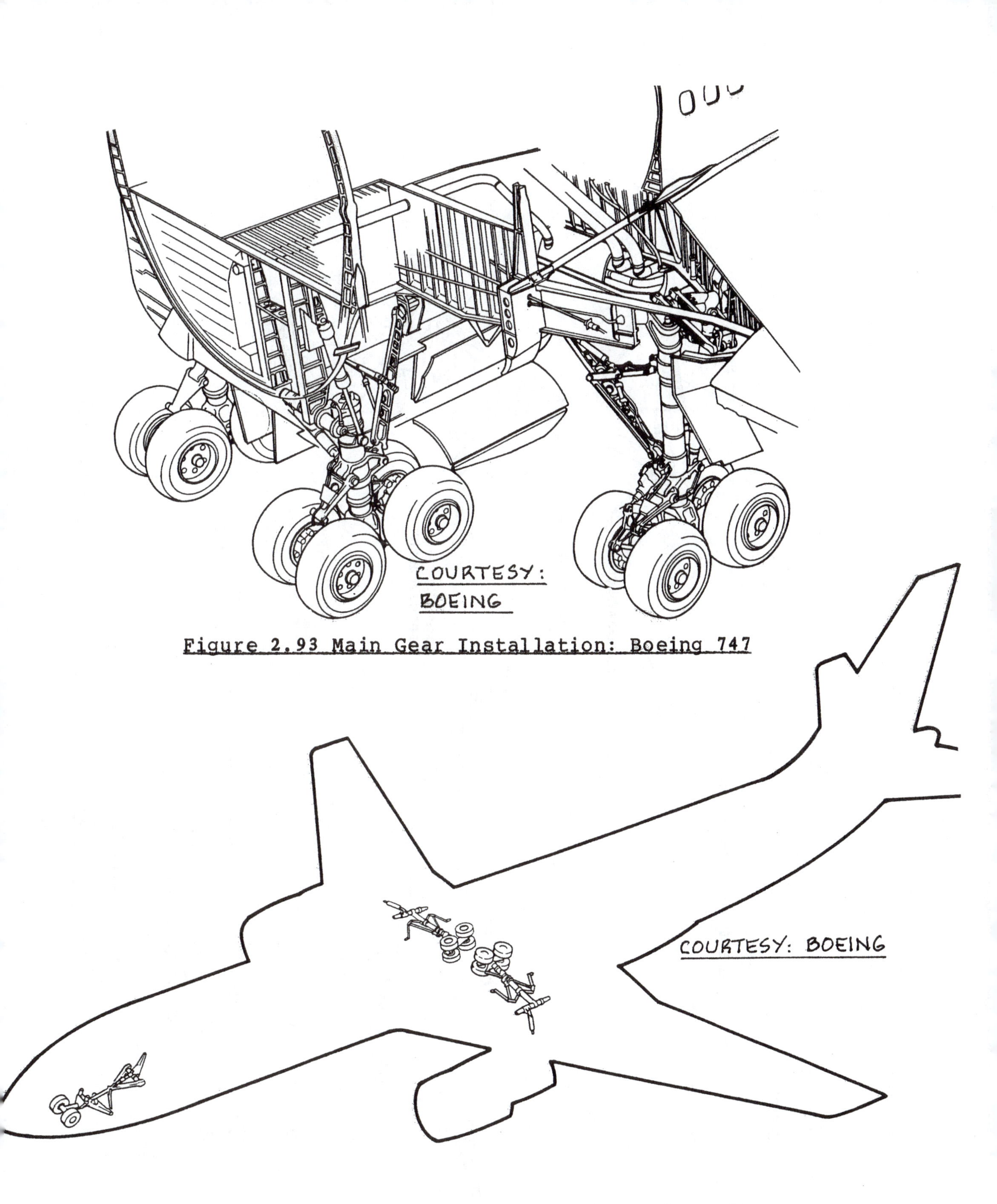

Figure 2.93 Main Gear Installation: Boeing 747

Figure 2.94a Landing Gear Overview: Boeing 767

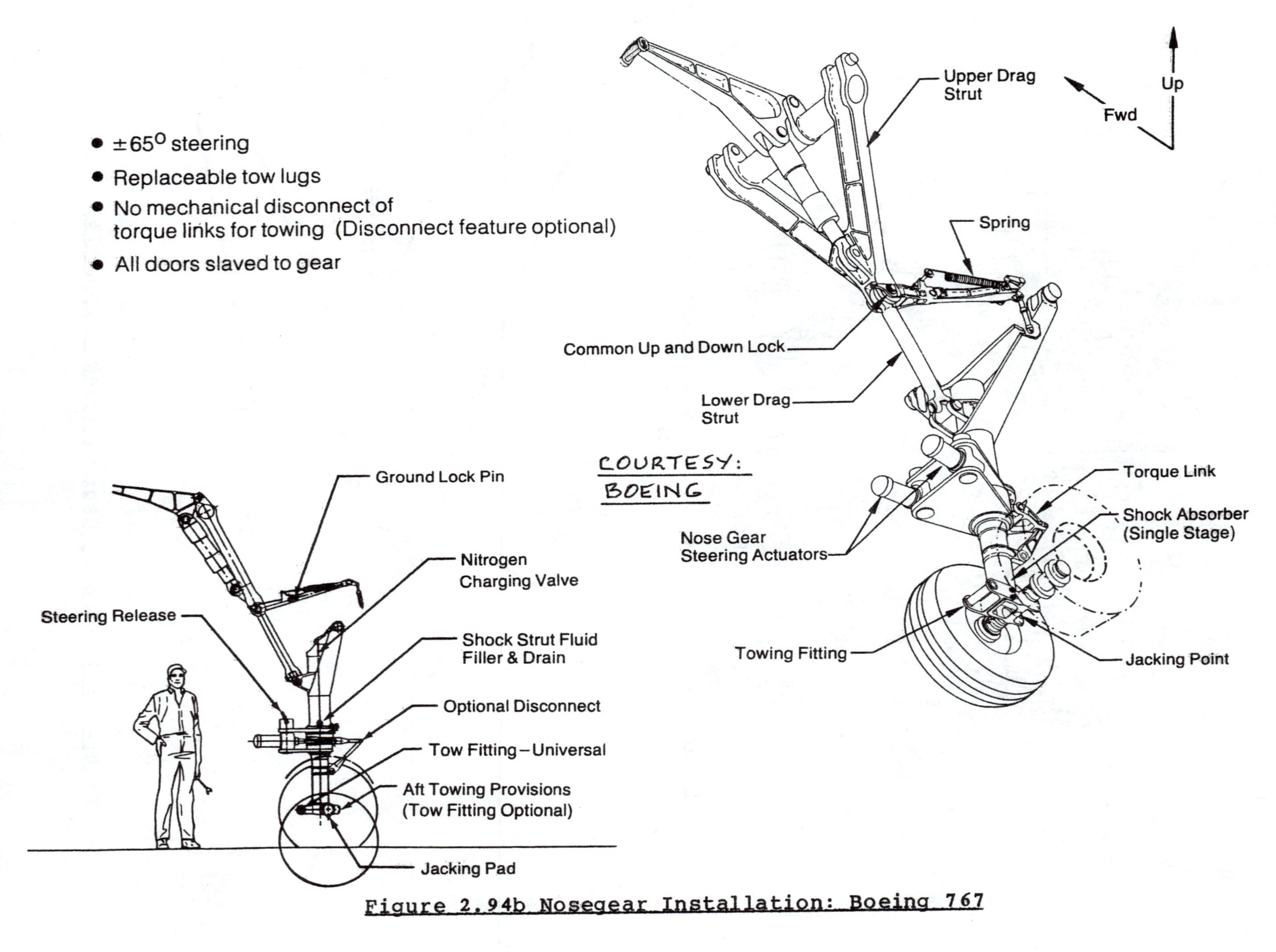

Figure 2.94b Nosegear Installation: Boeing 767

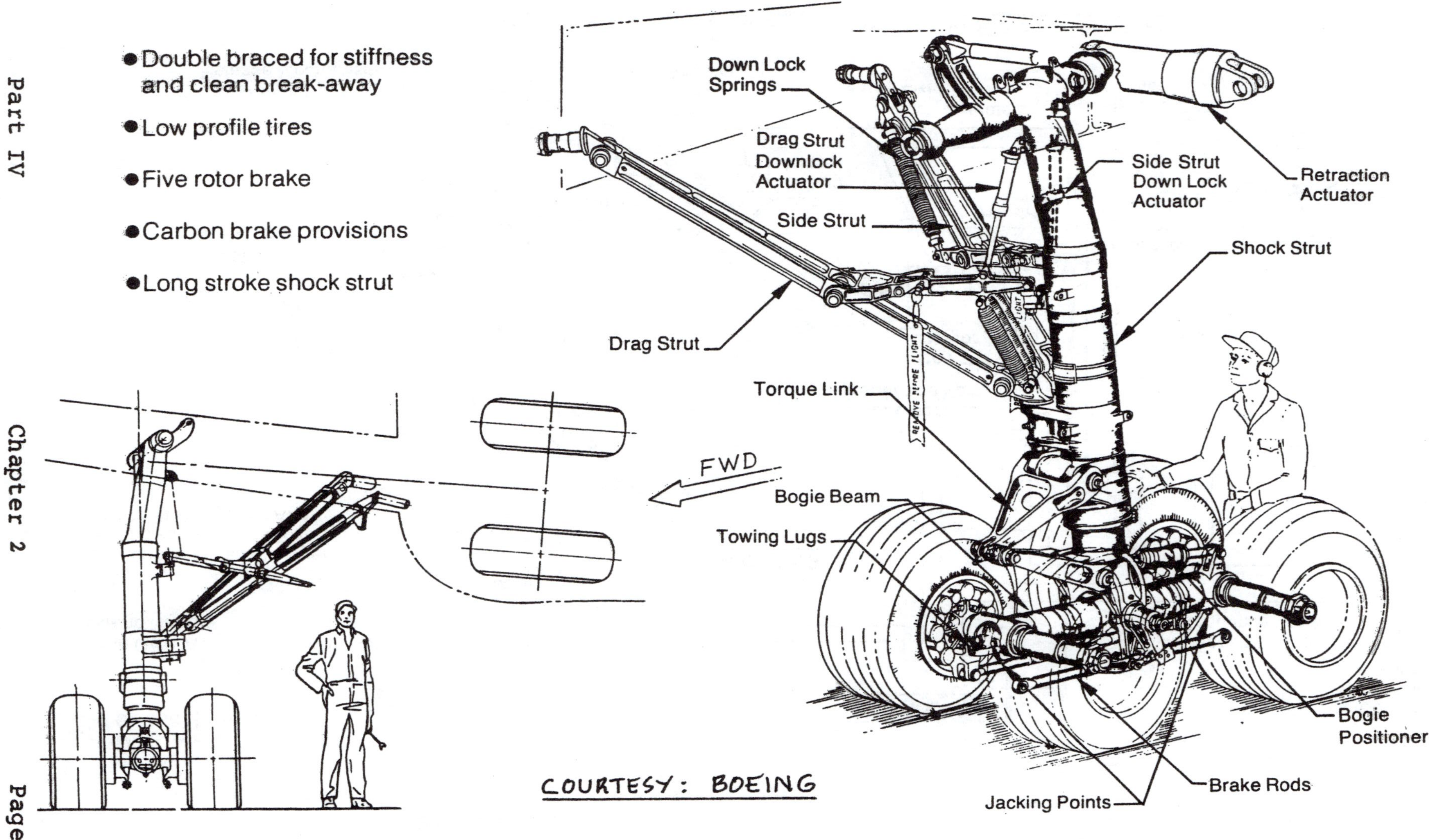

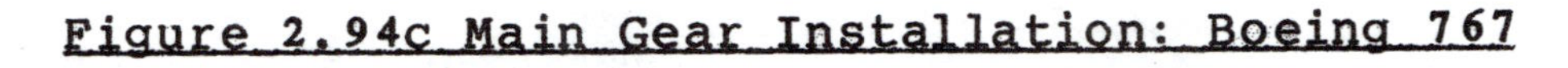
Figure 2.94c Main Gear Installation: Boeing 767

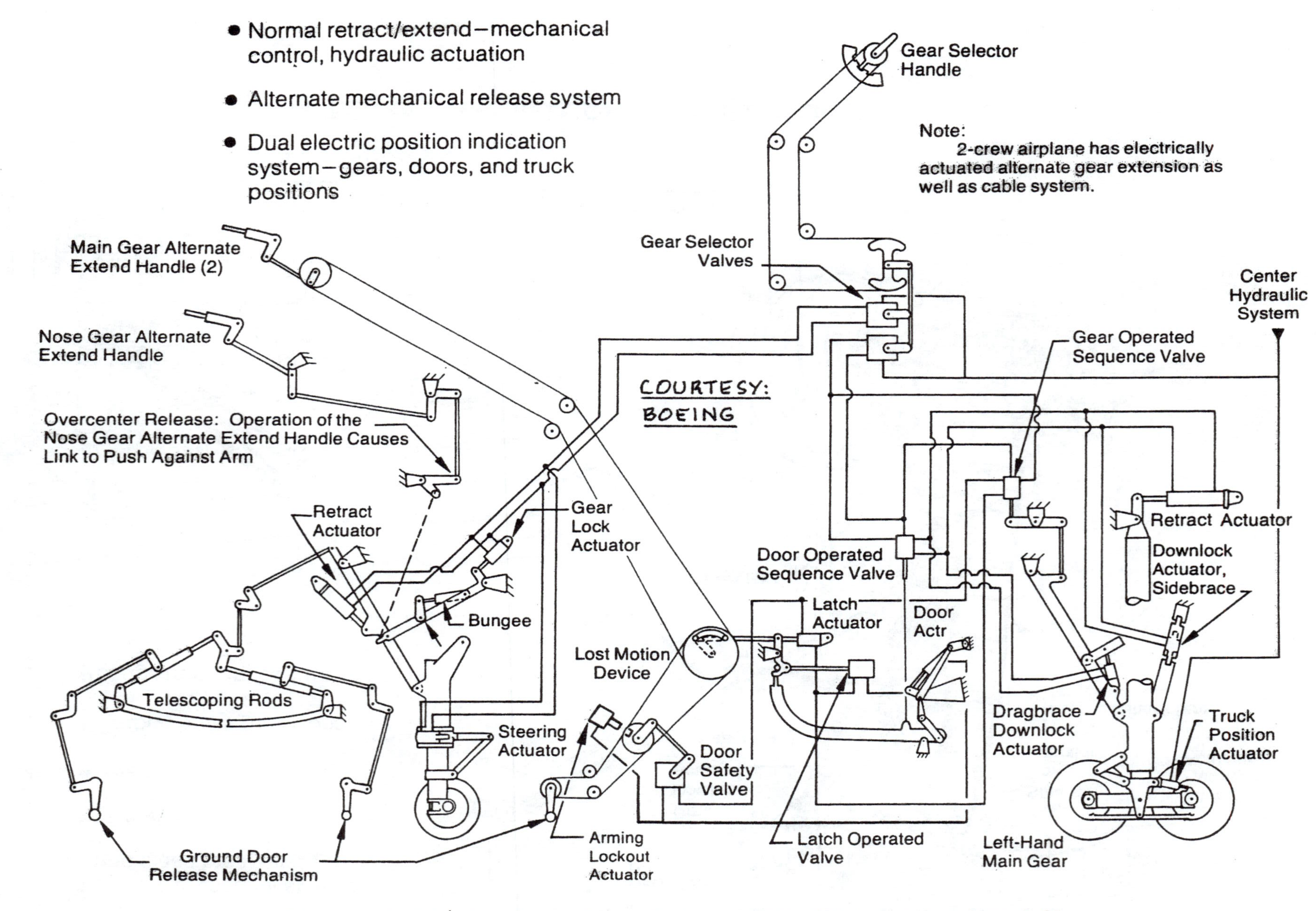

Figure 2.94d Landing Gear Retraction System: Boeing 767

a gear. The Hotol will however still have a conventional retractable gear which is designed for the landing weight only.

An interesting possibility with a droppable gear is that the method for ground and take-off propulsion can also be built into the droppable gear system, allowing lift-off without any fuel having been used.

2.12.5 Beaching Gears

Flying boats are often equipped with beaching gears. These gears are attached to the flying boat after landing on the water. After the beaching gears have been attached, the flying boat can taxi up a ramp onto terra firma and then maneuver like any other land-plane. Figure 2.101 shows an example of such a beaching gear.

2.12.6 Skis

Skis are used to allow airplanes to land on snow, ice and water. Examples of skis are shown in Figs. 2.102 and 2.103.

2.12.7 Floats

Floats are added to some airplanes to enhance their operational capability in areas with large bodies of water. Canada and Alaska are examples of such areas. Figure 2.104 shows an example of an airplane on floats. A structural detail of a typical float attachment is given in Figure 2.105.

The hydrodynamic performance of floats depends strongly on their cross sectional shape. Figure 2.106 summarizes typical float geometries.

2.12.8 Air Cushion Landing System (ACLS)

The air cushion landing system was developed to make an airplane independent of prepared surfaces. Figures 2.107 and 2.108 show examples of such a system. References 2, 14 and 15 should be consulted for data on the design of air cushion landing systems.

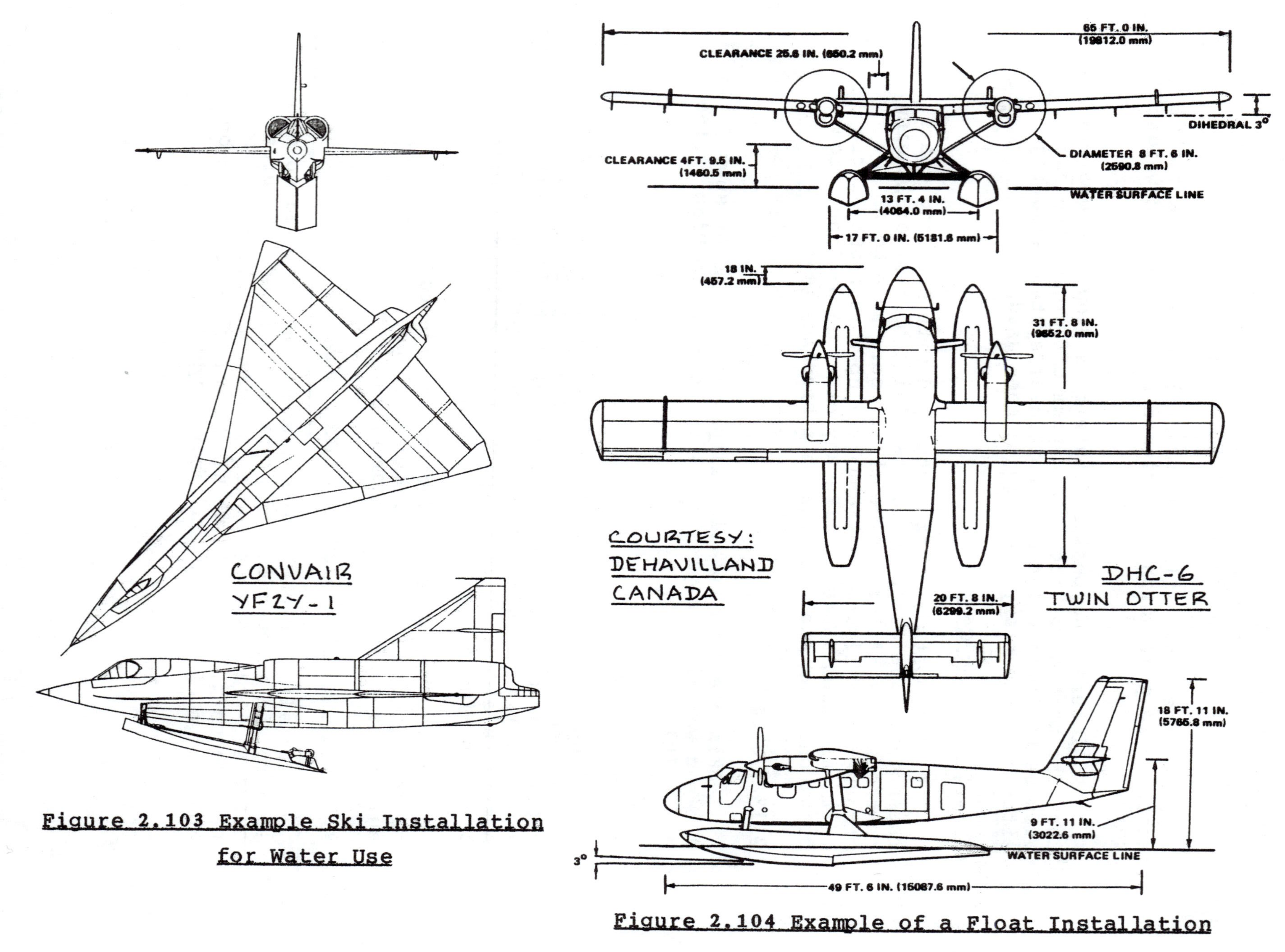

Figure 2.103 Example Ski Installation for Water Use

Figure 2.104 Example of a Float Installation

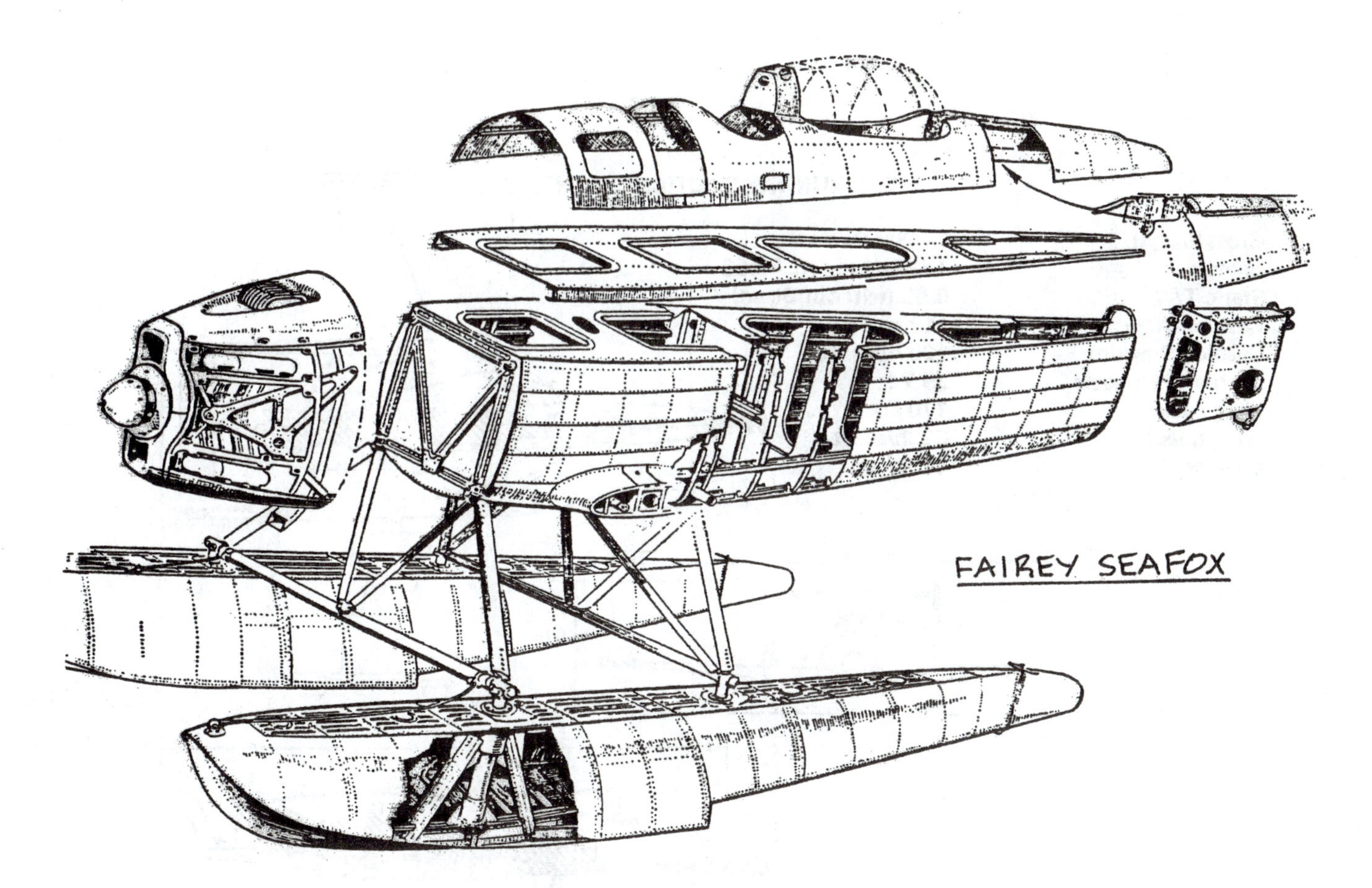

Figure 2.105 Example of a Float Structural Attachment

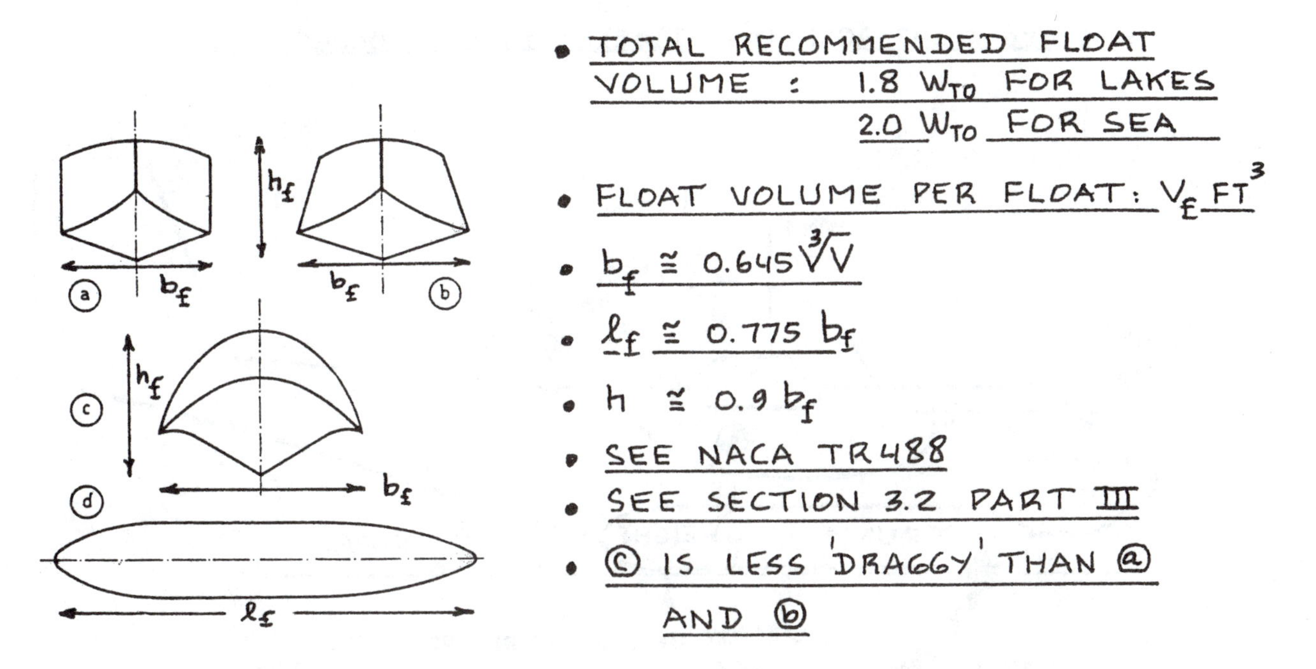

Figure 2.106 Summary of Float Geometries

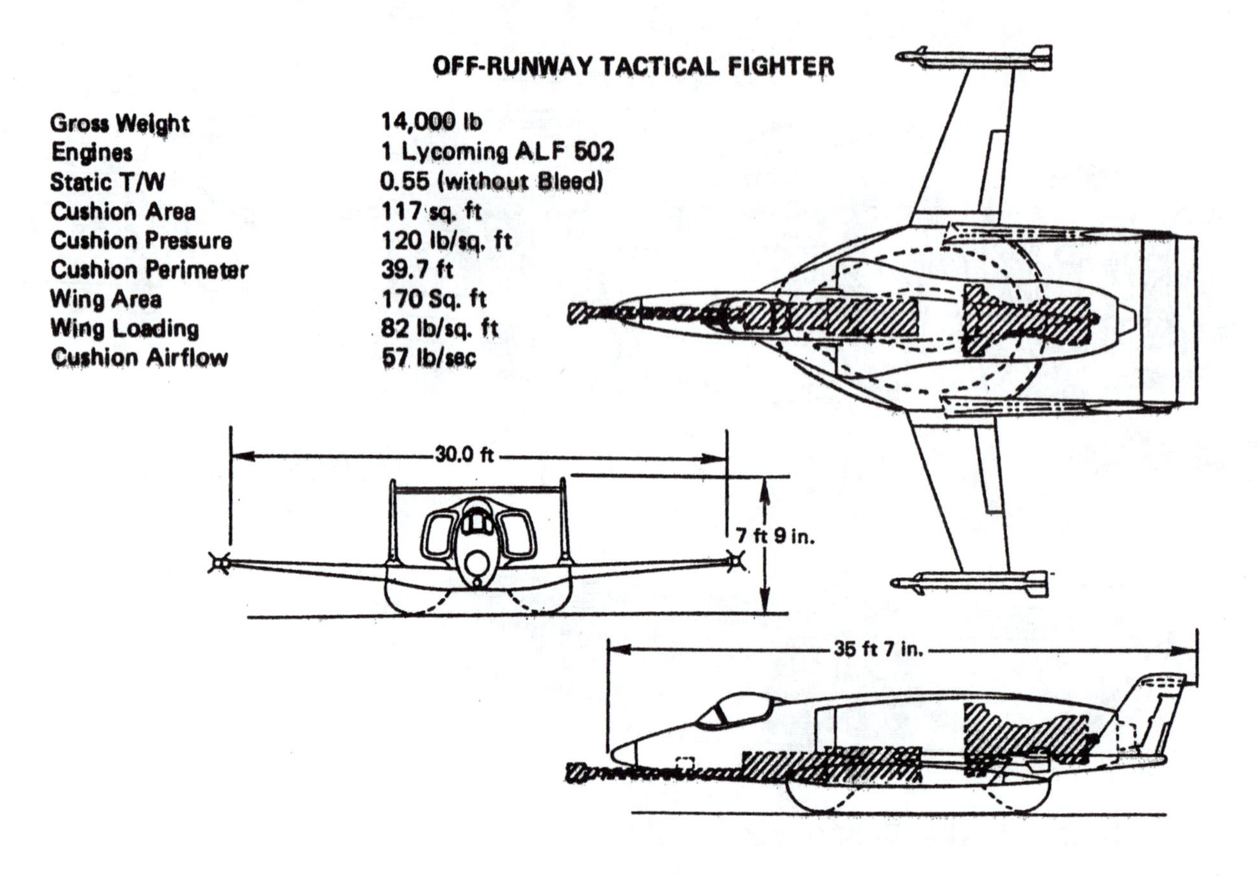

Figure 2.107 ACLS Installation in a Proposed Fighter

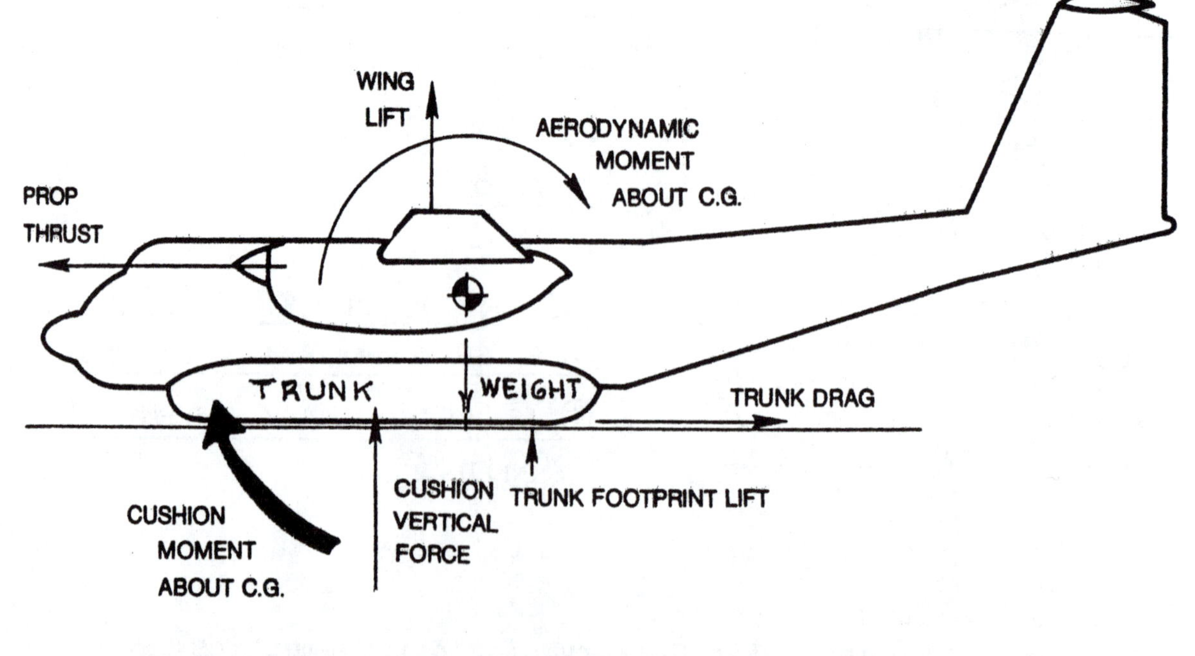

Figure 2.108 ACLS Installation in the XC-8A Transport

3. WEAPONS INTEGRATION AND WEAPONS DATA

The purpose of this chapter is to discuss a number of special design considerations which arise when integrating weapons and stores into military airplanes. In addition, data on typical weapons, stores and military cargo are presented.

The material is organized as follows:

3.1 Aerodynamic design considerations

3.2 Structural design considerations

3.3 Design for low radar and infrared detectability

3.4 Examples of weapon installations

3.5 Weapon and military payload data

3.1 AERODYNAMIC DESIGN CONSIDERATIONS

The following problem areas need to be considered when integrating weapons and/or stores into an airplane:

1. Drag 2. Stability and control 3. Separation

4. Gun exhaust gasses entering into engine inlets

3.1.1 Drag Considerations

When mounting stores under a wing or fuselage additional friction, profile and interference drag will be generated. Figures 3.1 and 3.2 show several 'rack' or 'pylon' mounted stores. For typical 'rack' geometries, the reader is referred to sub-section 3.5.2.

The incremental drag caused by the stores shown in Figures 3.1 and 3.2 can be estimated with the drag estimating methods of Part VI. The following general rules must be observed if drag increases due to stores are to be kept to a minimum:

1. For rack or pylon mounted stores which are roughly bodies of revolution, the distance from the store exterior to the wing or fuselage exterior should be larger than one half of the store diameter. This recommendation is illustrated in Figure 3.3.

Figure 3.1 Example of Rack and Pylon Mounted Stores

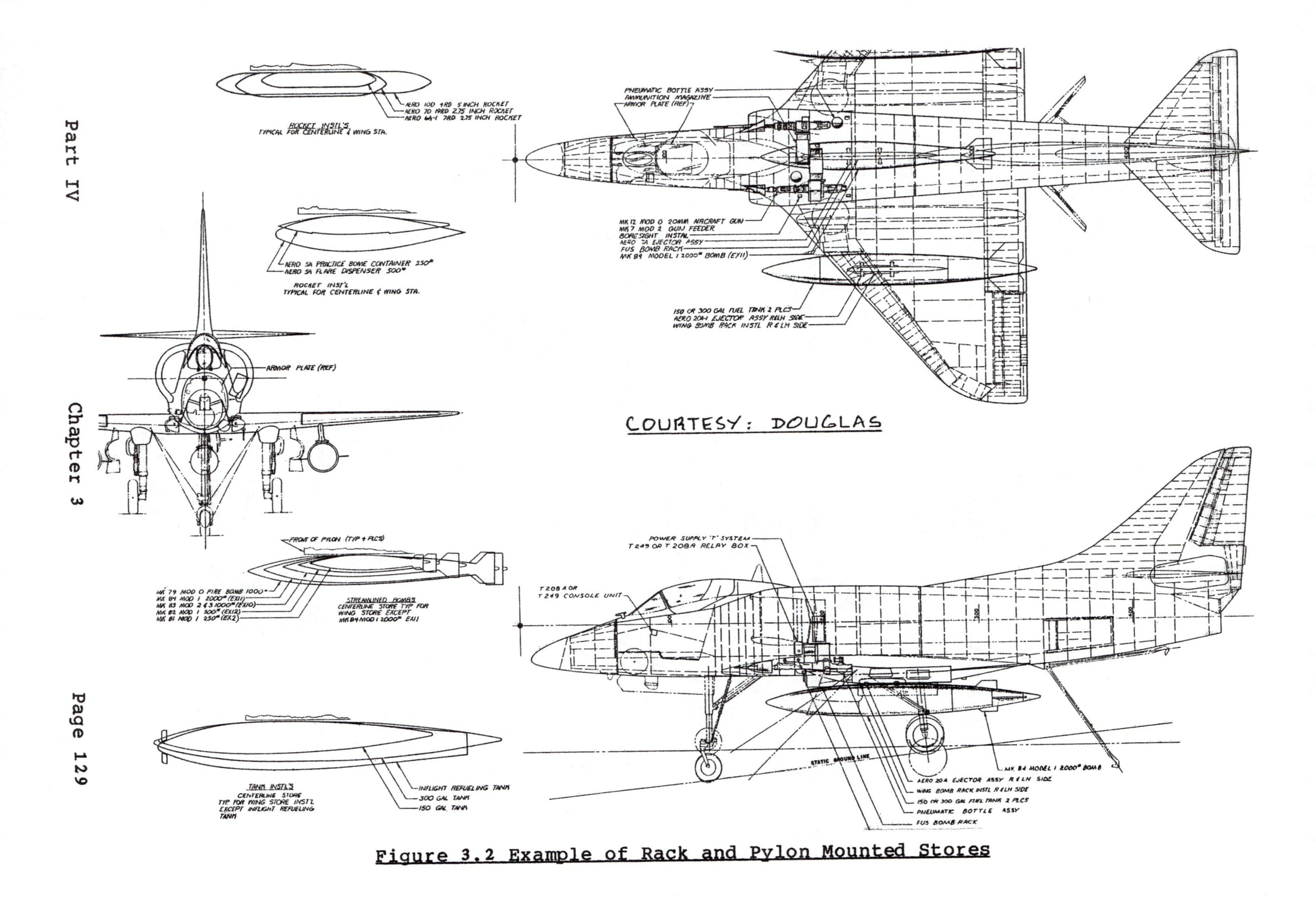

Figure 3.2 Example of Rack and Pylon Mounted Stores

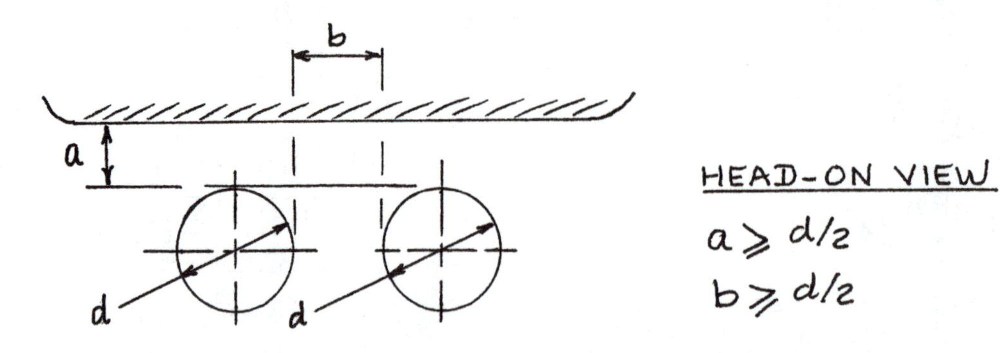

Figure 3.3 Recommended Mounting for Pylon and Rack Mounted Stores

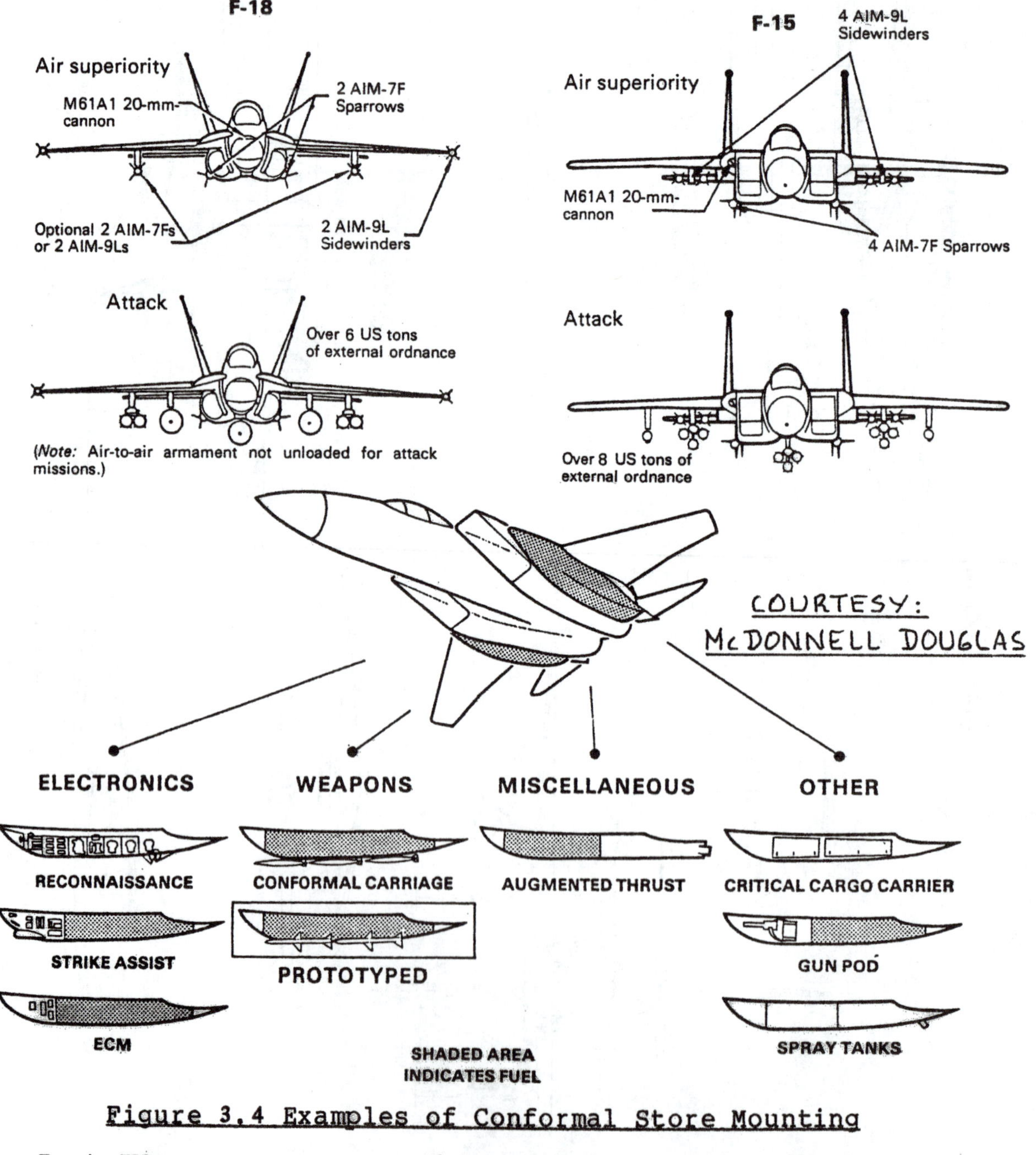

Figure 3.4 Examples of Conformal Store Mounting

Significant reductions in store drag increments can be achieved by mounting the stores 'conformally'. Examples of conformal and pylon mounted store arrangements are shown in Figure 3.4.

2. Figure 3.5 illustrates the drag reductions which can be achieved with a 'conformal' arrangement.

3.1.2 Stability and Control Considerations

Stores can alter the static longitudinal and static directional stability of an airplane. Figure 3.6 shows examples of store arrangements with such effects. The methods of Part VI and Part VII may be used to estimate the effects of stores on stability.

Stores and guns can add large moments about the center of gravity. These moments must be controllable without significant increase in pilot workload.

Store induced moments are primarily caused by drag. These drag induced moments need to be 'trimmed' in steady state flight. Upon release of the store a moment transient will occur. This moment transient should be minor.

Gun induced moments are caused by gun recoil forces. The data in Section 3.5 show that these recoil forces can be considerable.

In both cases the problem centers around the estimation of the moments imposed on the airplane by these recoil forces or/and by these drag forces.

Figure 3.7 shows a typical fuselage gun installation. If the recoil force imposed on the airframe is called: F_r and the moment arms are as shown in Figure 3.7, then it is possible to estimate the resulting recoil moments from:

$$\Delta C_{m_r} = F_r \bar{z}_r / \bar{q} S \bar{c} \quad (3.1)$$

$$\Delta C_{n_r} = F_r y_r / \bar{q} S b \quad (3.2)$$

The required control deflections to compensate for these moments follow from:

$$\Delta i_{H_r} = \Delta C_{m_r} / C_{m_{i_H}} \quad (3.3)$$

$$\Delta \delta_{R_r} = \Delta C_{n_r} / C_{n_{\delta_R}} \quad (3.4)$$

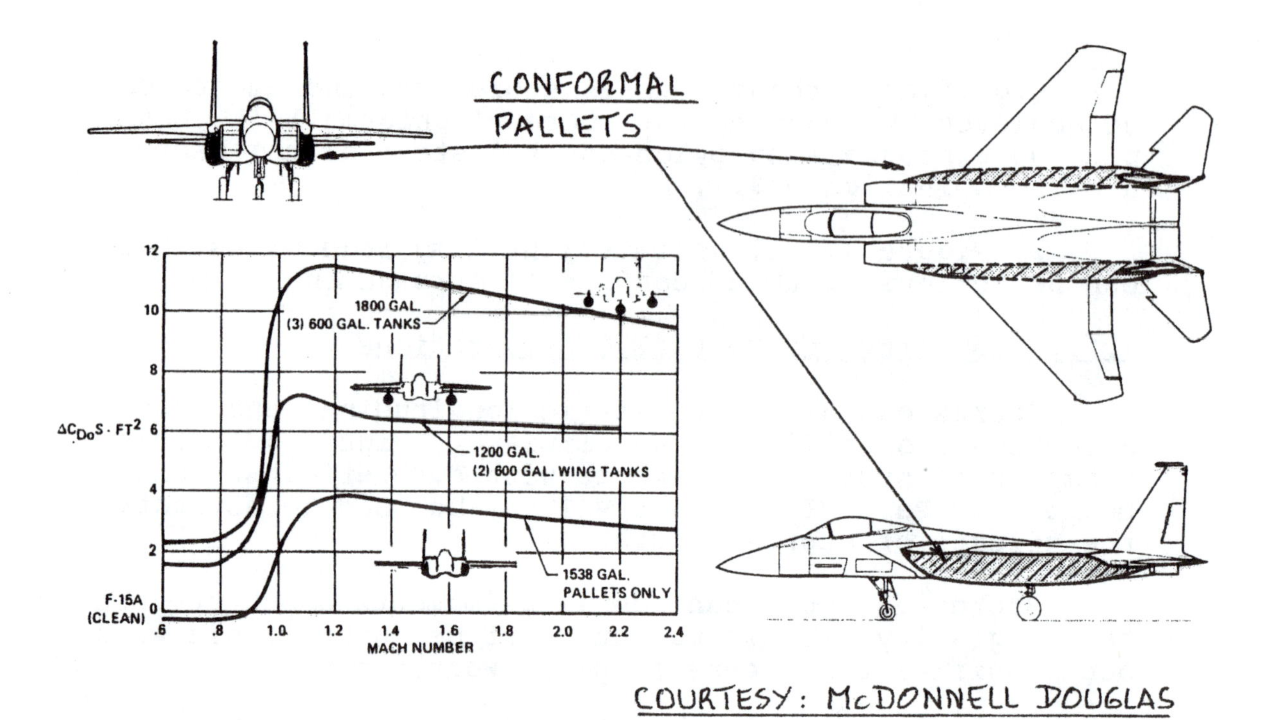

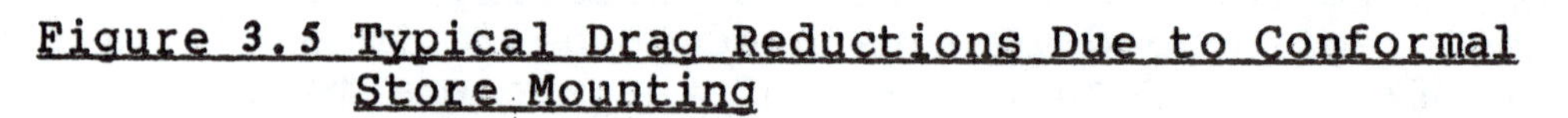

Figure 3.5 Typical Drag Reductions Due to Conformal Store Mounting

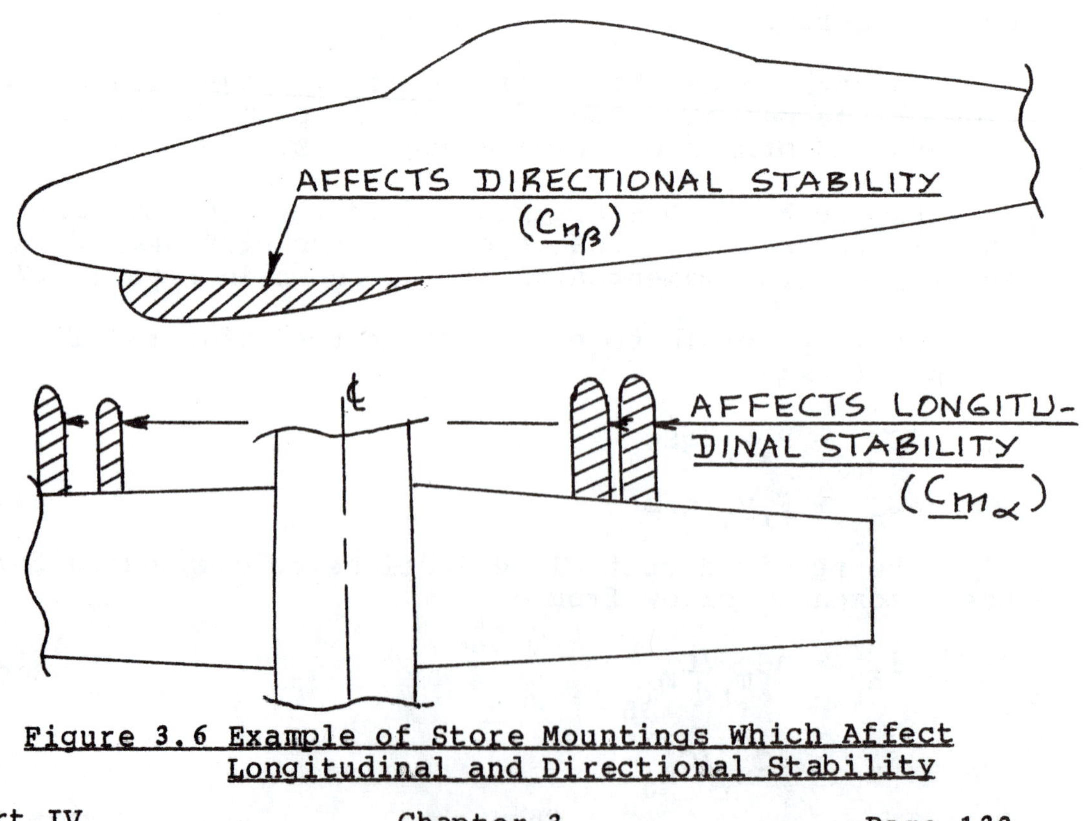

Figure 3.6 Example of Store Mountings Which Affect Longitudinal and Directional Stability

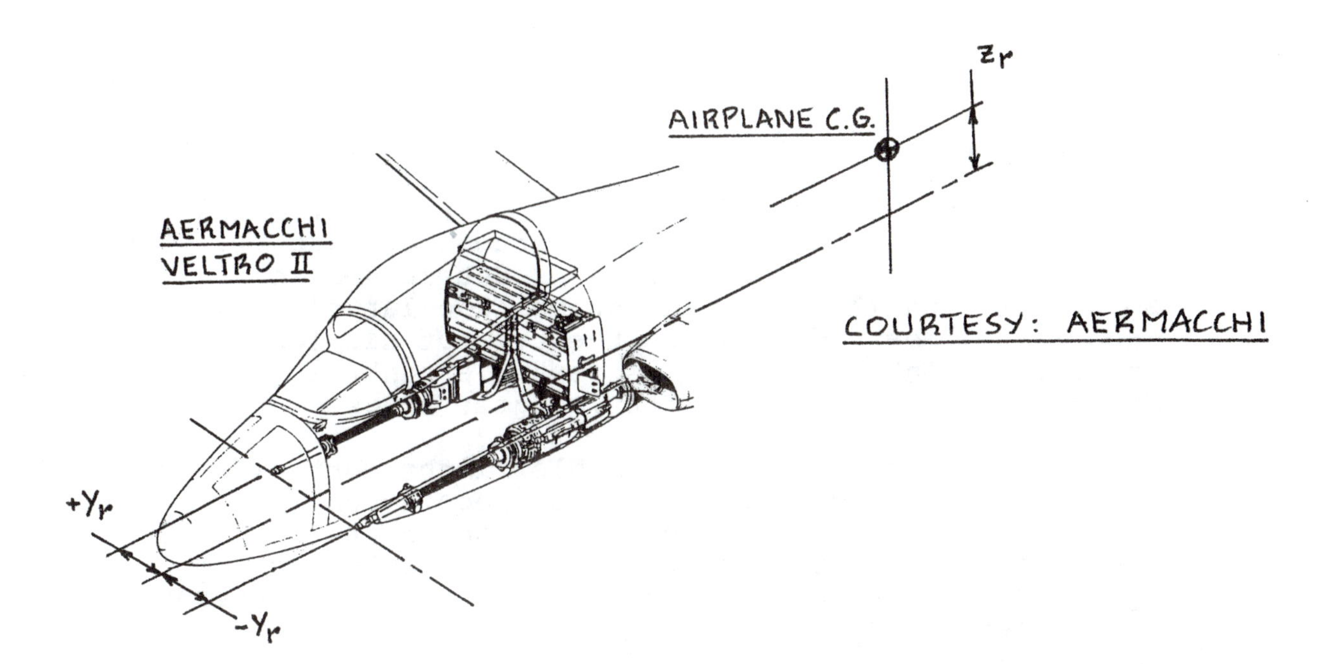

Figure 3.7 Example of a Fuselage Mounted Gun Installation

CANON RECOIL EFFECT ON CONTROL SYSTEM DESIGN

- THE SHOULDER MOUNTED GUN WAS THE OPTIMUM F-15 LOCATION:
  - PERMITTED ELEVATED BORE LINE FOR TRACKING STABILITY
  - ELIMINATED GAS INGESTION IN INLETS
  - MINIMIZED CROSSECTION AREA AND PERMITTED READY ACCESS
- THE LOCATION CAUSES A RECOIL MOMENT ON THE AIRCRAFT, WHICH IS NEGLIGIBLE WITH THE PRESENT 20MM GUN
- THE F-15 WAS DESIGNED TO ACCOMMODATE THE GAU-7A, 25MM GUN WHICH HAS NOT BEEN DEVELOPED, SIMULATION SHOWED COMPENSATION TO BE NEEDED TO ACCOMMODATE THE FACTOR OF 3 INCREASE IN RECOIL
- THE FOLLOWING SYSTEM WAS DESIGNED VIA SIMULATION AND WOULD BE ADDED WITH THE GAU-7A GUN

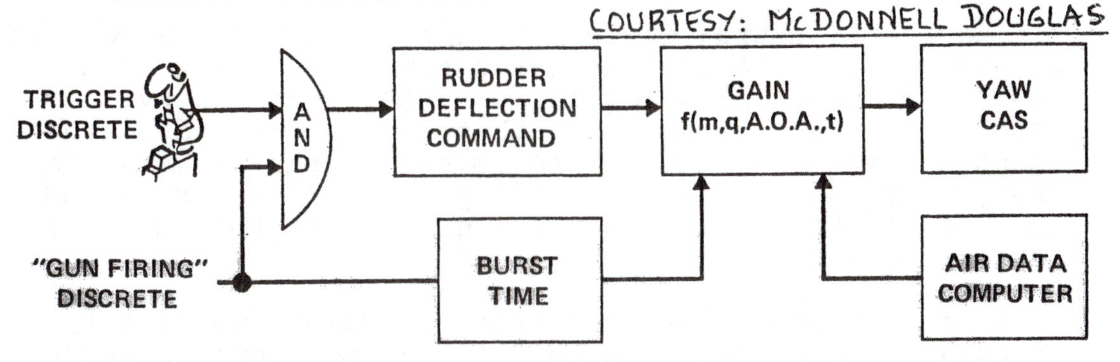

Figure 3.8 Automatic Compensation of Recoil Effects

The control power derivatives in Eqns.(3.3) and (3.4) may be estimated with the methods of Part VI. The incremental control deflections as determined in Eqns.(3.3) and (3.4) should not be 'too large'. What 'too large' means depends on the type of airplane and on its flight control system. Acceptable control deflections are in the 0.5 to 2 degree range.

For airplanes with reversible flight control systems the control deflections due to gun recoil must be obtained through a pilot induced force on the cockpit controls. These pilot control forces should not be excessive. Methods for computing pilot control forces are presented in Part VII.

For airplanes with irreversible flight control systems it is relatively easy to arrange for automatic compensation of the recoil moments. This can be done via the flight control computer system signalled by the pilot's triggering of a fire control switch. An example of such a system is shown in Figure 3.8.

3.1.3 Separation Considerations

Any store or missile which needs to be dropped or fired from an airplane must have 'clean' separation characteristics relative to the airplane as well as relative to other ordnance released at the same time.

To assure positive separation an analysis or test must be conducted which shows conclusively that upon release the now 'free' store or missile will not hit the airplane. This is a very difficult problem to analyze because the store or missile, upon release is not at first in a uniform flowfield but is instead in the complicated flowfield surrounding the airplane.

Separation trajectories are normally calculated with the help of finite element aerodynamic programs. The results of these calculations are then verified by windtunnel tests arranged so that the forces and moments on both airplane and store (or missile) can be measured. The US Navy David Taylor facility in Corduroc, Maryland is probably the best equipped facility for this type of testing.

After these windtunnel tests in-flight verification of positive separation is normally carried out at weapons test ranges, before the store or missile is released for operational deployment.

Figure 3.9 shows an example of a positive separation system as used frequently in separating fuel tanks from airplanes.

3.1.4 Gun Exhaust Gas Considerations

When installing a gun or cannon in an airplane, make sure that the gun exhaust gasses cannot enter the inlet of a jet engine or gas generator. Most gun exhaust gasses are highly corrosive to jet engine compressor and/or turbine blades.

Figure 3.10 shows examples of 'good' and 'poor' installations from this viewpoint.

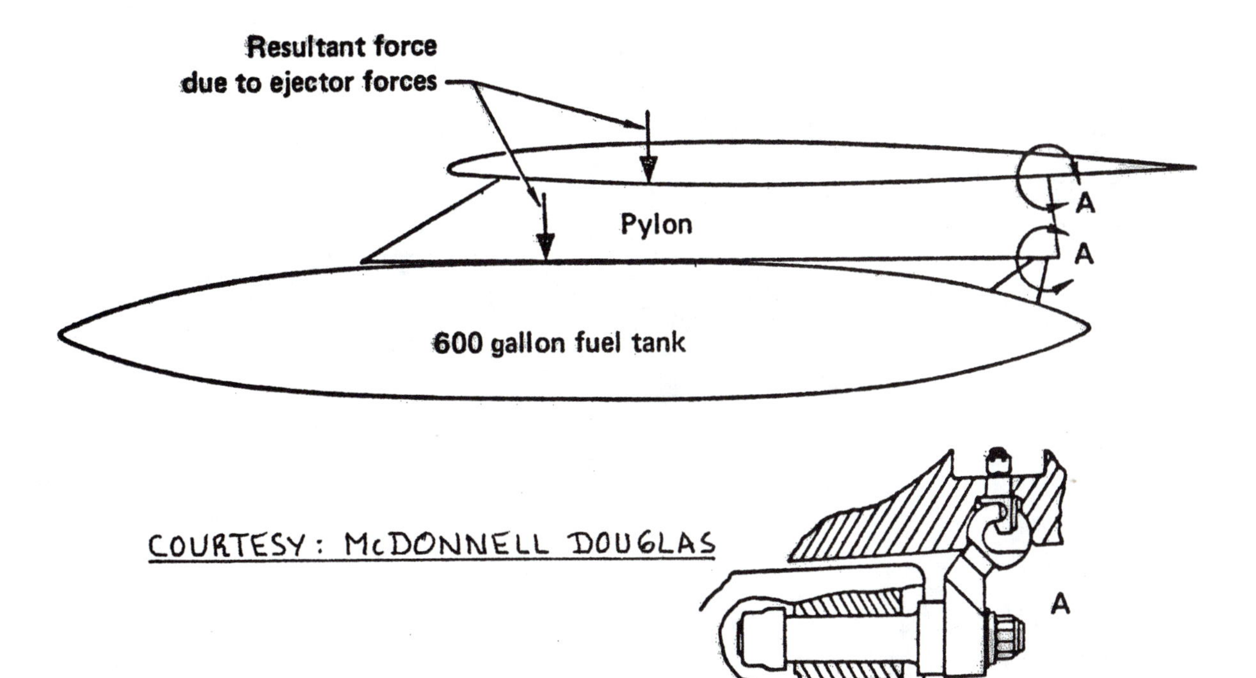

Figure 3.9 System for Positive Store Separation

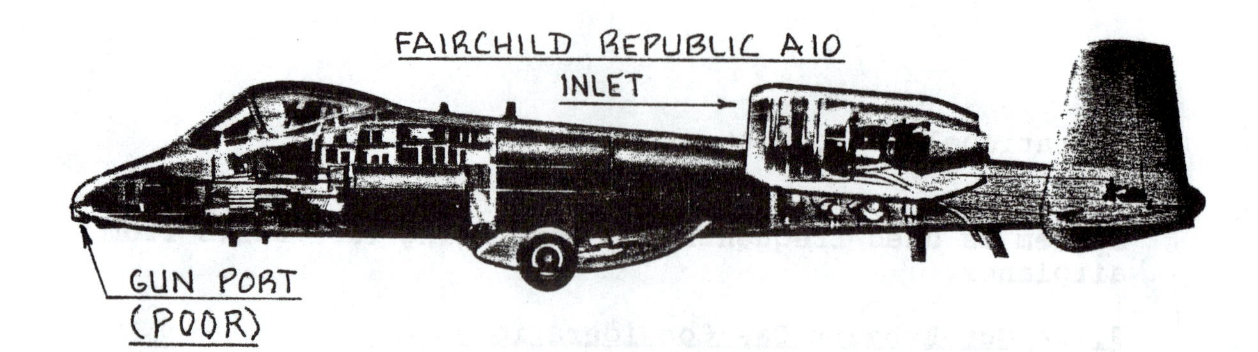

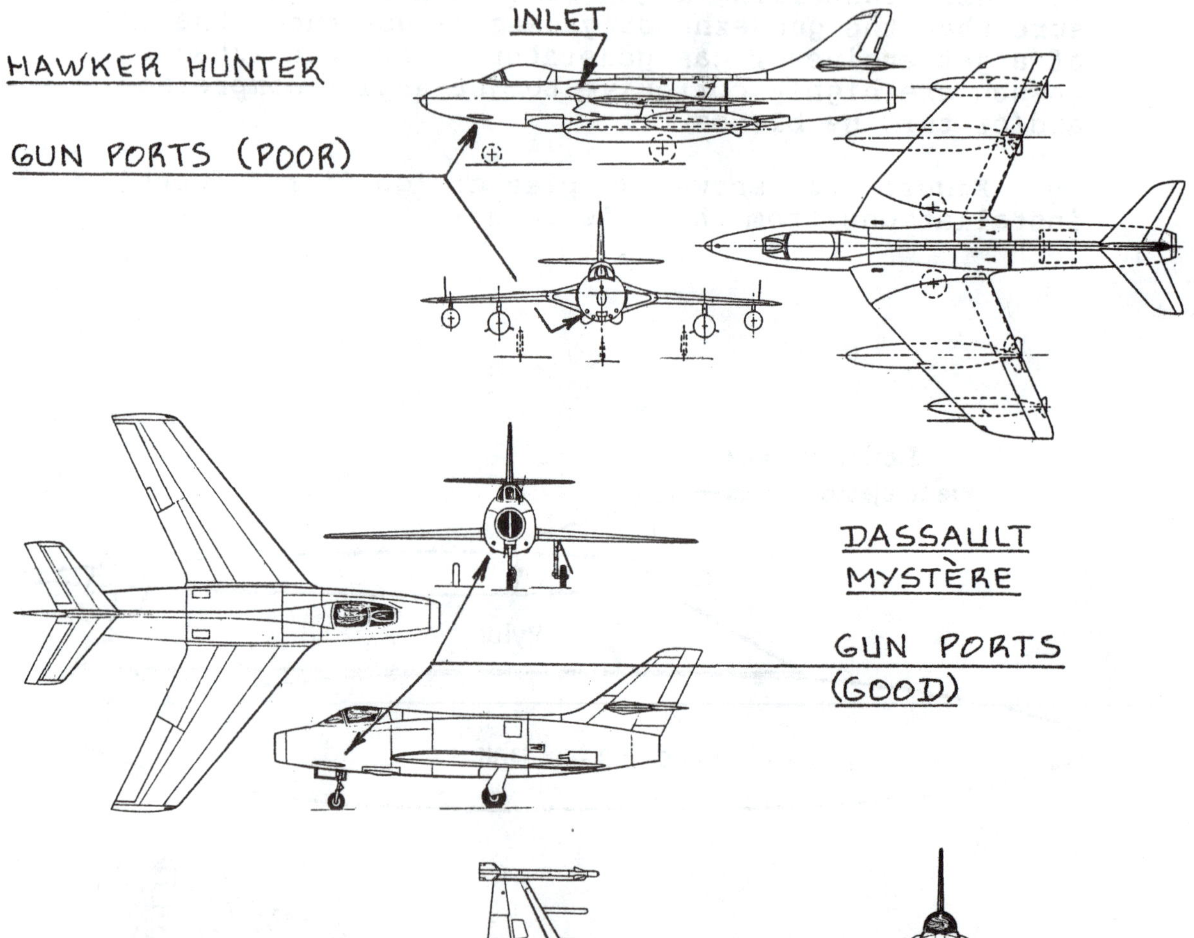

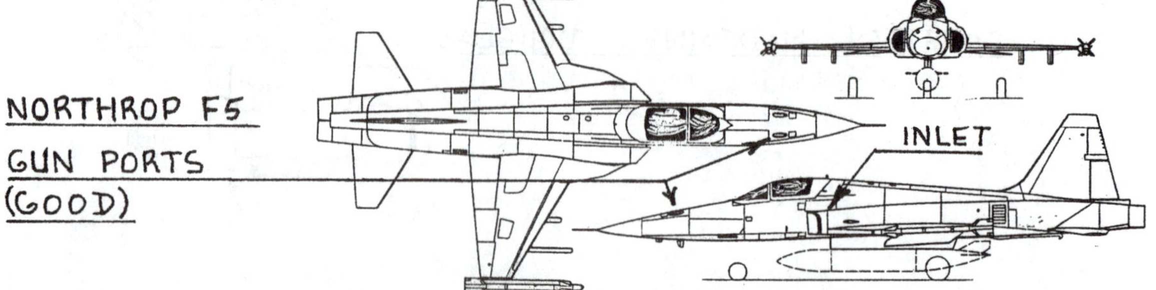

Figure 3.10 Example of Good and Poor Gun Locations From a Viewpoint of Gun Exhaust Versus Engine Inlet Locations

3.2 STRUCTURAL DESIGN CONSIDERATIONS

The following structural design considerations can be important in weapon and store integration:

1. Effect of gun recoil forces on the structure.

These recoil forces can be considerable (12,000 lbs in the case of some multi-barrel cannons!). These forces must be transitioned in to the airframe structure without causing undue deformations or fatigue problems.

2. Effect of gun induced vibrations on the structure and on sensors located close by.

Gun induced vibrations can cause ride quality problems at the pilot station. The gun attachment structure needs to be sufficiently stiff to prevent this.

Gun induced vibrations can cause local accelerations at places where flight crucial sensors are located. This can have two effects:

a) damage to the sensors

b) the sensors will interpret these accelerations as 'rigid' body signals thereby causing improper operation of the flight control system. Particularly this latter effect has been happening frequently during early weapons testing on several fighter programs. Relocating of sensors or the addition of notch filters can solve these problems.

3. Inertial load of stores and aerodynamic loads on stores need to be transmitted to the structural 'hard points' to which the stores are attached.

These hard points need to be designed so they do not deform excessively during maneuvering flight and so they do not induce unduly large stresses in the surrounding structure. Figure 3.11 shows an example of typical hard point installations. Another example was shown in Figure 4.80 in Part III.

4. Stores may cause flutter.

During flight testing in many fighters it is found that excessive vibration and/or flutter problems arise with certain store configurations. To prevent this from occurring it is necessary to conduct realistic flutter calculations (and sometimes flutter model tests)

Figure 3.11 Example of 'Hard Point' Locations

before committing to flight test and certainly before committing to production. Required 'fixes' late in a production program tend to be very expensive.

Figure 3.12 recalls an incident of store flutter (in this case a relatively benign 'limit cycle' oscillation) and indicates the 'fix'.

STORES EFFECT ON MECHANICAL CONTROL SYSTEM DESIGN

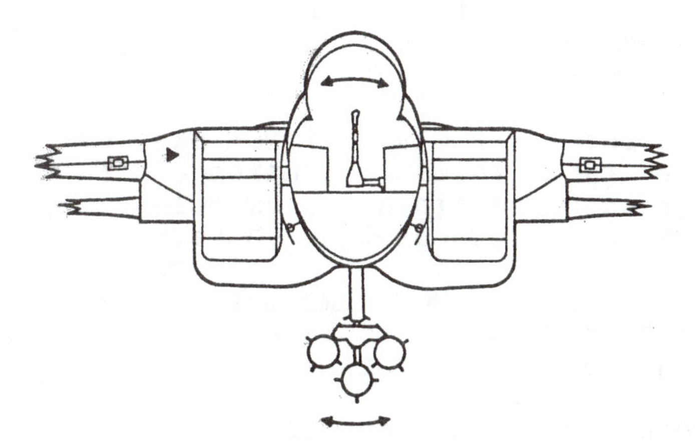

PROBLEM
- A 5.5hz DIVERGENT OSCILLATION WAS DISCOVERED AT 0.9M, 5000 FT DURING FLIGHT FLUTTER TESTING WITH MK-82 BOMBS. REACHED $\pm 1g$ LATERAL.
- PILOT HAD TO JETTISON STORES

CAUSE
- COUPLED AEROELASTIC MODE INVOLVING: PYLON YAW, RACK BENDING, FUSELAGE SIDE BENDING AND LATERAL MECHANICAL CONTROL SYSTEM.
- REQUIRED HIGH EXCITATION LEVEL TO INDUCE.

SOLUTION
- ADDED EDDY-CURRENT DAMPER TO LATERAL CONTROL SYSTEM

Figure 3.12 Example of a Store Flutter Incident

3.3 DESIGN FOR LOW RADAR AND INFRARED DETECTABILITY

When an airplane is difficult to detect with radar and/or with infrared sensors its combat survivability is improved. Combat survivability is also determined by the vulnerability of critical airplane systems to hostile action. Reference 16 contains a thorough treatise on analysis and design for combat survivability.

3.3.1 Design Considerations for Low Radar Detectability

The following design considerations for low radar detectability (low radar cross section or low RCS) were obtained from Dr. Alan E. Fuhs of the US Naval Postgraduate School in Monterey, California. Ref.17 presents the corresponding theories and methods for determining the radar cross section of an airplane. This sub-section summarizes those recommendations for low RCS which are important to the preliminary layout designer.

What makes an airplane visible to radar is a phenomenon called radar wave reflection or radar wave scattering. To understand scattering, it is useful to think of an airplane as a 'porcupine' with quills pointing outward as normal vectors to the surface. If the airplane is to be 'stealthy' it should not point any quills in the direction of a radar antenna.

To shape an airplane for low RCS a decision must be made regarding the primary radar threat: it makes a difference whether the radar sees an airplane from below, from the side up, from the side down or from above.

In shaping an airplane for low RCS some unfavorable trades will have to be accepted in regard to aerodynamic drag, inlet efficiency and cockpit visibility.

The following factors tend to promote radar wave scattering and therefore increase the RCS of an airplane:

1. Fuselages with square cross sections: round the fuselage cross section as gently as possible. Cant the fuselage sides away from radar looking at it from the side: a fuselage with a rounded triangular cross section tends to have a low RCS.

Figure 3.13 illustrates the good and the bad.

2. Inlets with sharp flat surfaces: locate the inlets above the airplane when the primary radar threat is from below. If propulsion considerations allow it,

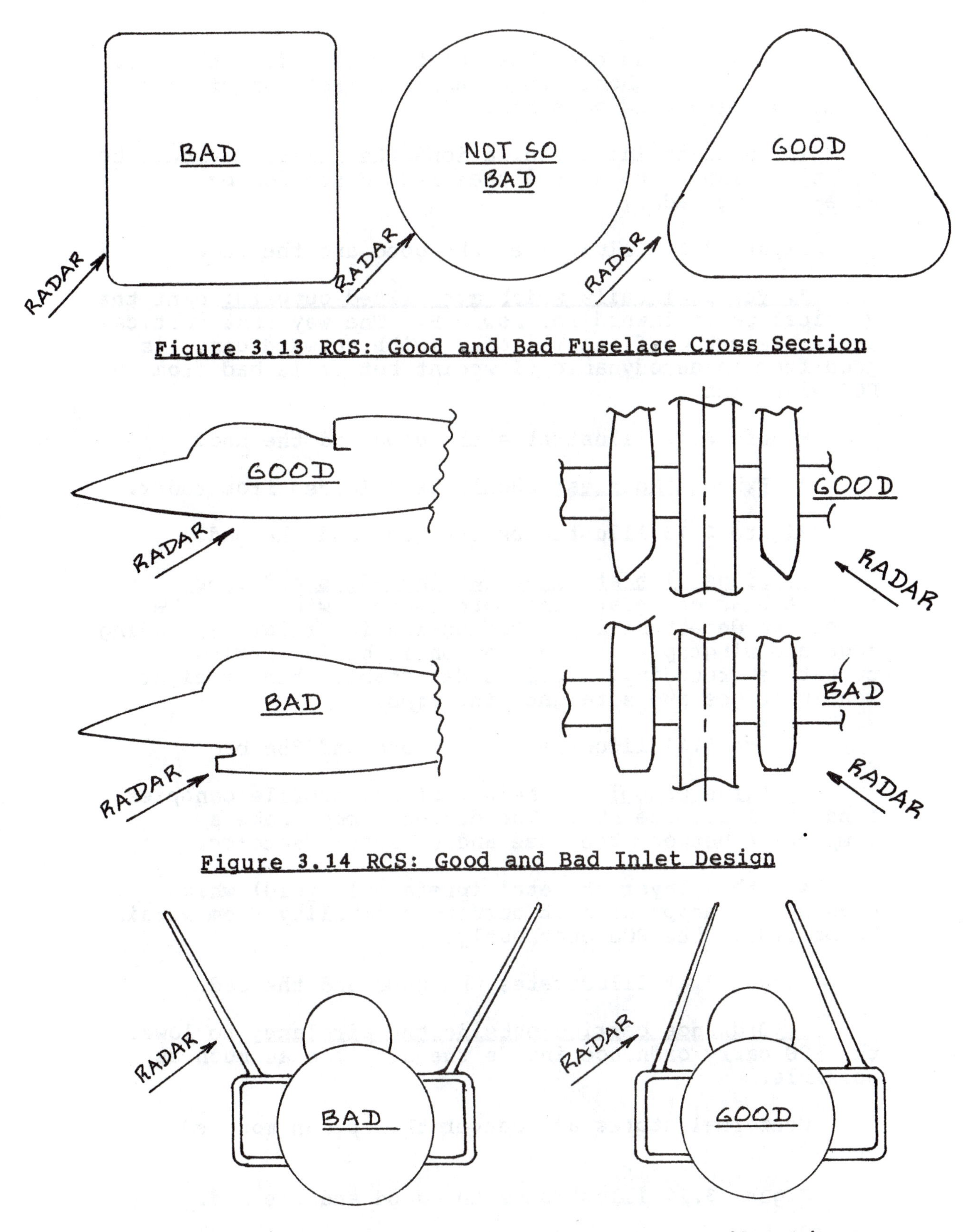

Figure 3.13 RCS: Good and Bad Fuselage Cross Section

Figure 3.14 RCS: Good and Bad Inlet Design

Figure 3.15 RCS: Good and Bad Vertical Tail Design

placing a wire mesh over the inlet helps reduce the RCS. The mesh spacing should be as small a fraction of the radar wavelength as possible.

In multi-engine installations the inlets may have to be angled inward to shield them from detection by side-looking radar.

Figure 3.14 illustrates the good and the bad.

3. Vertical tails which are tilted outward: cant the vertical tails inward for low RCS. The way most vertical tails are put on fighters today (with outward cant) is good from an aerodynamic viewpoint but it is bad from an RCS viewpoint.

Figure 3.15 illustrates the good and the bad.

4. Exhaust Nozzles: should be shielded from radar.

Figure 3.16 illustrates the good and the bad.

5. Wings with straight leading edges and tips: against head-on radar, the more sweep a wing has, the lower its detectability. For up-looking radar, a leading edge and wingtip which have a continuously varying curvature (rounding) are less detectable than straight leading edges and straight wing tips.

Figure 3.17 illustrates the good and the bad.

6. Large canopies: the use of low profile canopies tends to lower the RCS. The designer must make a compromise between the 'see and be seen' features.

Any thin layer of metal (preferably gold) which covers the canopy without hurting visibility from within helps reduce the RCS enormously.

Figure 3.18 illustrates the good and the bad.

7. Ordnance carried outside the airplane: to lower the RCS carry ordnance inside the airplane as much as possible.

Conformal stores are better than pylon mounted stores.

Figure 3.19 illustrates the good and the bad.

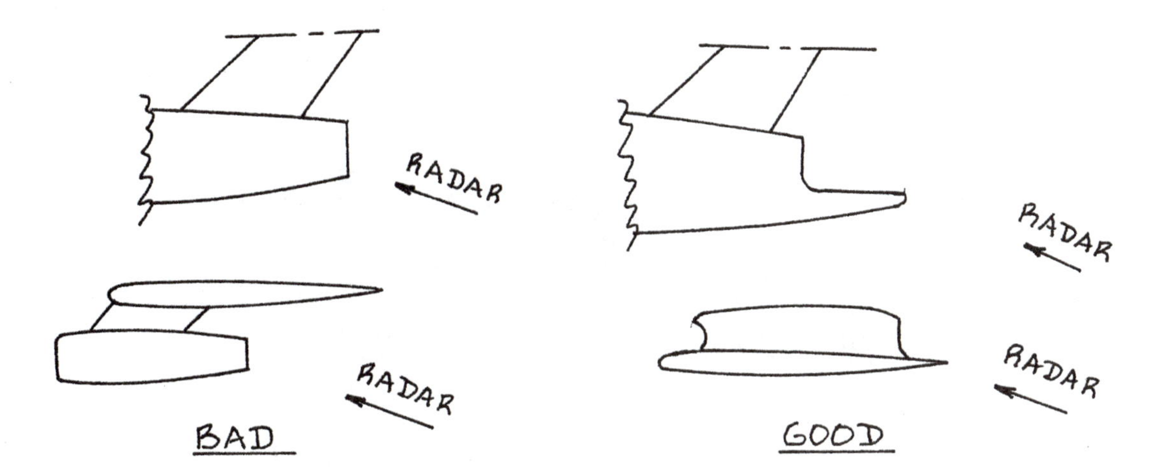

Figure 3.16 RCS: Good and Bad Exhaust Nozzle Design

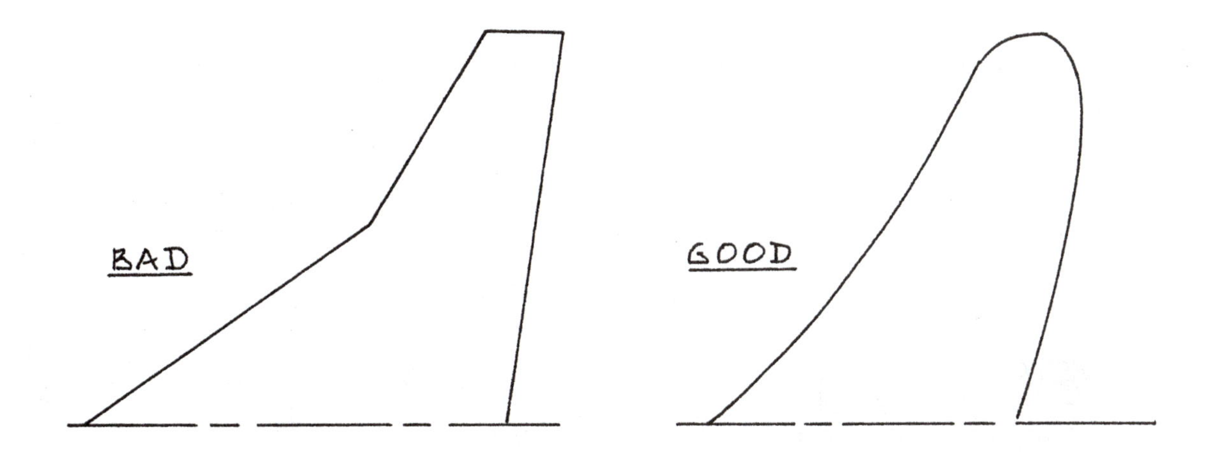

Figure 3.17 RCS: Good and Bad Wing Planform Design

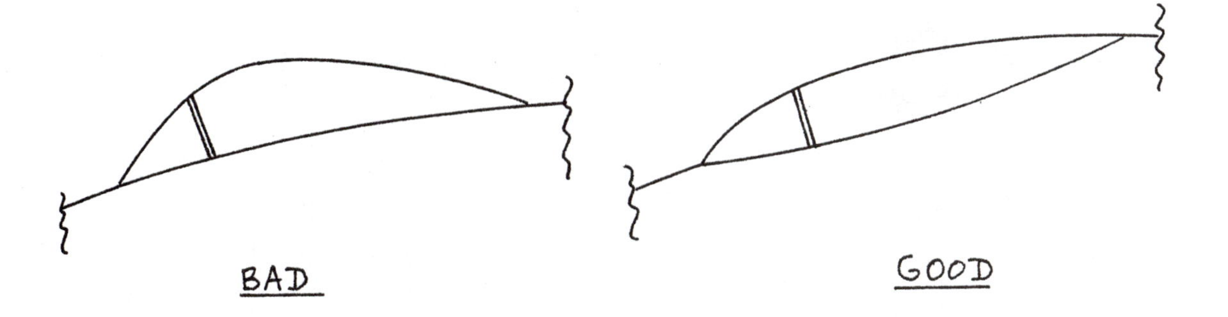

Figure 3.18 RCS: Good and Bad Canopy Design

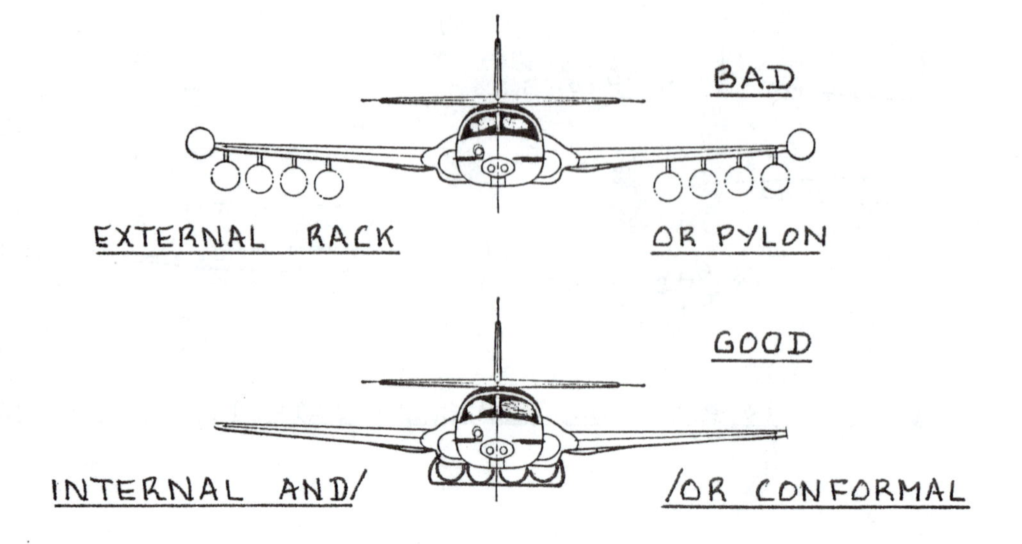

Figure 3.19 RCS: Good and Bad Ordnance Installations

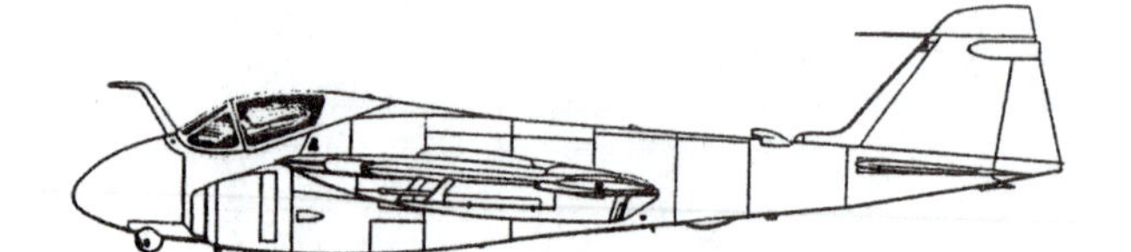

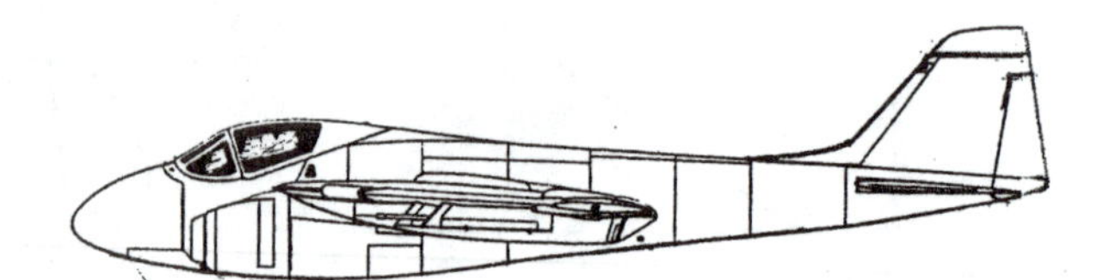

BAD

GOOD
ALL BUMPS AND PROTRUSIONS "INTERNALIZED"

Figure 3.20 RCS: Good and Bad Bumps and Protrusions

8. Bumps and protrusions: eliminate these as much as possible to achieve a low RCS.

Figure 3.20 illustrates the good and the bad.

9. Radar antennas: A radar antenna when 'looked at' by enemy radar looks like the reflection of light from a cat's eyes at night. Making the radar dome (radome) opaque at the search radar wavelength cuts down on RCS.

Finally: anything that can be done to smoothly blend surfaces and bodies helps lower the RCS of an airplane.

10. Radar absorbing materials: Not much can be said about this subject other than that this can be rather effective.

Reference 18 contains some interesting discussions of the characteristics of 'stealthy' aircraft.

3.3.2 Design Considerations for Low Infrared Detectability

All bodies emit infrared radiation. The 'warmer' a body is relative to its surroundings, the easier it can be detected by infrared sensors. Contrary to popular opinion, infrared seeking missiles do not home in on exhaust plumes but on the surrounding metal which, although cooler than the exhaust plume, emits 95 percent of the infrared energy.

To suppress infrared detectability the following guidelines are offered:

1. Shield all exhaust nozzles from any direction from which a threat can be expected.

2. High by-pass ratio engines are much more difficult to detect by infrared sensors than straight turbojet engines. Carrying the by-pass air duct all the way aft also serves to reduce the infrared signature.

3. Carry heat generating flares and release these whenever the threat of infrared seeking weapons exists. Heat generating flares can confuse infrared seeking weapons and thus lead them to the wrong target.

Figure 3.21 illustrates a method for infrared shielding. It is clear that a significant increase in

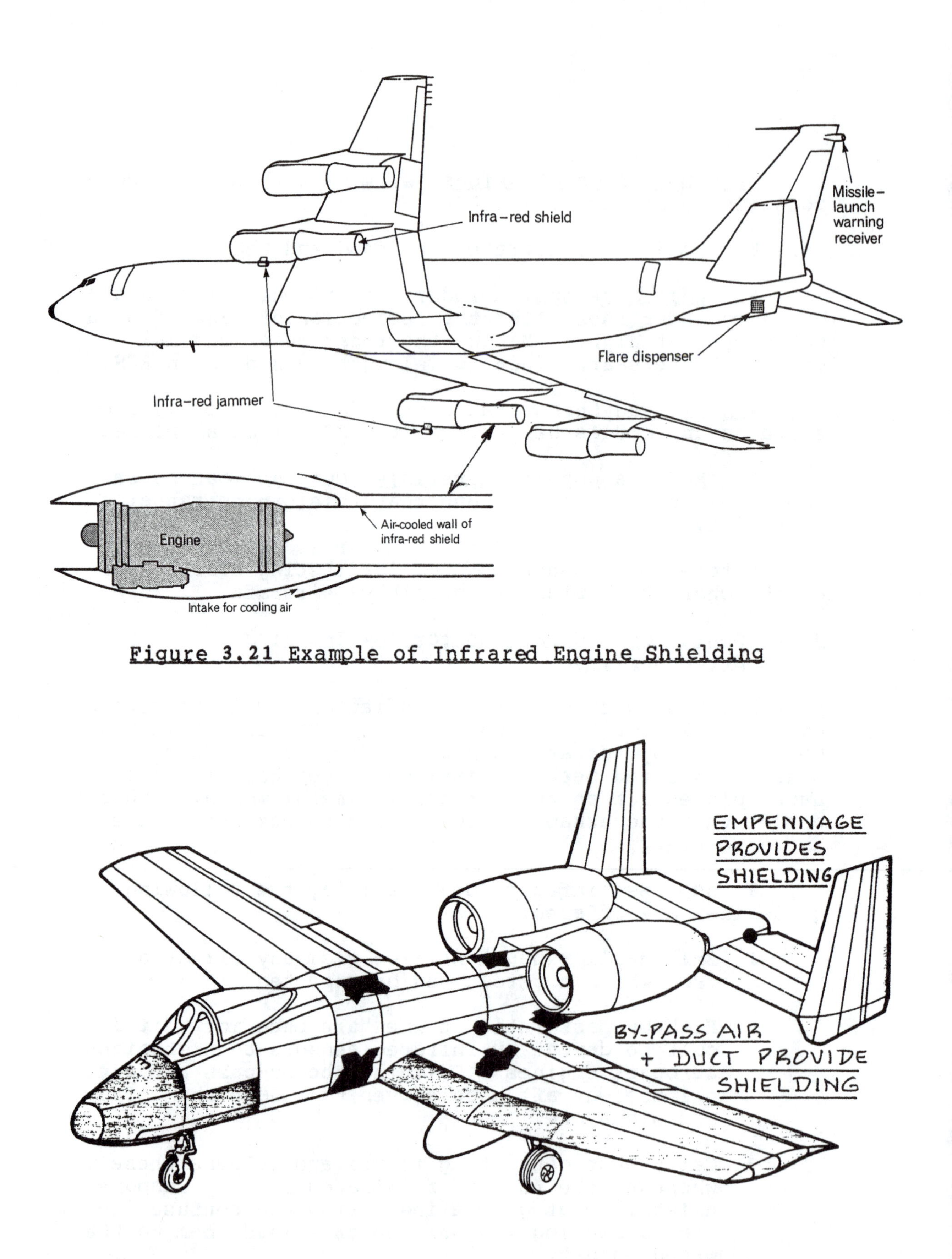

Figure 3.21 Example of Infrared Engine Shielding

Figure 3.22 Example of an Engine Installation with Low Infrared Signature

empty weight will be incurred.

Figure 3.22 shows the Fairchild Republic A-10 with its engines very well placed to avoid a large infrared signature from its area of greatest vulnerability: missile attack from below and behind.

Figure 3.21 also shows infrared jammers. A typical infrared jammer is the Xerox ALQ-123 system. This is a 378 lbs externally mounted pod, powered by a ram-air turbine with a modulated caesium lamp as its infrared source. Since infrared seekers produce an error signal for any infrared source which is not on its optical axis, a modulated, high intensity signal in its field of vision will produce errors in its steering commands causing the missile to stray off course.

Finally, Figure 3.23 shows a Pratt and Whitney concept of a fighter designed with low RCS and low infrared signature.

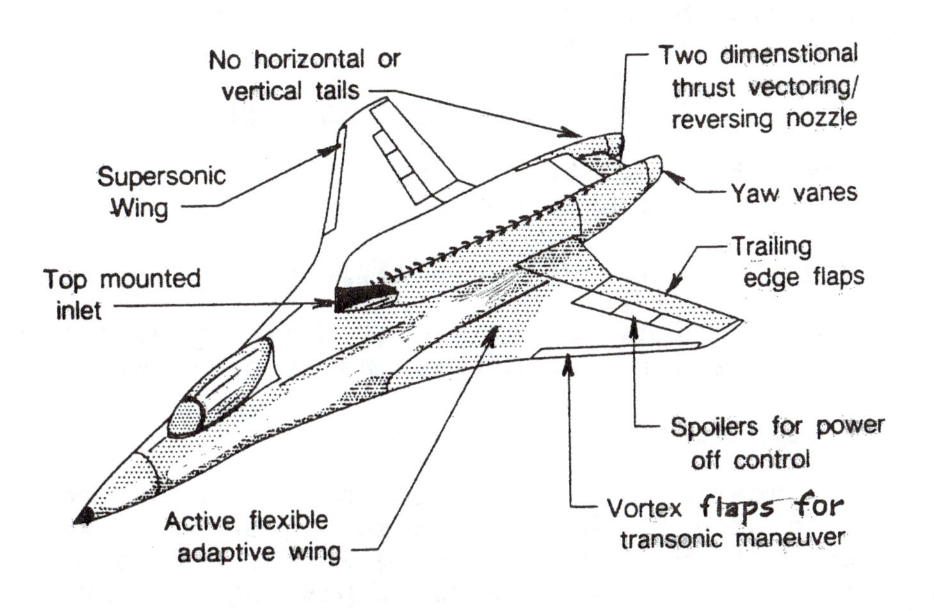

Figure 3.23 Fighter Concept for Low RCS and Low Infrared Detectability

3.4 EXAMPLES OF WEAPON INSTALLATIONS

In this section several examples are presented of airplane weapon installations.

3.4.1 Examples of Gun Installations

Typical gun/cannon installations are shown in the following figures:

Figure 3.7: Aermacchi Veltro II
Figure 3.24: Douglas A4D-2N Skyhawk (Also Fig.3.2)
Figure 3.25: Grumman F-14
Figure 3.26: General Dynamics F-16
Figure 3.27: Fairchild Republic A-10

3.4.2 Examples of External Store Arrangements

Examples of regular and conformal external store arrangements are shown in the following Figures:

Figure 3.28 Cessna YAT-37D
Figure 3.29 Grumman F-14
Figure 3.30 General Dynamics F-16
Figure 3.31 Sidewinder Installation on a Wing Tip
Figure 3.32 Panavia Tornado

3.4.3 Example of an Internal Store Installation

Figure 3.33 shows the internal weapons bay configuration of the B-1B bomber.

3.4.4 Example of Avionics Installations

Figure 3.34 shows a variety of military avionics installations used in the British Aerospace Hawk 200.

3.4.5 Example of Armor Plating

In a number of military airplanes it is deemed necessary to provide the pilot some degree of protection against enemy fire. Armor plating is the method normally used. Figure 3.24 shows the armor plating used in the Skyhawk. Armor plating is heavy (the armor is made of special steels) and provides only limited protection.

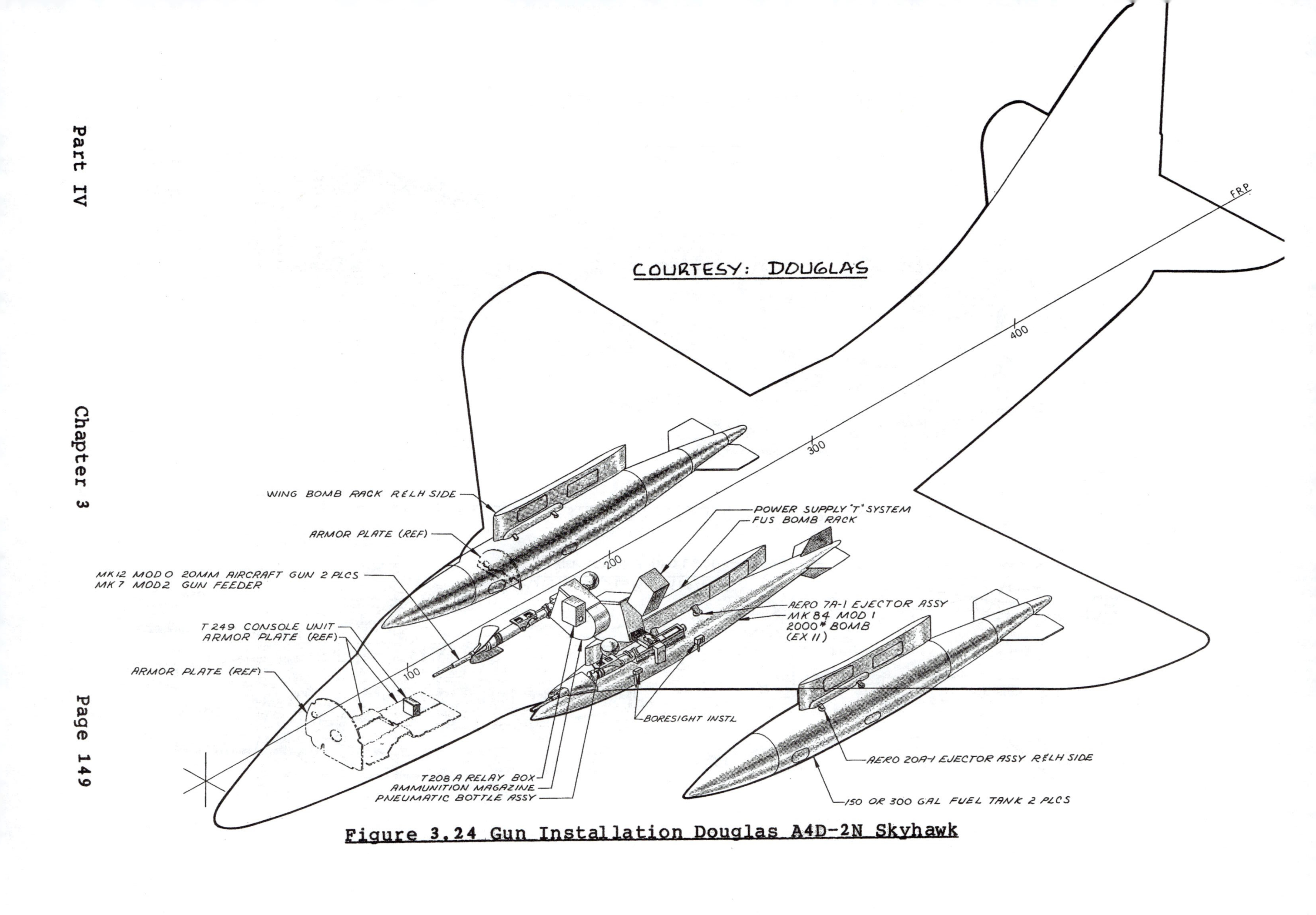

Figure 3.24 Gun Installation Douglas A4D-2N Skyhawk

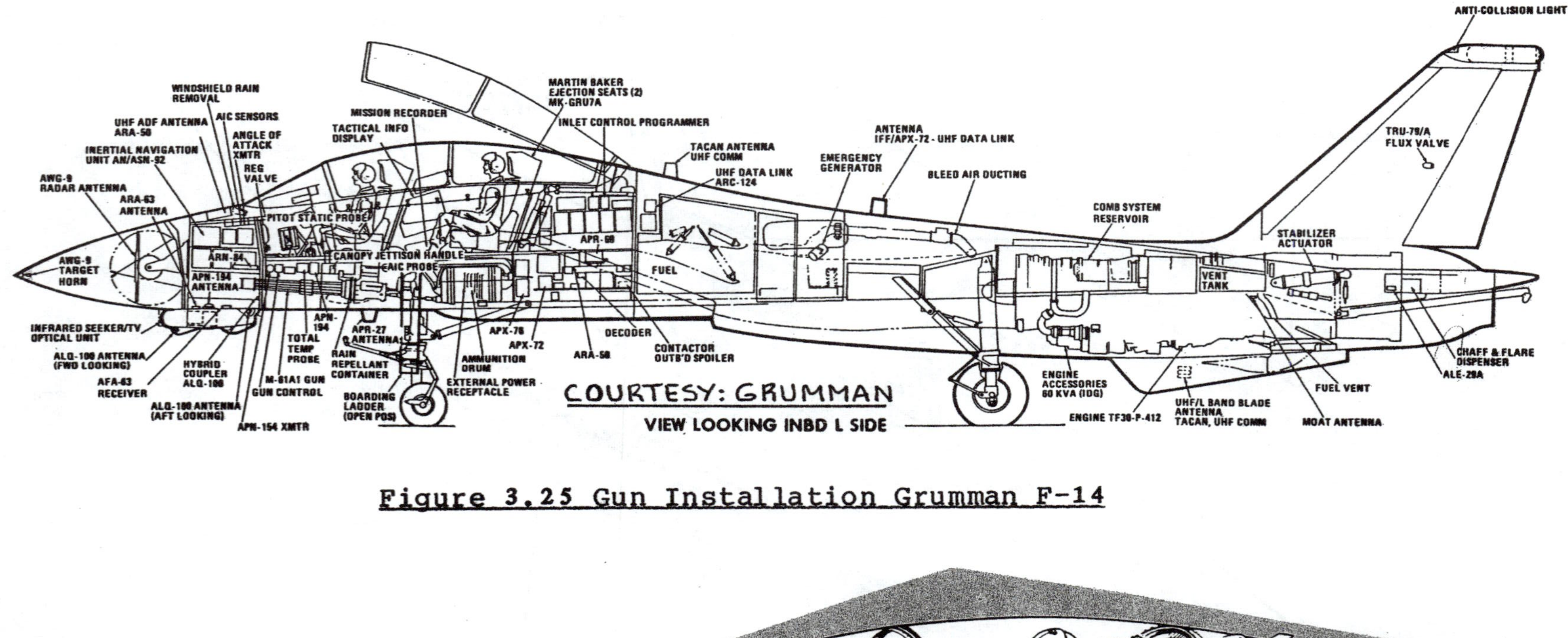

Figure 3.25 Gun Installation Grumman F-14

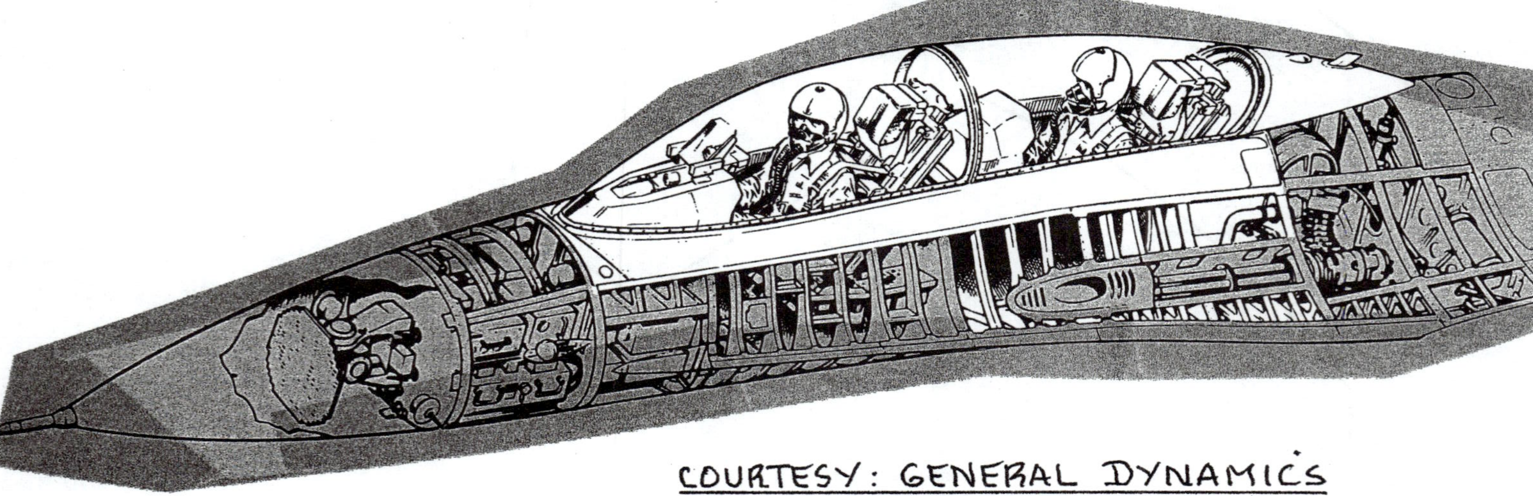

Figure 3.26 Gun Installation General Dynamics F-16

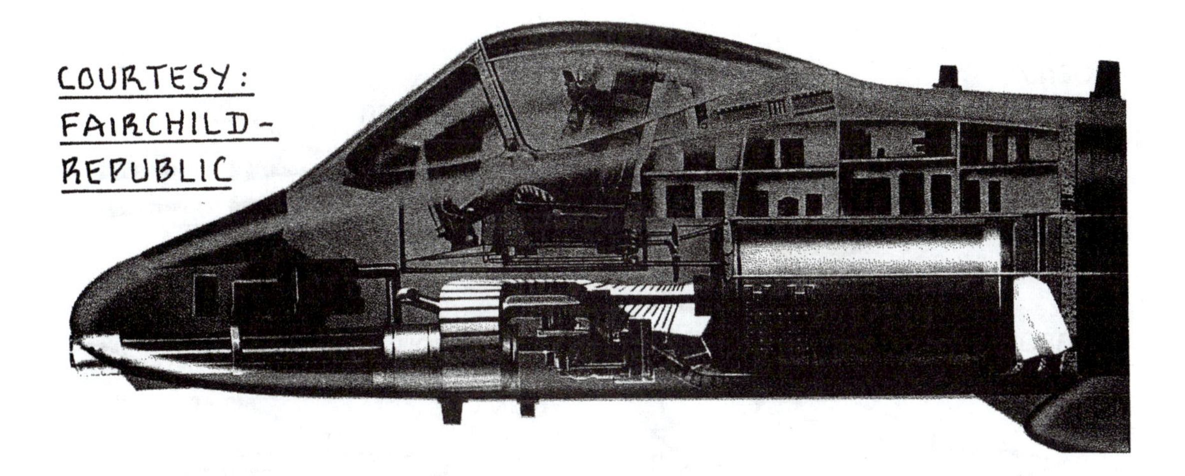

Figure 3.27 Gun Installation Fairchild Republic A-10

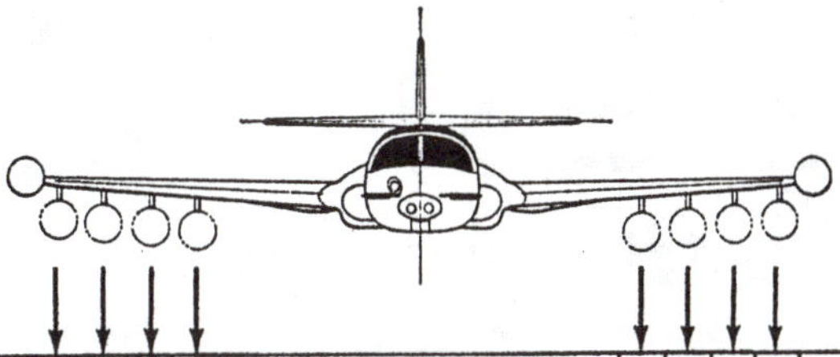

Store										Weight
100 Gal. Drop Tank			O	O		O	O			760 Lbs.
SUU-7A Bomblet Dispenser			O	O		O	O			
BLU-1/B (M-116) Fire Bomb			O	O		O	O			750 Lbs.
SUU-12/A Gun Pod (50 Cal.)			O	O		O	O			500 Lbs.
M-117 G. P. Bomb			O	O		O	O			750 Lbs.
BLU-11/B Fire Bomb	O	O	O	O		O	O	O	O	500 Lbs.
MK-81 Low Drag Bomb	O	O	O	O		O	O	O	O	250 Lbs.
MK-82 Low Drag Bomb	O	O	O	O		O	O	O	O	500 Lbs.
M-64/A1 G. P. Bomb	O	O	O	O		O	O	O	O	500 Lbs.
LAU-3A 19 Tube Launcher		O	O	O		O	O	O		470 Lbs.
AERO-6A 7 Tube Launcher	O	O	O	O		O	O	O	O	150 Lbs.
SUU-11/A Gun Pod (7.62mm)	O	O	O	O		O	O	O	O	235 Lbs.
XM-75 Grenade Launcher	O	O	O	O		O	O	O	O	
GAR-8 Sidewinder Missle	O	O						O	O	155 Lbs.

COURTESY: CESSNA

Figure 3.28 External Store Arrangement Cessna YAT-37D

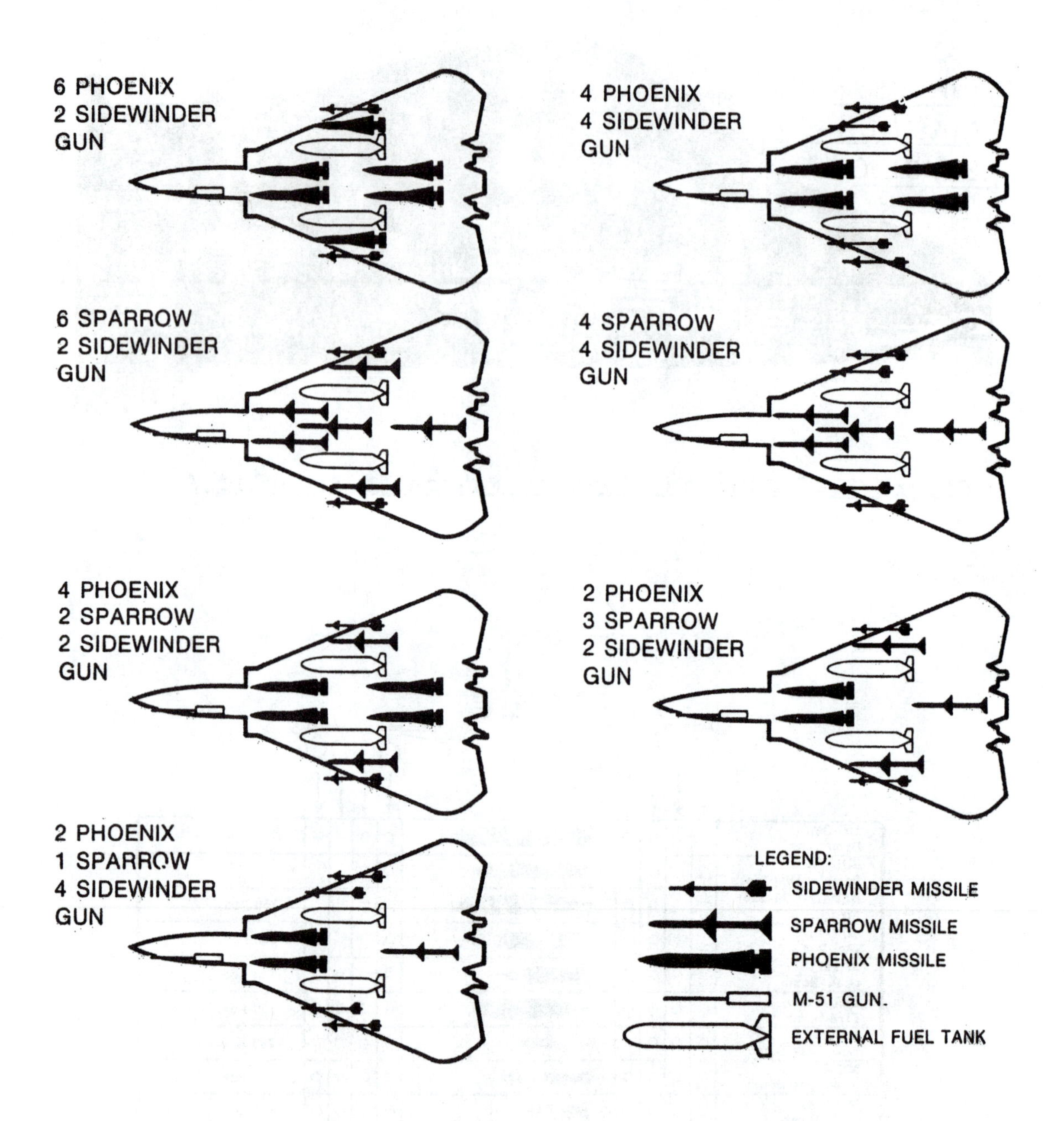

COURTESY: GRUMMAN

Figure 3.29 External Store Arrangement Grumman F-14

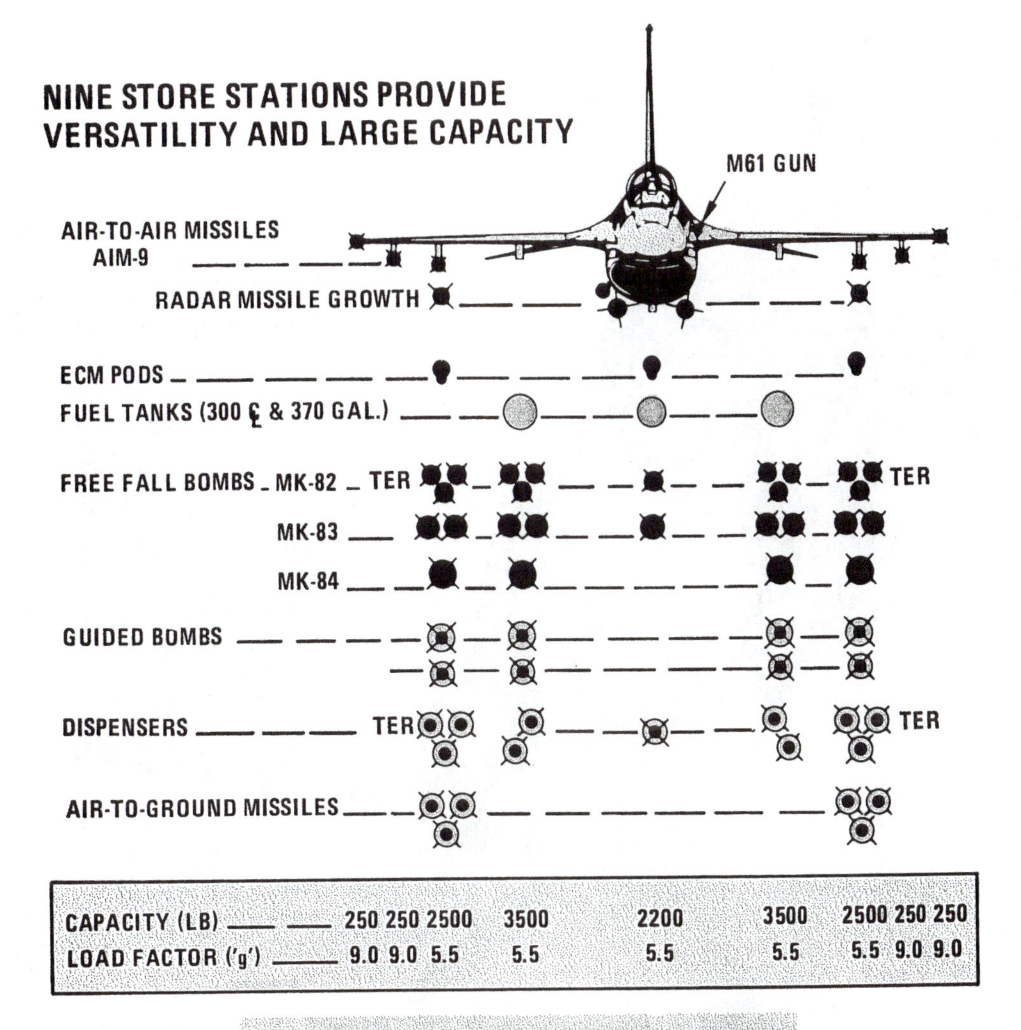

COURTESY: GENERAL DYNAMICS

Figure 3.30 External Store Arrangement Gen. Dynamics F-16

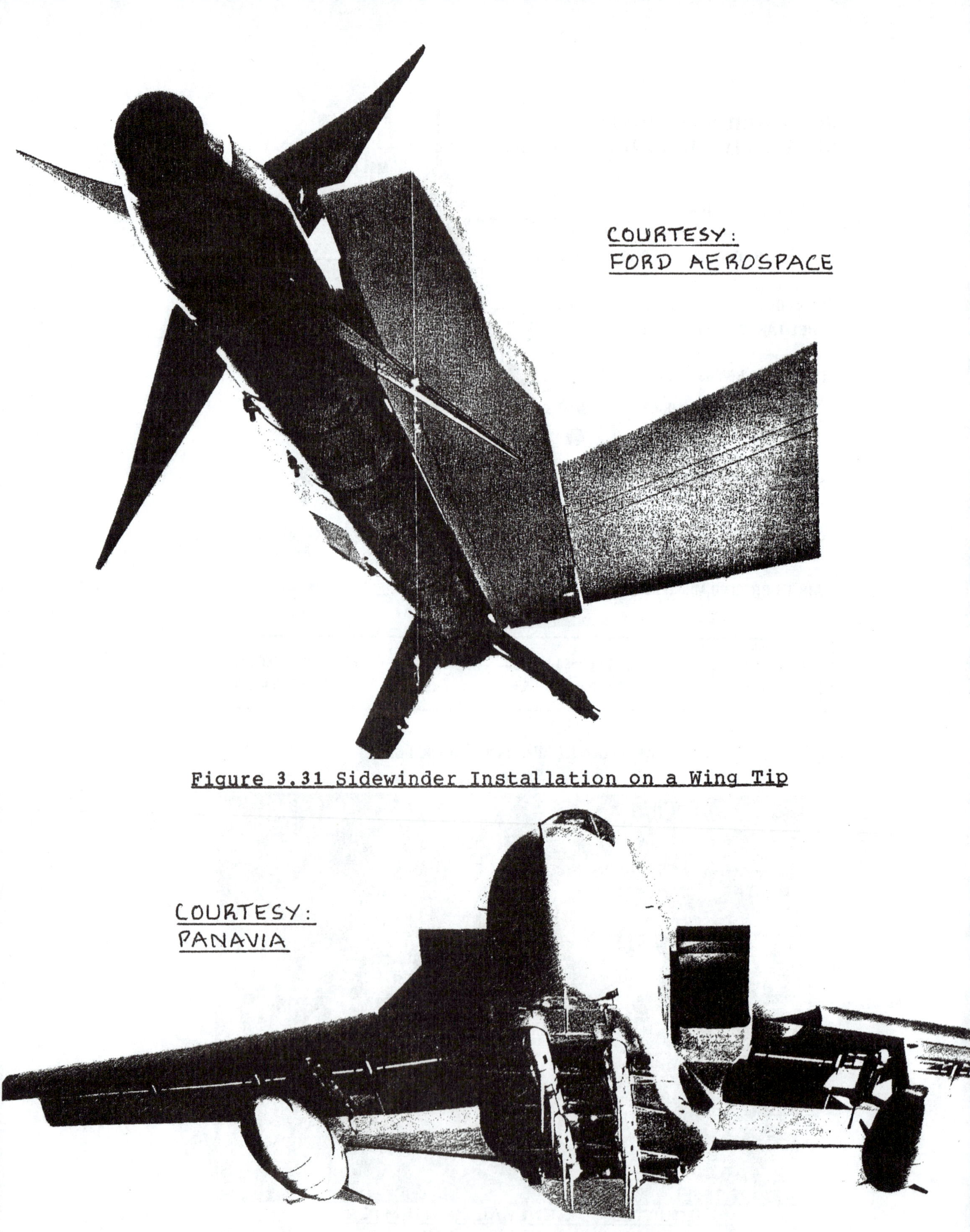

Figure 3.31 Sidewinder Installation on a Wing Tip

Figure 3.32 Conformal and Regular External Store Arrangement Panavia Tornado

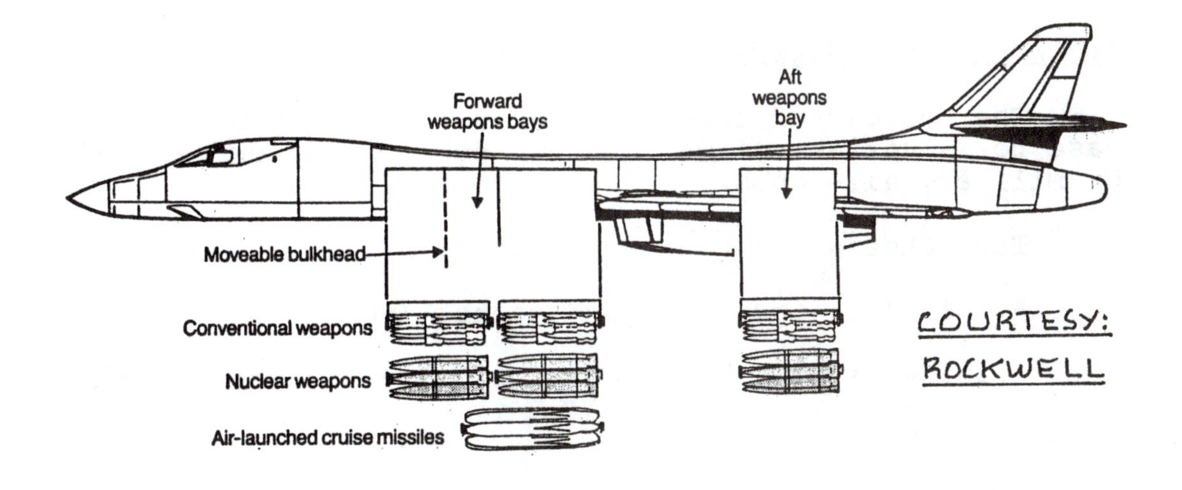

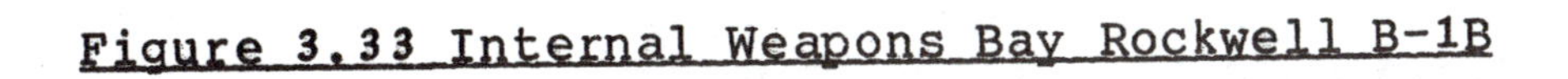
Figure 3.33 Internal Weapons Bay Rockwell B-1B

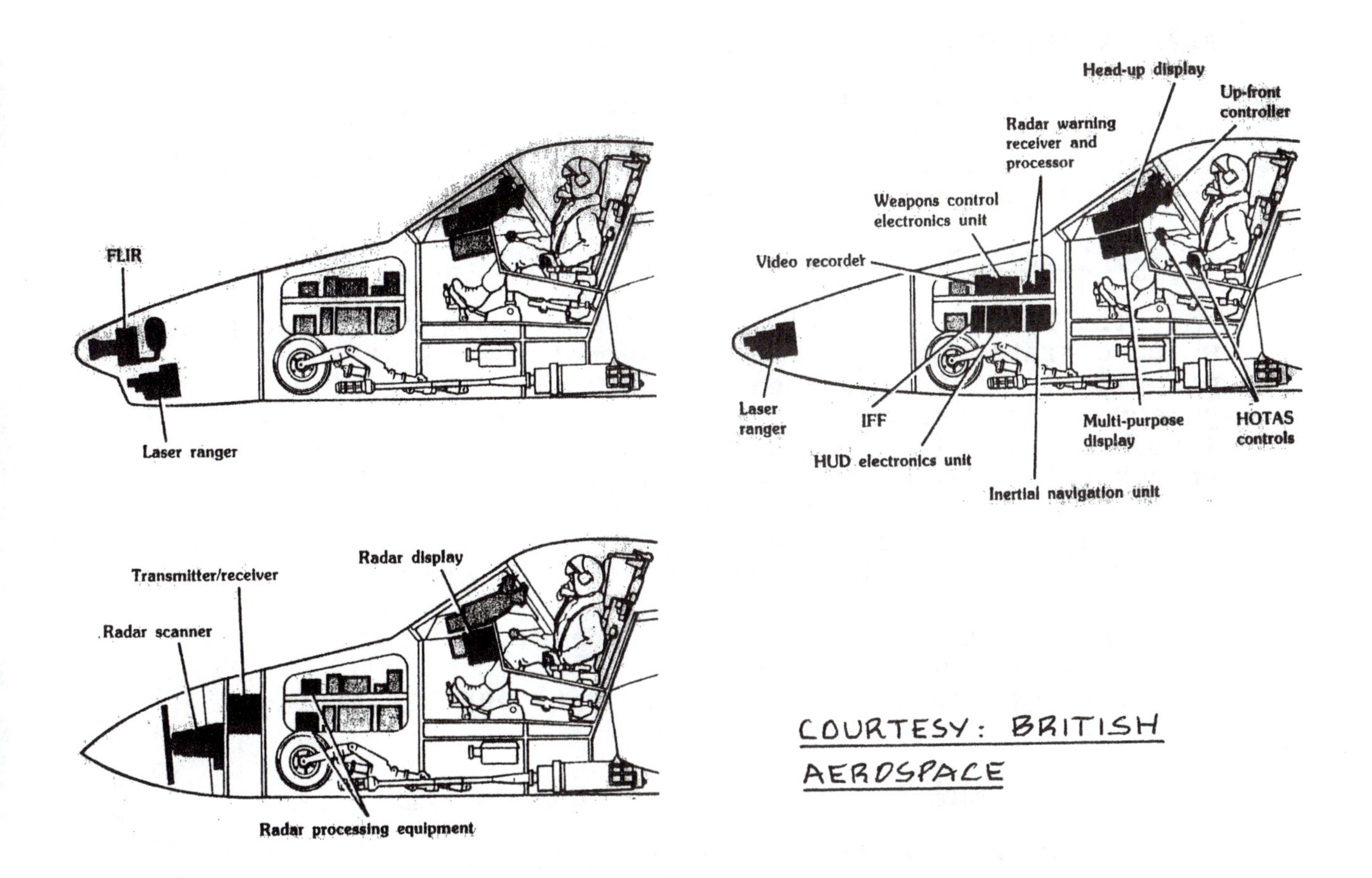

Figure 3.34 Avionics Options British Aerospace Hawk 200

3.5 WEAPON AND MILITARY PAYLOAD DATA

The purpose of this section is to provide a data base for a variety of weapons and other payloads carried by military airplanes.

The material is organized as follows:

3.5.1 Guns, gun pods, and rocket launchers

3.5.2 Free-fall munitions (bombs) and ejector racks

3.5.3 Missiles

3.5.4 External fuel stores

3.5.5 Special purpose stores

3.5.6 Military vehicles

3.5.1 Guns, Gun Pods, and Rocket Launchers

Figures 3.35 and 3.36 show dimensional data for rotating barrel guns used in military airplanes. A summary of the corresponding weight and ammunition data for these guns is given below:

GAU-2B/A 7.62mm Minigun (Fig.3.35a)

The General Electric GAU-2B/A is a six-barrel weapon which operates on the rotating barrel (Gatling) principle. It is intended for use primarily against personnel, trucks and other relatively light vehicles ('soft' targets). This gun is used on Cessna OA-37 light attack airplanes as well as on helicopters.

Weight: 67 lbs uninstalled.
Rate of fire: 4000-6000 RPM (rounds per minute), selectable.
Ammunition: NATO 7.62 mm standard.
Unit ammo weight: 0.053 lbs.
Average recoil force: 300 lbs at 6000 RPM firing rate.

GAU-4/A and M61A1 Vulcan 20mm Cannon (Fig.3.35b)

The General Electric Vulcan cannon operates on the same principle as the previously described Minigun. Ammunition is fed into the weapon chamber via a machanism that pulls the rounds from a drum. The rounds may be fed into the chamber by electrical or by hydraulic power.

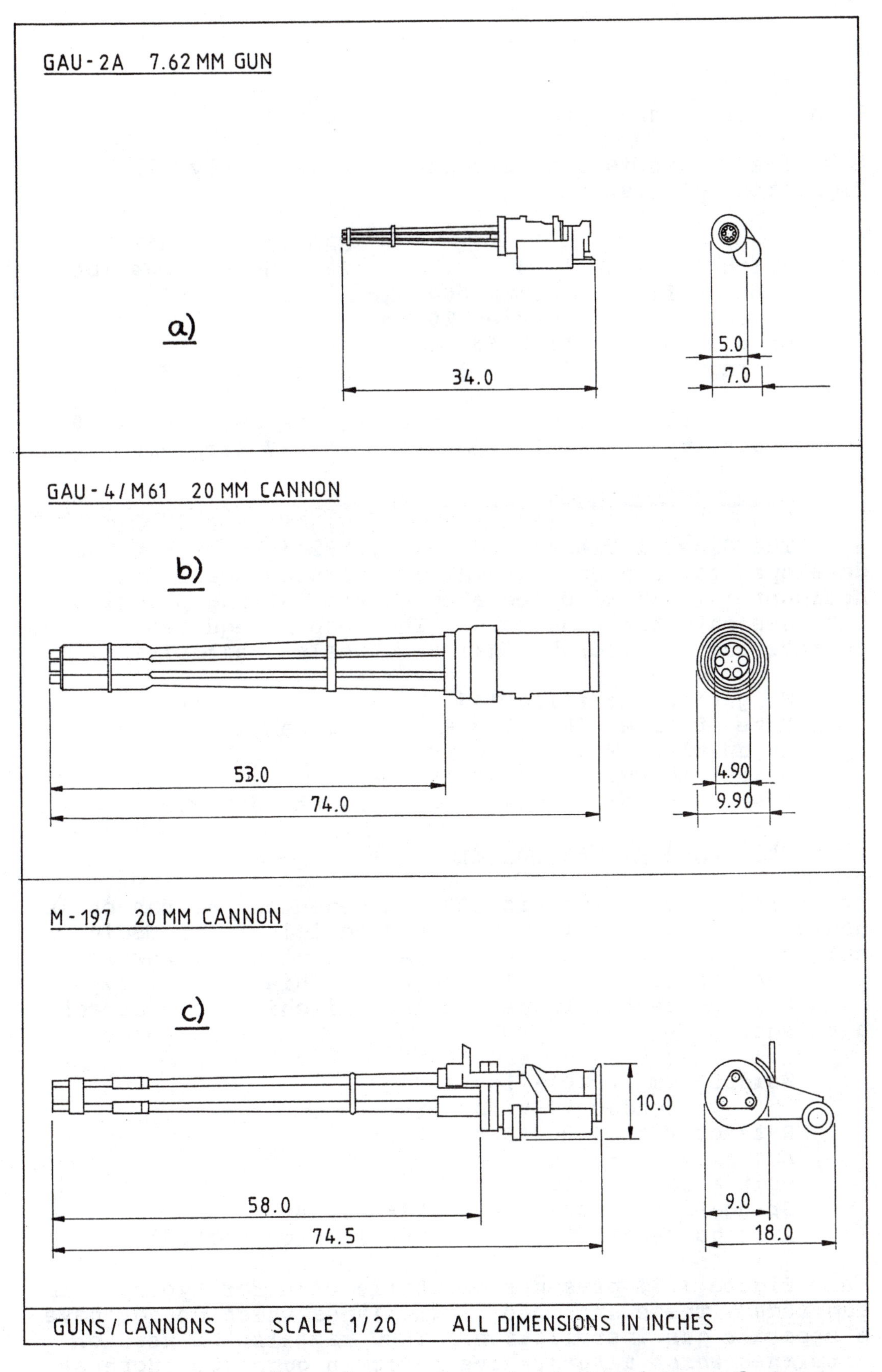

Figure 3.35 Dimensions of Typical Multi-Barrel Guns

Spent cases can be jettisoned or retained.

The Vulcan is standard armament in nearly all US fighter airplanes.

	GAU-4/A	M61A1
Weight: uninstalled	275 lbs	264 lbs

Rate of fire: maximum 6000 RPM
Ammunition: M50 series 20 mm
Unit ammo weight: 0.55 lbs
Average recoil force: 3,980 lbs at 6000 RPM.

A typical ammunition feed system weight with 1020 round capacity is 418 lbs for the LTV A-7 airplane.

M-197 20mm Cannon (Fig.3.35c)

The General Electric M-197 lightweight cannon was developed for use on light attack airplanes and for helicopters. It also operates on the Gatling principle but uses only three barrels. The weapon requires 3 hp to operate and can accept linkless or belted ammunition.

Weight: uninstalled 145 lbs
Rate of fire: 400-6000 RPM, selectable
Ammunition: M50 series 20 mm
Unit ammo weight: 0.55 lbs
Average recoil force: 1,500 lbs at 1500 RPM

GAU-8/A 30mm Cannon (Fig.3.36)

The General Electric GAU-8/A seven barrel cannon is designed for use against armoured vehicles with medium thickness armour. The cannon is hydraulically powered from the airplane hydraulic system. This Gatling type cannon is currently in use on the Fairchild A-10 attack airplane.

Weight: empty 2014 lbs
loaded 3799 lbs
Rate of fire: 6000 RPM
Ammunition: 30 mm
Unit ammo weight: 1.521 lbs
Standard ammo capacity: 1174 rounds
Average recoil force: 12,000 lbs (estimated)

Figure 3.37 presents geometric data for typical gun-pods. These are used in airplanes which do not have a separate gun installation. They may also be used on airplanes which already have built-in guns, to increase the firepower.

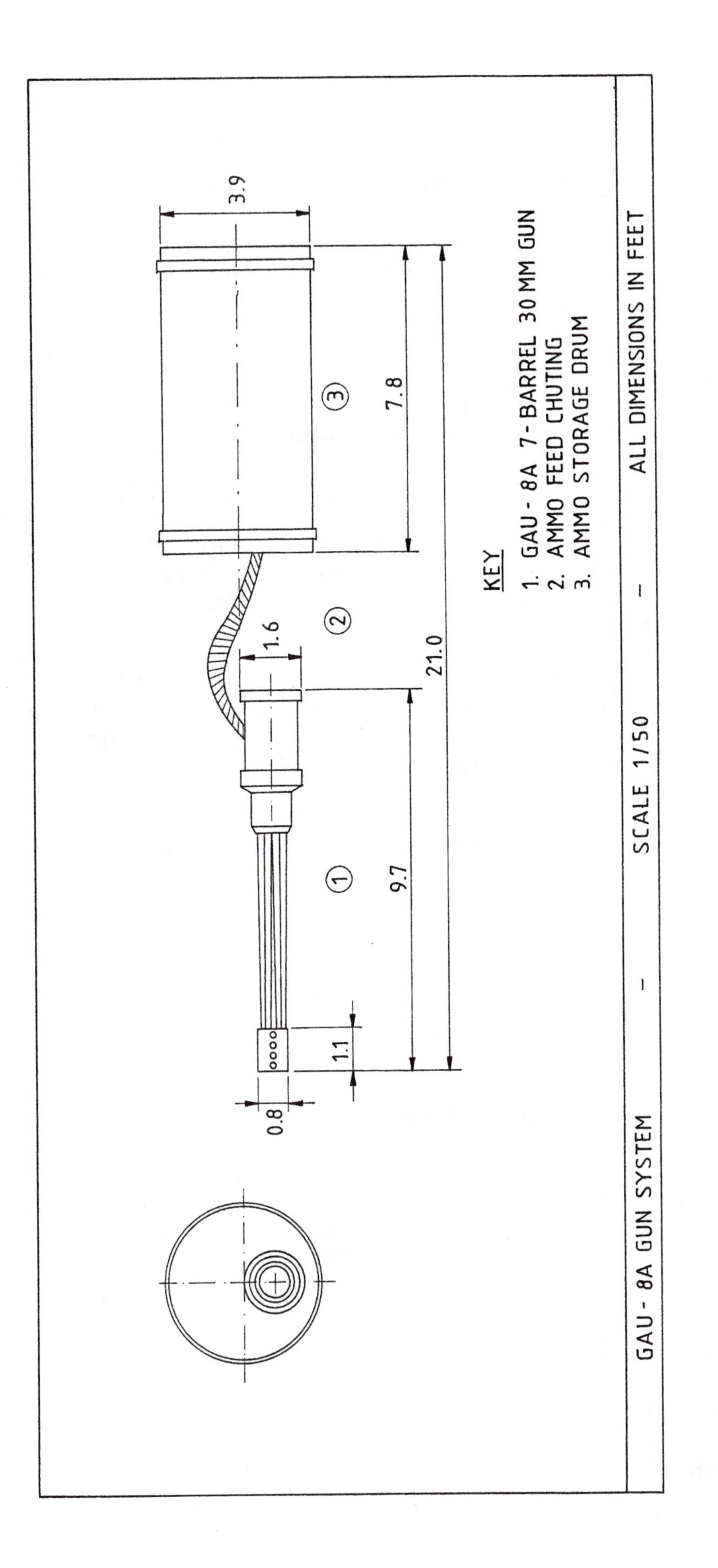

Figure 3.36 Dimensions of GAU-8A Gun System

SUU-11B/A 7.62mm Minigun Pod (Fig.3.37a)

This pod contains the GAU-2 minigun. An external power source is required. The pod is mounted on standard hard-points or on a pylon.

Weight: empty 245 lbs
loaded 324 lbs
Ammo: 7.62 mm NATO standard
Standard ammo capacity: 1,500 rounds

SUU-16/A and SUU-23/A 20mm Gun Pods (Fig.3.37b)

These pods are geometrically similar. The aft end of these pods contains a cylindrical ammo drum. The difference between these pods is that they carry different types of guns.

The SUU-16/A uses the externally powered M61A1 gun. The external power source can be in the form of a ram-air turbine which generates the required electrical power. An airspeed of 350 kts is needed for achieving full firing performance.

The SUU-23/A uses the self-powered GAU-4 gun.

	SUU-16/A	SUU-23/A
Weight: empty	1,067 lbs	1,078 lbs
loaded	1,720 lbs	1,731 lbs

Ammo: 20 mm
Ammo unit weight: 0.55 lbs
Standard ammo capacity: 1,500 rounds

Mark-4 Mod 0 20mm Gun Pod (Fig.3.37c)

This gun pod is in use with the US Navy on F-4, A-4 and A-7 airplanes. The Mk.4 employs a twin barrel (fixed) Mk.5 gun which has a selectable rate of fire from 750 to 4,200 RPM. Ammo is belt-fed into the weapon. The pod is powered by the gun exhaust gas.

Weight: empty 797 lbs
loaded 1,390 lbs
Ammo: 20 mm
Ammo unit weight: 0.55 lbs
Ammo capacity: 750 rounds
Ammo weight including links: 593 lbs

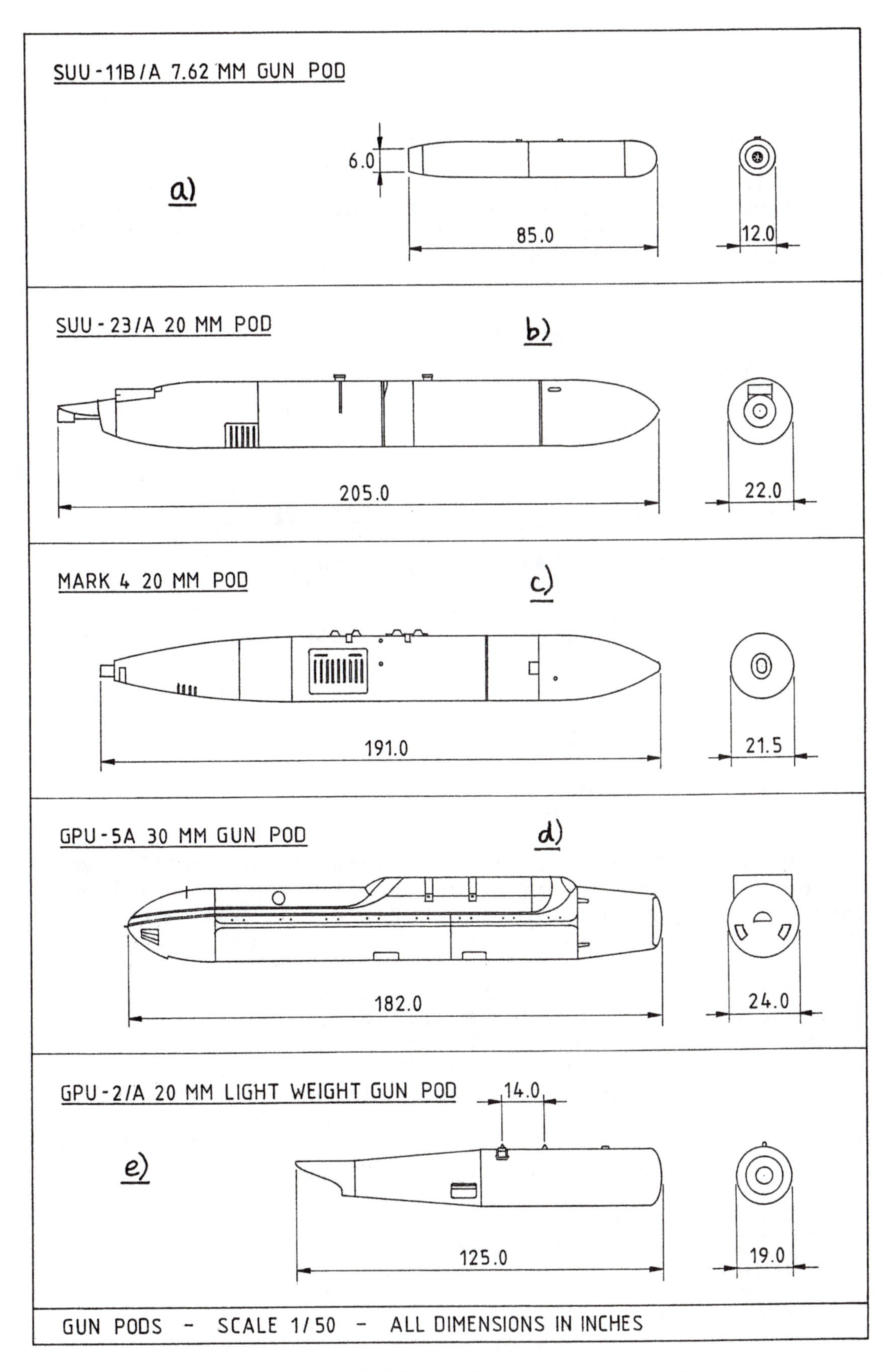

Figure 3.37 Dimensions of Gun Pods

GPU-5/A 30mm-Gun Pod (Fig.3.37d)

This pod is based on the GAU-13/A 5-barrel gun. The GAU-13 gun is a derivative of the GAU-8/A weapon. Anti-armor capability is similar to that of the GAU-8/A. This pod has been designed to be compatible with nearly all US attack airplanes.

Weight: empty 1,369 lbs
loaded 1,900 lbs
Ammo: 30 mm
Ammo unit weight: 1.79 lbs per round
Ammo capacity: 300 rounds
Firing rate: 2,400 RPM

GPU-2/A 20mm Lightweight Gun Pod (Fig.3.37e)

This pod is powered by a NiCad battery which provides power sufficient for three reloads (900 rounds of total ammo) before needing a recharge. The pod contains the M-197 gun which can be set to fire 750 or 1,500 RPM.

Weight: empty 433 lbs
loaded 595 lbs
Ammo: 20 mm
Ammo unit weight: 0.54 lbs
Ammo capacity: 300 rounds

Rockets fired from airplanes can be very effective in concentrating a large amount of fire power onto ground targets. In recent years most of these rockets are of the 'folding fin' type. The rocket fins are spring loaded and deployed immediately following their launch. The folding fin feature permits denser storage and lower installed drag. The rockets are normally installed in rocket launchers which can be dropped or retained.

Figure 3.38 shows two rocket launchers used on USAF airplanes.

LAU-68 7 Round 2.75 in. Rocket Launcher

The LAU-68 is the standard 7 round rocket launcher in the USAF. Figure 3.38a shows the geometry of this launcher.

Weight: empty 67 lbs
loaded 218 lbs

LAU-3 19 Round 2.75 in. Rocket Launcher

The LAU-3 is the standard 19 round rocket launcher for the USAF. Fighter airplanes may carry several of these rocket launchers. The 19 rounds can be fired within 2 seconds. Figure 3.38b shows the geometry of this rocket launcher.

Weight: empty 78 lbs
loaded 415 lbs

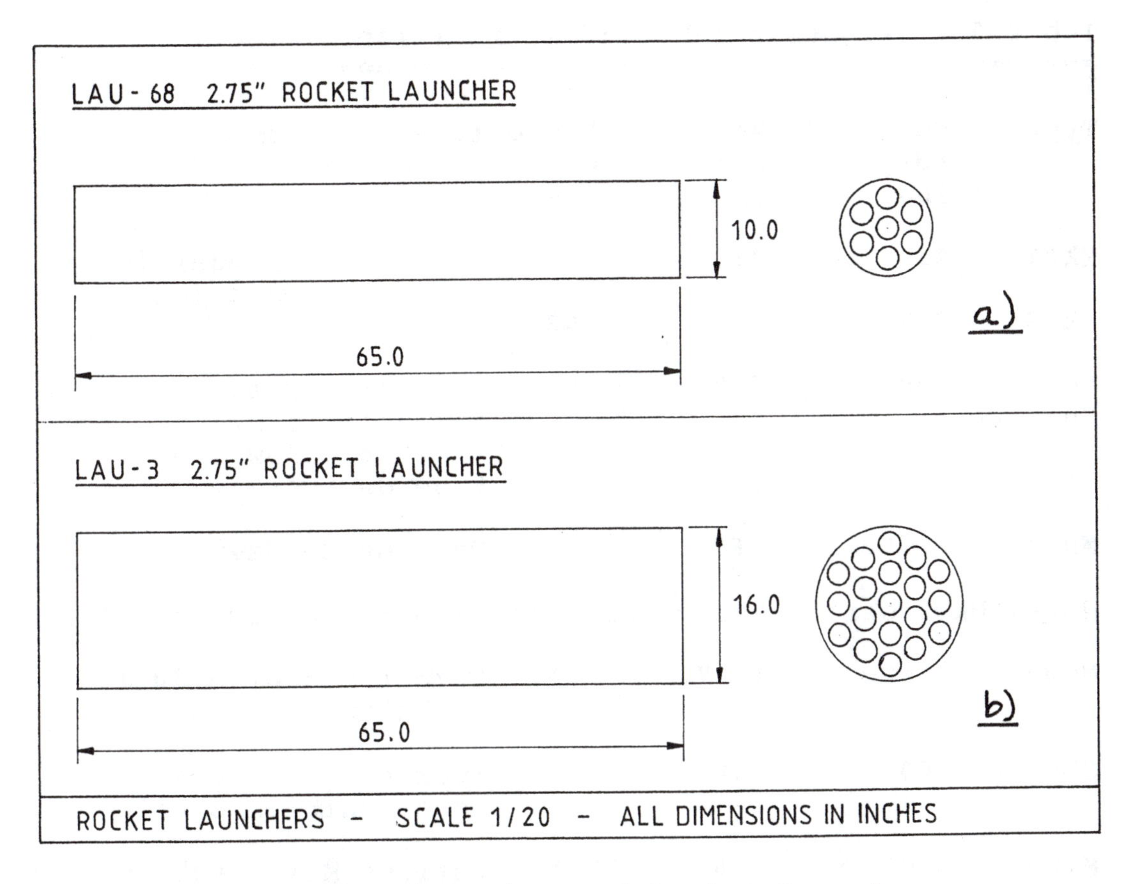

Figure 3.38 Dimensions of Rocket Launchers

3.5.2 Free-Fall Munitions (Bombs) and Ejector Racks

This class of munitions is subdivided into unguided and guided free-fall types. Both types are normally released from ejector racks. These racks can be mounted externally or internally.

Unguided Free-Fall Munitions

Free-fall munitions (also called bombs) are steel cased explosives designed for blast and fragmentation effects. A more recent version is the so-called cluster bomb, designed to disperse a large amount of 'bomblets'.

Figure 3.39 shows the geometries for these weapons. Table 3.1 lists the corresponding weights.

Table 3.1 Weights for Free-fall Munitions

Type	Nominal Weight lbs	Actual Weight lbs	Explosives Weight lbs	Comment
Mk81	250	270	96	No longer in production
Mk82	500	531	192	
Mk82-Snakeye	500	560	192	This high drag bomb uses the Mk15 retardation device to allow low altitude delivery.
Mk83	1,000	985	N.A.	Used by US Navy
Note: Mk80 series are compatible with 14 in. racks only.				
Mk84	2,000	1,970	N.A.	Compatible with 30 in. racks only
SUU-30	500	500*	N.A.	*varies with type of sub-munition carried
Mk20 Rockeye	500	476	N.A.	carries Mk118 sub-munition which can penetrate light to medium armor

Bombs, when carried internally or externally are mounted below racks from which they may be released by the pilot following an electrical command.

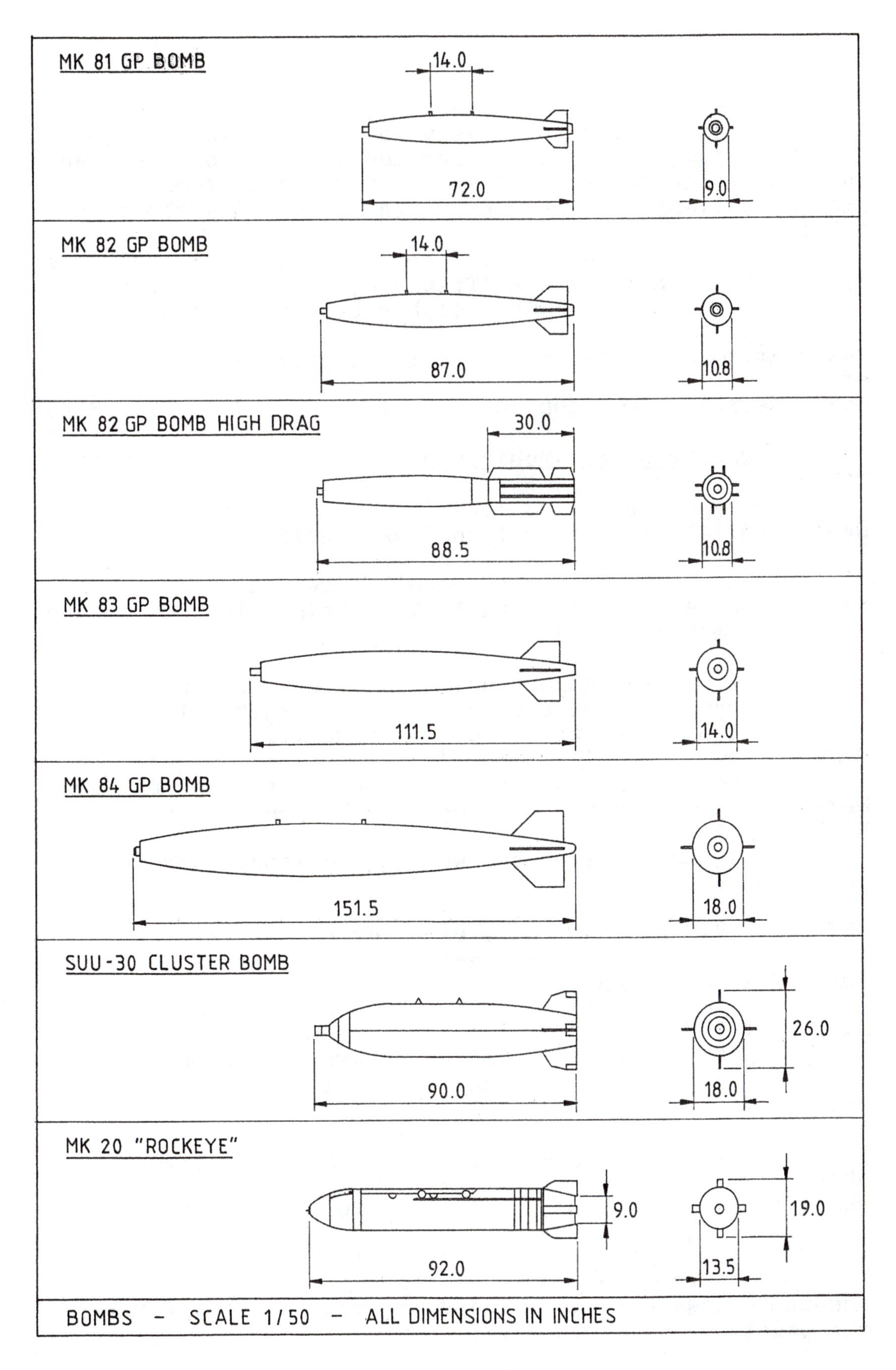

Figure 3.39 Dimensions of Free Fall Munitions (Bombs)

Bomb racks tend to be very 'draggy' as well as to present a large radar cross section (RCS). Conformal and internal installations tend to reduce these problems. Figures 3.4 and 3.34 show examples of such installations.

The most frequently used racks are the MER (Multiple Ejector Rack) and the TER (Triple Ejector Rack), both shown in Figure 3.40. The weights of these racks are:

MER weight: 220 lbs   TER weight: 93 lbs

Fig.3.41 shows EDO built 14 in. and 14/30 in. racks.

Guided Free-Fall Munitions

Two types of guided free-fall weapons will be presented: electro-optical and laser guided weapons.

Electro-optical guided weapons use one or more television cameras and data links to visually lock-on and home-to the target.

Laser guided weapons use the target reflection of a laser beam to lock-on and home-to the target. A device which 'illuminates' the target is required.

Figure 3.42 shows the geometries of several of these weapons. Most are based on Mk 80 series bombs.

Table 3.2 provides the weight and range data for these weapons.

Table 3.2 Weights and Range Data for Guided Free-Fall Munitions

Type	Based on	Weight lbs	Range miles	Comment
GBU-15	Mk 84 or SUU-54 Cluster	2,515	N.A.	Rockwell
AGM-62-Walleye I		1,100	10	Martin
AGM-62-Walleye II		2,340	10	Martin
GBU-10/B Paveway-II	Mk 84 series	2,052	N.A.	Texas Instr.
GBU-12/B	Mk 82	627	N.A.	Texas Instr.

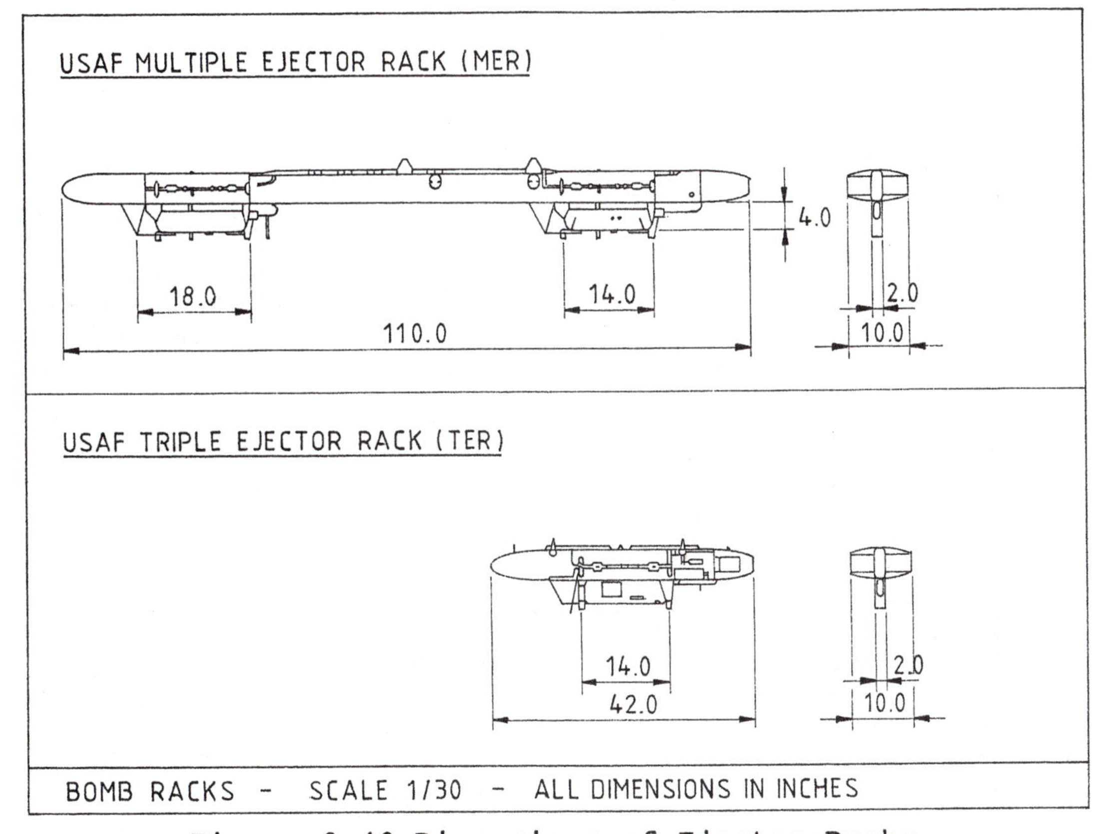

Figure 3.40 Dimensions of Ejector Racks

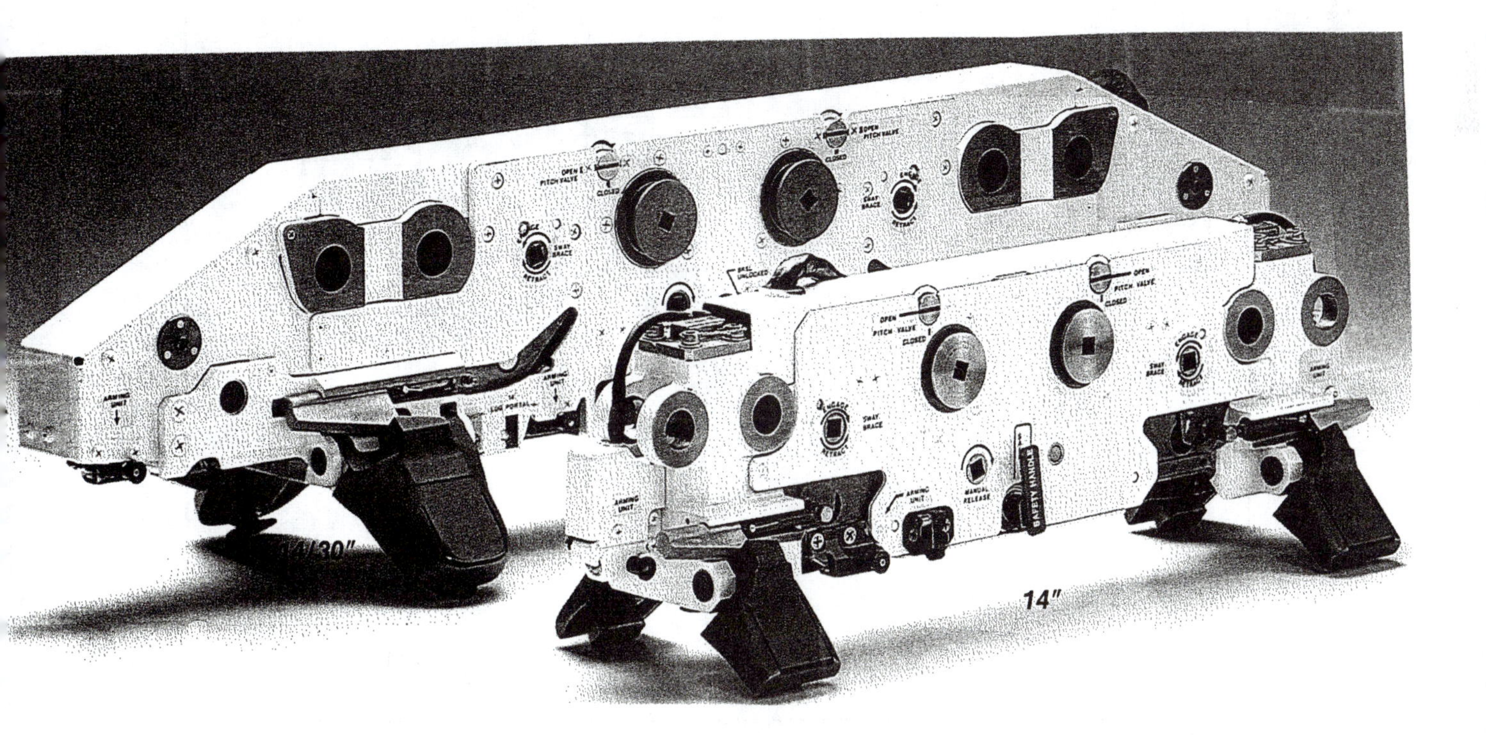

Figure 3.41 Perspective View of EDO Ejector Racks

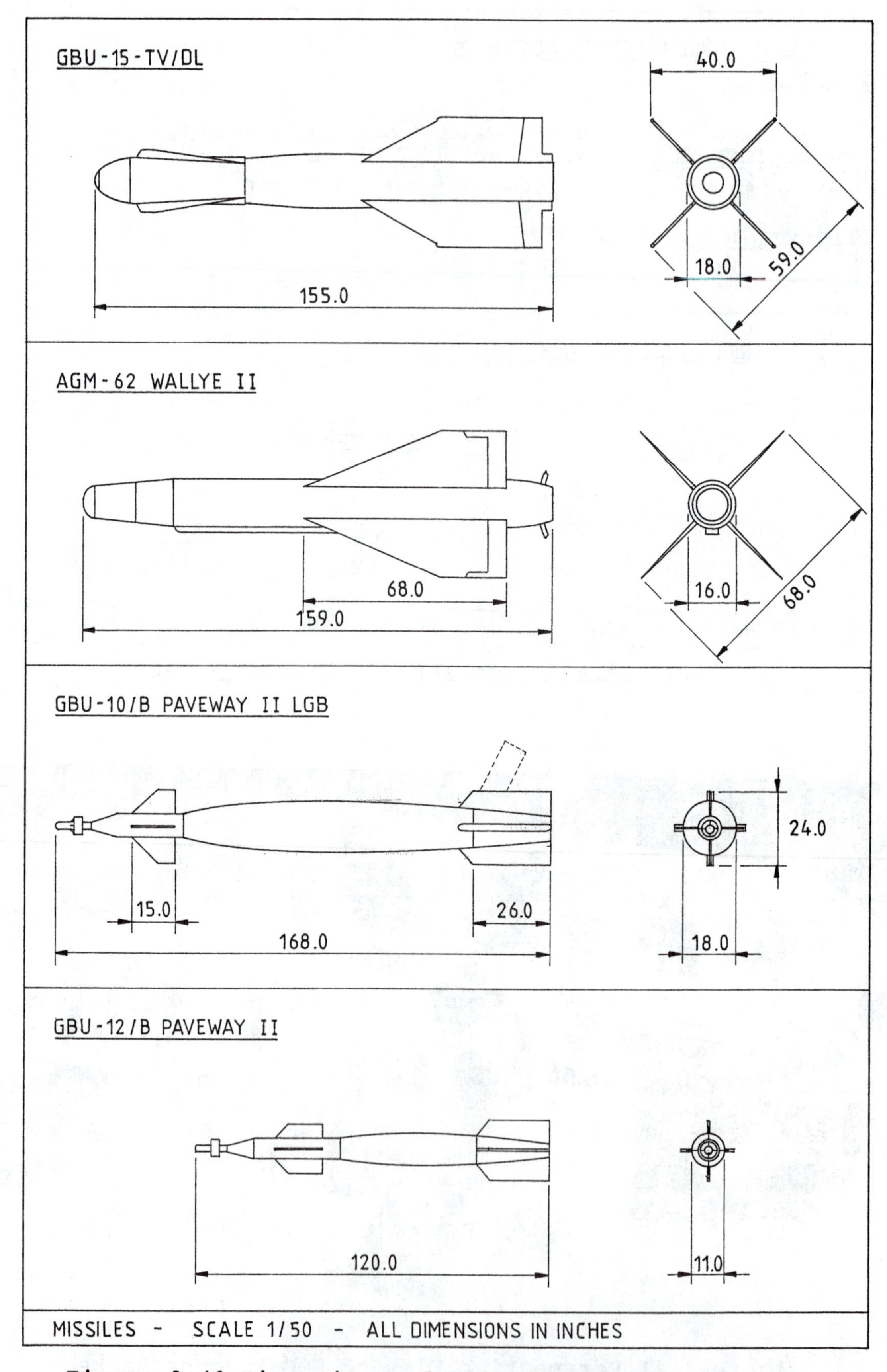

Figure 3.42 Dimensions of Guided Free Fall Munitions

3.5.3 Missiles

The data presented in this sub-section cover two types of missiles: Air-to-air missiles and Air-to-surface missiles.

Air-to-air Missiles

Air-to-air missiles are normally used in an air intercept role against enemy fighters, bombers and reconnaissance airplanes. For that reason they are also referred to as Air-Intercept-Missiles (AIM).

Air-to-air missiles are classified in terms of range and in terms of the guidance systems used:

Short range: less than 10 miles. These are mostly infrared seekers: passive guidance.

Medium range: 10-30 miles. These are mostly guided via radar illumination of the target: semi-active guidance.

Long Range: 30 miles or more. These are also mostly guided via radar illumination of the target: semi-active guidance.

Figure 3.43 contains geometric information for five air-to-air missiles. Table 3.3 provides weight and range data for these missiles.

Table 3.3 Weight and Range Data: Air-to-air Missiles

Type	Weight (lbs)	Range (miles)	Warhead (lbs)	Launch Method
Infrared Guided AIM's				
AIM-9J Sidewinder	170	6	N.A.	Rail
AIM-9L Sidewinder	191	10	25	Rail
Radar Guided AIM's				
AIM-7F Sparrow	514	24	66	Rail or eject.
AIM-54C Phoenix	1,000	120	132	eject.
AIM-120 AMRAAM	327	>100	N.A.	eject.

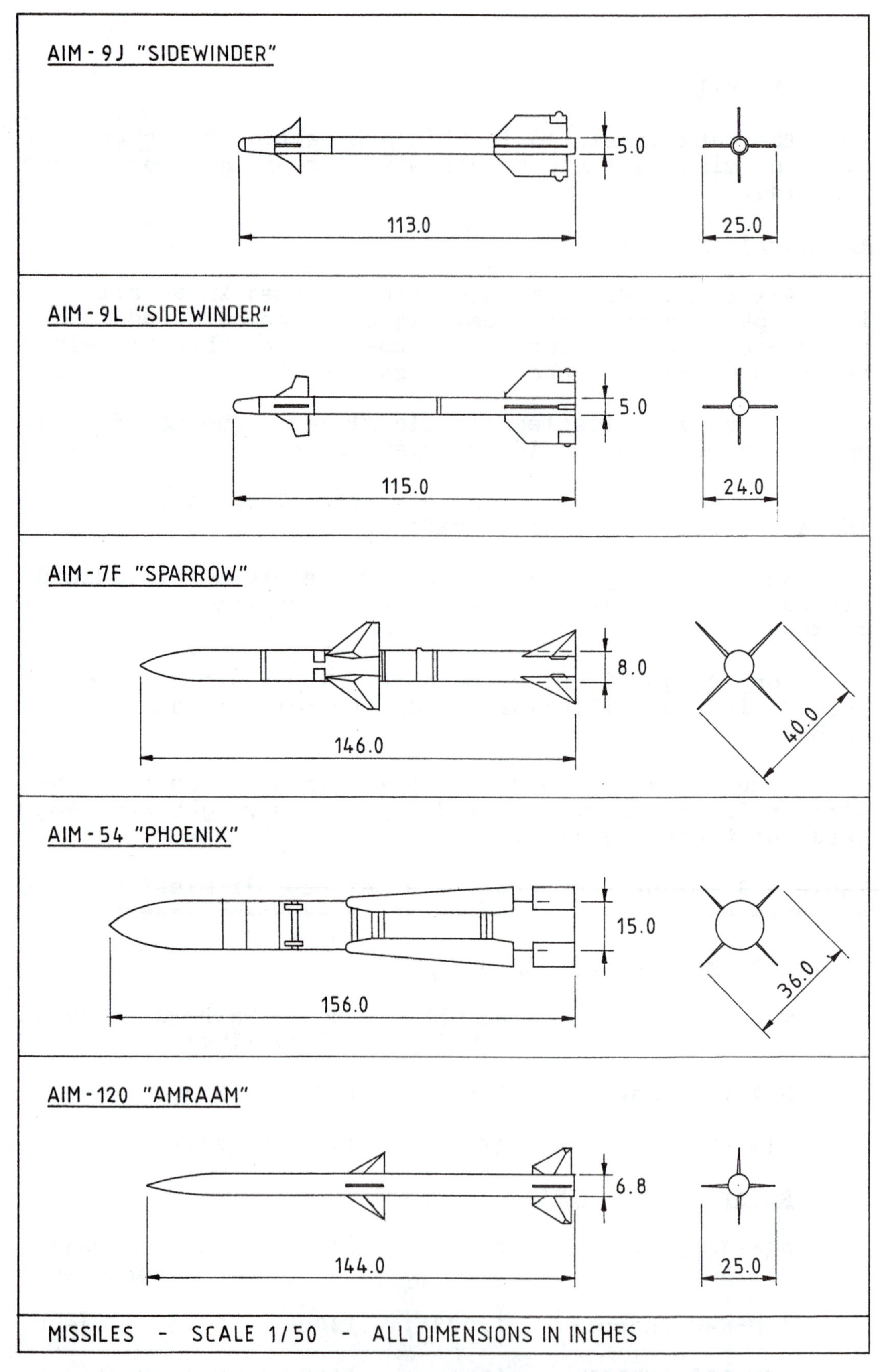

Figure 3.43 Dimensions of Air-to-air Missiles

Air-to-surface Missiles

Air-to-surface missiles are also referred to as Air-to-ground (AGM) missiles.

AGM's or air-to-ground missiles use a wide variety of guidance systems depending on range and purpose. Figure 3.44 presents the geometries of five such missiles. Table 3.4 contains weight and range data for these missiles.

Table 3.4 Weight and Range Data: Air-to-surface Missiles

Type	Guidance	Purpose	Weight	Warhead	Range
Maverick AGM-65A/B	TV	Anti-tank+ Anti-ship	463	125	N.A.
Maverick AGM-65C	Laser	same	463	not produced	
E			634	299	N.A.
Maverick AGM-65D	IFR	same	463	N.A.	N.A.
F			634	N.A.	N.A.
Standard ARM AGM-78D	Radar	Anti-radar	1,397	N.A.	>15.5
Shrike AGM-45	Radar	Anti-radar	400	145	10
HARM AGM-88A	Radar	Anti-radar	807	145	13
Hellfire AGM-114A	Laser	Anti-tank	95	N.A.	N.A.

For further information on missile systems the reader should consult Reference 19.

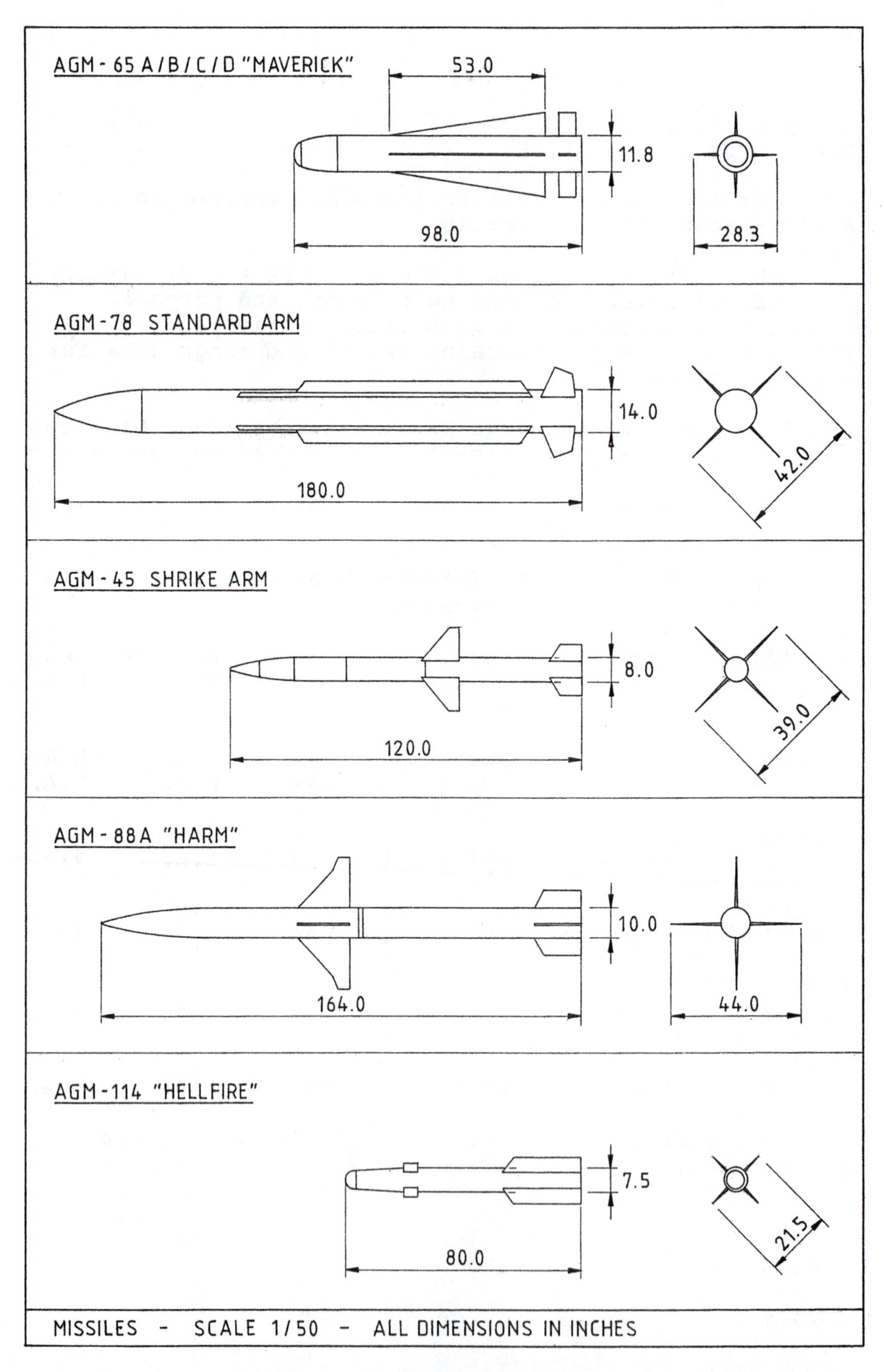

Figure 3.44 Dimensions of Air-to-surface Missiles

3.5.4 External Fuel Stores

For aerodynamic, structures and mission oriented reasons there is not much interchangeability between external fuel stores. As a general rule these external fuel stores are developed uniquely for each airplane.

In preliminary design it is useful to have data relating tank weights (empty) to usable tank volume. This information is given in Fig.3.45. Figures 3.46 and 3.47 define dimensions for five external fuel stores (tanks).

3.5.5 Special Purpose Stores

Most combat airplanes can be equipped with external stores to enhance their effectiveness in combat zones. These external stores can have several functions:

electronic warfare, navigation, offensive avionics and reconnaissance.

Figures 3.48-3.50 contain information on the geometries of special purpose stores. Table 3.5 provides weight and function data for these stores.

Table 3.5 Special Purpose Stores

Type	Purpose	Figure	Weight gross	empty
AN/ALQ-101 V-6	ECM	3.48a	400	
V-10	ECM	3.48b	540	
AN/ALQ-119 V-15	ECM	3.48c	580	
AN/ALQ-131 V	ECM	3.48d	675	
Pave Penny Sensor	Targetting (via HUD)	3.49a	32	
LANTIRN Pod 1	Targeting	3.49b	431	
Pod 2	Navigation	3.50a	544	
AN/ASQ-153 Pave Spike	Targeting (Laser)	3.49c	425	
AN/ALE 43	Chaff Pod	7.49d	534	184
Lundy Conformal	Chaff Pod+ IR decoys	7.50b	31.3	12

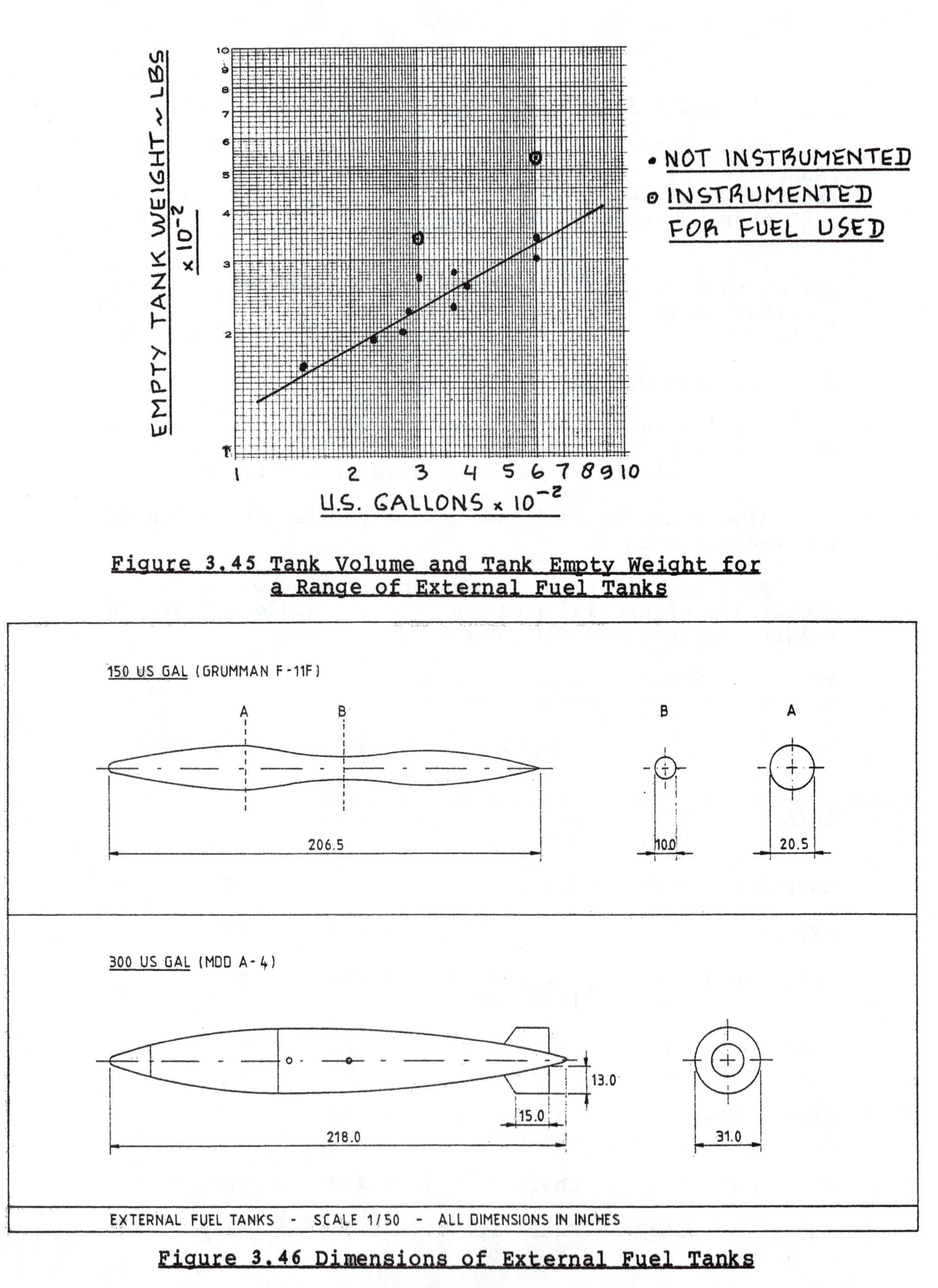

Figure 3.45 Tank Volume and Tank Empty Weight for a Range of External Fuel Tanks

Figure 3.46 Dimensions of External Fuel Tanks

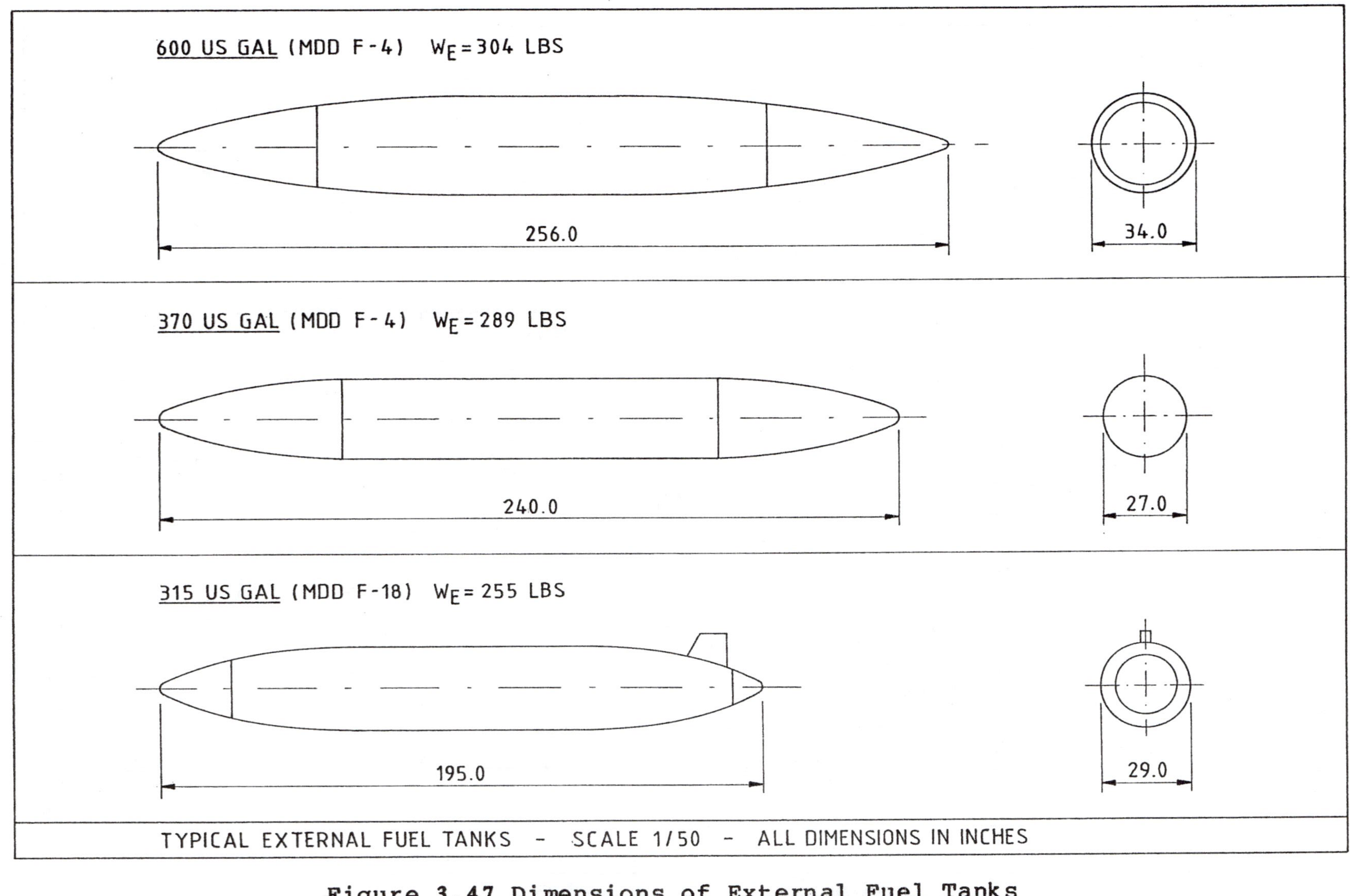

Figure 3.47 Dimensions of External Fuel Tanks

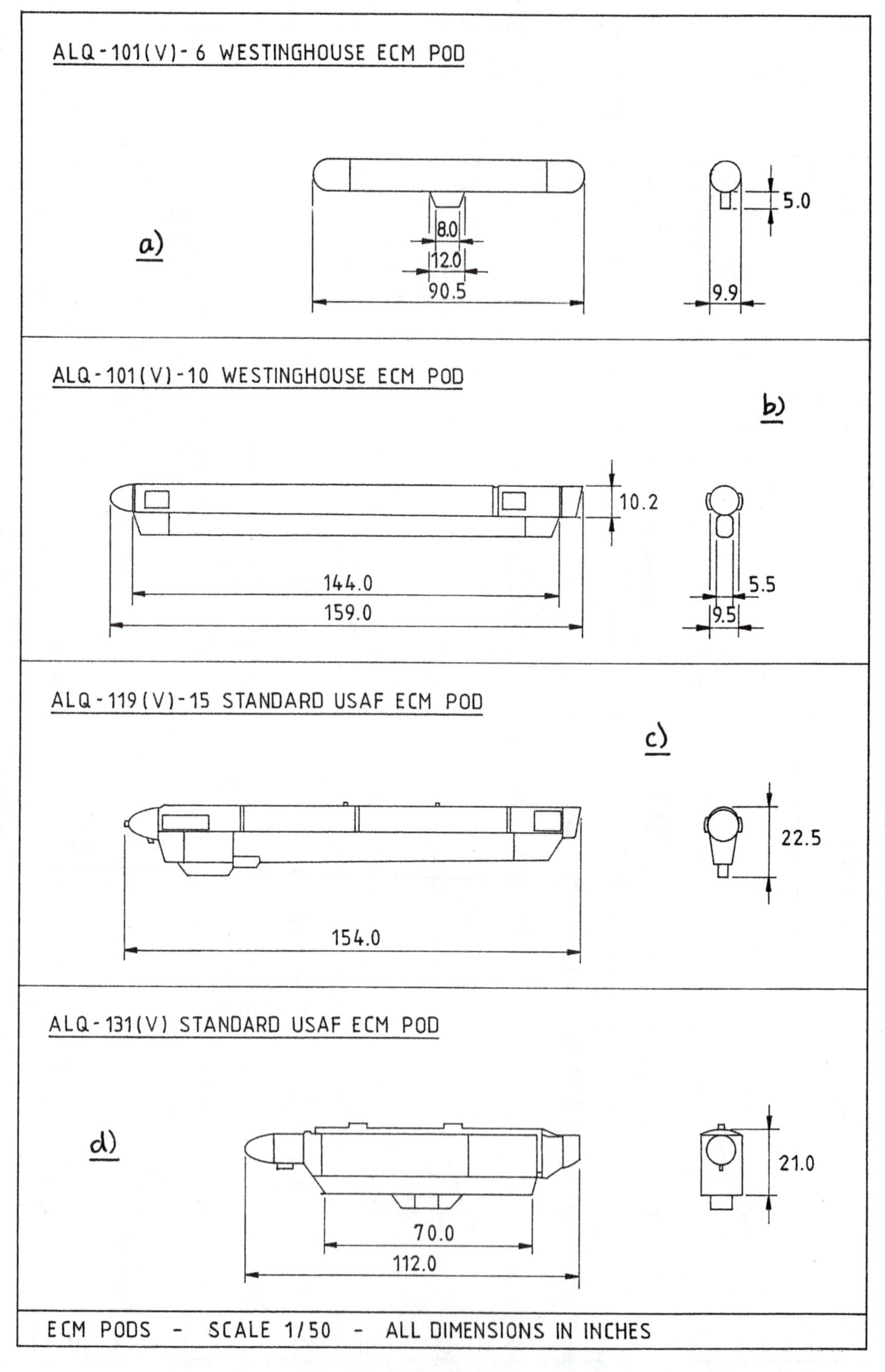

Figure 3.48 Dimensions of Special Purpose Pods

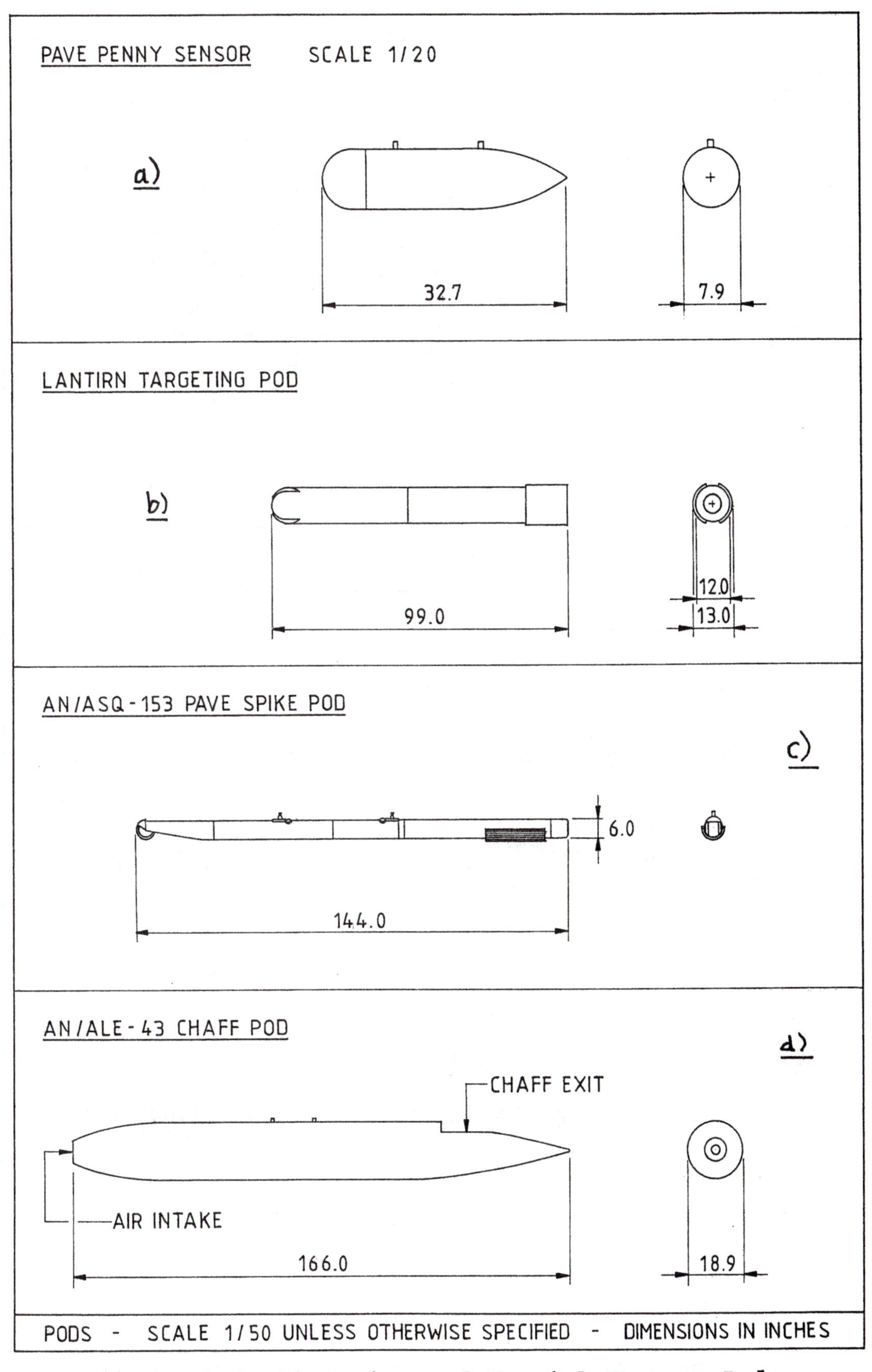

Figure 3.49 Dimensions of Special Purpose Pods

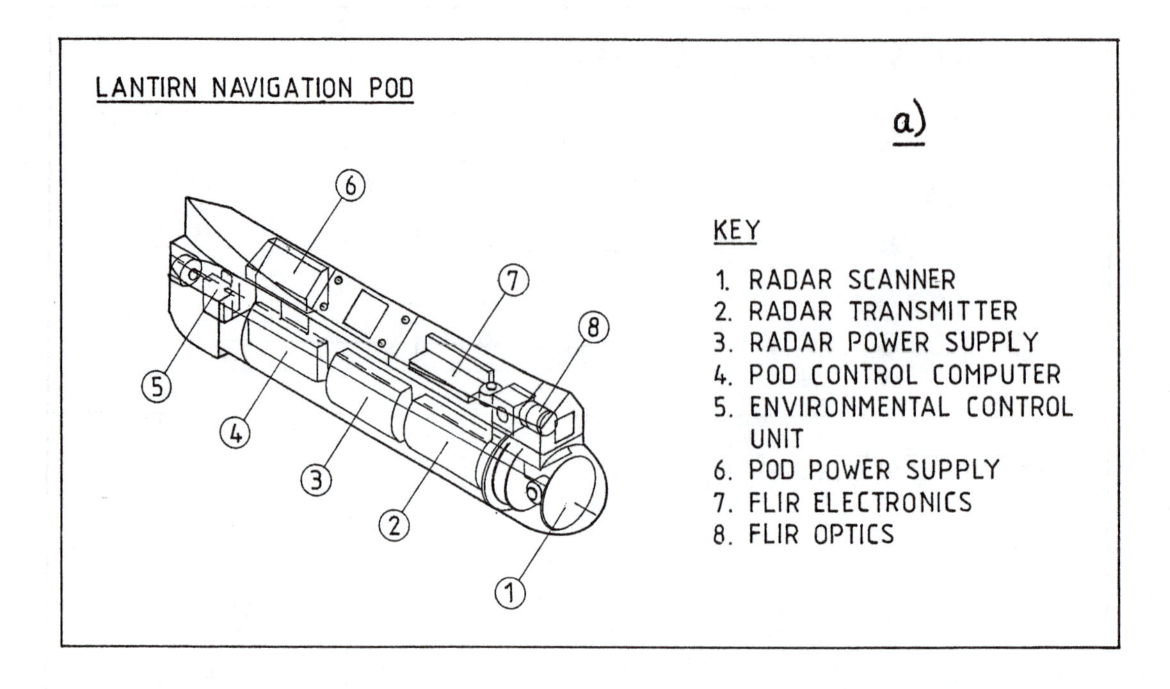

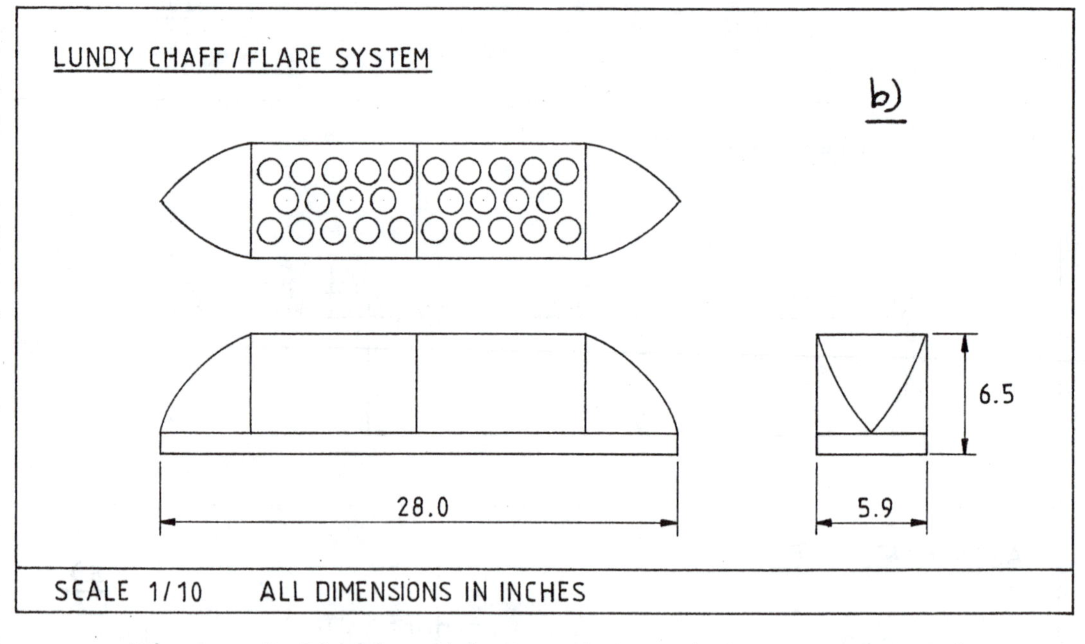

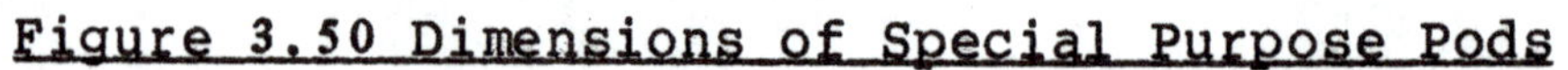

Figure 3.50 Dimensions of Special Purpose Pods

3.5.6 Military Vehicles

Military transport airplanes must be designed so that they can load, carry and unload a wide range of military vehicles. The number and type of vehicles to be carried are normally given in the mission specification of the airplane. The purpose of this sub-section is to present weight and geometric data on a range of large military vehicles. Reference 19 should be consulted for further data on military vehicles.

Table 3.6 lists the type, purpose and weight of each vehicle. Figures 3.51 through 3.58 provide geometric information for these vehicles.

Table 3.6 Typical Military Vehicles

Type	Purpose	Weight Loaded	Figure
M1 Abrams	Main battle tank	117,600*	3.51
M60	Battle tank	108,000	3.52
M2 Bradley	Fighting vehicle	50,000	3.53
M113	Personnel Carrier	22,520	3.54
M107(175mm)	Self-propelled Gun	62,027	3.55
M108(105mm)	Self-propelled Gun	49,458	3.56
M109(155mm)	Self-propelled Gun	52,428	3.57
M110(203mm)	Self-propelled Gun	62,370	3.58

*Empty weight: 100,000 lbs

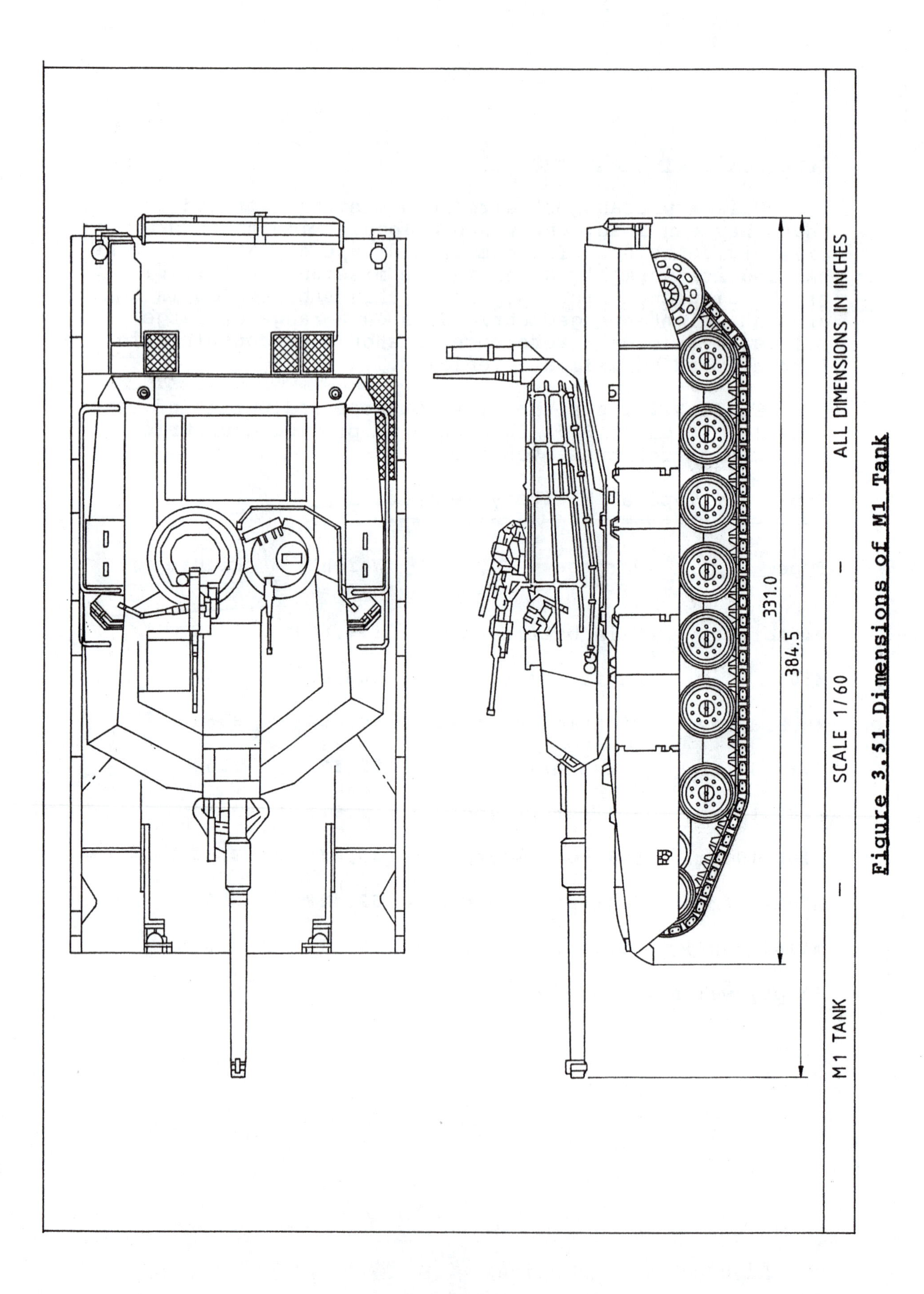

Figure 3.51 Dimensions of M1 Tank

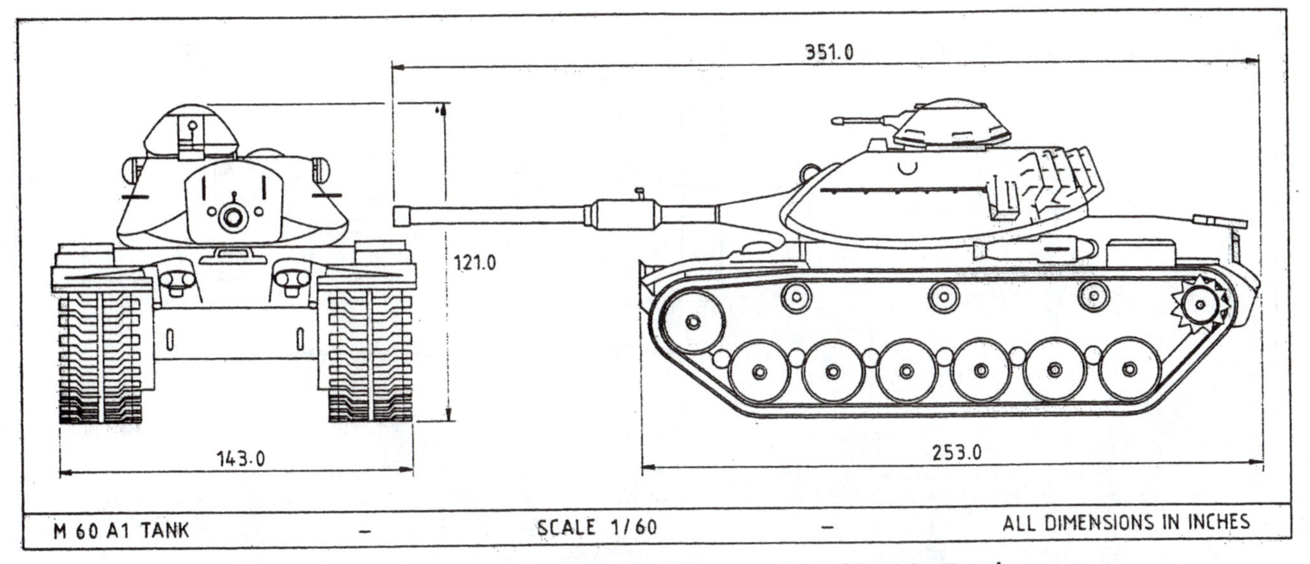

Figure 3.52 Dimensions of M60-A1 Tank

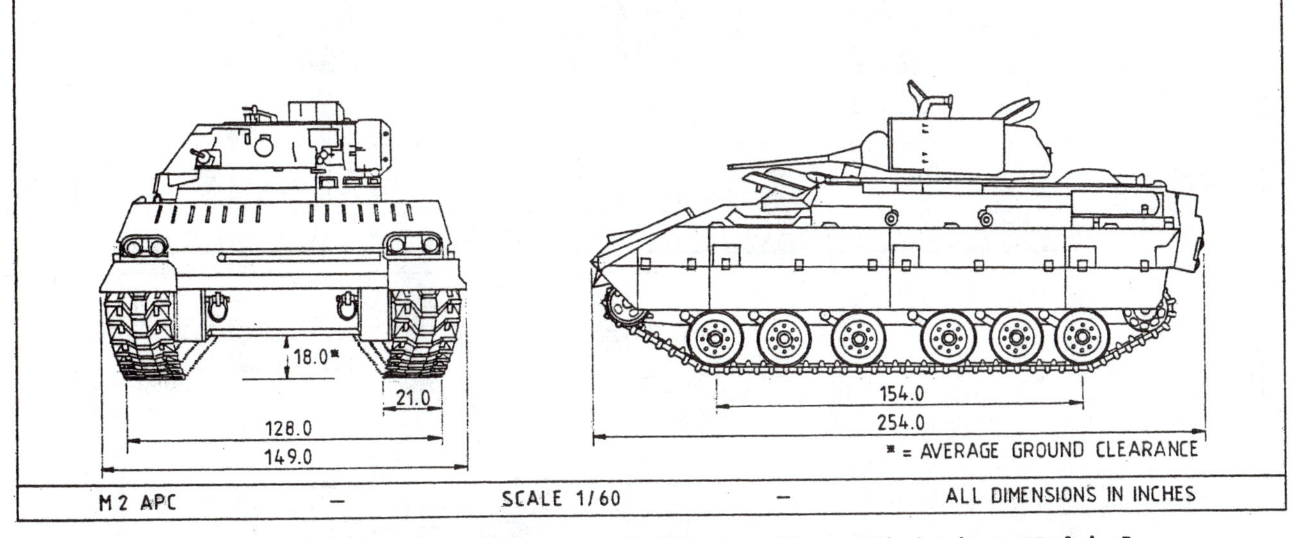

Figure 3.53 Dimensions of M2 Bradley Fighting Vehicle

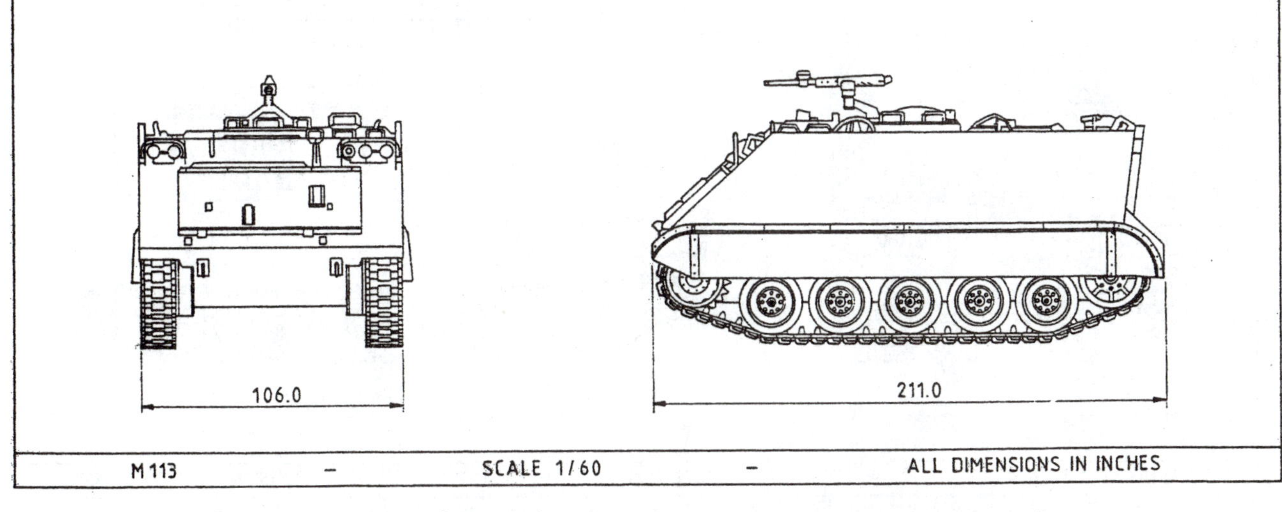

Figure 3.54 Dimensions of M113 Personnel Carrier

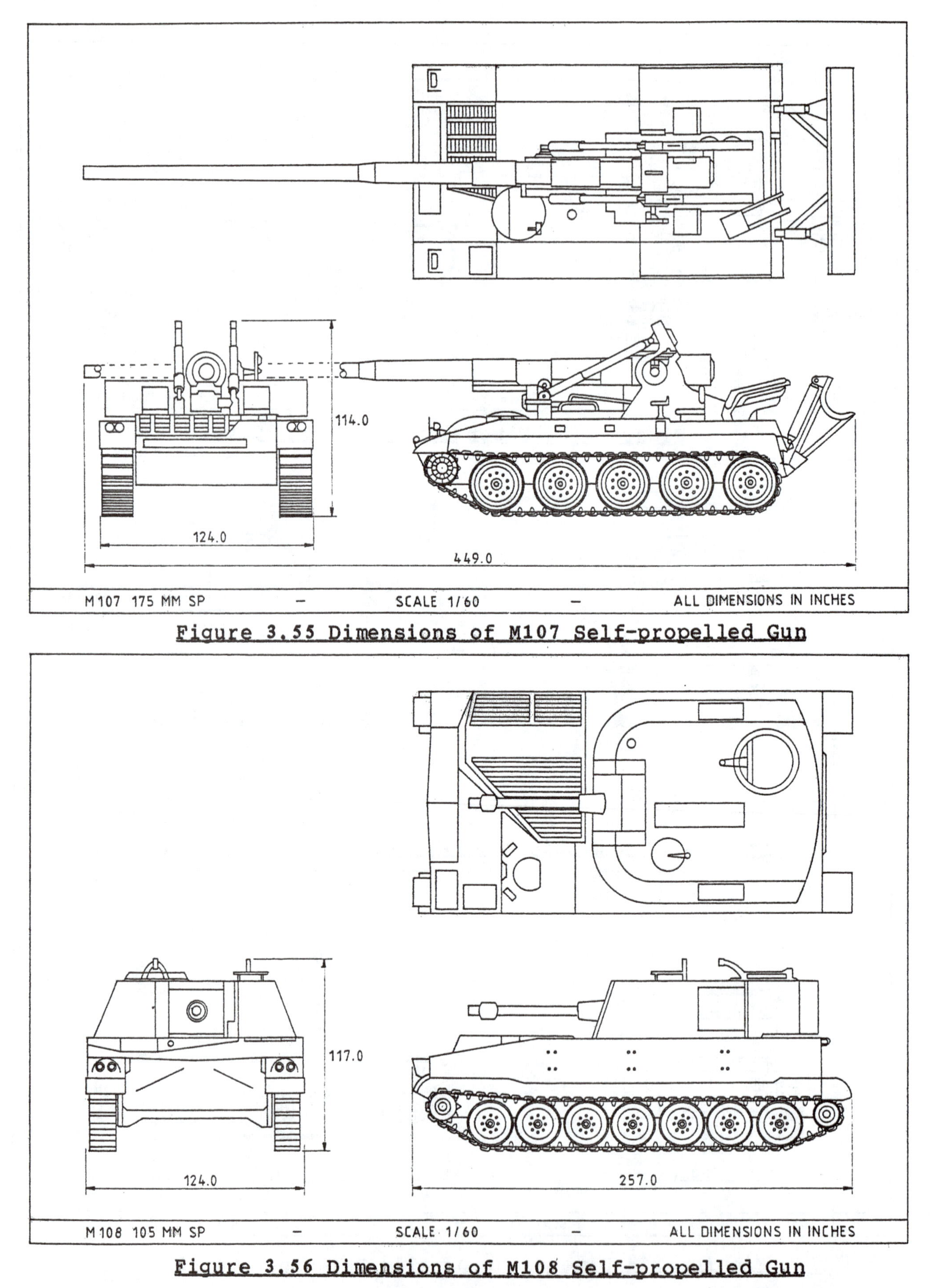

Figure 3.55 Dimensions of M107 Self-propelled Gun

Figure 3.56 Dimensions of M108 Self-propelled Gun

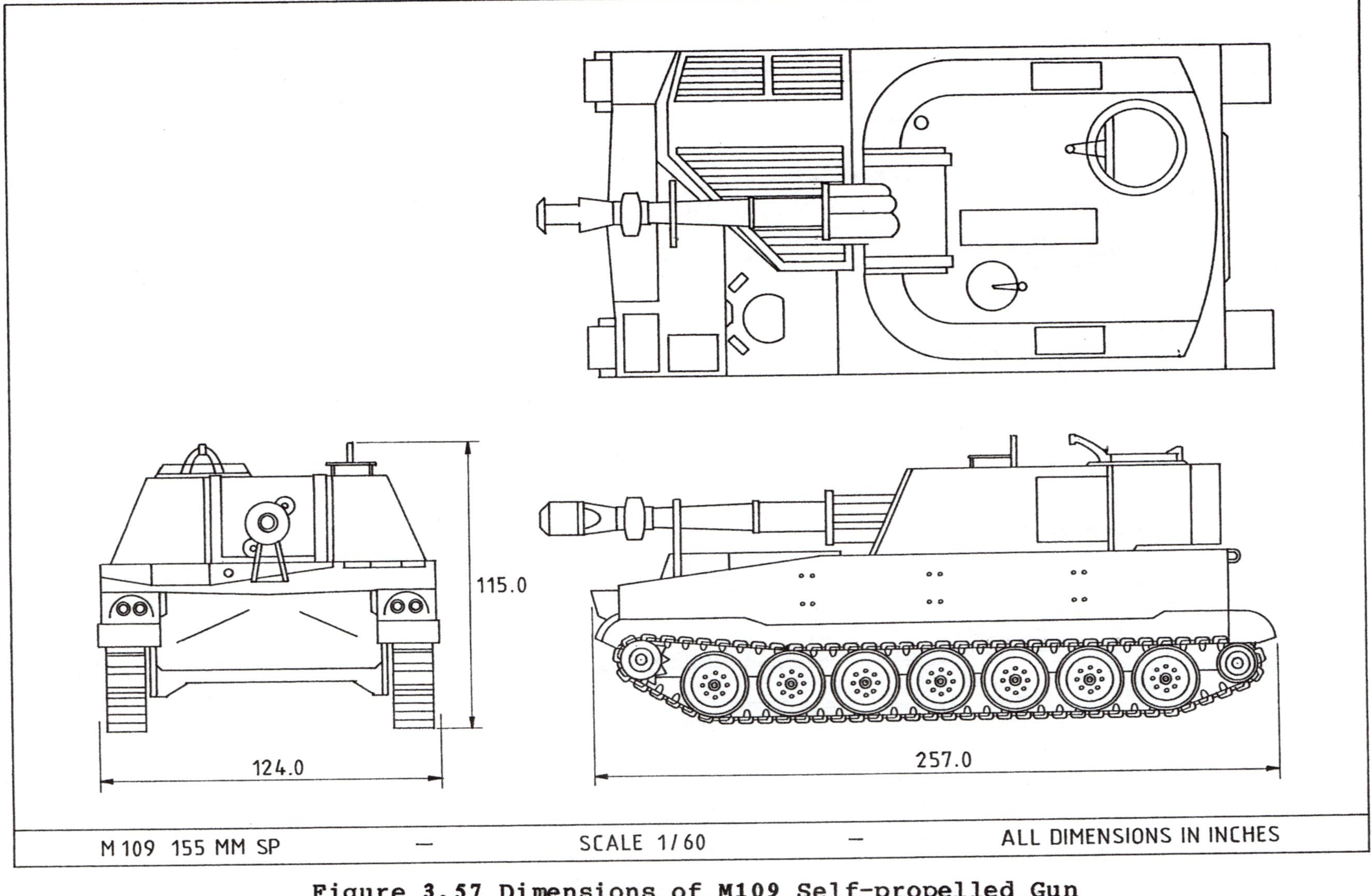

Figure 3.57 Dimensions of M109 Self-propelled Gun

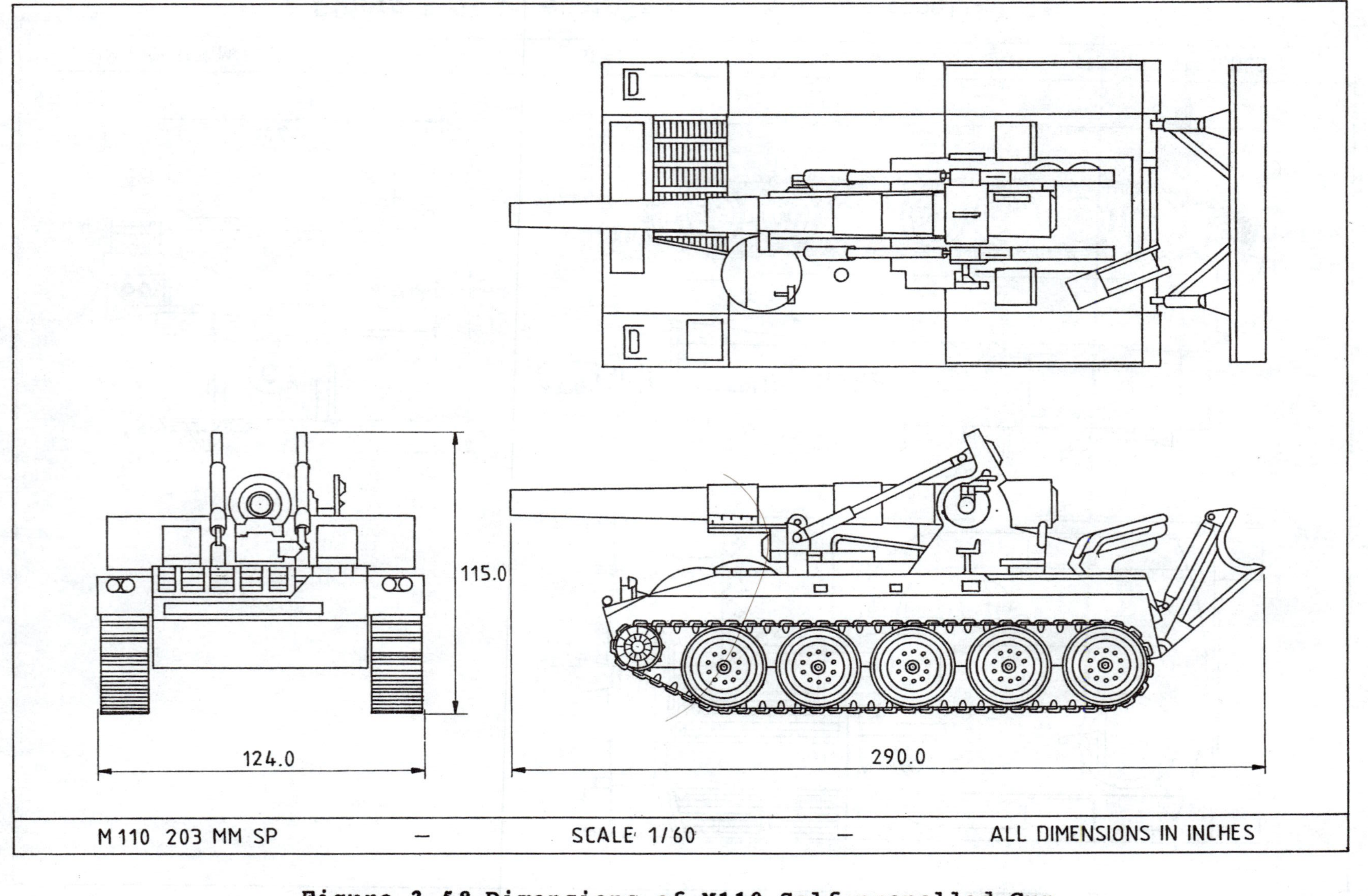

Figure 3.58 Dimensions of M110 Self-propelled Gun

4. FLIGHT CONTROL SYSTEM LAYOUT DESIGN

The purpose of this chapter is to discuss the preliminary layout design of flight control systems.

Flight control systems can be divided into: Primary and Secondary flight control systems.

Examples of primary flight control systems are:

Lateral Controls to:	Longitudinal Controls to:	Directional Controls to:
Ailerons, Spoilers Differential Stabilizer	Elevator Stabilizer Canard	Rudder

Examples of secondary flight control systems are:

Trim controls Controls to:	High lift Controls to:	Thrust (or Power) Controls to:
Lateral, Longitudinal and Directional primary flight controls	Trailing and/ or leading edge flaps	Engine fuel controls (Throttles), Manifold gates, Propeller blade incidence

In the case of automated and powered flight control systems the split between primary and secondary flight controls is not clear. In several fighter airplanes the thrust controls and the flight controls are in fact already integrated or combined with the primary (aerodynamic) flight controls. Examples are : the Harrier AV8B (reaction controls use bleed air from the compressor) and the F15 SMTD demonstrator have integrated aerodynamic and propulsion controls. Such integration of primary and secondary controls into one flight control system will probably be commonplace in most high performance airplanes.

Preliminary sizing of the aerodynamic flight controls is discussed in Chapters 6 and 8 of Part II. Preliminary sizing of high lift devices is discussed in Chapter 7 of Part II.

The preliminary layout design of flight control systems is presented in the following manner:

4.1 Layout of reversible flight control systems

4.2 Examples of reversible flight control systems

4.3 Layout of irreversible flight control systems

4.4 Examples of irreversible flight control systems

4.5 Trim systems

4.6 High lift control systems

4.7 Power control systems

4.1 LAYOUT OF REVERSIBLE FLIGHT CONTROL SYSTEMS

Definition: In a reversible flight control system, when the cockpit controls are moved, the aerodynamic surface controls move and vice versa.

Reversible flight control systems are typically mechanized with cables, push-rods or a combination thereof.

Major design problems associated with this type of flight controls are:

1. Friction 2. Cable stretch 3. Weight
4. Handling qualities 5. Flutter

Major advantages associated with this type of flight controls are:

1. Simplicity (= reliability) 2. Low cost
3. Relatively maintenance free

4.1.1 Reversible Lateral Flight Control Systems

Figure 4.1 shows several possible layouts for reversible lateral flight control systems.

Note: the lateral cockpit control movement must be consistent with the ranges specified for commercial and for military airplanes in Section 2.2 of Part III.

Credit: Most figures in this section were adapted from: Bok, F.P. and Beulink, D.G., Aircraft Mechanisms (in Dutch), Report: VTH-D22, June 1974, Delft Technological University, The Netherlands.

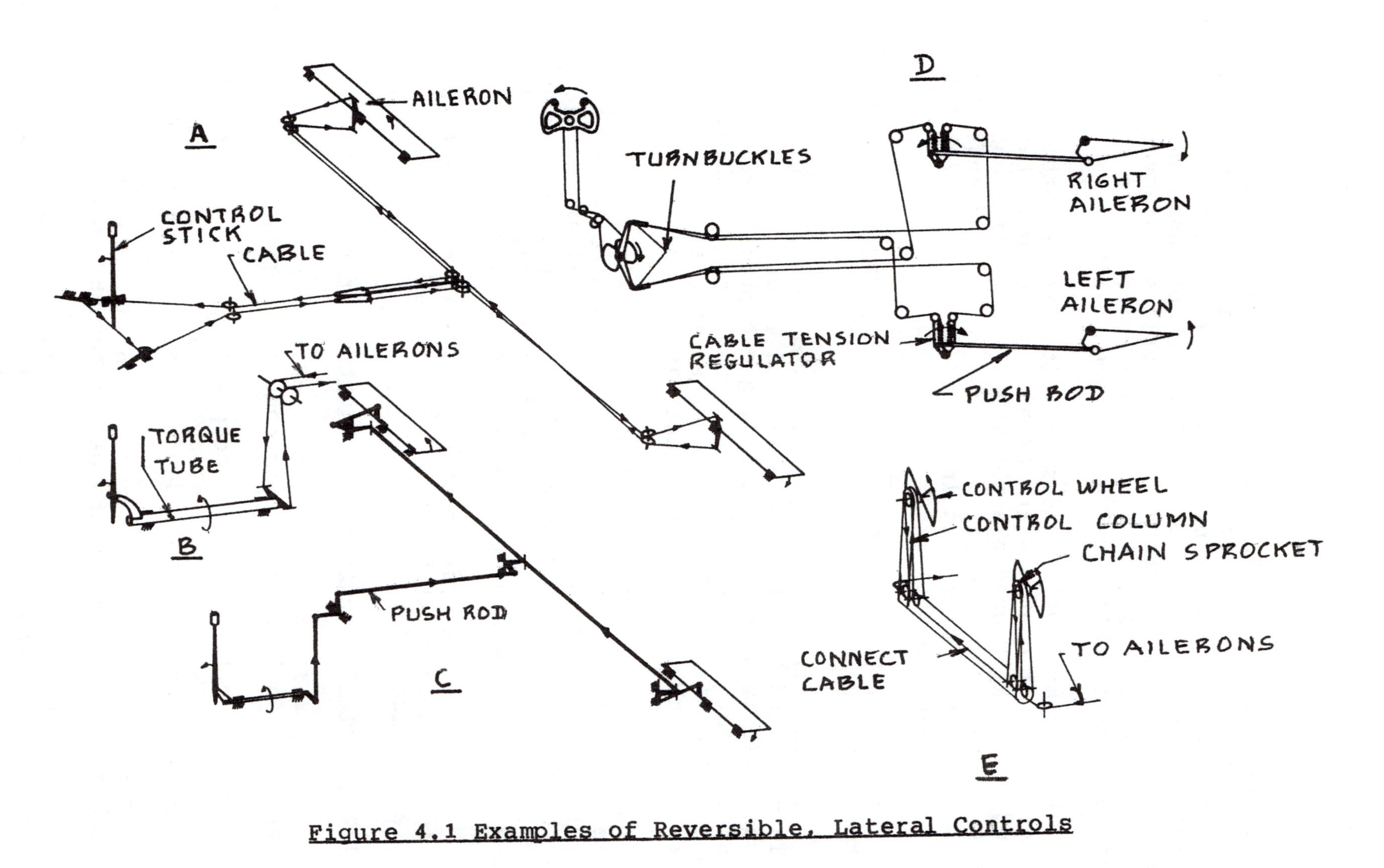

Figure 4.1 Examples of Reversible, Lateral Controls

System type A in Figure 4.1 is commonly used in light airplanes. The lateral stick motion is translated directly into cable motion. Figure 4.2 shows the stick-to-cable arrangement in more detail. Experience shows that this direct attachment method works satisfactorily only as long as: $(l/s) > 6$. If this criterion is violated, excessive cable stretch problems result.

In airplanes with tandem cockpit arrangements the lateral cockpit controls are often combined with the longitudinal cockpit controls through a concentric cable/tube arrangement. Figure 4.1B shows such an arrangement with more detail presented in Figure 4.3. The $(l/s) > 6$ criterion applies here also.

Figure 4.1C shows an example of a push-rod driven lateral control system.

In many airplanes a wheel mounted on a control column is used to actuate the lateral flight controls. Figure 4.1D is an example of a single control wheel installation. Figure 4.1E shows a dual control wheel installation. The latter are frequently used in airplanes with a side-by-side cockpit.

Most ailerons when deflected asymmetrically over the same angle, induce an adverse yawing moment onto the airplane. Ref.20 (Ch.4) contains a physical explanation of this effect. To counteract this adverse yawing moment the ailerons may be deflected 'differentially': Fig.4.4 shows a simple method for achieving this.

Note that the systems of Figures 4.1B through 4.1E avoid the cable stretch problem associated with the system of Figure 4.1A.

4.1.2 Reversible Longitudinal Flight Control Systems

Figure 4.5 presents a number of layouts used in reversible longitudinal flight controls systems.

Note: The longitudinal cockpit control movement must be consistent with the ranges specified for commercial and for military airplanes defined in Ch.2 of Part III.

The cable-crossing feature shown in Fig.4.5A is acceptable only as long as the cables do not rub together.

The system of Figure 4.5B employs a mixture of cables and push-rods. The system of Figure 4.5C uses only push-rods. Push-rod systems tend to have less

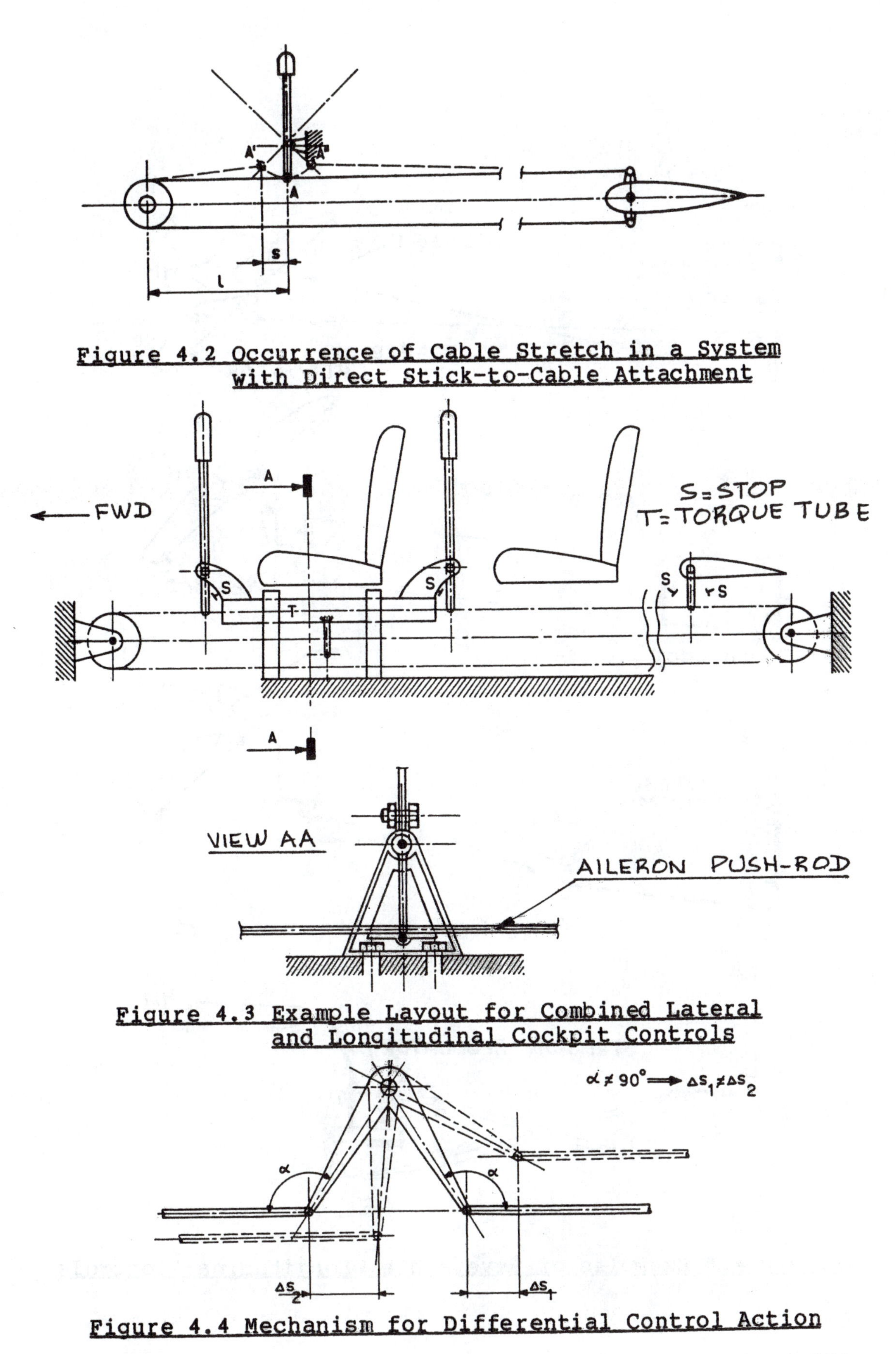

Figure 4.2 Occurrence of Cable Stretch in a System with Direct Stick-to-Cable Attachment

Figure 4.3 Example Layout for Combined Lateral and Longitudinal Cockpit Controls

Figure 4.4 Mechanism for Differential Control Action

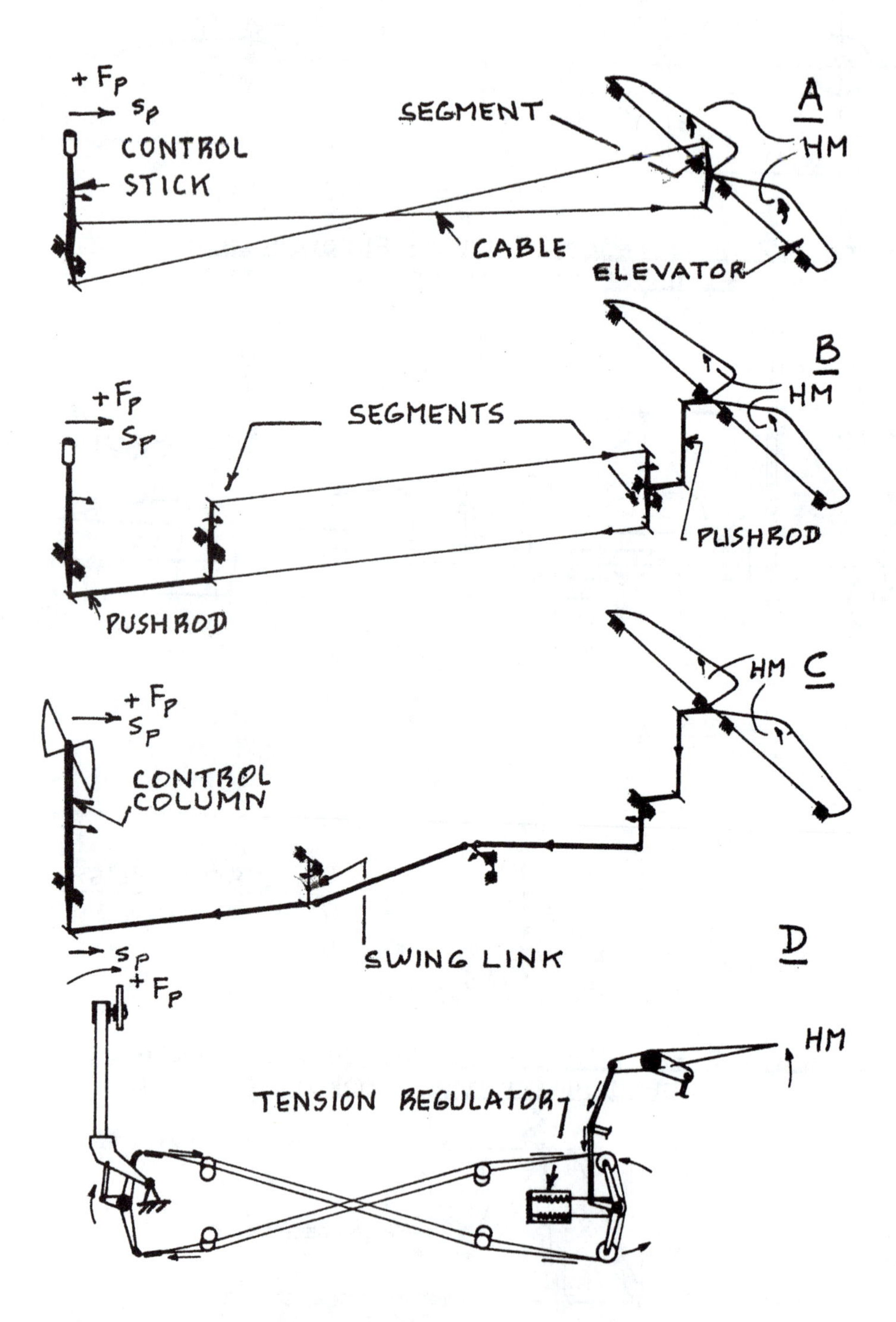

Figure 4.5 Examples of Reversible, Longitudinal Controls

friction than cable systems. They also tend to be a bit heavier.

Note the redundant cable system shown in Fig.4.5D: this type of cable redundancy is required in FAR25 certified airplanes only. In FAR23 airplanes single cable routings are acceptable.

4.1.3 Reversible Directional Flight Control Systems

Figure 4.6 shows examples of reversible directional flight control systems. The system of Figure 4.6A is found in many light airplanes.

Note: The directional cockpit control movement must be consistent with the ranges specified for commercial and for military airplanes defined in Section 2.2 of Part III.

Whenever dual cockpit controls are needed, the layouts shown in Figures 4.6B or C are employed. When redundancy is required, a cable layout such as shown in Figure 4.6D may be used.

In airplanes with V-tails (butterfly tails) the tail-mounted control surfaces normally serve both longitudinal and directional control functions. The separation of these functions is accomplished with a so-called mechanical mixer unit. Figure 4.7 shows example schematics of such mixer units.

4.1.4 Important Design Aspects of Reversible Flight Control Systems

In laying out a reversible flight control system, the following important design aspects need to be kept in mind:

*Mechanical design requirements for cable and for push-rod systems

*Efficiency considerations

*Cable and push-rod control force levels

*Control surface types and hinge moments

*Aerodynamic balance requirements

*Mass balance requirements

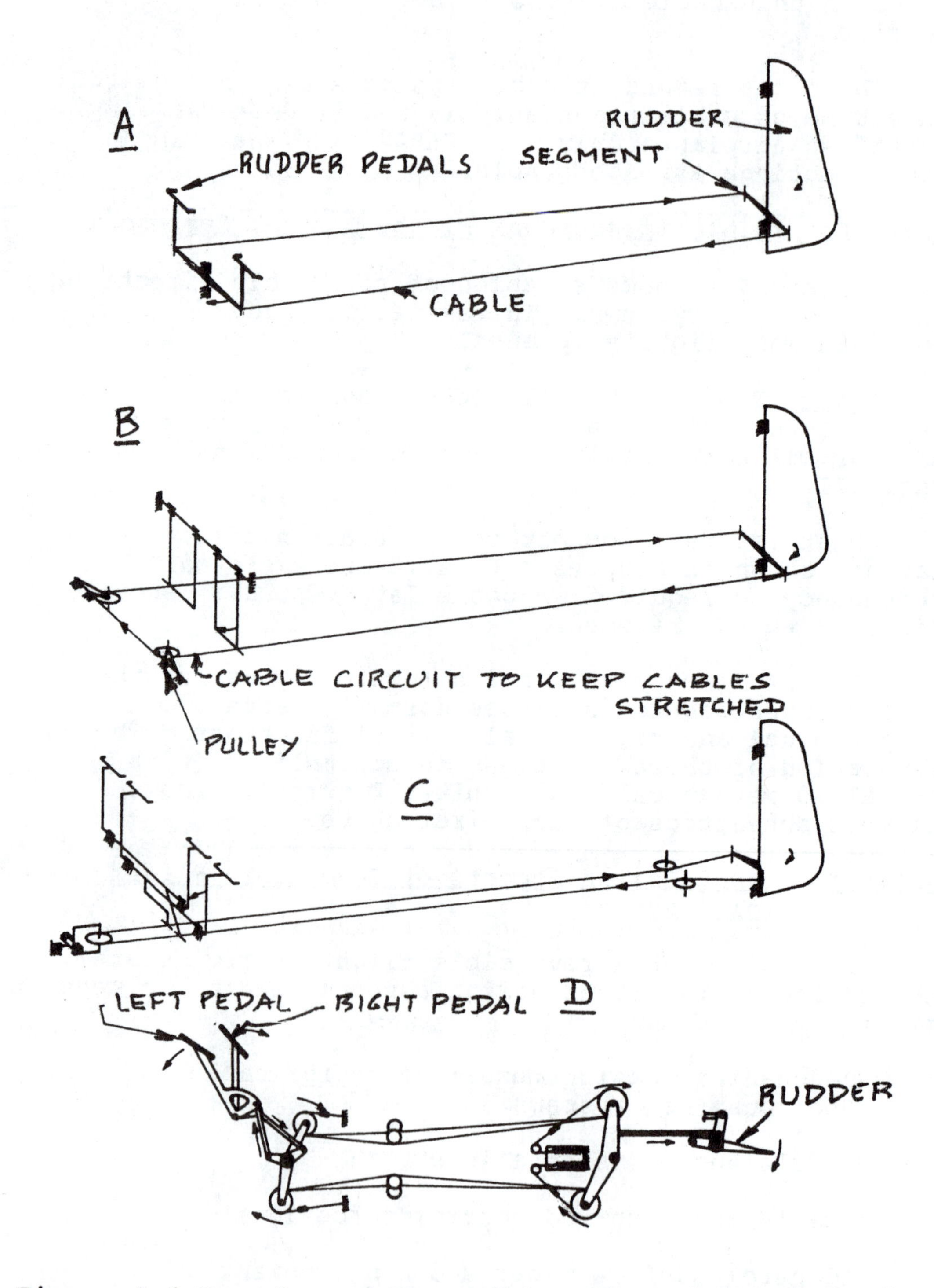

Figure 4.6 Examples of Reversible, Directional Controls

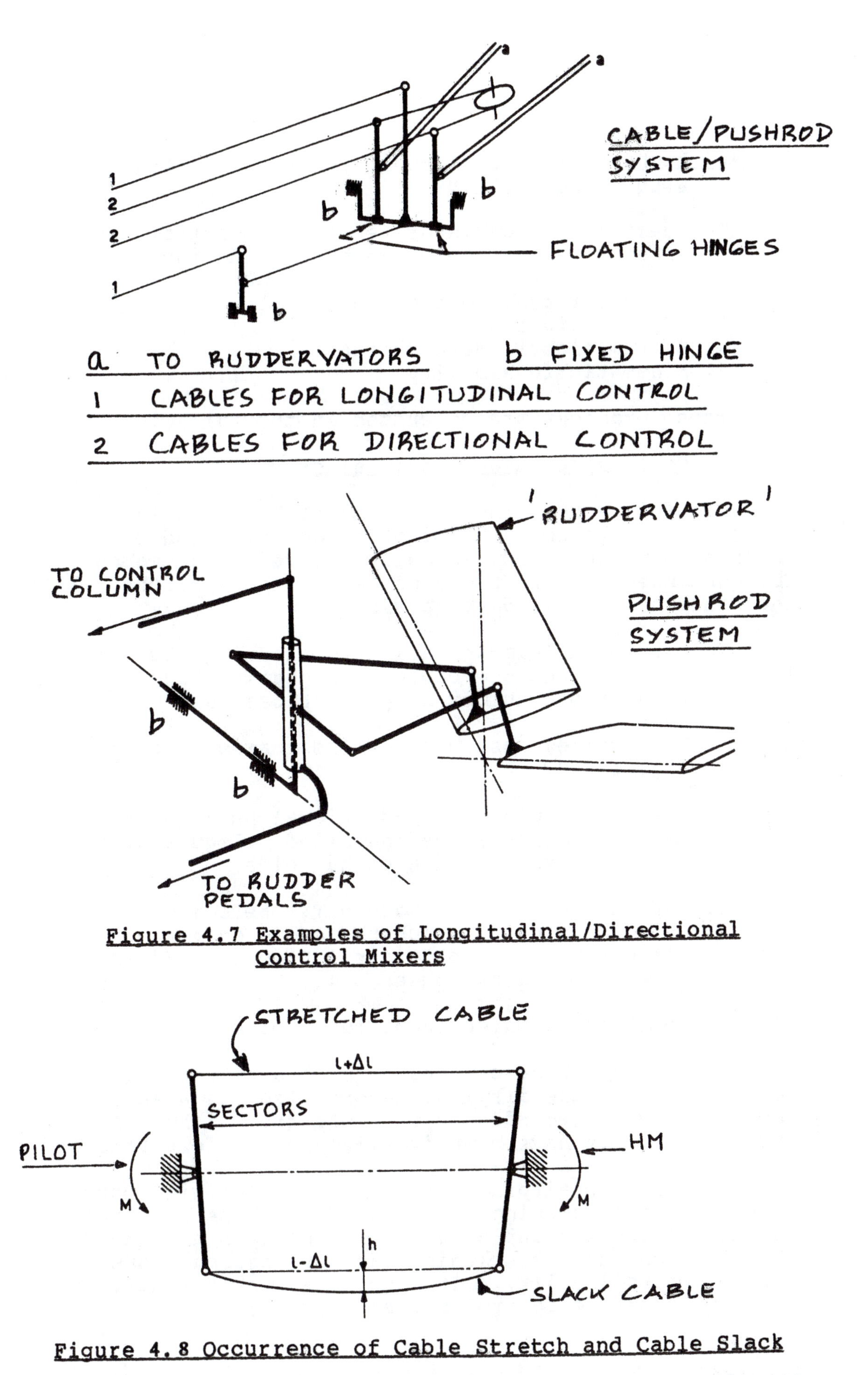

Figure 4.7 Examples of Longitudinal/Directional Control Mixers

Figure 4.8 Occurrence of Cable Stretch and Cable Slack

4.1.4.1 Mechanical design requirements associated with cable systems

The designer of cable driven flight control systems is confronted with a range of mechanical design problems:

1. Cable stretch and cable slack
2. System friction
3. System elastic deformation
4. Kinematic feasibility

These problems are discussed in the following.

1. Cable stretch and cable slack

Figure 4.8 shows how cable stretch occurs in a simple cable system. The pilot's control force (moment) is opposed by an aerodynamically induced hinge moment. The consequence is that one cable will be stretched while the other cable will develop slack.

To prevent a 'slack' cable from leaving its pulley(s) or to prevent the cable from getting tangled into some other system a number of precautions must be taken:

1. Each pulley must have cables guards as shown in Figure 4.9.

2. For long cable runs, additional pulleys and/or cable guides must be installed. Figure 4.10 shows an example of a cable guide.

3. Turnbuckles and/or cable stretchers may have to be used to pre-stretch the cable system so that slack cables do not occur. Figures 4.11 and 4.12 show examples of these.

Cable slack can also occur for thermal reasons:

When an airplane cruises at high altitude for long periods of time the airplane becomes 'cold-soaked'. If the cruise was preceeded by take-off from a 'hot' field, the airplane may have been 'warm-soaked' before take-off.

An aluminum airplane with steel cables will experience cable slackening in cruise due to the differences in thermal coefficient of expansion between aluminum and steel. Such cable slackening can have serious consequences to controllability. In some cases cable slackening can contribute to control surface flutter.

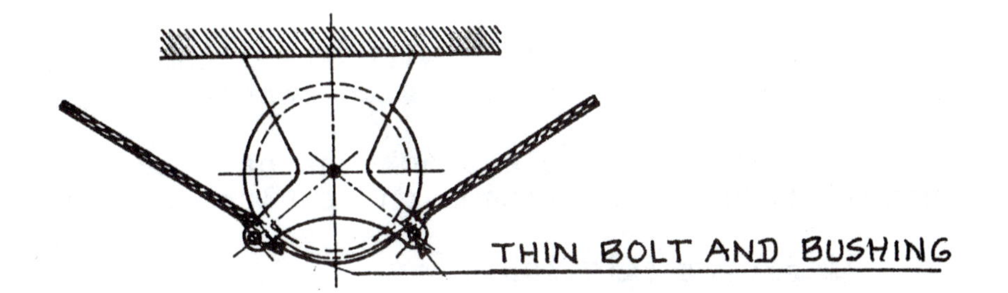

Figure 4.9 Example of a Cable Guard Installation

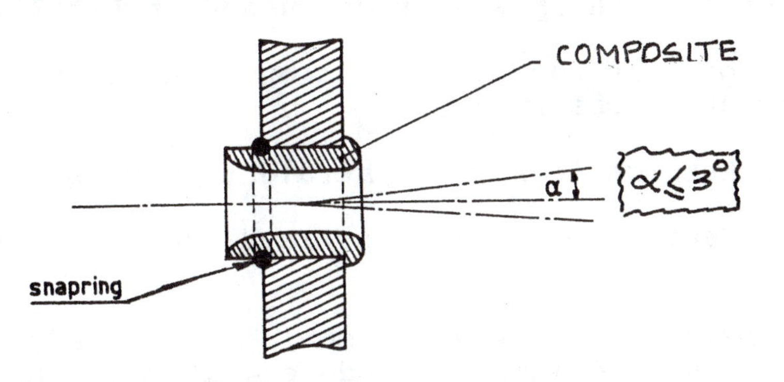

Figure 4.10 Example of a Cable Guide Installation

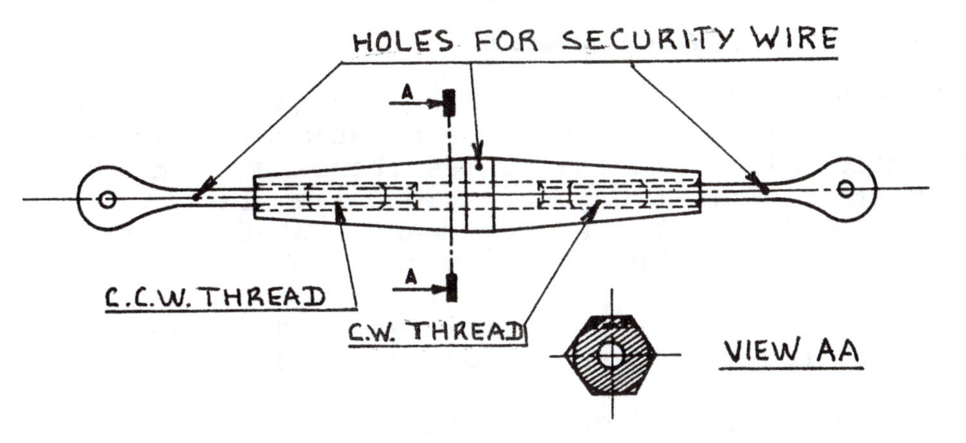

Figure 4.11 Example of a Turnbuckle

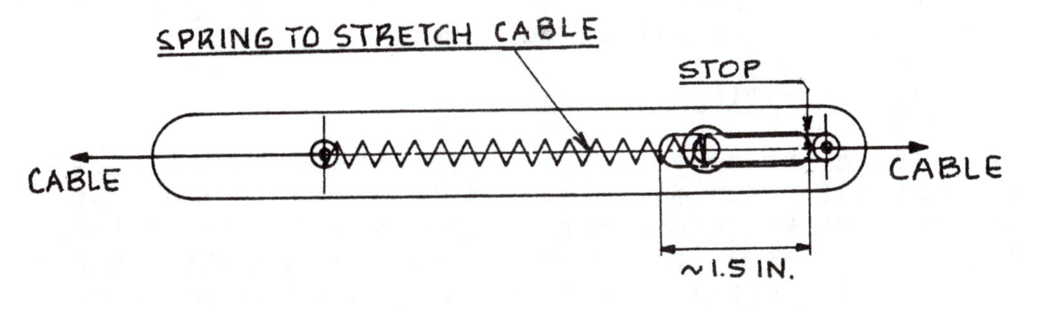

Figure 4.12 Example of a Cable Stretcher

A composite airplane with steel cables will experience exactly the opposite scenario.

This type of cable slack can be prevented by the installation of cable tension regulators: see Fig.4.13.

2. System friction

A major cause of handling quality problems with airplanes is control system friction. The effect of control system friction on the so-called 'return-to-trim-speed' behavior of an airplane is discussed in Ref.20, Ch.5.

To prevent too much friction, the following ground rules must be observed:

1. Keep cable runs as straight as possible.

2. Keep the number of pulleys and guides as small as possible.

Remember that at every cable turn an additional pulley is needed which introduces extra friction and extra weight.

3. System elastic deformations

Another major problem in the design of cable systems is the occurrence of elastic deformations. Sources for these deformation are the cables themselves and the pulley attachment structure. Excessive elastic deformation in a cable system means that full control surface travel will not be attained.

To prevent elastic control system deformation the following groundrules should be observed:

1. Use oversized cables (This will 'cost' weight)

2. Make sure pulleys are attached to 'stiff' structural components. Note: do NOT attach pulleys to flat plates: they deform too easily.

4. Kinematic feasibility

For a cable system to be acceptable it must be kinematically sound. Figure 4.14 shows a number of possible arrangements for cable systems. Mechanisms A and B are kinematically sound: the quadrangles ABFE and CDFE remain parallelograms when the system is used. System A is better than system B because of its higher

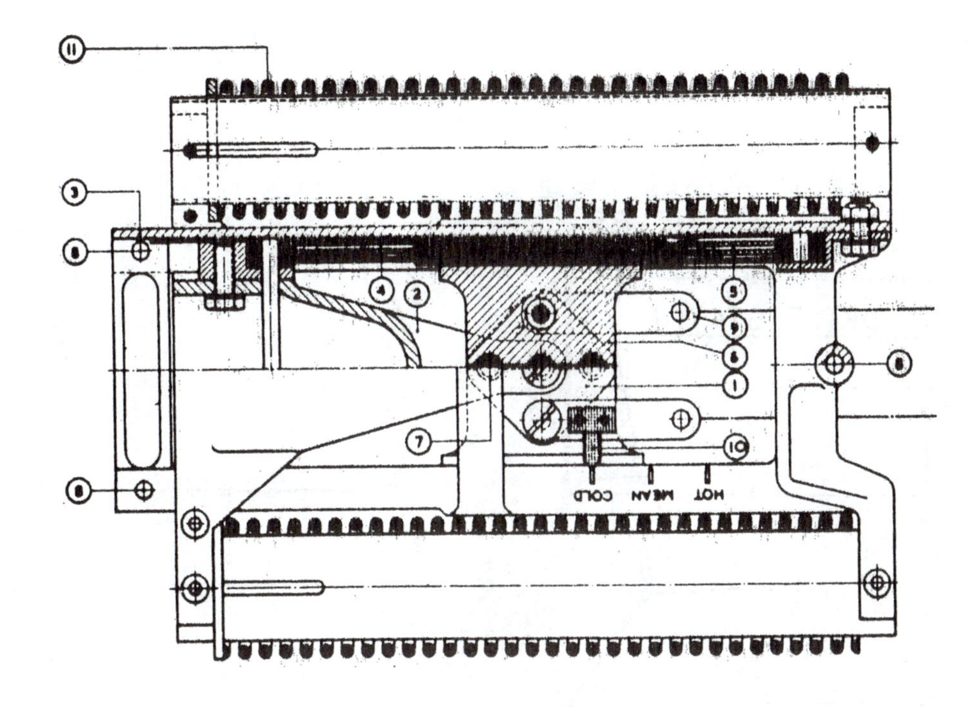

1 Rotating plate

2 Beam, attached to tension reg.

3 Beam, attached to airplane

4 Brake plate attached to 2

5 Brake plate attached to 3

6 Brake lining

7 Pins

8 Attachment points to airplane structure

9 Clevis for control cable attachment

10 Pointer

11 Springs

Airplane Temperature		Distance from HOT to pointer	
deg.C.	deg.F.	mm.	in.
52	125	0	0
38	100	4.6	0.18
24	75	9.1	0.36
10	50	13.7	0.54
-4	25	18.2	0.72
-18	0	22.5	0.89
-32	-25	27.2	1.07
-46	-50	31.8	1.25

Figure 4.13 Example of a Cable Tension Regulator

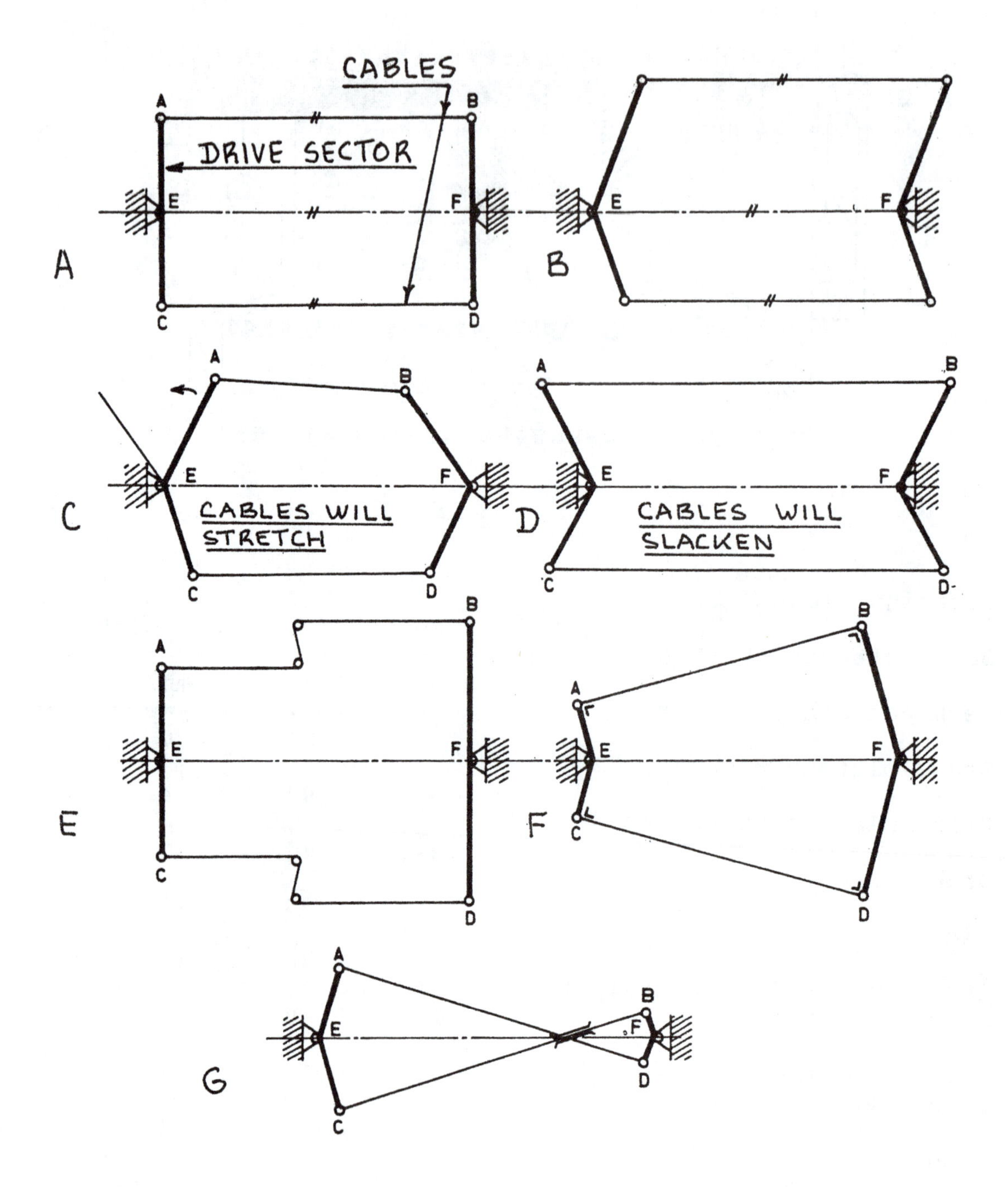

Figure 4.14 Examples of Cable Mechanisms

rotation efficiency. Rotation efficiency is highest if the angle between the cable and the driving sector is 90 degrees with the system in its neutral position.

Mechanisms C and D in Figure 4.14 are unworkable because the cable lengths AB and CD do not remain constant after some rotation: this is undesirable. The reader will recall that the system of Figure 4.2 also violates the 'constant-cable-length' requirement. Despite this shortcoming the system is used in several light airplanes. It is important that the $l/s > 6$ criterion be met to prevent excessive cable stresses!

Mechanism E is kinematically sound but it requires more pulleys: increased weight, complexity, friction and maintenance.

Mechanism F is kinematically sound (it meets the previously stated 90 degree requirement).

Note that the output rotation in both systems E and F is not the same as the input rotation. This effect is sometimes desirable.

Mechanism G employs 'crossing' cables but is kinematically sound. In detail design it is essential that the 'crossing' cables do not in fact touch each other: that would cause extra friction and chafing.

One way to always maintain the 90 degree angle between a cable and its sector is to use a cable-quadrant. Figure 4.15 shows an example of a cable-quadrant.

Cables have 'built-in' redundancy because of the multiple strands from which they are made. This built-in redundancy is only as good as the reliability of the devices used to connect cables to sectors, to quadrants, to turnbuckles and to cable tension regulators. These connections often take the form of 'swaged ends'. Fig.4.16 shows an example of a swaged end in the form of a clevis. The quality control used in manufacturing such swaged ends must be very good.

For detailed requirements which flight control systems must meet the reader should read Ref.21, Subpart D.

Primary flight control system cables should have a diameter of greater than 0.15 in.

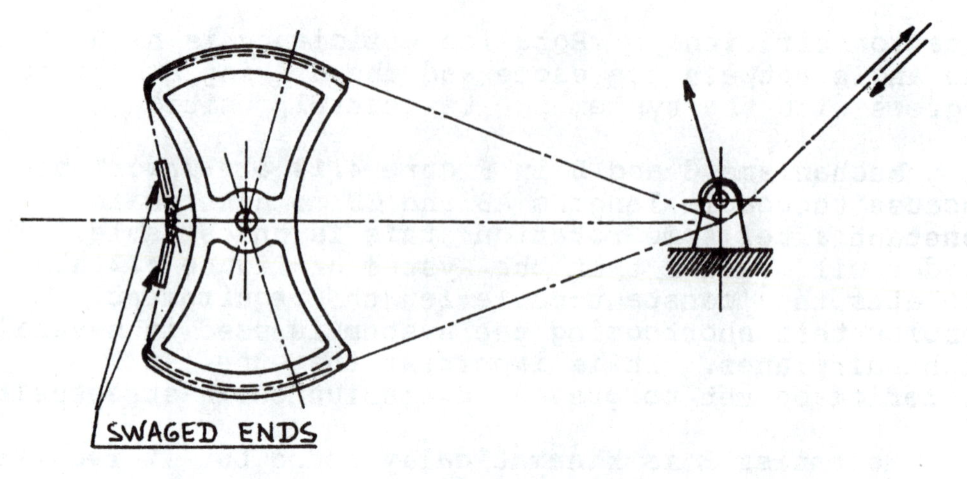

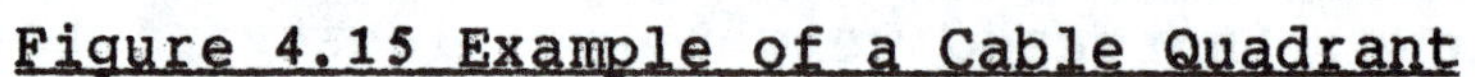
Figure 4.15 Example of a Cable Quadrant

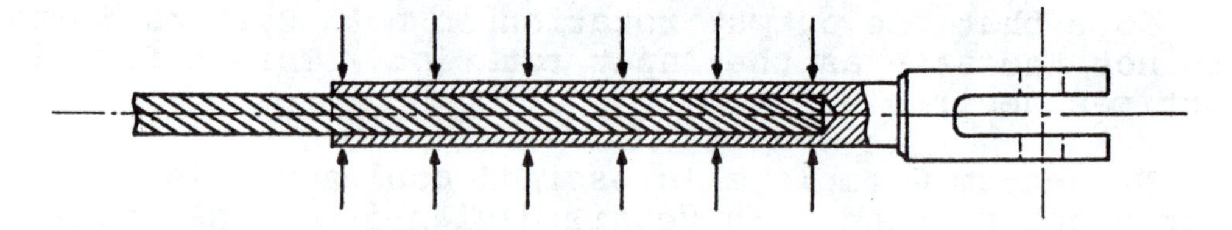

Figure 4.16 Example of a Swaged Fork End

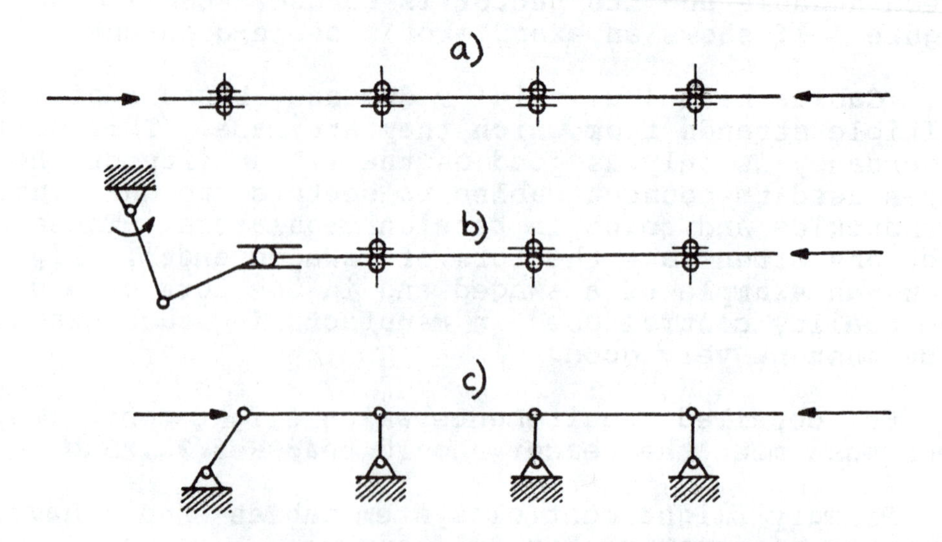

Figure 4.17 Prevention of Buckling in Push-rod Systems

4.1.4.2 Mechanical design requirements associated with push-pull (push-rod) systems

Figures 4.1C and 4.5C show examples of applications of push-rod flight control systems. A major problem associated with push-rod systems is buckling of the push-rods. Clearly, such buckling is unacceptable.

Figure 4.17 shows what can be done to guard against the buckling of push-rods. The unsupported length of each push-rod must be selected so that under a push load of 1.5 times the maximum expected control force application in each rod, no buckling will occur.

Typical push-rod dimensions which normally will prevent buckling are:

Push-rod diameter: > 0.5 in.

Push-rod wall thickness: > 0.08 in.

Push-rod length: < 45 in.

For accurate dimensioning of pushrods, Ref.7 should be consulted.

Whether or not a push-rod system is lighter or heavier than a cable system depends on the controls layout of an airplane.

The comments made under 4.1.4.1 about friction, kinematic feasibility and elastic deformations also apply to push-rod systems.

4.1.4.3 Efficiency considerations

The efficiency of a mechanical flight control system is defined as:

$$\eta_{cs} = W_e/W_i \qquad (4.1)$$

where: W_e = work done by the control surface against the aerodynamic hinge moments

W_i = work done by the pilot on the cockpit control(s)

The work done by the pilot consists of three contributions:

$$W_i = W_e + W_v + W_w \qquad (4.2)$$

where: W_v = the work done to overcome elastic deformation in the control system

W_w = the work done to overcome system friction

A well designed mechanical flight control system has efficiency values in the range of 0.85 to 0.90.

The work done by the pilot can be expressed as:

$$W_i = \int F_p ds_p \tag{4.3}$$

The pilot control force, F_p and the cockpit control travel, s_p are defined in Figure 4.5.

Using considerations of virtual work it can be shown that:

$$F_p = G(HM), \tag{4.4}$$

where: HM is the control surface hingemoment defined in Eqn.(4.7) and

G is the control system gearing ratio defined as:

$$G = \delta/ds_p \tag{4.5}$$

where: δ is the control surface deflection.

The gearing ratio is normally expressed in rad/ft. Table 4.1 contains typical values for G.

4.1.4.4 Calculation of cable and/or push-rod forces from control surface hingemoments

To determine the minimum required sizes of control cables and/or control push-rods, the maximum operating forces must be determined. These operating forces follow from equilibrium considerations between control surface hinge moments and the control force needed to oppose it. Figure 4.18 depicts a typical layout.

It is assumed that the control surface hingemoment, HM (in ftlbs) is known. The cable control force, F_c required to oppose this hingemoment follows from:

$$F_c = (HM/a)(b/c) \tag{4.6}$$

Table 4.1 Control Deflections and Gearing Ratios

Airplane Type	Elevator			Aileron			Rudder		
	Surface Travel degrees	Wheel Travel	Gearing Ratio rad/ft	Surface Travel degrees	Wheel Travel	Gearing Ratio rad/ft	Surface Travel degrees	Pedal Travel	Gearing Ratio rad/ft
Cessna 172	28 up 23 dwn	6.6 in. total	1.62	20 up 15 dwn	+/-90 deg	0.50	+/-16 deg	4.0 in. total	1.68
Cessna 210	23 up 17 dwn	7.5 in. total	1.12	25 up 15 dwn	+/-90 deg	0.50	+/-24	2.8 in. total	3.59
Cessna 303	28 up 15 dwn	6.0 in. total	1.50	20 up 15 dwn	+/-85 deg	0.50	+/-30 deg	4 in. total	3.14
SIAI-M S211		stick	0.70		stick	0.67			1.42
GL M36			0.86			0.39			2.29
Transp. Jets			0.72			0.35			1.30

Note: The gearing ratios are all defined so that:

$$F_p = (G) * (HM)$$

(lbs) (rad/ft) (ftlbs)

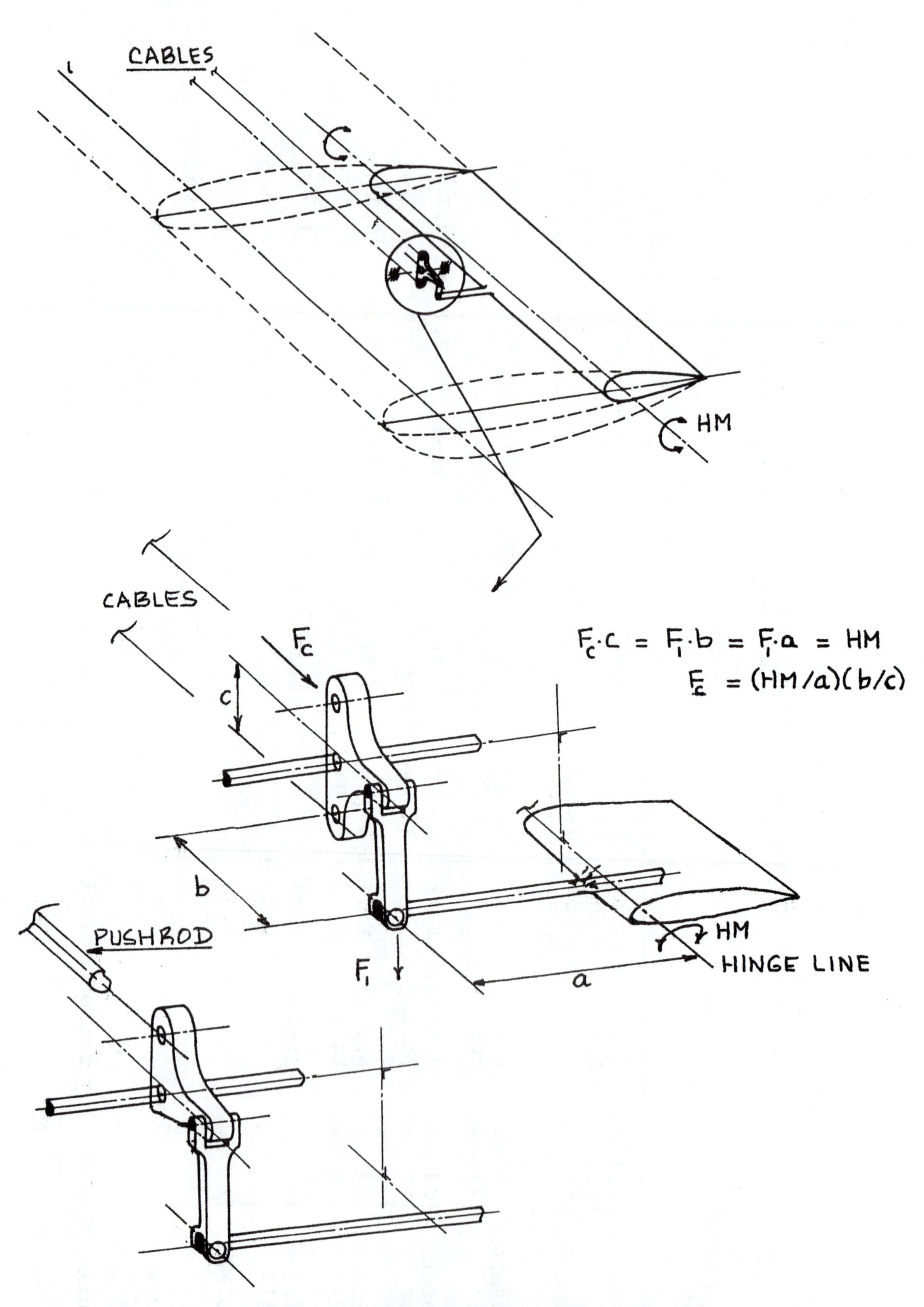

Figure 4.18 Layout for Translating Cable Motion or Push-rod Motion into Control Surface Deflection

Note from Figure 4.18 that for the control force in a push-rod system the same equation (4.6) applies.

The value of the control surface hinge moment, HM depends on the control surface deflection, on the flight condition and on the control surface geometry and design. Reference 21 defines the conditions for determining control system limit and ultimate loads. A brief discussion on the dependence of control surface hinge moments on parameters such as control surface geometry and design is provided next.

4.1.4.5 Control surface hinge moments and control surface tabs and types

Figure 4.19 shows a typical control surface cross section. The hingemoment acting about the control surface hinge line can be computed from:

$$HM = C_h\bar{q}(S\bar{c})_{\text{control surface}} \tag{4.7}$$

The hinge moment coefficient, C_h depends on such factors as:

1. hinge line location
2. gap size
3. nose shape
4. trailing edge angle
5. overhang
6. horn size and shape
7. tab configuration
8. surface deflection
9. Reynold's Number
10. Mach Number

Figure 4.19 also shows a sealed gap and two control surface horn configurations.

It is shown in Reference 20 (Ch.5) and in Part VI that the hinge moment coefficient, C_h can be expressed as:

$$C_h = C_{h_o} + C_{h_\alpha}\alpha + C_{h_\delta}\delta + C_{h_{\delta_t}}\delta_t \tag{4.8}$$

Part VI contains methods for estimating the control surface hinge moment derivatives used in Eqn.(4.8).

Figure 4.20 shows a number of control surface tab configurations. An indication of their potential applications is given below. For a discussion of tab

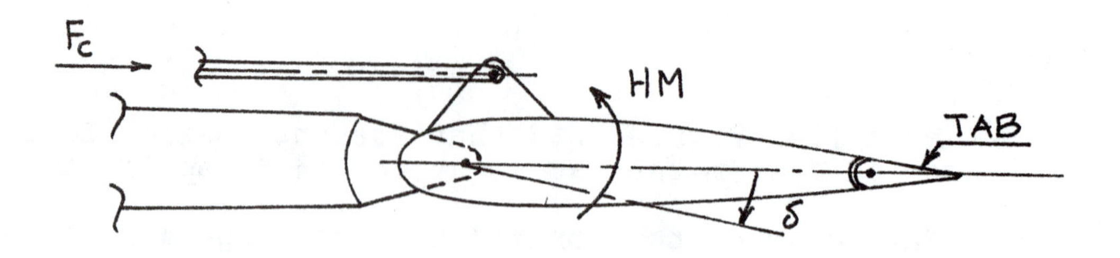

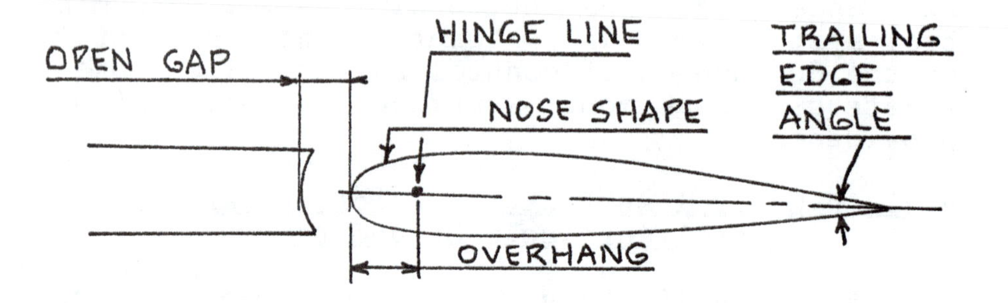

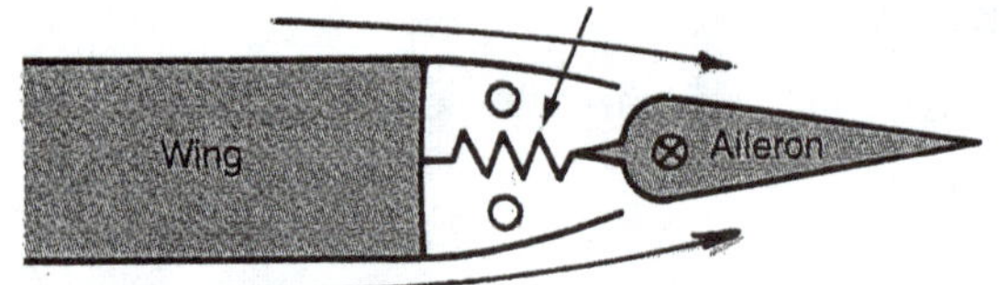

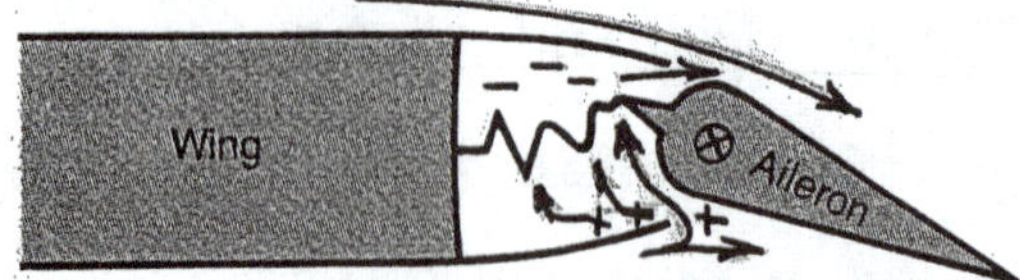

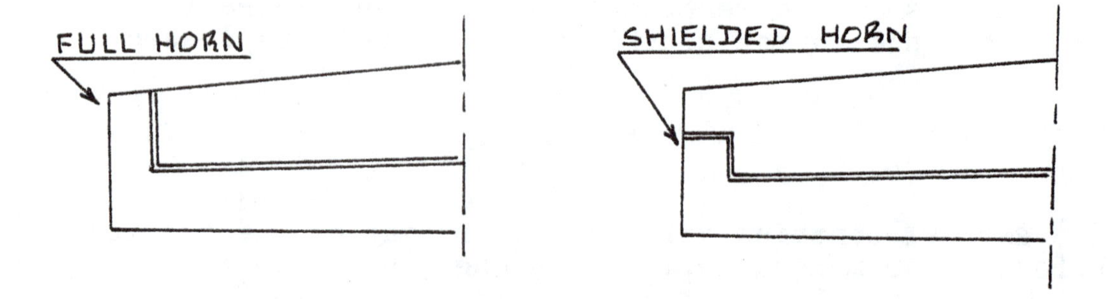

Figure 4.19 Typical Control Surface Cross Section and Control Surface Horn Arrangements

effects on pilot control forces (and thus on handling qualities) the reader should refer to Ref.20 and/or to Part VI.

Ground Adjustable Tabs: Figures 4.20A and 4.20B show potential applications.

These tabs allow the pilot control forces to be 'trimmed' in one flight condition only. These tabs are light, cheap and simple.

Flight Controllable Trim Tabs: Figure 4.20C shows an application.

These tabs allow the pilot control forces to be 'trimmed' in any flight condition. An additional control system connecting the trim tab with the cockpit must be provided. This increases weight, cost and complexity.

Geared Tabs (also called Balance Tabs): Figure 4.20D shows an application.

These tabs allow the hinge moment derivative C_{h_δ} to be increased or decreased depending on the sign and magnitude of gearing used. In some instances the gearing ratio may be varied in flight. These tabs also increase weight, cost and complexity.

Blow Down Tabs: Figure 4.20E shows an application.

These tabs can be used to alter the apparent stability level in airplanes. Ref.20 contains a detailed discussion of such tabs.

Servo Tabs and Spring Tabs: Figures 4.20F and 4.20G show applications.

These tabs are used to allow reversible flight control systems to be used in airplanes of such size and performance that direct control over the flight control surfaces would yield pilot control forces which are too high.

Control surface types: Figure 4.21 shows a range of control surface types.

When the elevator and the rudder are mounted in close proximity (Figure 4.21A) it is necessary to arrange for a 'cutout' in one of these surfaces to prevent mechanical interference.

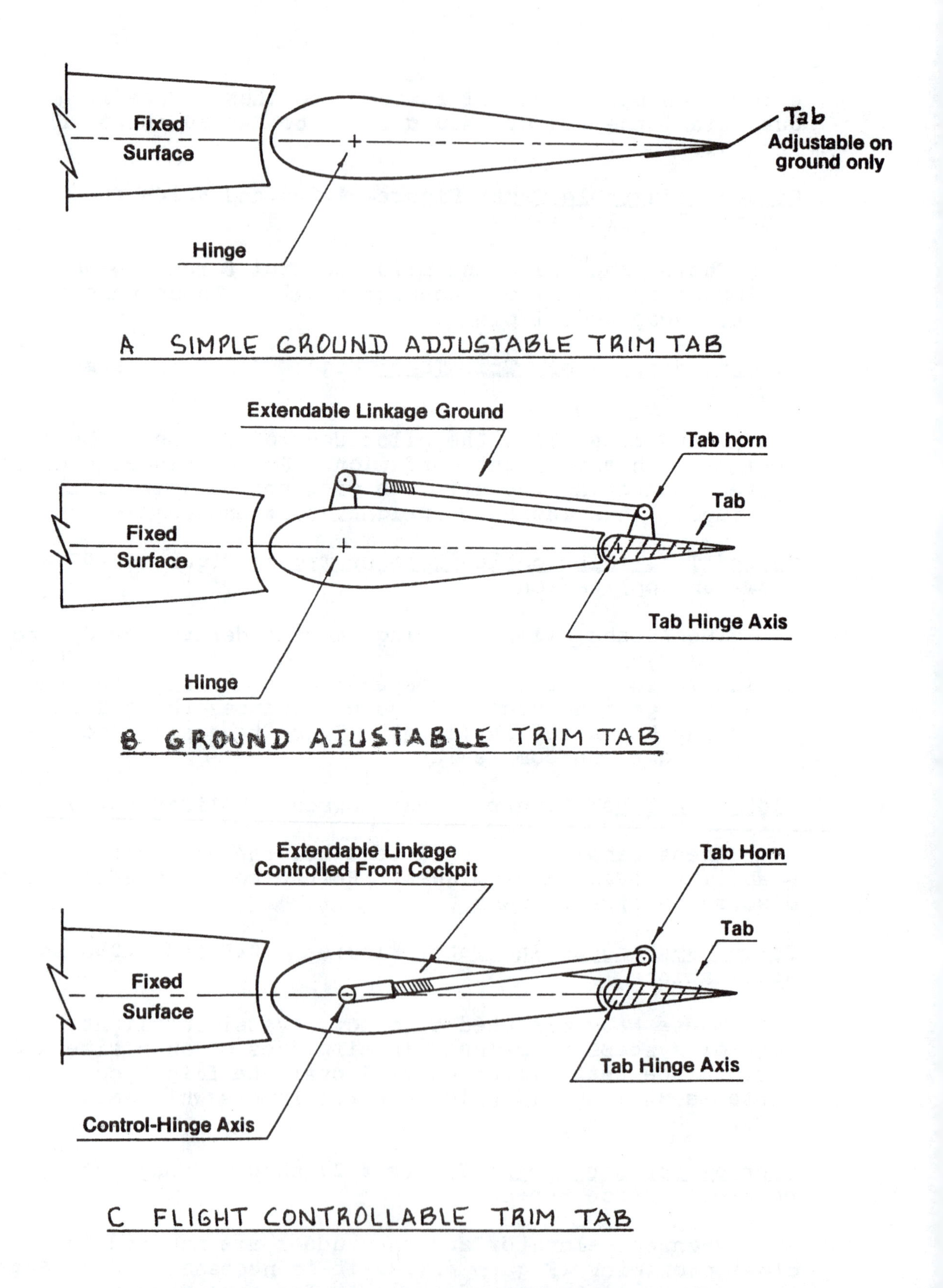

Figure 4.20 Example Tab Configurations

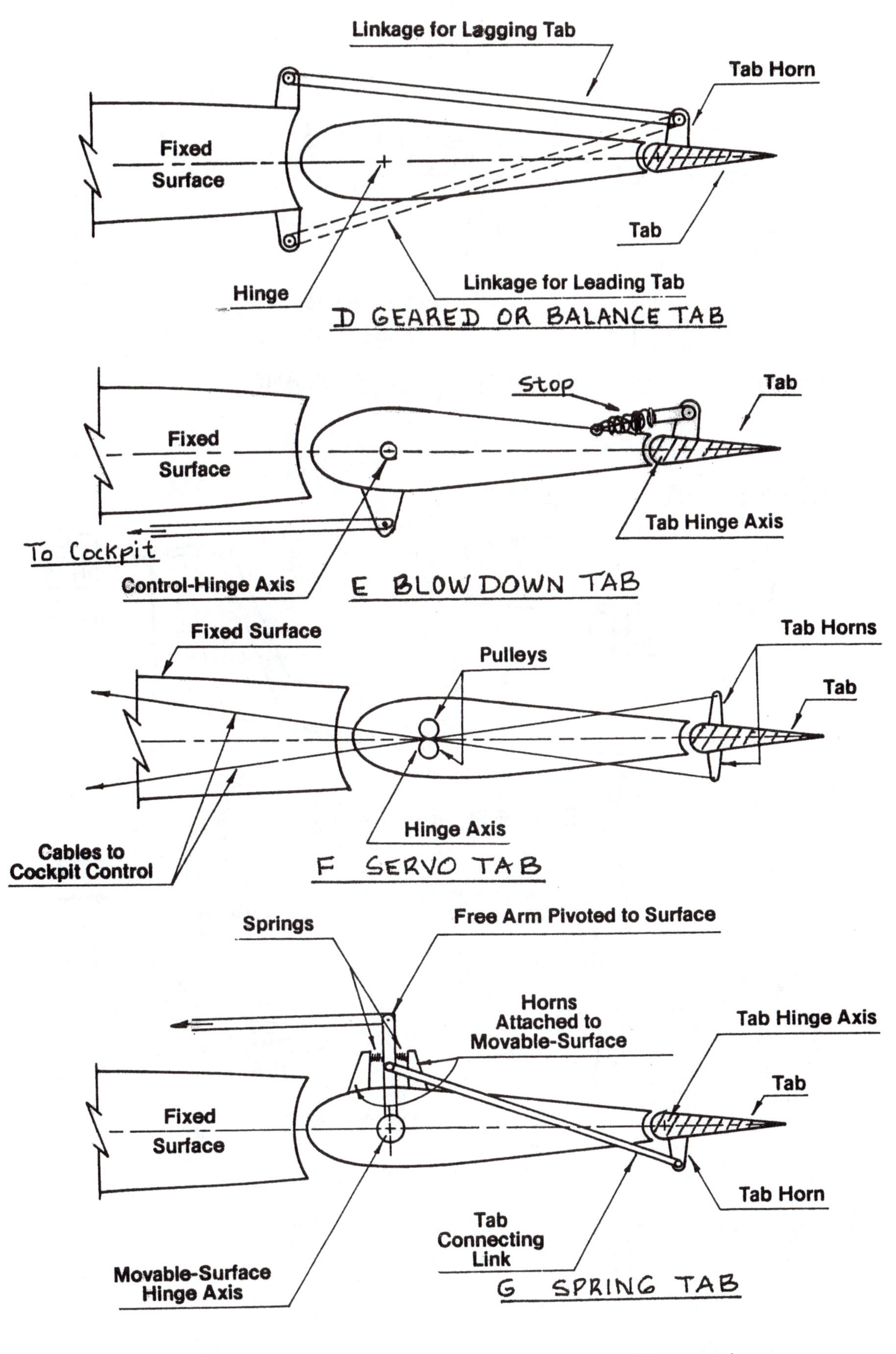

Figure 4.20 (Cont'd) Example Tab Configurations

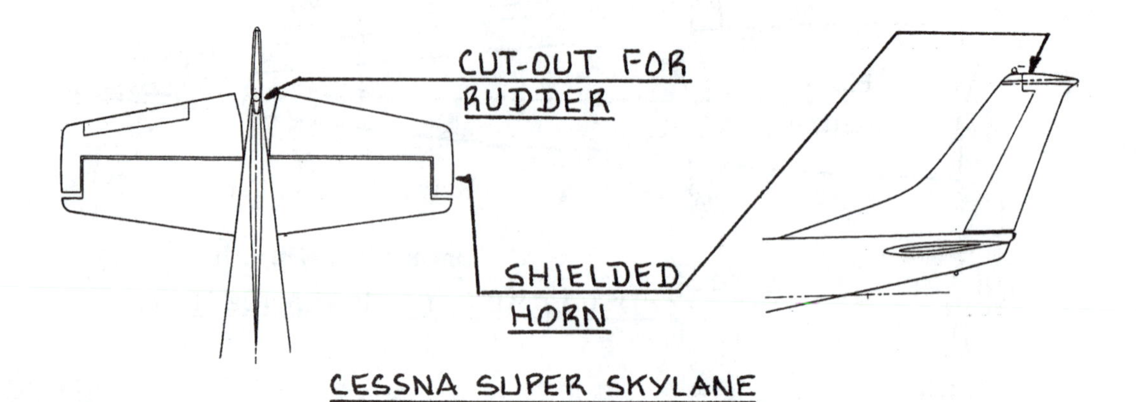

CESSNA SUPER SKYLANE

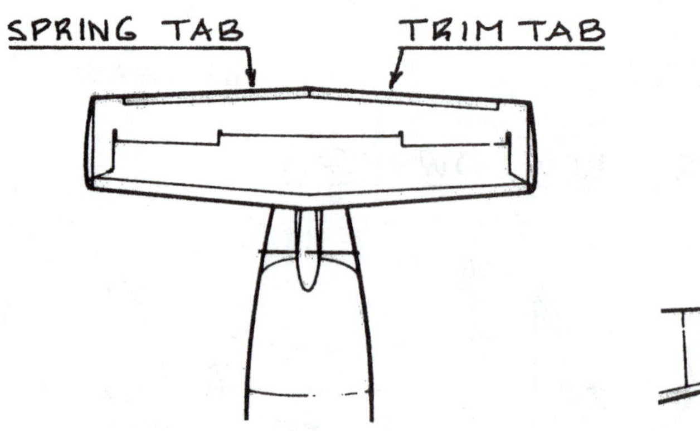

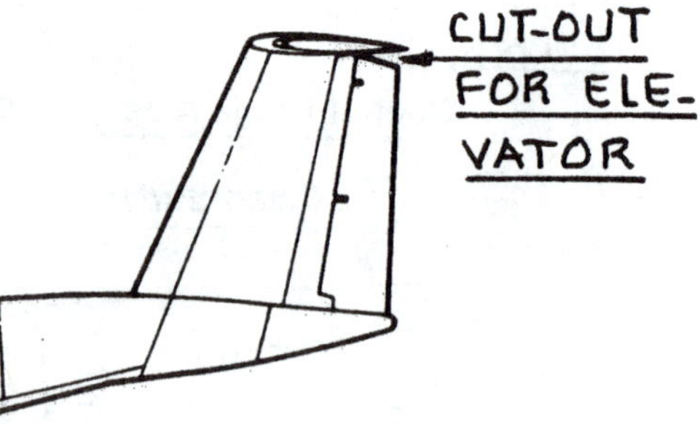

DHC 5D BUFFALO

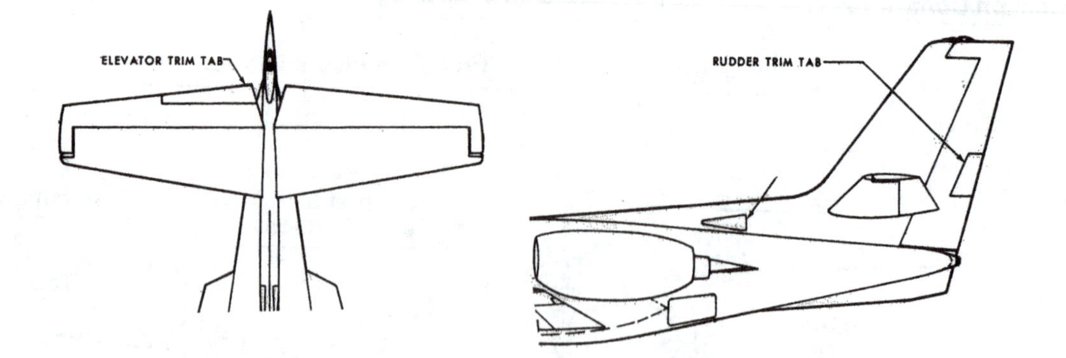

CESSNA CITATION

NOTE: WITH ANY TYPE HORN BALANCE WATCH OUT FOR ICING

Figure 4.21 Examples of Control Surface Configurations

Most control surfaces in airplanes with reversible flight control systems need special design provisions for aerodynamic and for mass balancing. Sub-sub-sections 4.1.4.6 and 4.1.4.7 contain more information on these balancing requirements.

4.1.4.6 Aerodynamic balance requirements and control system gadgets

To keep the pilot control force, F_p and the pilot control force gradients with respect to speed and load factor, $\partial F_p/\partial V$ and $\partial F_p/\partial n$ within acceptable limits, the hinge moment derivatives in Eqn.(4.8) must be kept within certain bounds. These bounds are dictated by handling quality requirements laid down in Refs 21 and 22.

Part VI addresses the relationship between the hingemoment derivatives and the handling quality requirements of Refs 21 and 22 from a design viewpoint.

The process used to 'tailor' the aerodynamic hinge moments to achieve certain handling quality objectives is referred to as aerodynamic balancing of the flight controls.

Aerodynamic balancing can be achieved by careful selection of items 1 through 7 mentioned in sub-sub-section 4.1.4.5.

Control surface tabs (Figure 4.20) and control surface horns (Figure 4.21) are important tools used to achieve satisfactory aerodynamic balance.

<u>CAUTION:</u> The tab configurations of Figures 4.20B through 4.20G are prone to flutter if special precautions to the design and maintenance of their attachments are not taken. Failure of a tab attachment rod and/or a tab spring can have serious consequences. References 23-25 contain methods for analyzing the flutter characteristics of various control surface/tab arrangements.

In many instances it turns out that desirable handling qualities cannot be attained without the use of so-called control system gadgets. Typical of such gadgets are:

<u>Downspring:</u> Figure 4.22 shows an example of a downspring. The downspring adds a force to the pilot control force,

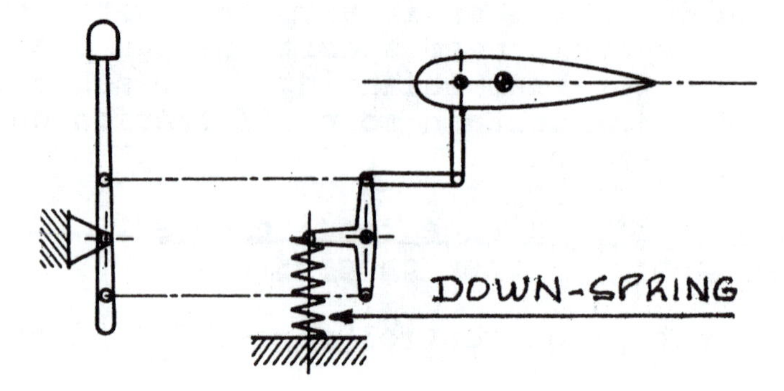

Figure 4.22 Example of a Down-spring

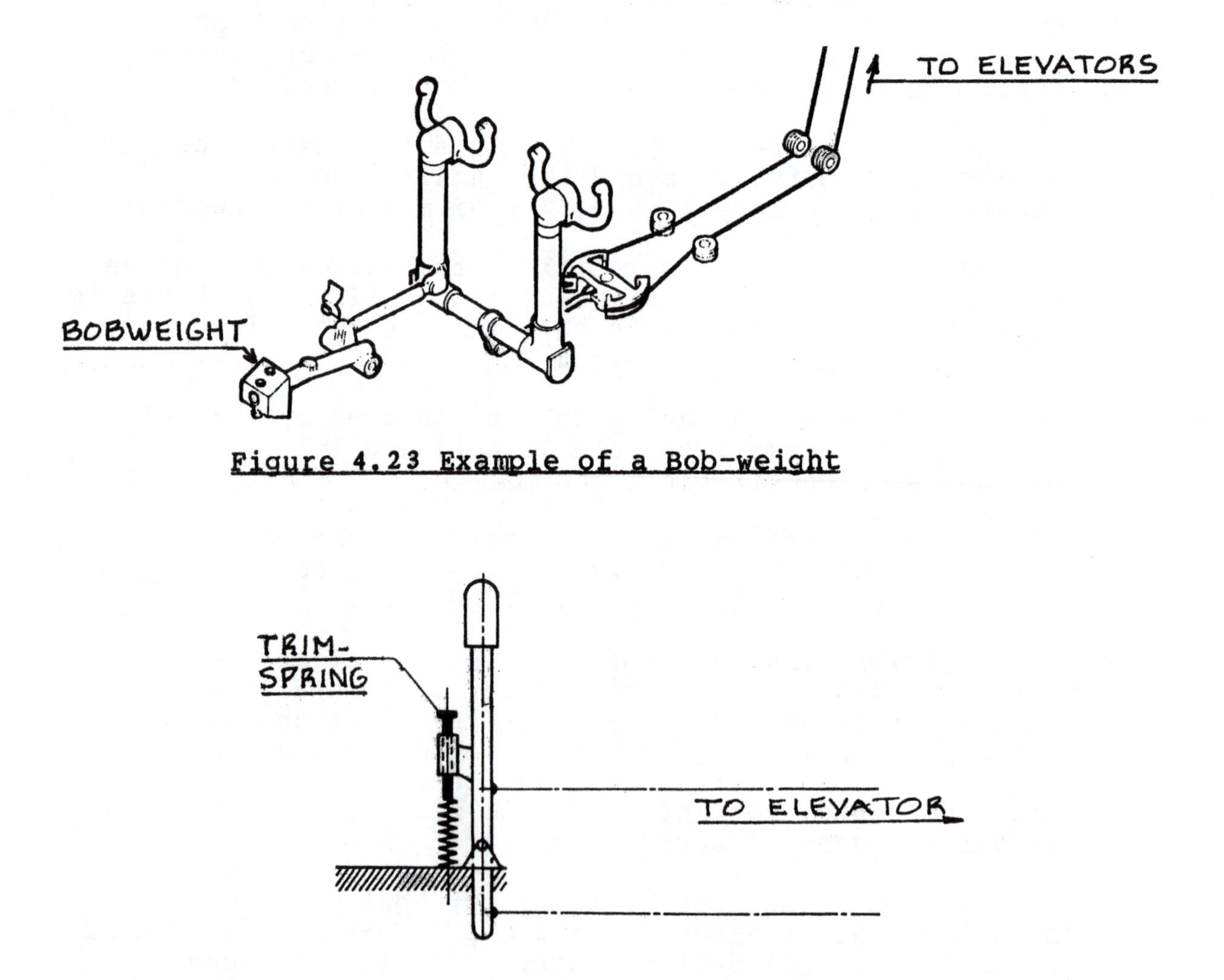

Figure 4.23 Example of a Bob-weight

Figure 4.24 Example of a Trim-spring

F_p. This added force is proportional to control stick or to control surface deflection.

Bob-weight: Figure 4.23 shows an example of a bob-weight. Such bob-weights are used to alter the stick-force-per-g gradients of an airplane.

Reference 20 (Chapter 5) contains detailed analyses of the influence of downsprings and bob-weights on airplane handling qualities.

Trim-spring: Figure 4.24 shows an example of a trim-spring. The trim-spring is a simple method for obtaining stick force trim. The trim spring does cause the stick to move and care must be taken that the stick motion does not fall outside the recommended ranges of Section 2.2 in Part III.

4.1.4.7 Mass balancing requirements

All mechanical flight controls and all reversible flight control surfaces must be mass balanced.

Figure 4.25 shows what is meant by the mass balancing of a mechanical flight control system. It is clear from Figure 4.25 what would happen to the flight control surface if the mechanical system itself is not balanced: the control surface may be forced to move when the airplane is perturbed by vertical accelerations. This could lead to undesirable oscillations and in some cases could also lead to flutter.

Figure 4.26 illustrates several potential methods for balancing of a mechanical control system.

Figure 4.27 shows what is meant by the mass balancing of a flight control surface. Under-balancing of a flight control surface can lead to flutter.

The attachment structure of mass balances as well as the torsional stiffnes of the control surfaces itself are extremely important detail design considerations.

Figure 4.28 illustrates several methods of mass balancing of a flight control surface.

CAUTION 1: Painting of flight control surfaces must be done BEFORE mass balancing. The distribution of paint over a flight control surface can seriously change its mass balance.

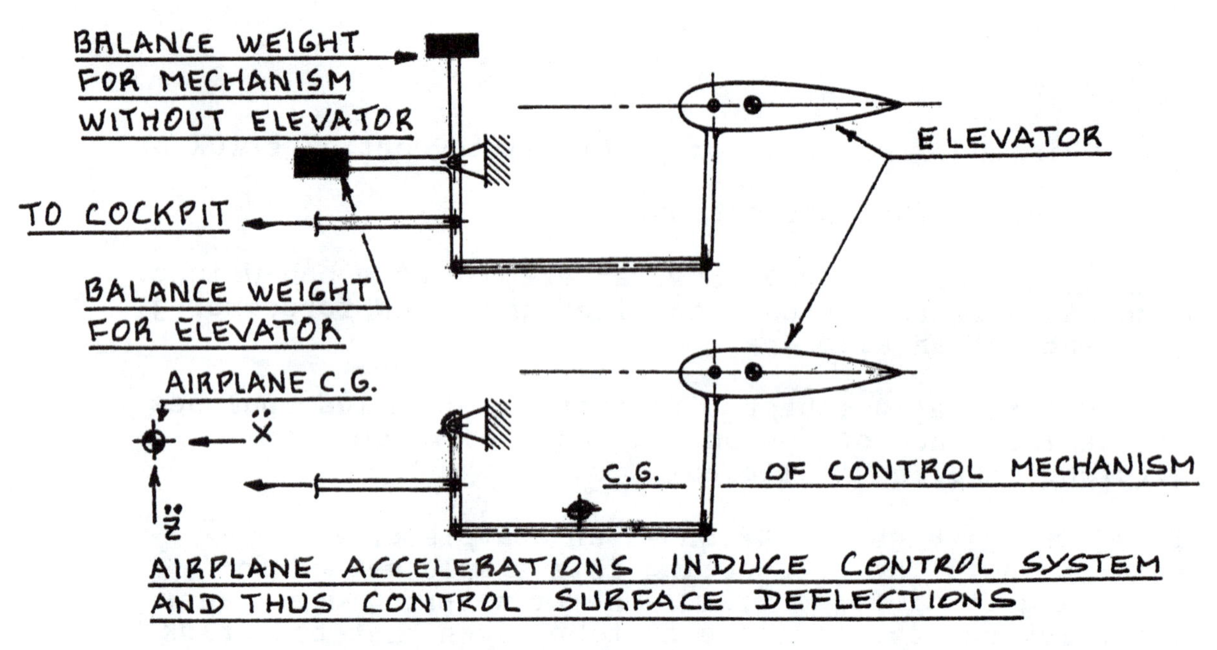

Figure 4.25 Effect of Underbalancing of a Control System

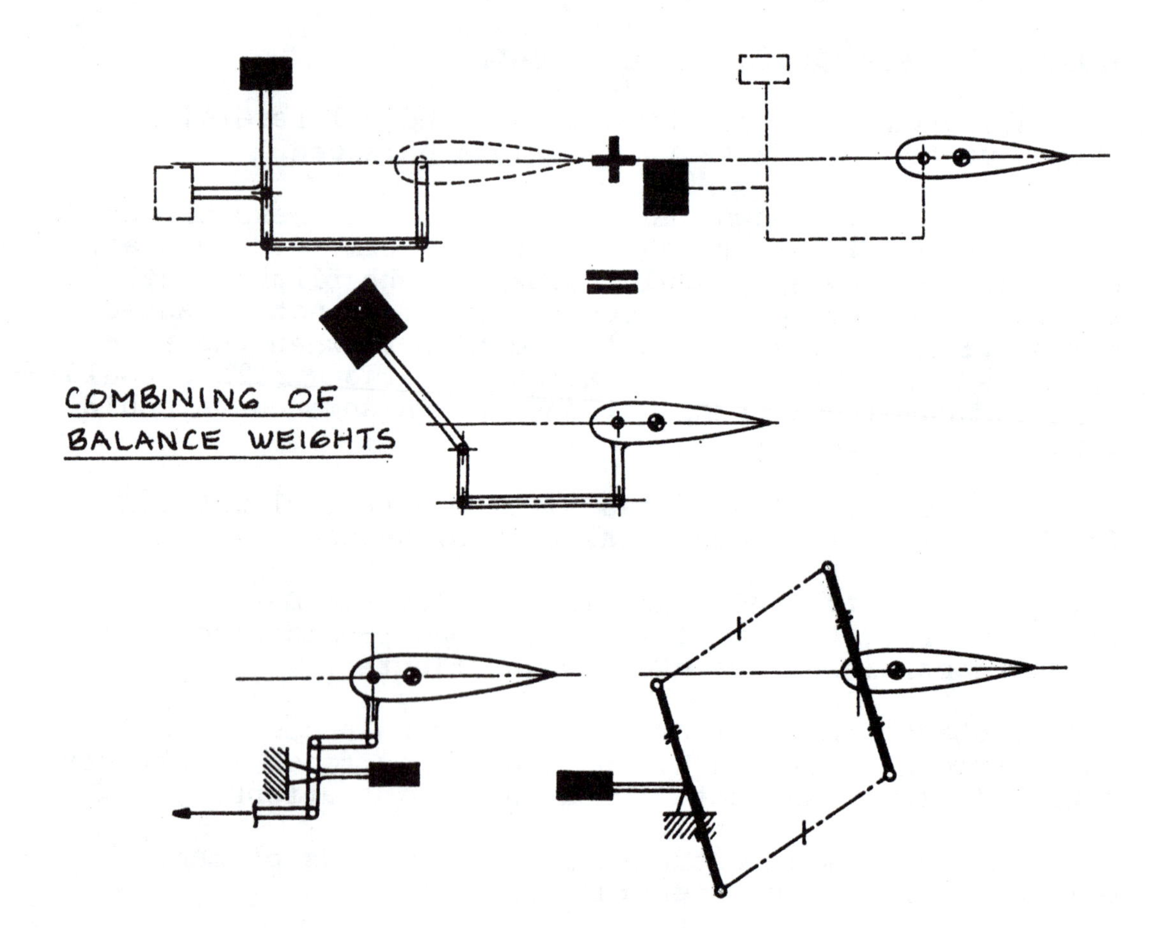

Figure 4.26 Examples of Balancing of a Control System

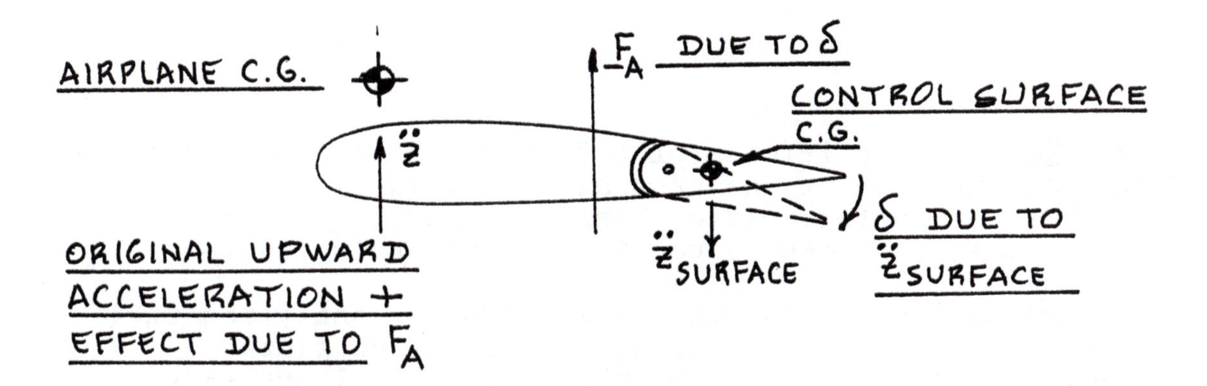

Figure 4.27 Effect of Underbalancing of a Control Surface

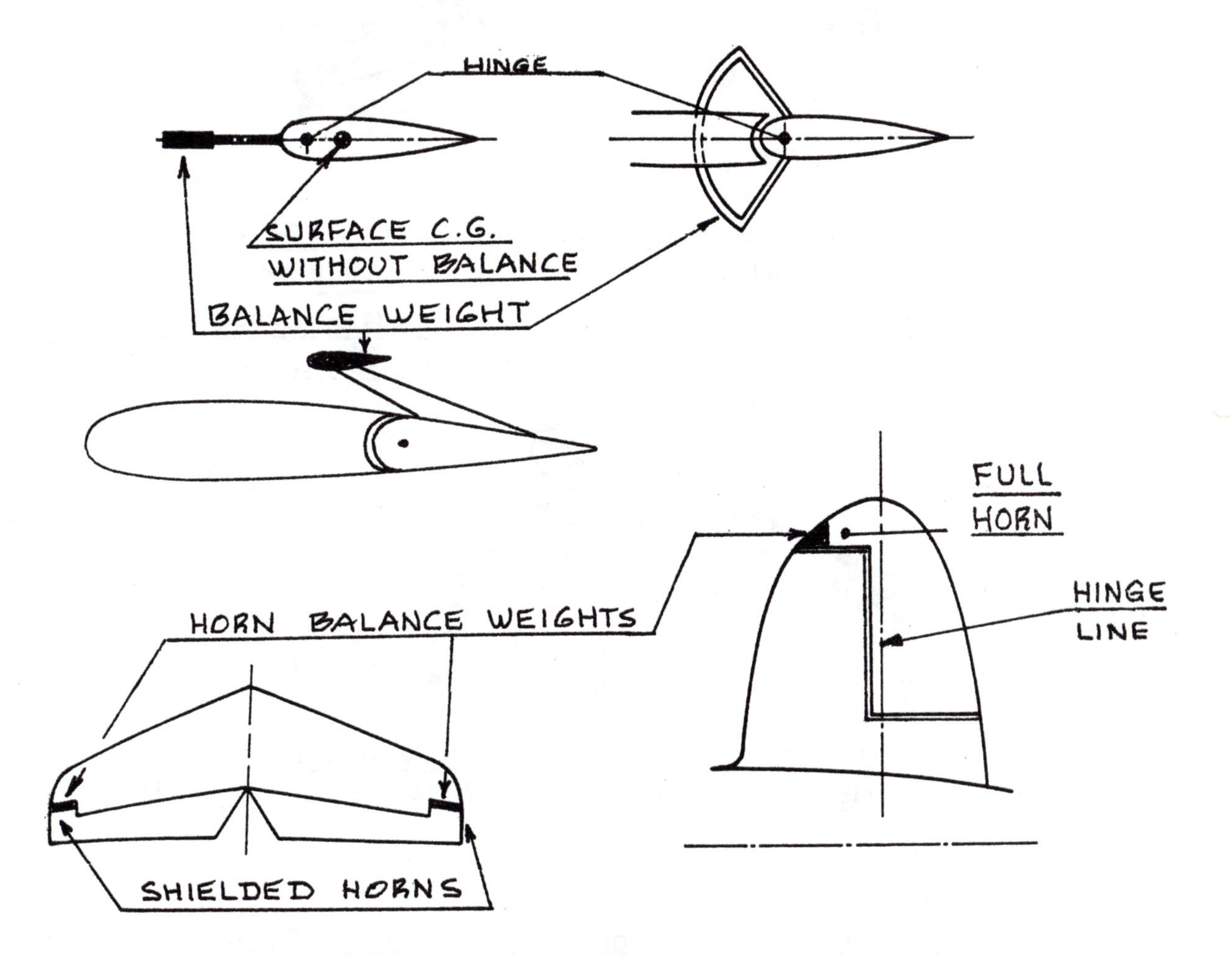

Figure 4.28 Examples of Balancing of a Control Surface

CAUTION 2: The stiffness of a flight control surface and the distribution of the mass balance weights are extremely important design considerations. Figure 4.29 illustrates what can happen if this point is ignored.

References 23-25 contain methods for analyzing the effect of mass balance and stiffness on the flutter characteristics of airplane plus its control system(s).

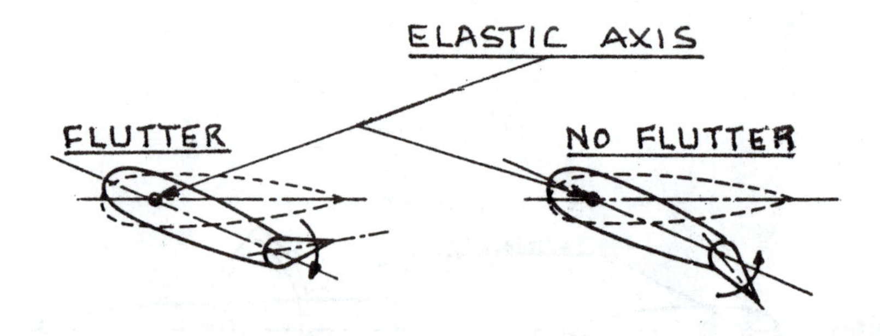

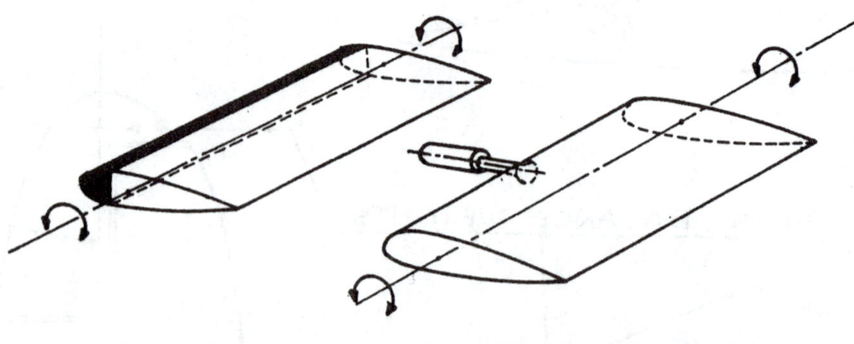

Figure 4.29 Effect of Control Surface Stiffness on the Distribution of Mass Balance Weights

4.2 EXAMPLES OF REVERSIBLE FLIGHT CONTROL SYSTEMS

In this section a number of example layouts of reversible flight control systems are presented in the form of '3-D' layout drawings. Such drawings are also referred to as 'ghost' layout drawings. The ghost layouts are presented in Figures 4.30 through 4.42. Commentary on these drawings is given below.

===

Figure 4.30: Lateral Controls Piper PA-38-112

This is a dual control installation. Note that lateral control motion is translated into the cable system via two sets of torque tubes. These torque tubes must have sufficient torsional stiffness to prevent significant loss of control authority!

Note also the chain-sprocket arrangement used to drive the aileron torque tube.

To reduce aileron induced adverse yaw the ailerons are deflected differentially. The bellcrank installation which provides this differential motion is shown in an insert in Fugure 4.30.

Figure 4.31: Control Column Assembly Piper PA-38-112

Observe that the control column is arranged in the form of a large T-bar. For longitudinal control the T-bar pivots about point 9. For lateral control the chain-sprocket drive rotates the aileron torque tube 7.

Figure 4.32: Longitudinal Controls Piper PA-38-112

Note that the bottom of the T-bar assembly is directly attached to one of the longitudinal control cables. This is similar to the arrangement discussed in Figure 4.2!

The longitudinal cables are routed via pulleys which use mountings common to the directional control pulleys. The longitudinal cables finally attach to a sector mounted on the elevator torque tube.

Figure 4.33: Directional Controls Piper PA-38-112

Rudder pedal motion in the cockpit is translated into rudder deflection via cables which use pulleys for guidance. Note that these pulleys use mountings common also to the longitudinal control pulleys.

The rudder cables attach to a sector mounted at the bottom of the rudder. This arrangement makes it necessary for the rudder to be fairly stiff to assure that rudder torsion does not reduce directional control effectiveness.

==

Figure 4.34 Lateral Controls Cessna 441

Aileron motion is generated through wing mounted cables which are actuated by a differential sector (See detail B) which itself is rigidly attached to a 'wheel' mounted in the wing. This 'wheel' in turn is controlled by cables operated by a chain/sprocket system mounted in the cockpit control column.

Observe the autopilot servo tie-in to the wing mounted 'wheel'.

Figure 4.35: Longitudinal Controls Cessna 441

Longitudinal cockpit control motion is translated into motion of cables which in turn move a sector mounted in the fuselage tail cone. This sector moves a push-rod which drives the elevator torque-tube. Note again the autopilot servo tie-in.

Figure 4.36 Directional Controls Cessna 441

Pedal motion is translated into cable motion which eventually drives a sector mounted at the bottom of the rudder torque-tube. The autopilot servo drives the rudder sector via separate cables.

Note the tie-in between the rudder and aileron system through springs in Detail A. This tie-in is to overcome a mild deficiency in rolling moment due to sideslip.

==

Figure 4.37 Flight Controls Canadair CL215

Note the extensive use of redundant cabling in this ghost view.

Because of the amphibious role of this airplane all control cables are routed up behind the cockpit and then routed aft.

Figure 4.38 Lateral Controls Canadair CL215

The lateral cockpit control system does not use the chain/sprocket method but instead a cable-drum system.

Note that the control yokes in the cockpit are not interconnected as they were in the previous examples.

Figure 4.39 Longitudinal Controls Canadair CL215

The cockpit yokes are directly attached to the elevator control cables. Note that the elevator is controlled directly with assist from a spring tab. Also note that the elevator halves are connected with a torque-tube.

Figure 4.40 Directional Controls Canadair CL215

Observe that the rudder pedals again control a sector at the bottom of the rudde torque-tube. This sector itself is controlled by a cable system. There is an interconnect between the lateral and the directional controls via an aileron tab.

==

Figure 4.41 Flight Controls Short Skyvan

Note that this system is a total push-rod system. The controls are routed up behind the cockpit. The high wing layout makes this a logical choice.

Note from detail A the extensive anti-buckling measures taken in this system.

==

Figure 4.42 Flight Controls Learjet M23

Despite the high performance capabilities of this airplane its flight control system is completely reversible. Cables are used throughout except for the final control run to the elevators. Here a redundant push-rod system is used.

Note the autopilot tie-ins in all three systems.

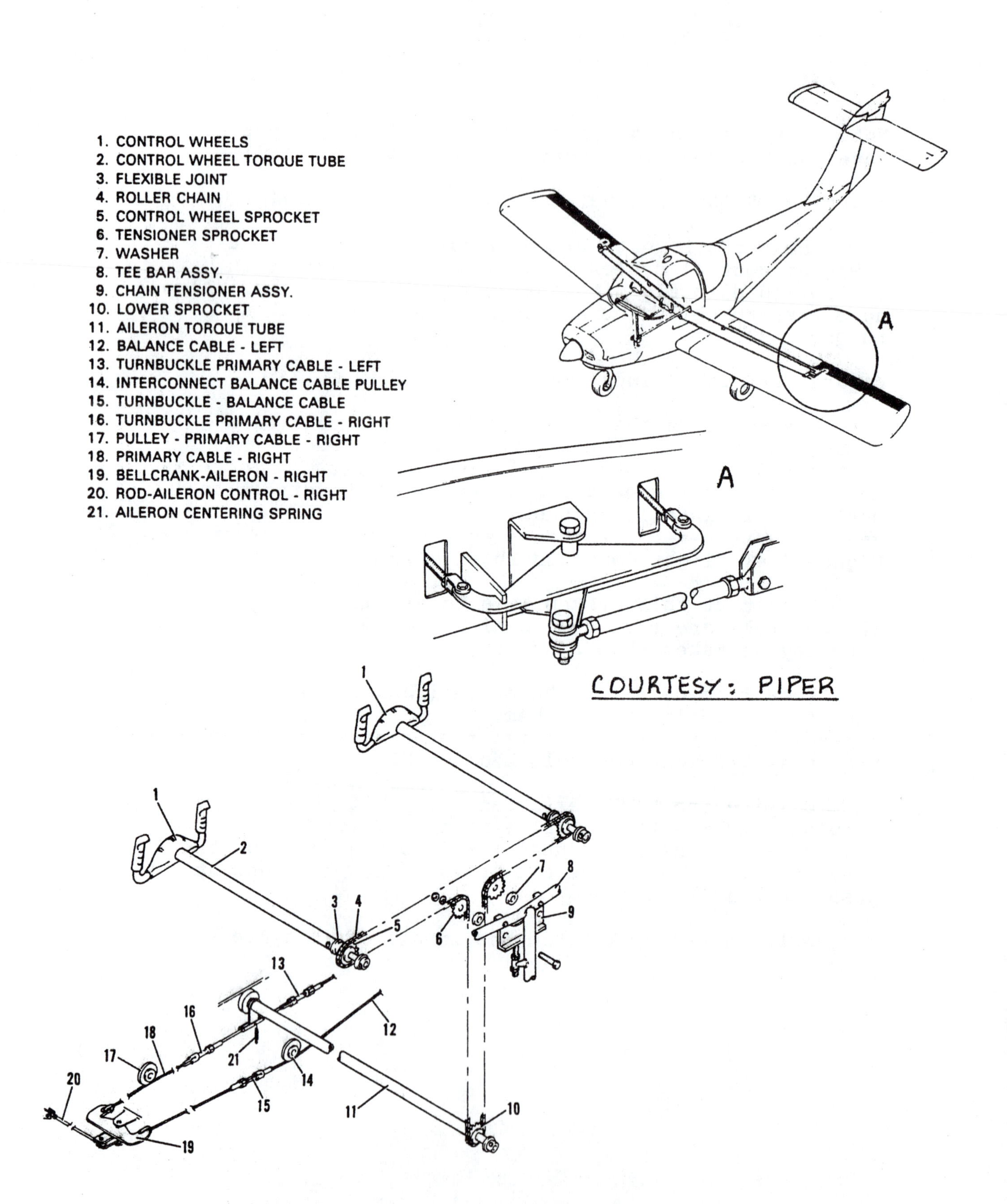

Figure 4.30 Lateral Controls Piper PA-38-112

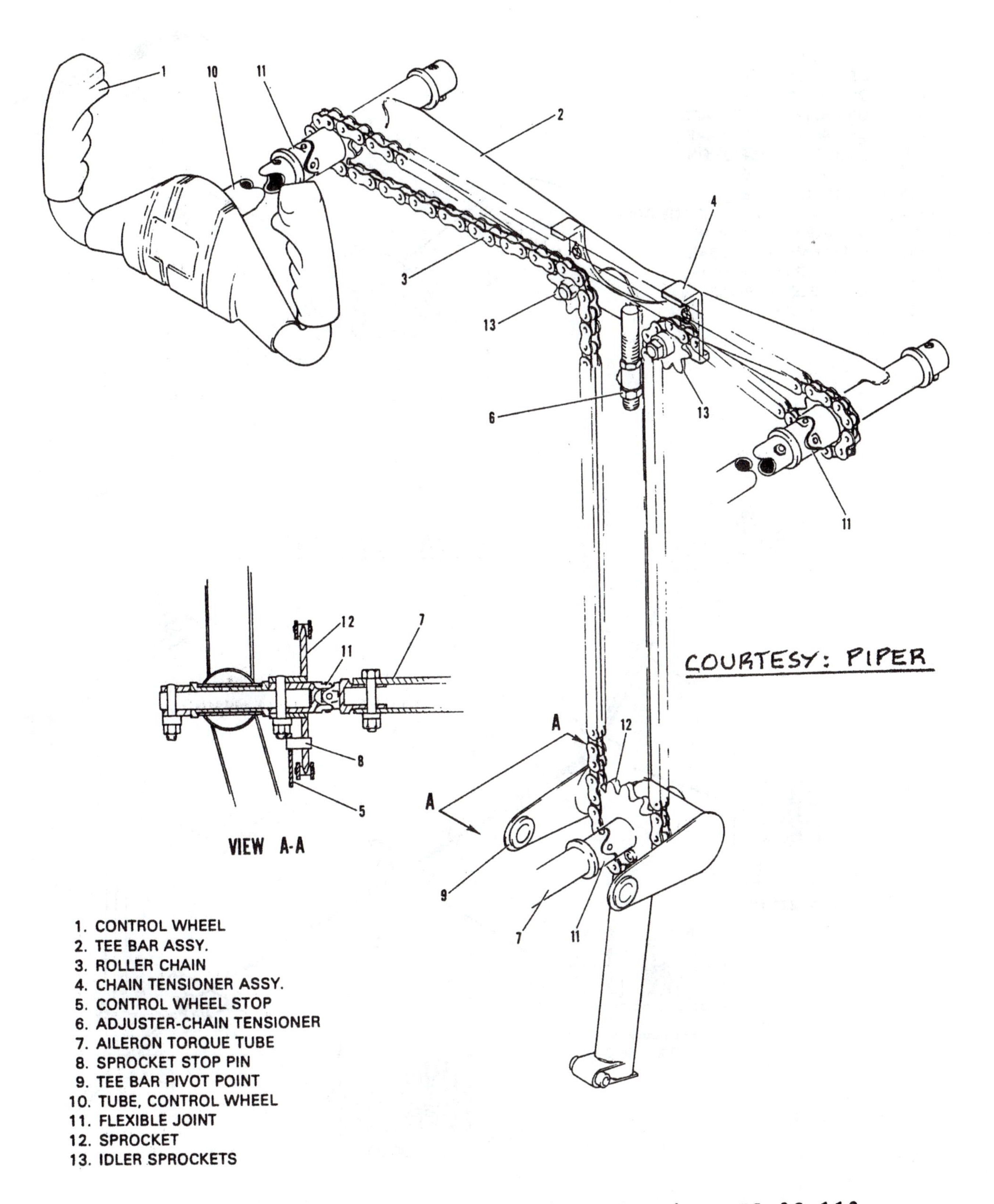

Figure 4.31 Control Column Assembly Piper PA-38-112

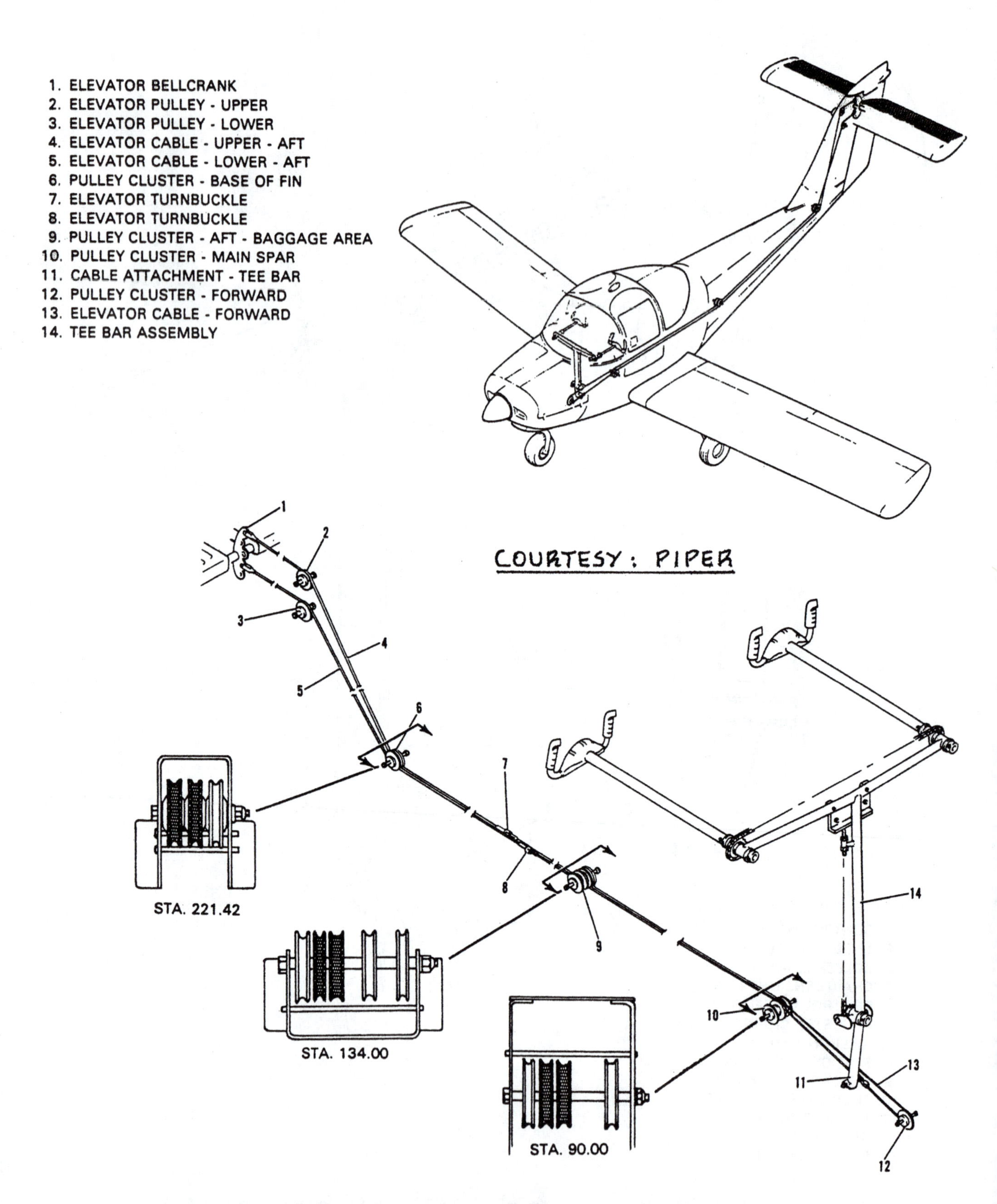

Figure 4.32 Longitudinal Controls Piper PA-38-112

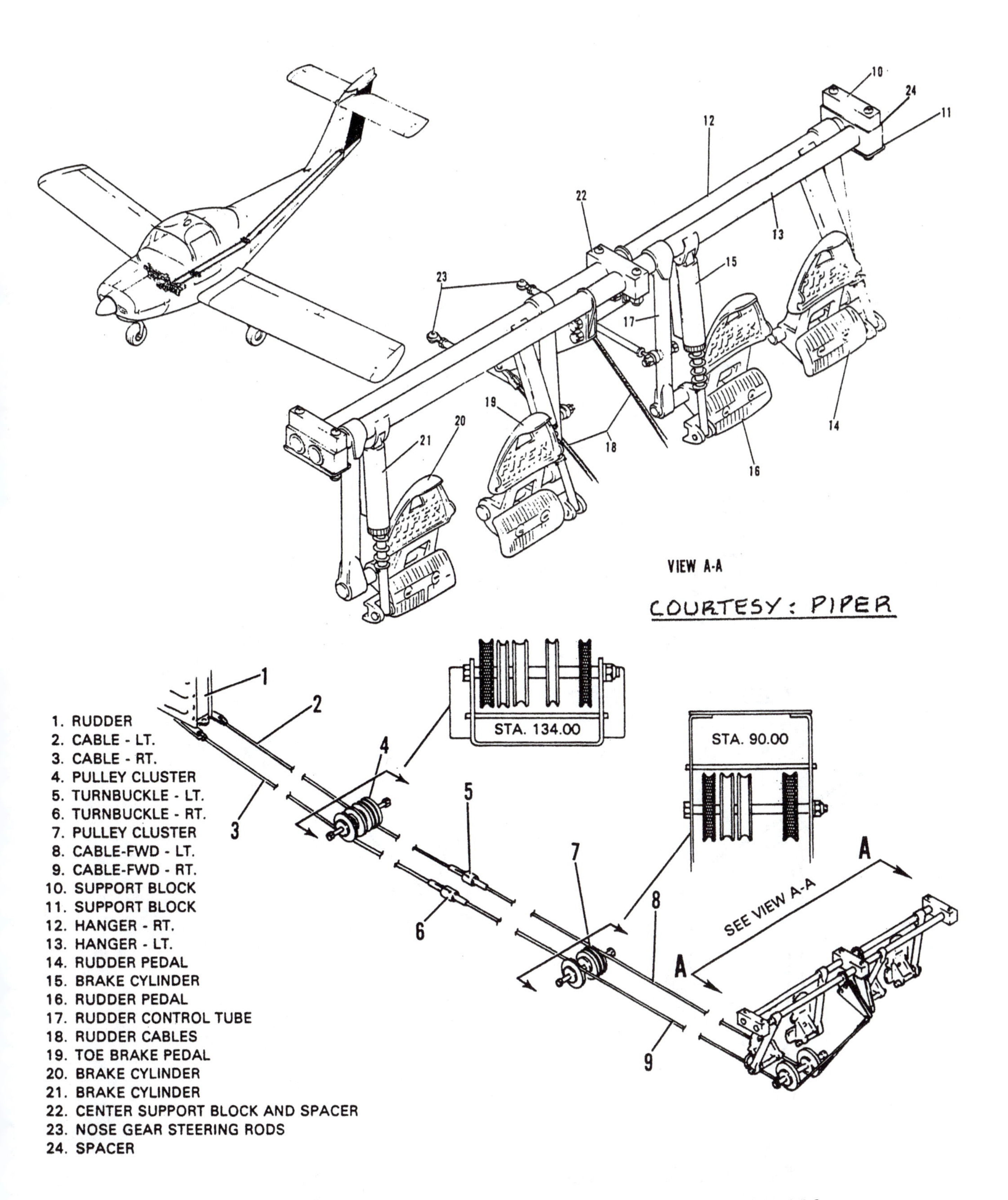

Figure 4.33 Directional Controls Piper PA-38-112

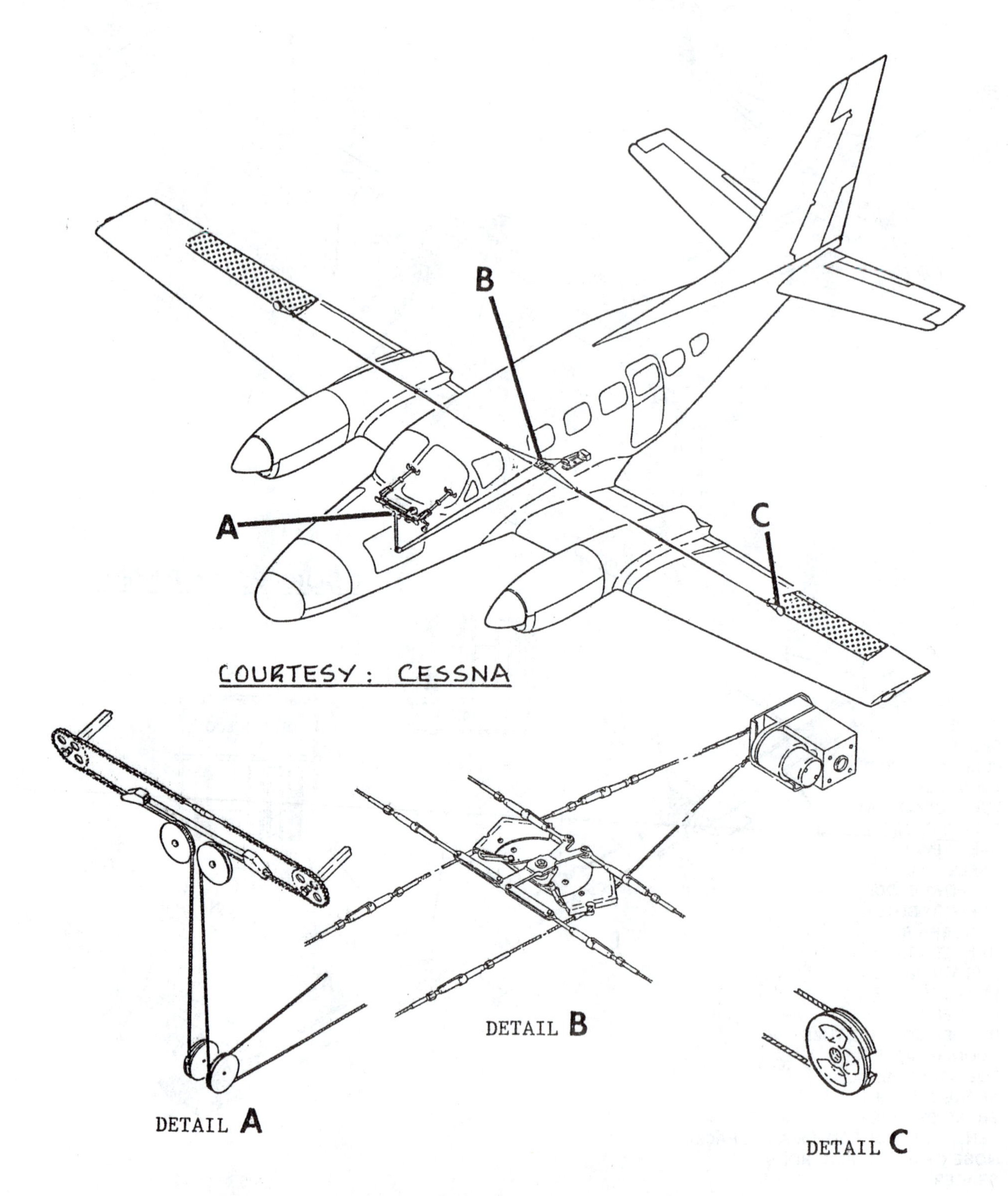

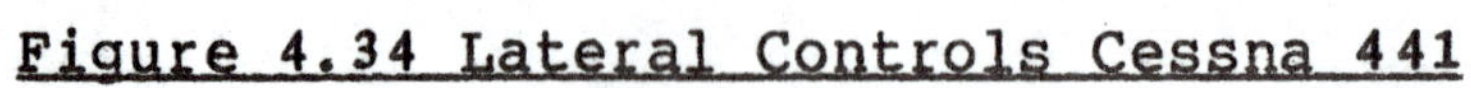

Figure 4.34 Lateral Controls Cessna 441

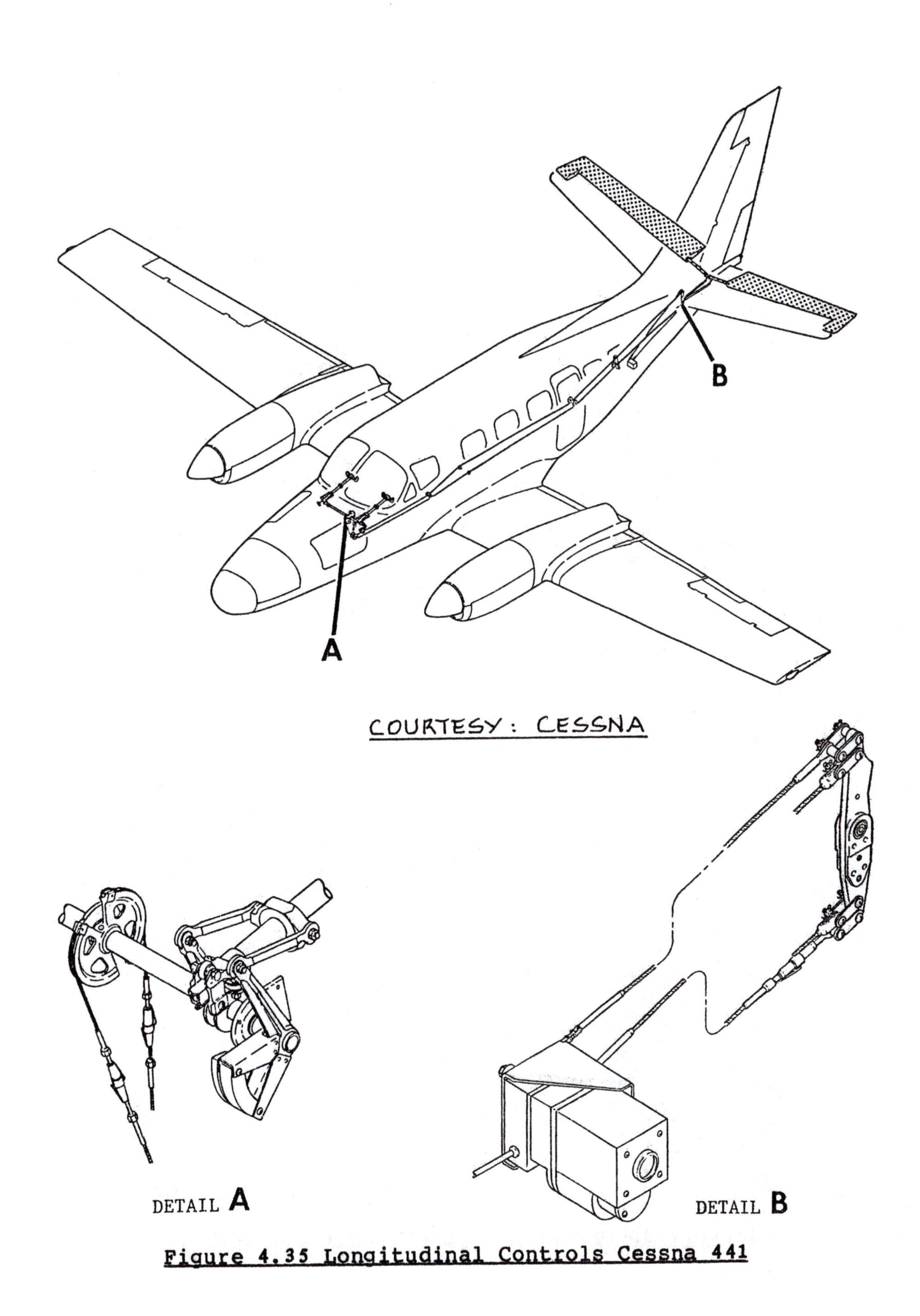

Figure 4.35 Longitudinal Controls Cessna 441

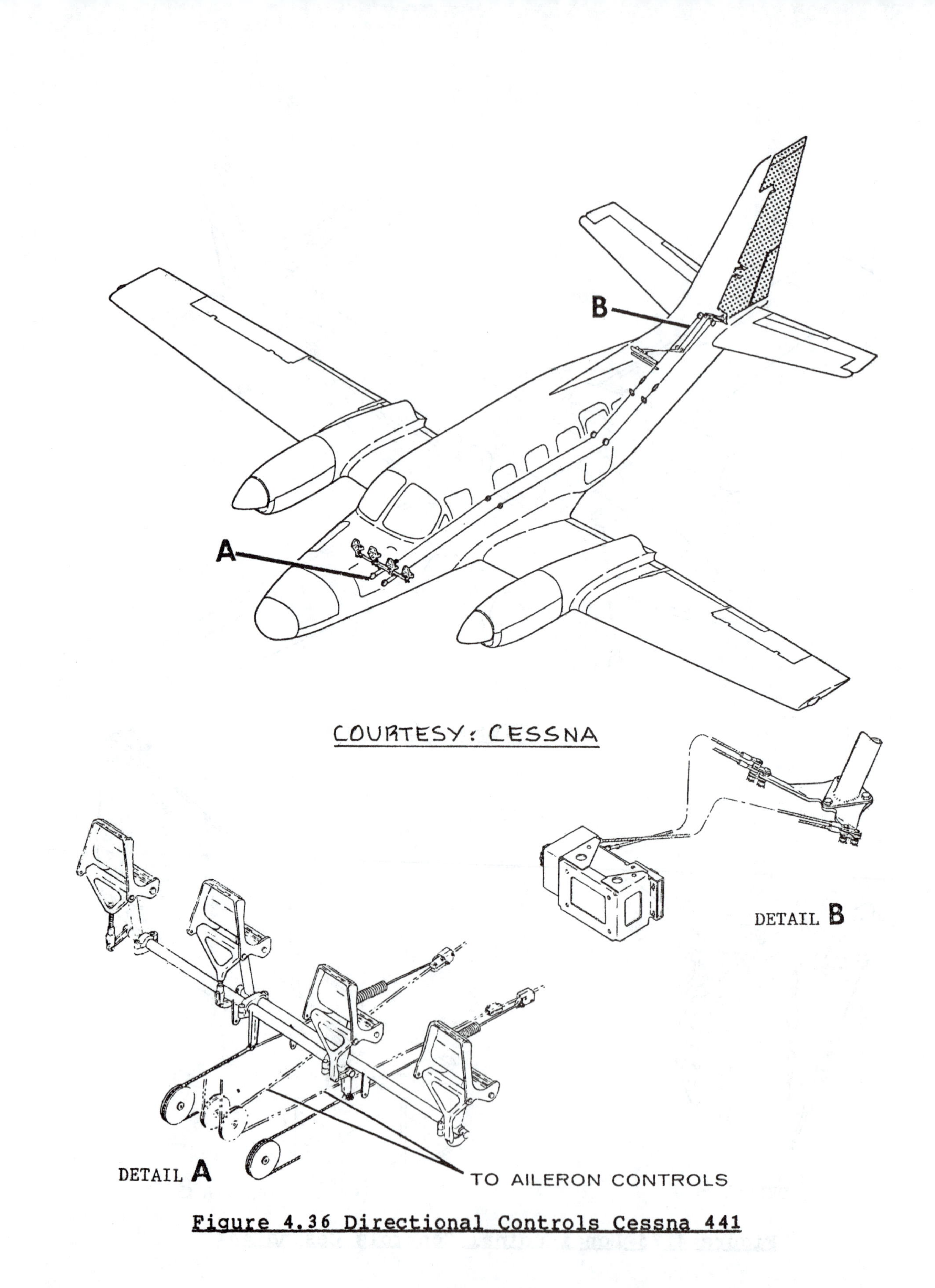

Figure 4.36 Directional Controls Cessna 441

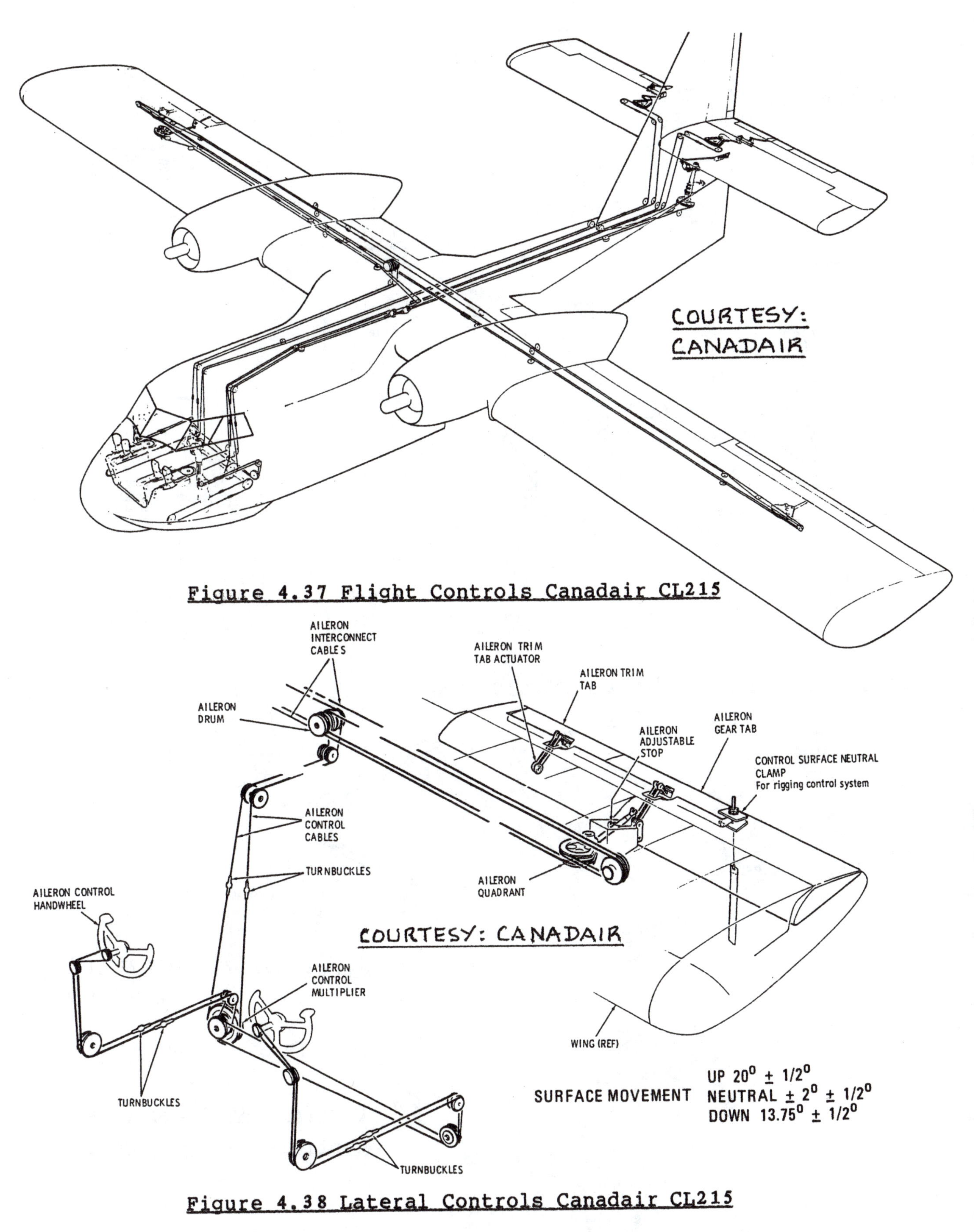

Figure 4.37 Flight Controls Canadair CL215

Figure 4.38 Lateral Controls Canadair CL215

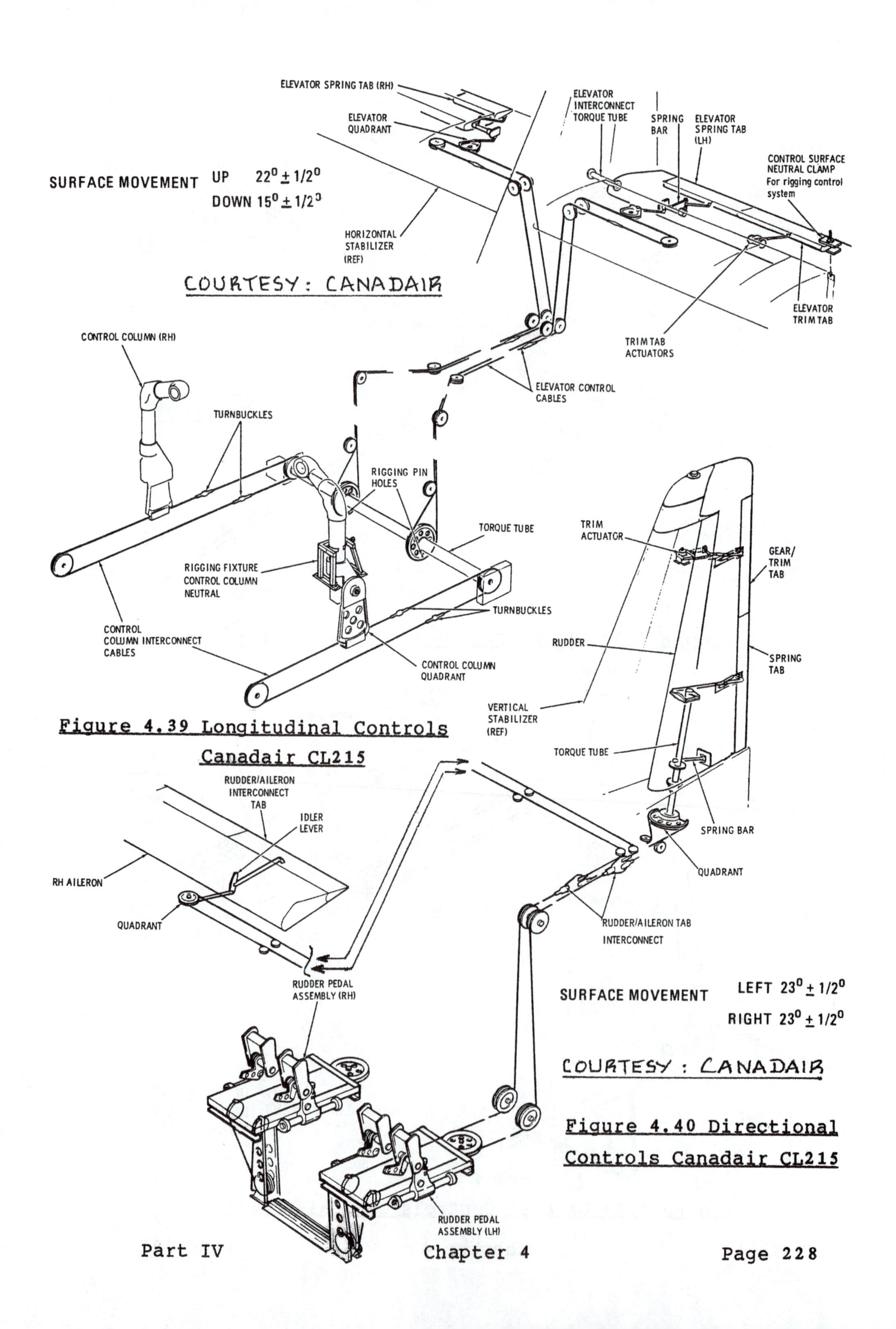

Figure 4.39 Longitudinal Controls Canadair CL215

Figure 4.40 Directional Controls Canadair CL215

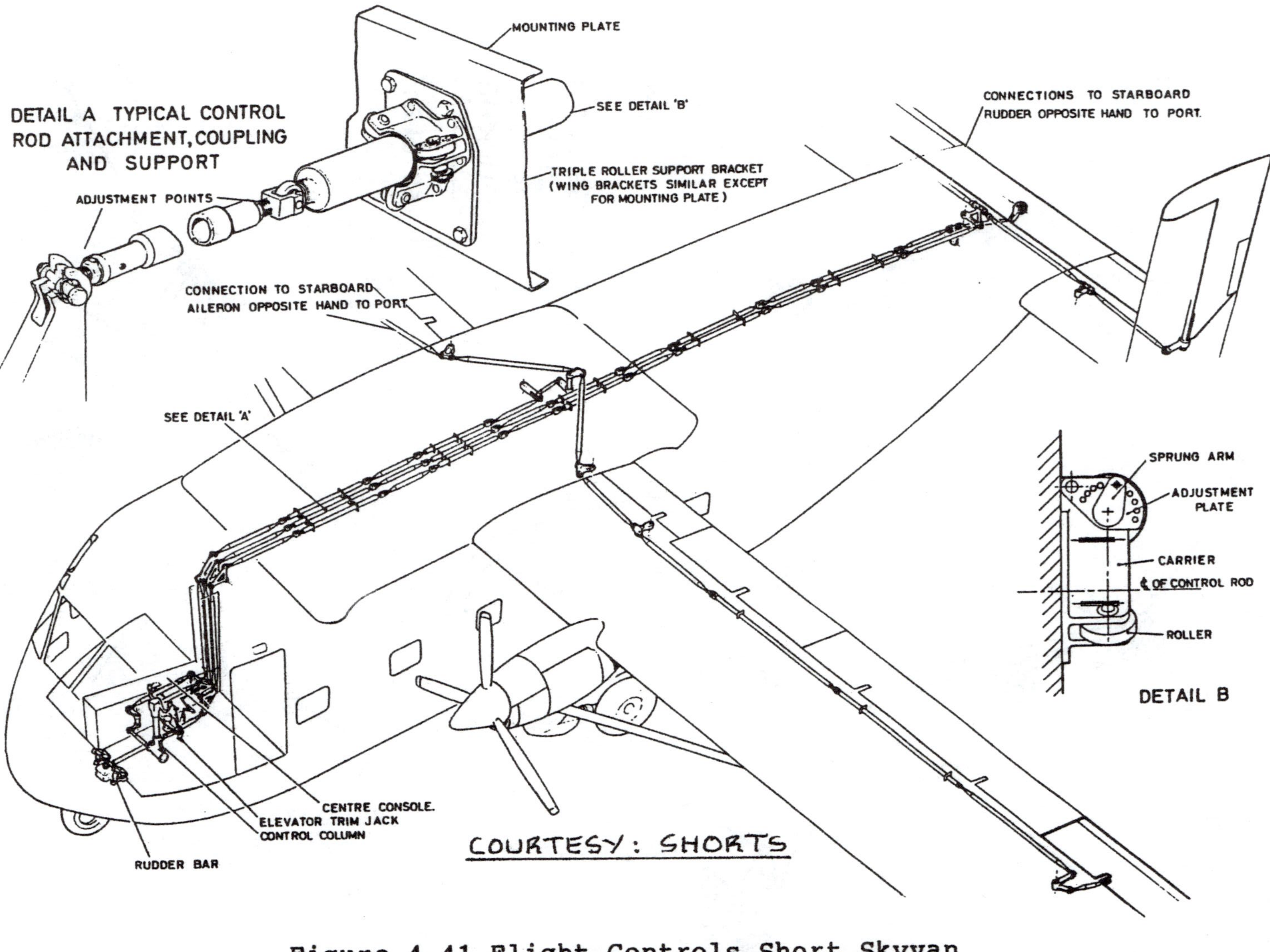

Figure 4.41 Flight Controls Short Skyvan

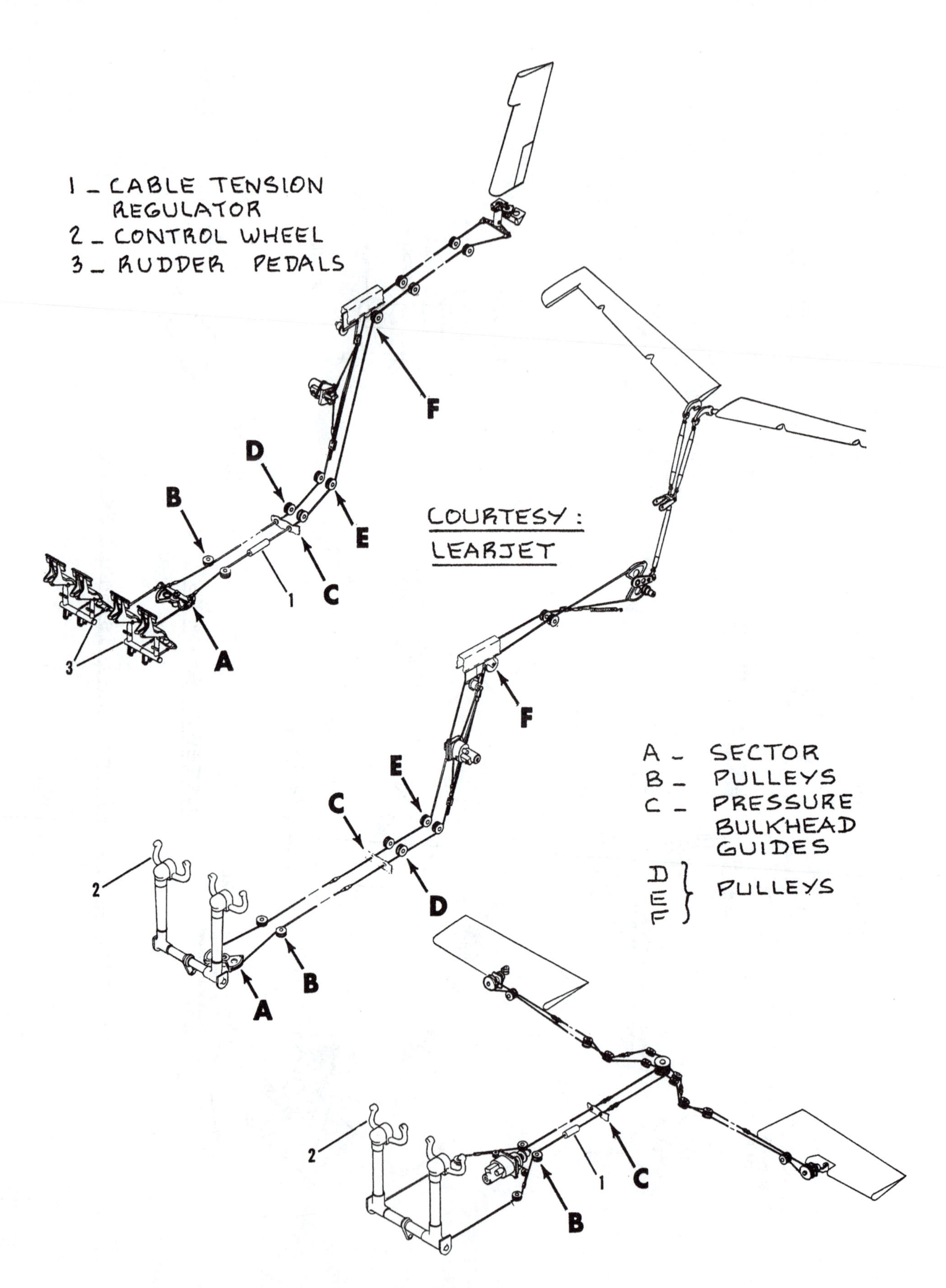

Figure 4.42 Flight Controls Learjet M23

4.3 LAYOUT OF IRREVERSIBLE FLIGHT CONTROL SYSTEMS

Definition: In an irreversible flight control system (hydraulic and/or electrical), when the cockpit controls are moved, the aerodynamic surface controls move and NOT vice versa.

Another way of stating this is to say that in an irreversible flight control system an actuator moves the aerodynamic surface controls. The pilot merely 'signals' the actuator to move. This signalling process is usually an irreversible process.

The material in this chapter is organized as:

4.3.1 Actuators (Servos)
4.3.2 Sizing of actuators
4.3.3 Basic arrangements of irreversible flight control systems
4.3.4 Design problems with irreversible flight control systems
4.3.5 Control routing through folding joints
4.3.6 Iron bird

4.3.1 Actuators (Servos)

In irreversible flight control systems the flight controls are moved by means of actuators (also called servos). Figure 4.43 shows a typical actuator-to-control-surface arrangement.

Actuators have two ends: a fixed and a moving end. The fixed end is attached to airplane structure. The moving end (also called the control rod) is attached to an aerodynamic surface control.

Two types of actuators are used most frequently: hydraulic and electromechanical. Figs 4.44 and 4.45 show examples of hydraulic and electromechanical actuators respectively. Sofar, in primary flight control systems the hydraulic actuator is the most frequently used type.

Actuators can be designed for linear or for rotary outputs. Figure 4.46 shows example installations.

Note: The structure at the fixed actuator end must be stiff. If this condition is not satisfied, the control surface will not deflect as much as the installation geometry would predict. This is called servo-elasticity.

For the actuator rod to move, the actuator must

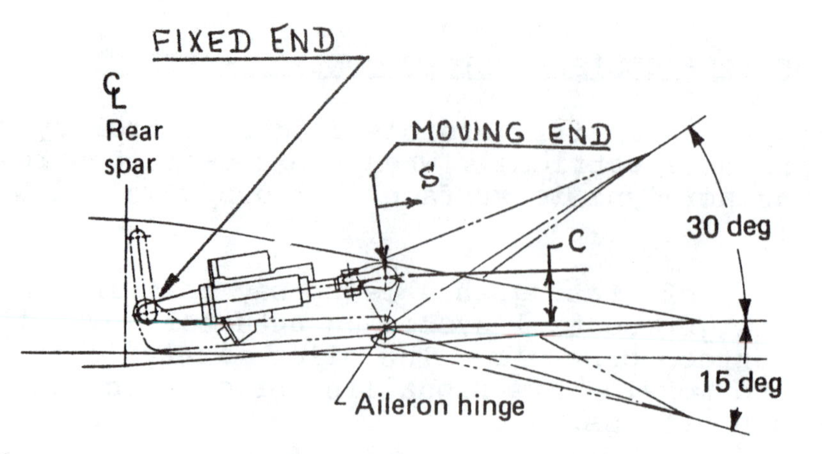

Figure 4.43 Typical Actuator-to-Surface Installation

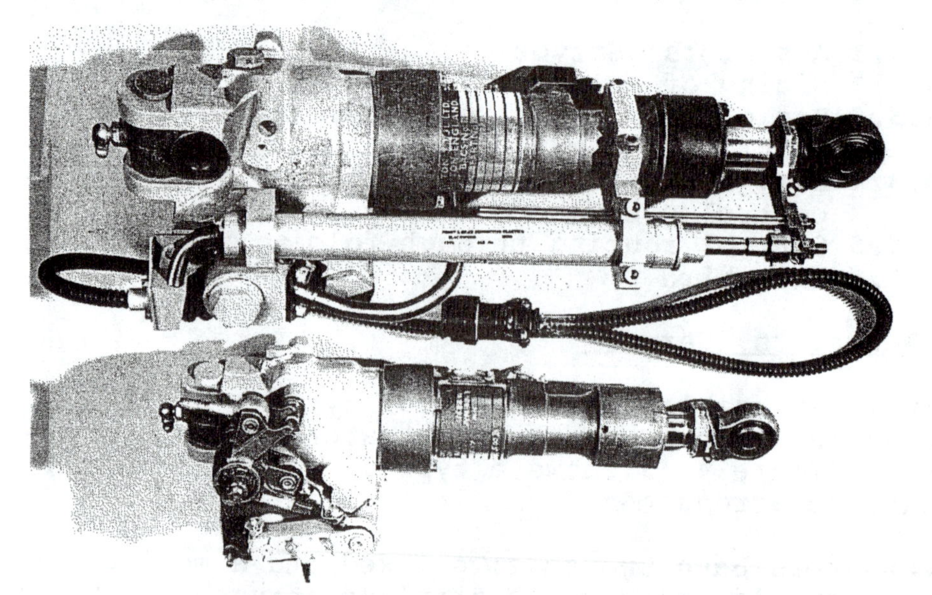

Figure 4.44 Hydraulic Actuator Examples

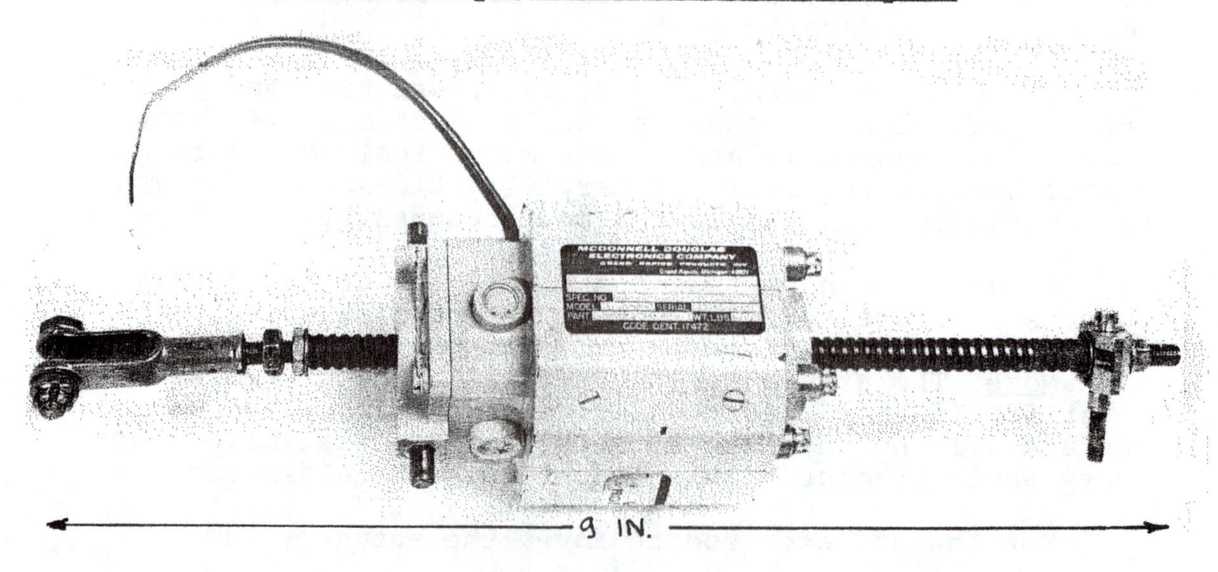

Figure 4.45 Electromechanical Actuator Example

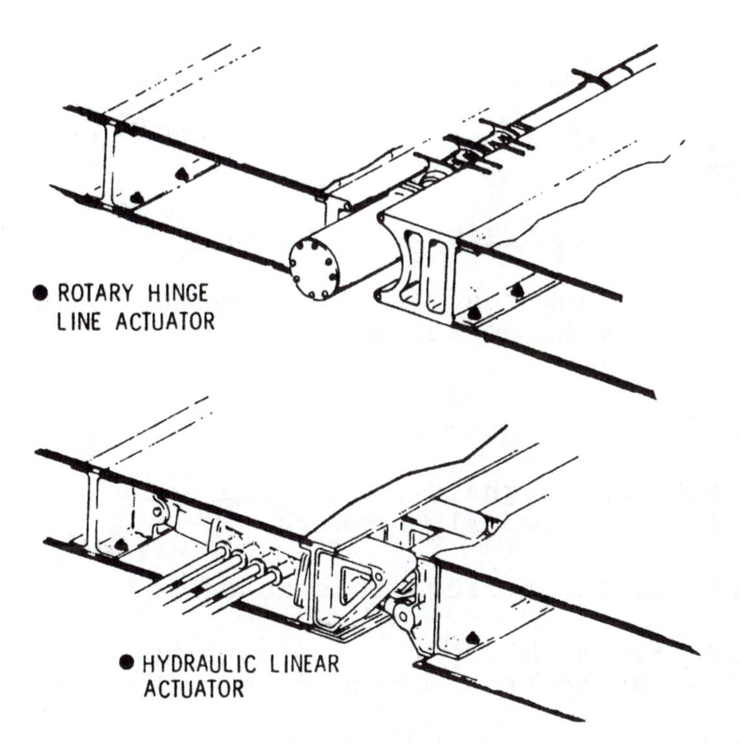

Figure 4.46 Rotary and Linear Actuator Installations

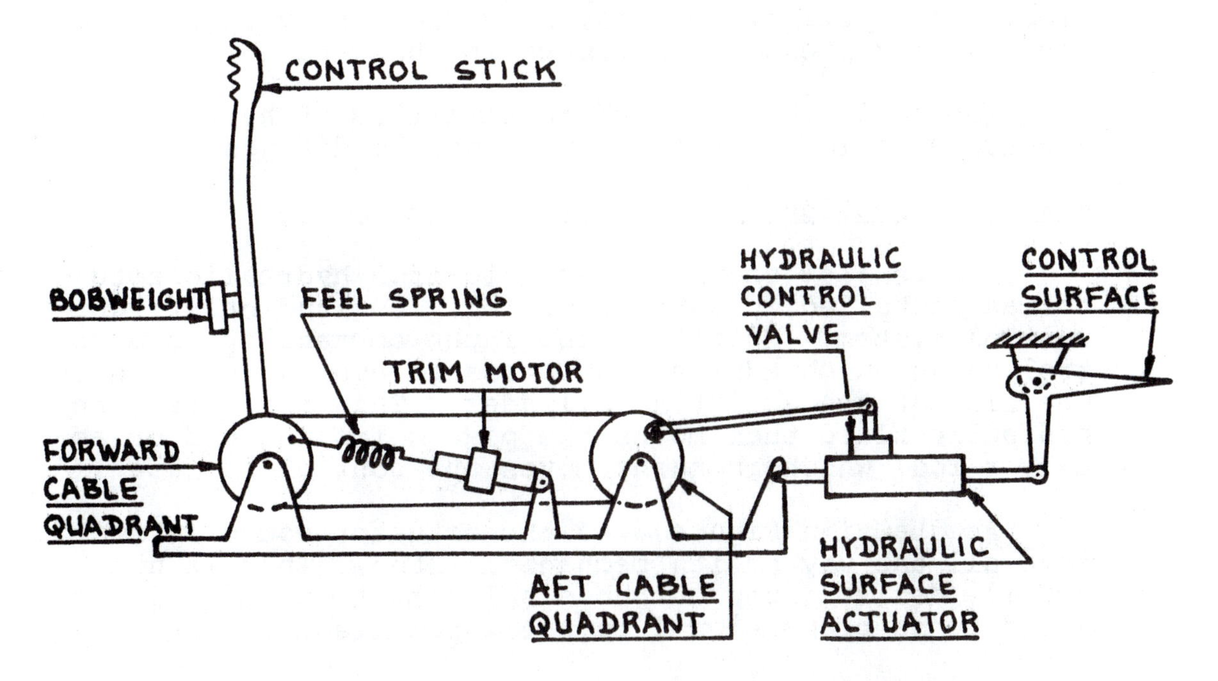

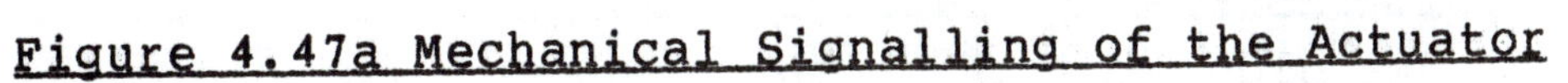
Figure 4.47a Mechanical Signalling of the Actuator

receive two inputs:

1) a command input signal, and 2) power

The general characteristics of these inputs will now be discussed.

1) Command input signal

The command input signal to an actuator can be sent by one or more of three methods:

a) mechanical signalling: this is done with cables and/or with push-rods

b) electrical signalling (Analog and/or digital): this is called fly-by-wire.

c) optical signalling: this is called fly-by-light

Figures 4.47a, b and c illustrate how actuator command signals may be transmitted from cockpit to actuator.

2) Power

Power to drive actuators (also called servos) is normally derived from a hydraulic system and/or from an electrical system*. The layout design of hydraulic and electrical systems is discussed in Chapters 6 and 7.

The basic operating characteristics of hydraulic and electromechanical actuators will now be discussed.

4.3.1.1 Operation of hydraulic actuators

Figure 4.48 shows a schematic of a hydraulic actuator as installed in an irreversible longitudinal flight control system. Note that the input command opens a control valve which admits high pressure hydraulic fluid to one side of the actuating cylinder. This high pressure hydraulic fluid then moves the piston (moving end or actuator rod) which in turns moves the control surface.

As drawn in Figure 4.48 the actuator would in fact move all the way to its mechanical stop. This is not usually desired: the output displacement is normally desired to be proportional to the input displacement.

---

* In some older installations pneumatic actuators are still used. These are not discussed in this text.

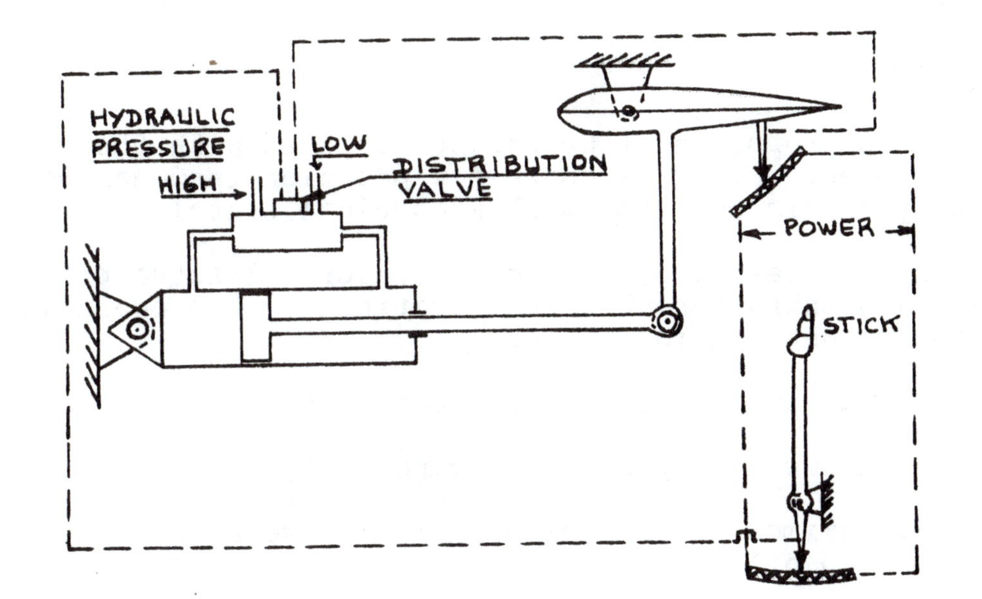

Figure 4.47b Electrical Signalling of the Actuator

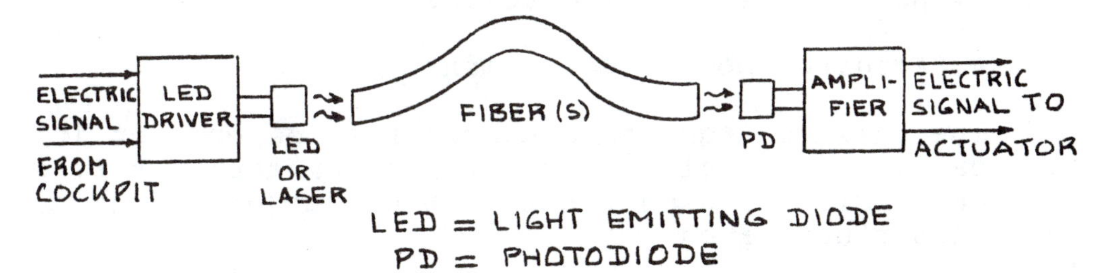

Figure 4.47c Optical Signalling of the Actuator

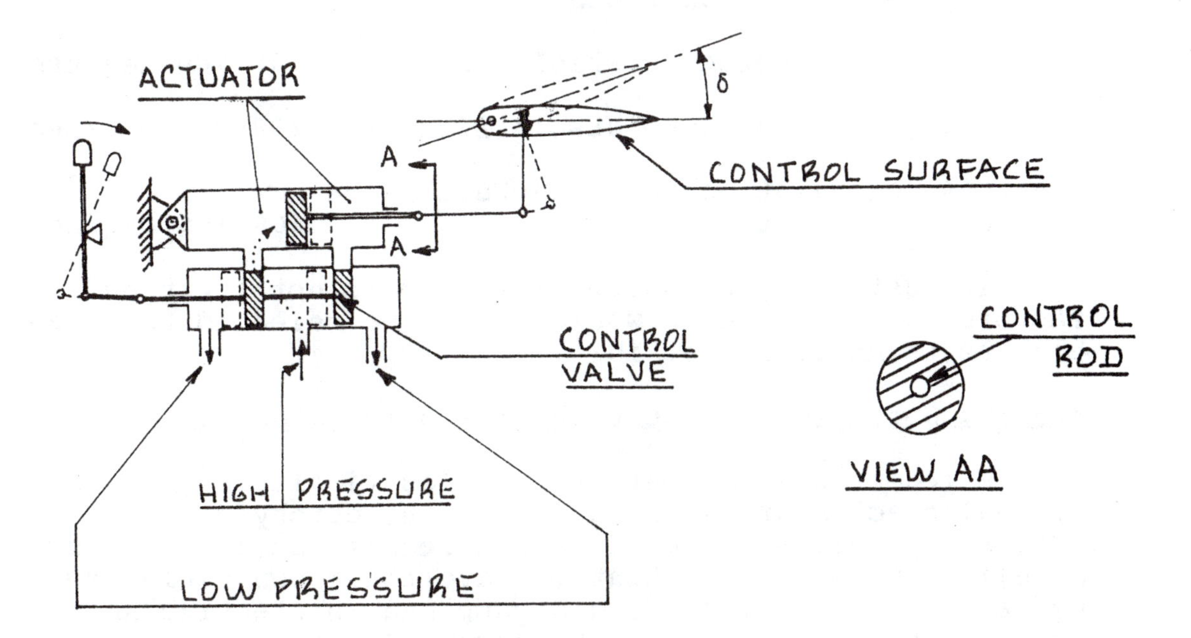

Figure 4.48 Mechanically Signalled Hydraulic Actuator Without Feedback

Such proportionality can be achieved with the help of feedback. Fig.4.49 shows how position feedback works in a hydraulic actuator with mechanical signalling.

In the design and/or selection of all actuators the following characteristics are critical to the operation of the flight control system:

1. Maximum output force capability

2. Maximum output stroke capability

3. Maximum output velocity (rate) versus load capability

4. Stall load magnitude

5. Actuator weight and volume

6. Actuator power requirements

The maximum required actuator rate depends on its application. Typical ranges for maximum actuator rates must be consistent with the following ranges of maximum control surface rates:

Transport flight controls: 100-200 deg/sec

Fighter flight controls:

inherently stable fighters: 100-300 deg/sec

de-facto stable fighters: 200-800 deg/sec

Gust alleviation and structural mode control systems: 500-800 deg/sec

If cost, weight and complexity were not limiting factors, actuator rates would always be selected to be as high as technically feasible.

4.3.1.2 Operation of electrohydrostatic actuators

A recent development of the electrically signalled hydraulic actuator is the so-called electrohydrostatic actuator. This actuator does not require an airplane hydraulic system. EHS actuators have their own miniature hydraulic system, including a pump and an electric motor (rare earth) which drives the pump. EHS actuators are driven by electric power and signalled by fly-by-wire or by fly-by-light systems. Figure 4.50 shows an example.

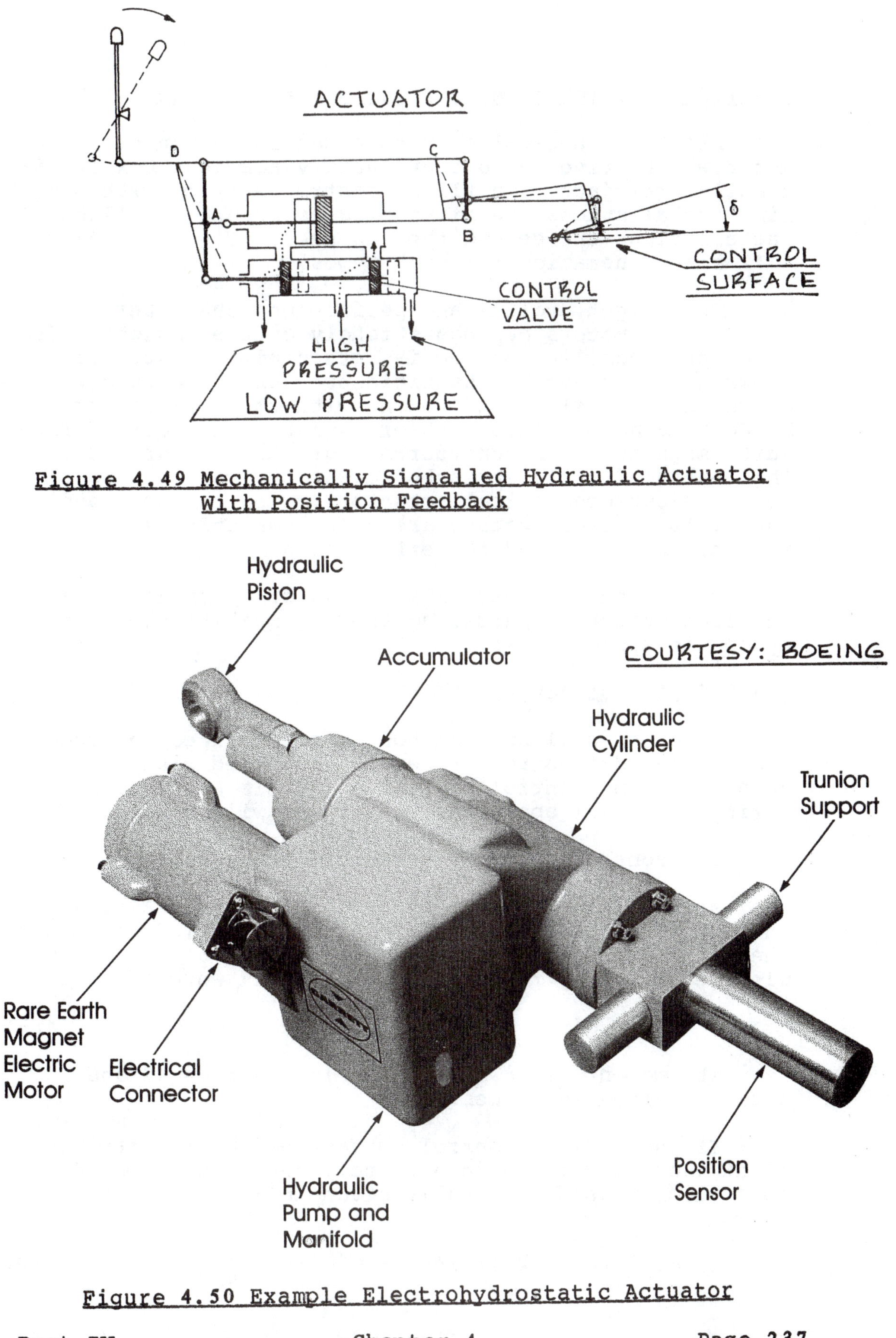

Figure 4.49 Mechanically Signalled Hydraulic Actuator With Position Feedback

Figure 4.50 Example Electrohydrostatic Actuator

4.3.1.3 Operation of electromechanical actuators

Electromechanical actuators consist of an electric motor which drives an output shaft via a gear box or via a ball-screw jack. In flight control systems with hinge-line installations the first type is used. For linear installations the second type is used. Figures 4.51 and 4.52 show schematics for both types.

The weight, volume and performance characteristics of electric motors depends strongly on the magnetic field strength capability of the magnets used. Figure 4.53 shows a comparison of the fieldstrength associated with conventional ALNICO magnets and with the more recent Nd-Fe-B magnets. These latter magnets (also called rare earth magnets) have considerably greater performance. This enhanced field strength capability makes it possible to use electromechanical actuators: with the new magnet materials these actuators are quite competitive in weight, in volume and in performance.

Reference 26 contains a design study of electromechanical systems, indicating their competitiveness with hydraulic systems.

4.3.2 Sizing of Actuators

References 21 and 27 should be consulted for specific maximum design requirements associated with actuators used in flight control systems. In this section only the sizing to normal operating conditions will be discussed.

The control force, F_c required to overcome the control surface hingemoment may be computed from Eqn.(4.6) on page 202. If the actuator stroke is called 's' and the control surface deflection is called 'δ' the following relationship follows from an 'equal work' condition:

$$F_c = \{C_{h_\delta} \bar{q}(S\bar{c})_{\text{control surface}}\}/c \qquad (4.9)$$

The moment arm c is defined in Figure 4.43 and is normally given in inches.

If the desired control surface deflection rate is dδ/dt (normally given in deg/sec), the required control rod velocity ds/dt (normally given in in/sec), follows from:

$$ds/dt = (1/57.3)(c)d\delta/dt \qquad (4.10)$$

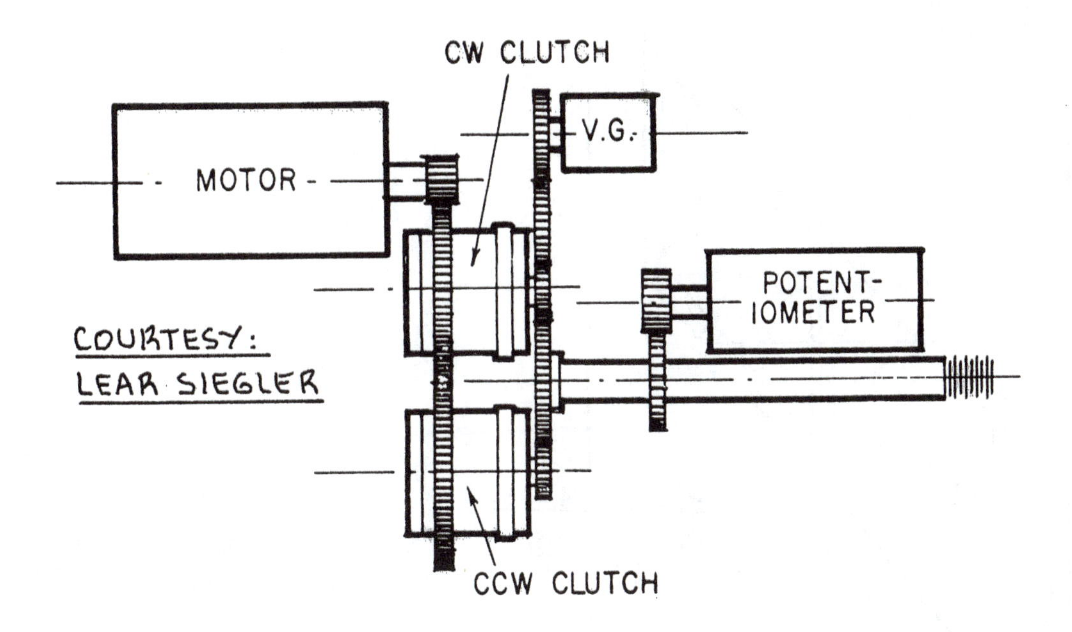

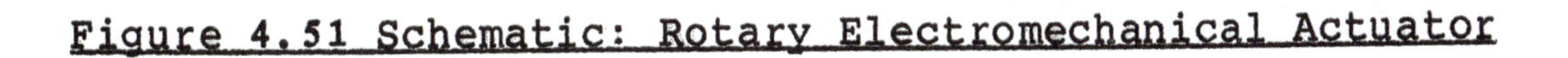
Figure 4.51 Schematic: Rotary Electromechanical Actuator

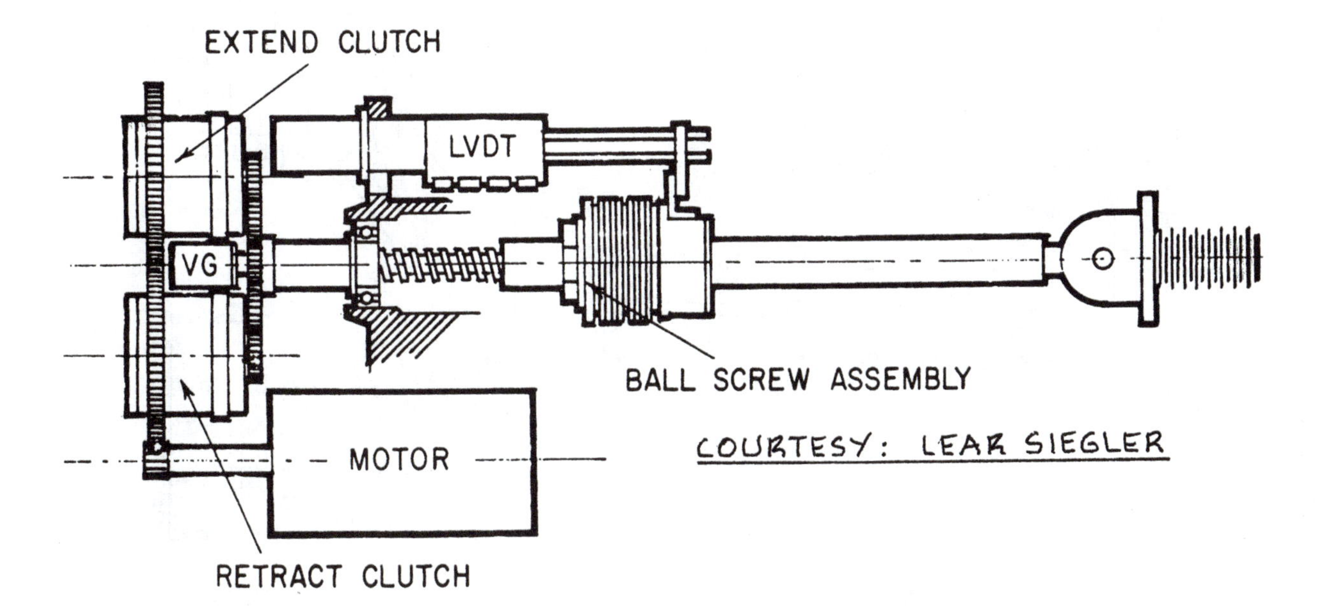

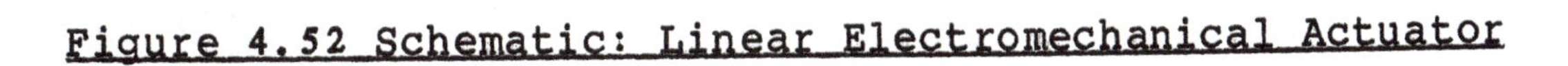
Figure 4.52 Schematic: Linear Electromechanical Actuator

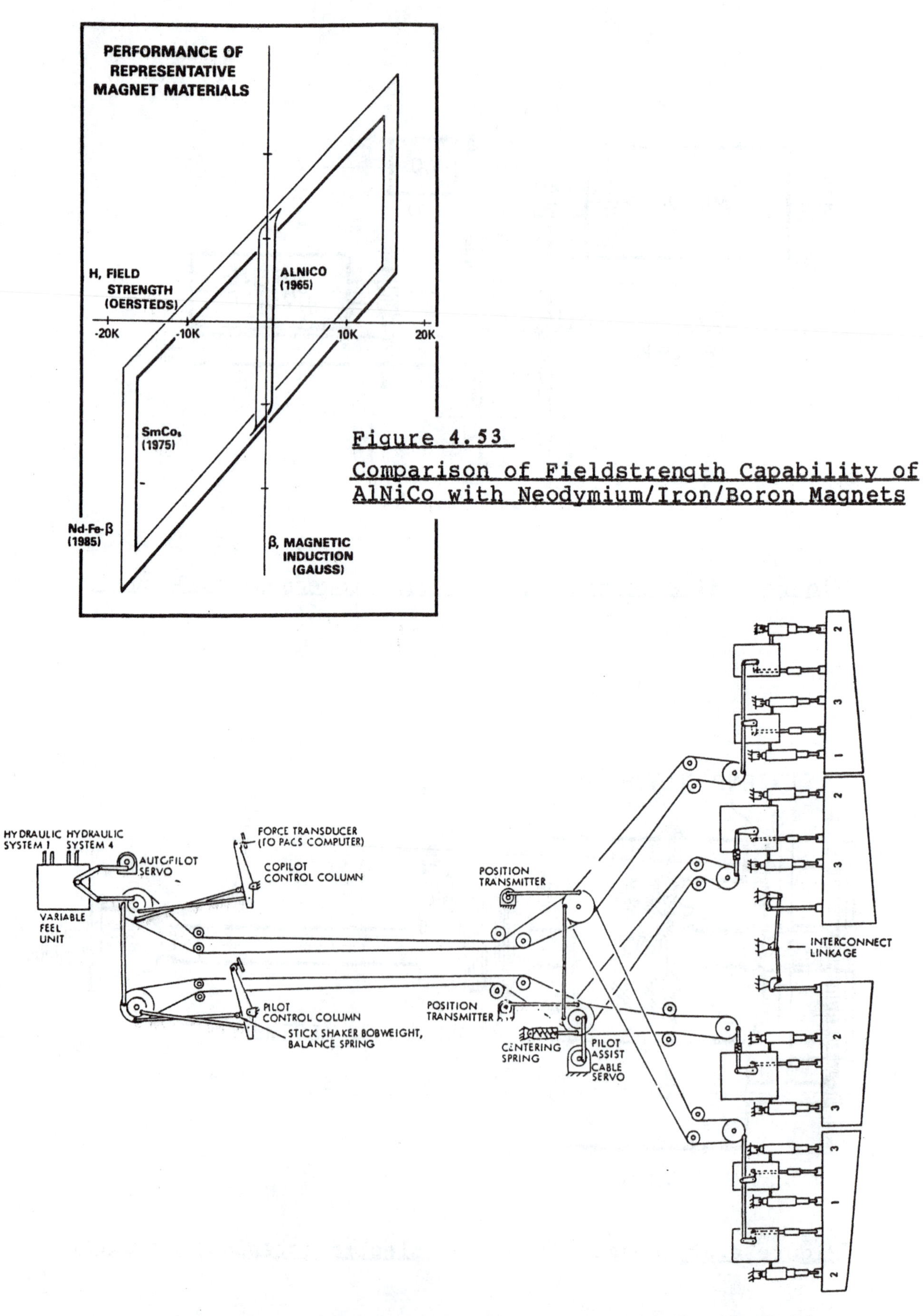

Figure 4.53

Comparison of Fieldstrength Capability of AlNiCo with Neodymium/Iron/Boron Magnets

Figure 4.54 Lockheed C5-A Elevator Control Schematic

Hydraulic Actuators

If the hydraulic system operating pressure is p_h (normally given in psi), while the effective piston area is S_p (normally given in in.^{2}), the following relation may be used to compute the required actuator size:

$$S_p = F_c/(p_h - 200) \tag{4.11}$$

Figure 4.48 defines the effective piston area, S_p.

The hydraulic system backport pressure is assumed to be typically 200 psi. The required hydraulic fluid flow follows from:

$$\text{Hydraulic fluid flow} = S_p(ds/dt) \text{ in}^3/\text{sec} \tag{4.12}$$

Typical operating pressures in current hydraulic systems are 3000-3500 psi for commercial applications and 5000 psi for military applications. Ref.28 states that systems with 8000 psi pressure may become feasible. Chapter 6 discusses the layout of hydraulic systems.

Electromechanical Actuators

These actuators are rated in terms of their normal operating load as given in Eqn.(4.9), their stroke velocity as given by Eqn.(4.10) and their stall load. By using an actuator efficiency of 80 percent the required electrical operating power follows once the basic operating voltage of the system is known. Typical operating voltages are 12-28 VDC and 60-90 KVA. Chapter 7 presents a discussion of electrical systems.

4.3.3 Basic Arrangements of Irreversible Flight Control Systems

Table 4.2 lists example applications of different types of irreversible flight control systems.

4.3.3.1 Hydraulic system with mechanical signalling

Figure 4.54 shows a schematic for the elevator control system of the Lockheed C5-A. It is a hydraulically powered, irreversible system with mechanical signalling.

This type of system is typical of that used in most commercial jet transports as well.

Table 4.2 Examples of Control System Applications

Airplane Type	Type of Signalling	Actuators	Redundancy	Reversibility	Trim System
Boeing 737					
Lateral	Mechanical	Hydraulic	2 systems	Full reversion	Through feel system
Longitudinal	Mechanical	Hydraulic	2 systems	Full reversion	El.Mech./Stabilizer
Directional	Mechanical	Hydraulic	3 systems	No reversion	Through feel system
Boeing 747					
Lateral	Mechanical	Hydraulic	4 systems	No reversion	Through feel system
Longitudinal	Mechanical	Hydraulic	4 systems	No reversion	Through feel system
Directional	Mechanical	Hydraulic	4 systems	No reversion	Through feel system
Rockwell B-1B					
Lateral	Mechanical Fly-by-wire	Hydraulic	4 systems	No reversion	Through feel system using diff. stabil.
Longitudinal	Mechanical	Hydraulic	4 systems	Fly-by-wire	Through feel system using stabilizers
Directional	Mechanical	Hydraulic	4 systems	Fly-by-wire	Through feel system
McDonnell-Douglas F/A-18A					
Lateral	Fly-by-wire	Hydraulic	2 systems	No back-up	Through feel system
Longitudinal	Quadriplex	Hydraulic	2 systems	Mechanical	Through feel system
Directional	Digital	Hydraulic	2 systems	No back-up	Through feel system
Aeritalia/Aermacchi/EMBRAER AM-X					
Lateral					
Spoilers	Fly-by-wire	Hydraulic	2 systems	None	Through feel system
Ailerons	Mechanical	Hydraulic	2 systems	Full reversion	Through feel system
Longitudinal	Mechanical	Hydraulic	2 systems	Full reversion	Through feel system
Directional	Fly-by-wire	Hydraulic	2 systems	No back-up	Through feel system

Figure 4.55 shows a schematic for the longitudinal control system of the YF-17 fighter. The hydraulic actuators control the left and right stabilizer.

Mechanical signalling is done with cable and/or with push-rod systems. Such systems are mechanically similar to those used in reversible control systems. Section 4.1 deals with the layout design of reversible flight control systems and section 4.2 gives example applications.

A problem with powered controls is: redundancy. If no mechanical reversion is provided it is necessary to design redundancy into the hydraulics, the electrics and the actuators so that complete system failure is an 'extremely' remote event. (Ch.13 defines event probability)

In powered flight controls designing for redundancy means the following:

In hydraulically driven systems: a large number of hydraulic pumps, at least three independent power sources for the hydraulic pumps and redundant actuators.

In electrically driven systems: a large number of electric generators, two independent standby sources for electric power and redundant actuators.

A fundamental advantage of mechanical signalling is that provisions for mechanical reversion can be incorporated for only a minor weight penalty.

Figures 4.56 through 4.58 show schematics for the Boeing 737 flight control system. Note that the rudder axis incorporates complete hydraulic system redundancy. The lateral and longitudinal axes incorporate complete mechanical reversion and thus do not need hydraulic redundancy. Observe that both aerodynamic balance and mass balance were added to make this certifiable.

Mechanical signalling involves a large weight penalty. Electrical and/or optical signalling avoid this.

4.3.3.2 Hydraulic system with electrical or optical signalling

Figure 4.59 shows a schematic for this system type. Note the triplicated signal paths and the use of three actuators per elevator.

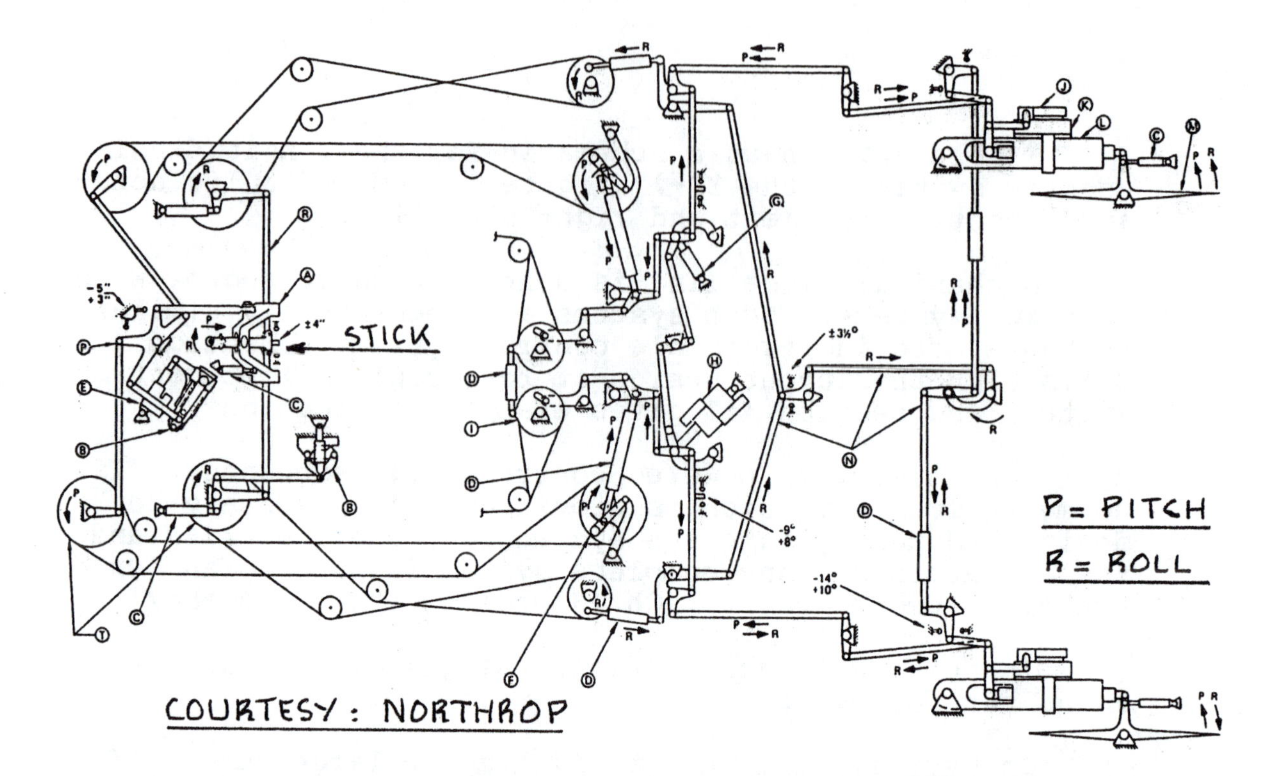

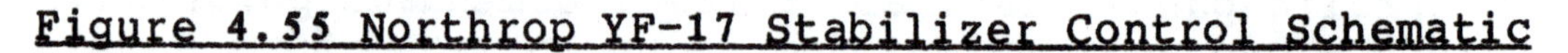

Figure 4.55 Northrop YF-17 Stabilizer Control Schematic

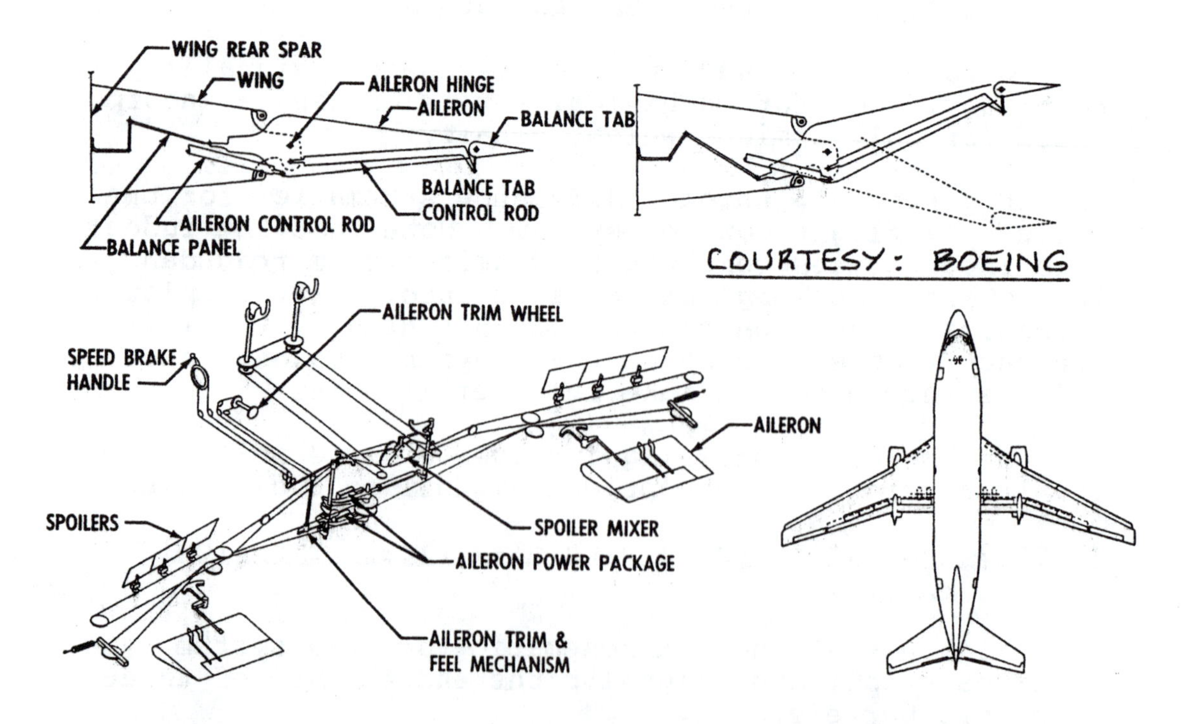

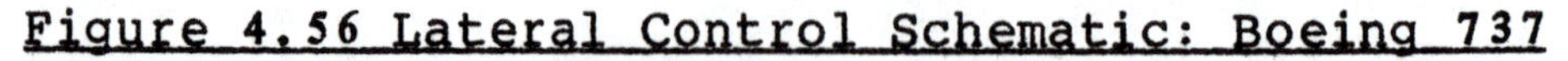

Figure 4.56 Lateral Control Schematic: Boeing 737

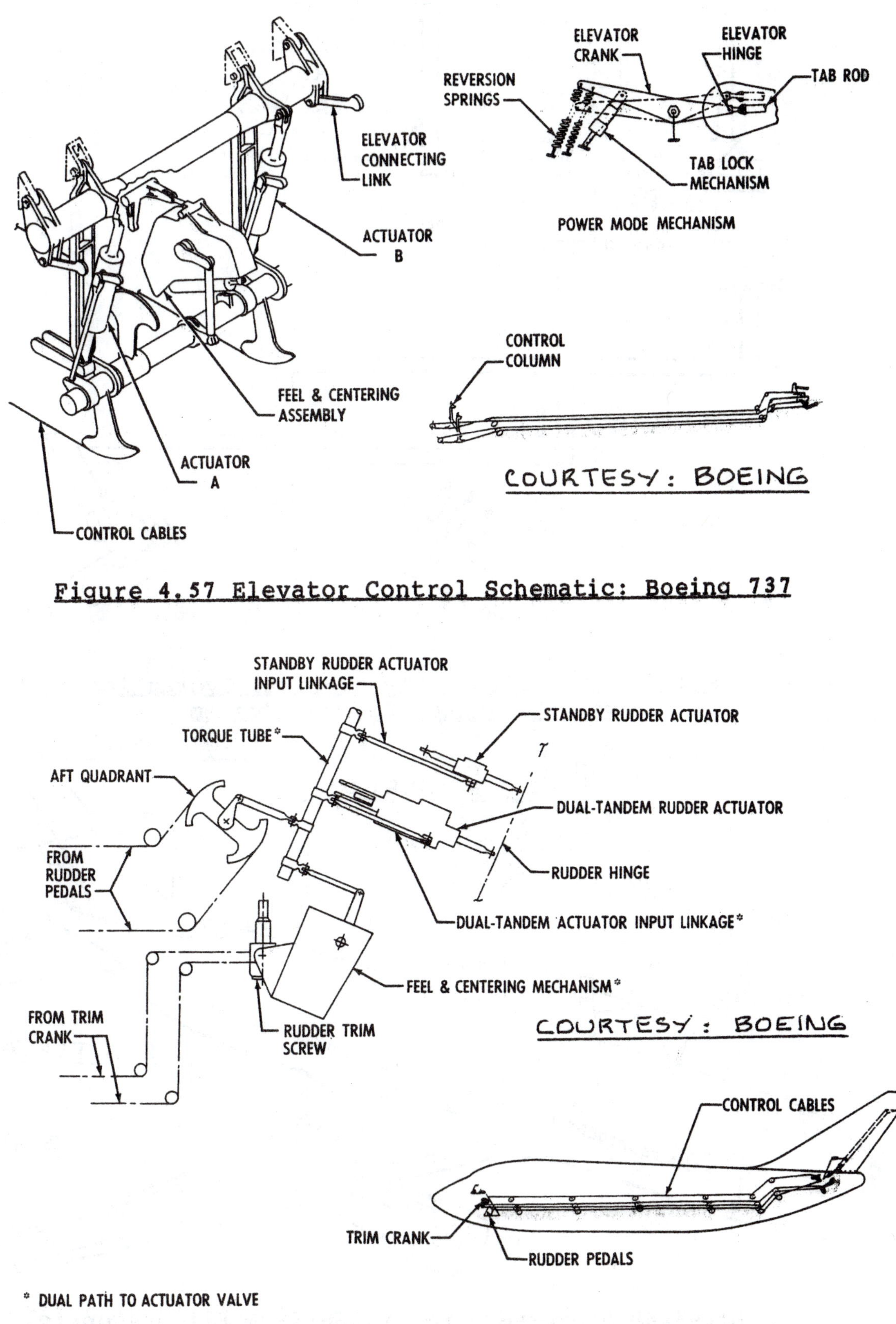

Figure 4.57 Elevator Control Schematic: Boeing 737

Figure 4.58 Rudder Control Schematic: Boeing 737

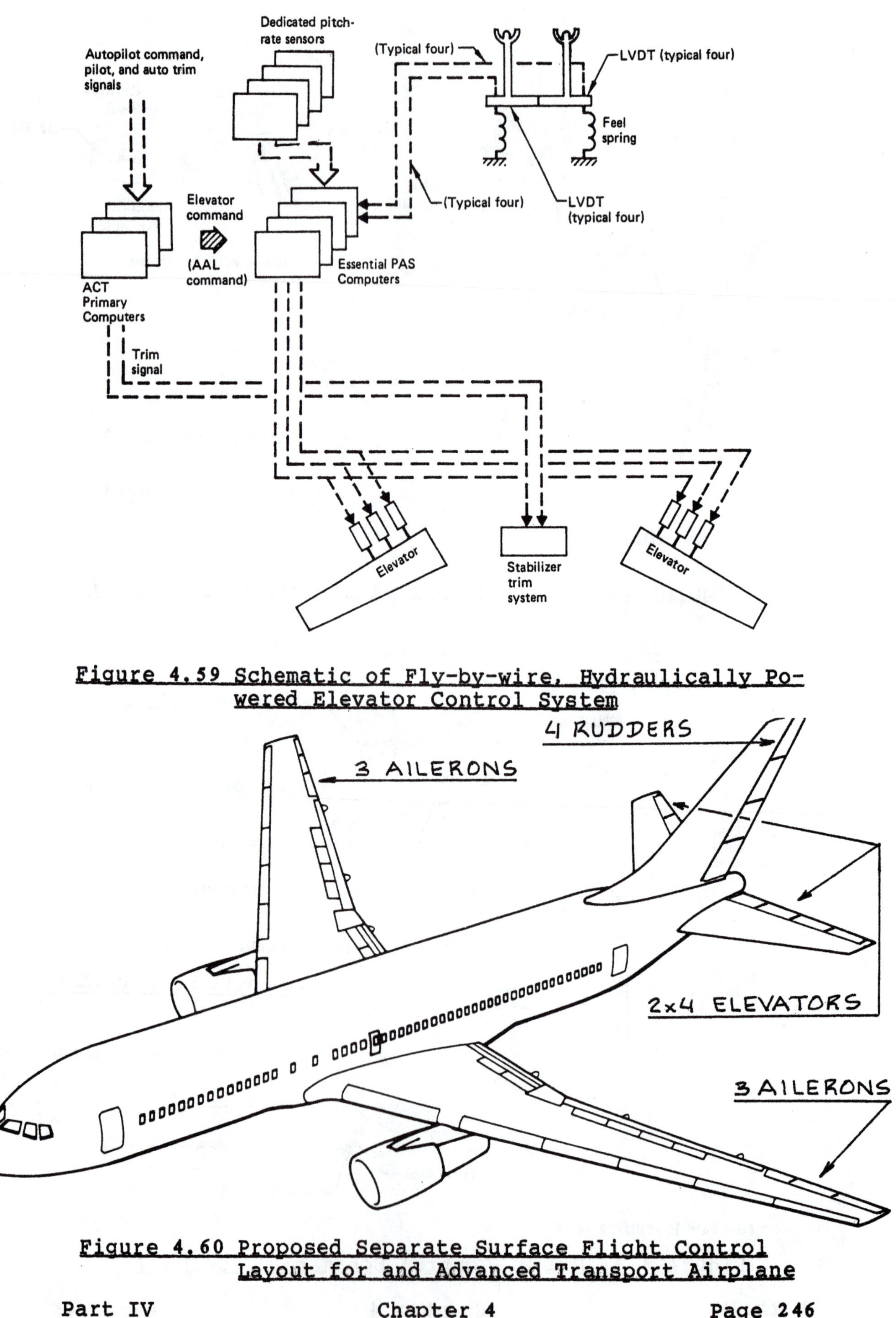

Figure 4.59 Schematic of Fly-by-wire, Hydraulically Powered Elevator Control System

Figure 4.60 Proposed Separate Surface Flight Control Layout for and Advanced Transport Airplane

4.3.3.3 Separate surface control systems with electrical or optical signalling

Because of requirements for redundancy and because primary flight control system actuators are normally sized to each control surface individually, these actuators tend to be large, complex and very expensive. In fighter airplanes these actuators contribute significantly to maintenance, cost and inventory problems.

An attractive way around these problems is a system where all flight control surfaces are split into smaller separate surfaces. These surfaces can be sized so that they all require the same type of actuator. This way, large numbers of these actuators can be produced which reduces unit cost. Also, because of the large number of actuators per airplane it won't be necessary to design redundancy into each actuator, again lowering its cost.

Figure 4.60 illustrates a possible separate surface control system configuration.

4.3.3.4 Electromechanical flight control systems

Application of these systems has sofar been restricted to secondary flight controls. Flaps and trim systems frequently employ electromechanical actuators. Examples are shown in Section 4.5.

Figure 4.61 shows a potential system layout for an electromechanical flight control system.

4.3.4 Design Problems With Irreversible Flight Control Systems

Major problems associated with reversible flight controls are:

1. Complexity
2. Reliability
3. Redundancy
4. Cost
5. Accessibility for repairs
6. Susceptibility to lightning strikes in the case of electrically signalled systems

Major advantages associated with reversible flight control systems are:

1. Flexibility in combining pilot control commands with automatic control commands

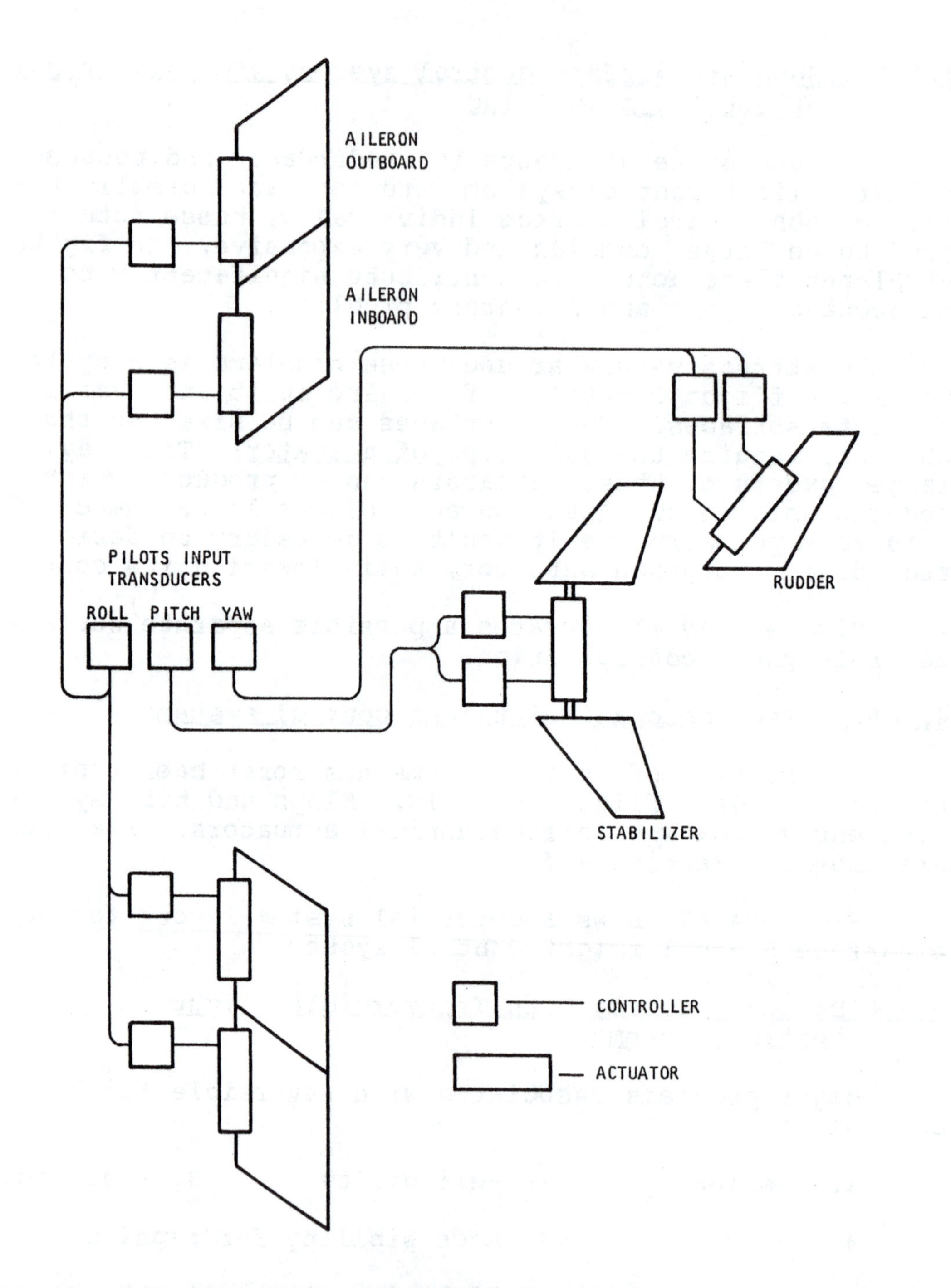

Figure 4.61 Proposed Schematic for an Electromechanical Flight Control System for a Fighter

2. Ability to tailor handling qualities

3. Potential of lower weight, particularly if electrical and/or optical signalling is used.

In the design and development of irreversible systems key roles are played by the following design considerations:

A. Reliability of all system components and the resulting need for redundancy.

B. Relative location of system components to assure that failures or damage induced by outside causes cannot cause total failure of the system.

C. Maintenance and accessibility.

It is useful to list a number of DO'S and DON'TS with regard to the layout design of irreversible flight controls:

<u>DO'S:</u>

Actuators: should be located at the aerodynamic surface controls such that maximum system stiffness is attained.

should be located such that inspection, maintenance and removal can be easily carried out.

<u>DON'TS:</u>

Actuator signal paths:

do not locate redundant signal paths close together. Closeness invites disaster due to such causes as: engine component disintegration, propeller blade separation, local structural failure due failure of another system, terrorist action or combat damage.

do not locate the signal paths so that lightning strikes can cause the system to fail.

4.3.5 Control Routing Through Folding Joints

Carrier based airplanes and some cargo airplanes require folding joints in wing, empennage or fuselage. Control routing through wing joints is discussed in Part III, Section 4.3. Figure 4.62 shows a typical folding

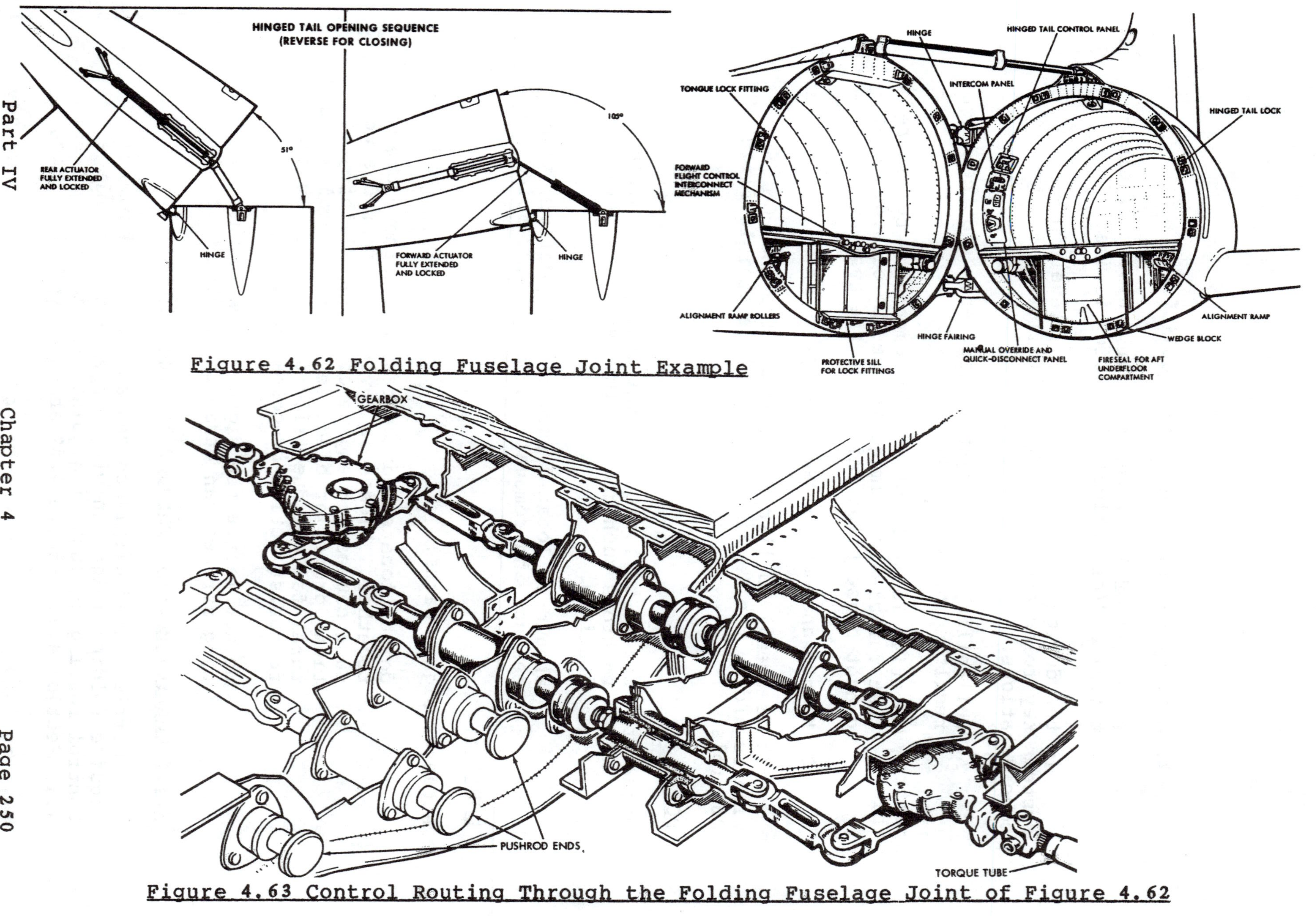

Figure 4.62 Folding Fuselage Joint Example

Figure 4.63 Control Routing Through the Folding Fuselage Joint of Figure 4.62

fuselage joint. The control routing through this joint is shown in Figure 4.63.

4.3.6 Iron Birds

To ensure the proper operation of complex flight control systems it is highly desirable to construct a so-called Iron Bird. Such an iron bird represents a full scale working model of the entire flight control system. Actual components used in the airplane are also installed in this iron bird. Figure 4.64 shows an example of an iron bird. Any defects in the design of the flight control system can be relatively easily 'ironed' out in this manner.

Figure 4.64 Iron Bird for Boeing 737

4.4 EXAMPLES OF IRREVERSIBLE FLIGHT CONTROL SYSTEMS

In this section a number of example layouts of irreversible flight control systems are presented as Figures 4.65 through 4.79. Commentary on these figures is given below.

==

Figure 4.65 Flight Control Features: Boeing 767

The Boeing 767 has a hydraulically powered, mechanically signalled flight control system for all aerodynamic controls except the spoilers. The spoilers are hydraulically powered but signalled by a fly-by-wire system.

Hydraulic power is provided by three independent systems. Two are powered by engine driven pumps, one by engine bleed-air. An emergency hydraulic power source is provided by a ram-air driven turbine (RAT). A description of the 767 hydraulic system is given in Chapter 6.

Figure 4.66 Control Surface Actuation: Boeing 767

Observe the redundancy in the actuator layout.

Figure 4.67 Aileron Control System: Boeing 767

The pilot lateral control inputs are translated into valve displacements in the central control actuators. If this system were to fail, an override is available: there is a direct mechanical link from the right control wheel to the aileron control actuators.

Each aileron is powered by two independent actuators. The distribution of hydraulic power to these actuators is indicated. Observe the lockout system for the outboard ailerons: this is necessary to prevent aileron/wing reversal at high speed. The lockout system is commanded by flap position.

The autopilot inputs to the aileron controls are routed to the central control actuators. These in turn signal the aileron actuators.

Figure 4.68 Spoiler Control System: Boeing 767

The spoilers are signalled by a fly-by-wire system using control wheel steering transducers. Logic circuits decide whether the spoilers are operated as lateral controls, as in-flight speedbrakes or as ground spoilers.

Note the emergency evacuation system spoiler override. This system lowers the spoilers when the emergency evacuation system is activated. This prevents the evacuation slides from being damaged by extended spoilers.

Figure 4.69 Elevator Control System: Boeing 767

Note that the primary elevator control circuit uses mechanical signalling. The autopilot flight control computer signals 3 autopilot actuators which in turn drive the mechanical system to signal the elevator actuators. Three autopilot actuators are required for Category 3A automatic landing operation.

Figure 4.70 Directional Control System: Boeing 767

The 767 uses a one-piece rudder driven by three independent actuators. The rudder pedals are mechanically connected to the actuator valves. The yaw dampers use two independent servos to signal the rudder servos.

Figure 4.71 Primary Flight Control System: McDD DC-10

The McDonnell-Douglas DC-10 employs a hydraulically powered, mechanically signalled flight control system. Note the split elevators and the split rudder. In the roll control axis, low speed and high speed ailerons are used (as in the 767) in addition to spoilers.

Hydraulic power is provided by three independent systems. Each system is powered by two engine driven pumps. One system also receives power from an electrically driven pump. This system allows control of the airplane if all engines were to fail.

Figure 4.72 Distribution of Hydraulic System Power to the Flight Controls

The distribution of hydraulic power to the flight controls was arranged to be able to cope with complete failure of two hydraulic systems.

Figure 4.73 Cockpit Flight Controls: McDD DC-10

Note the cable runs extending aft and below the cabin floor. Access to the control cables is from below.

Figure 4.74 Lateral Control System: McDD DC-10

Note that ALL lateral controls are mechanically signalled in this airplane. There is an electrical interconnect between the landing gear and the spoiler controls to provide for ground speedbrake operation.

The spoilers are used in a direct lift control mode with the flaps down. A mixing unit tells the spoilers which mode of operation is desired. This mixer unit is illustrated in Figure 4.75.

Figure 4.75 Aileron-Spoiler Mixer Unit: McDD DC-10

This unit decides the mode of operation of the aileron/spoiler combination. The unit is totally mechanical.

Figure 4.76 Elevator Control System: McDD DC-10

Note the direct link between the elevator cockpit control and the elevator control cables.

Pilot inputs to the elevator actuators are mechanical. Autopilot inputs to these actuators are electrical!

In an irreversible flight control system it is necessary to arrange for control force 'feel' artificially. This is done with the help of a control feel unit.

Figure 4.77 Control Feel Unit: McDD DC-10

This figure indicates schematically the types of input signals required to provide the pilot with the proper control-force/speed and control-force/'g' gradients.

Figure 4.78 Directional Control System: McDD DC-10

Note that the rudder is split into two parts, each driven by different hydraulic systems.

Figure 4.79 Control Surface Damper: McDD DC-10

In irreversible flight controls mass balancing of aerodynamic surface controls is not required as long as mechanical failure of the actuator is an extremely remote event. Whether or not it is depends on the design choices made in the actuator installation. If actuator failure were to occur, either mass balancing has to be included (to prevent flutter) or a control surface damper unit must be installed. The DC-10 uses control surface dampers. The Boeing 767 uses mass balancing as shown in Figure 4.69.

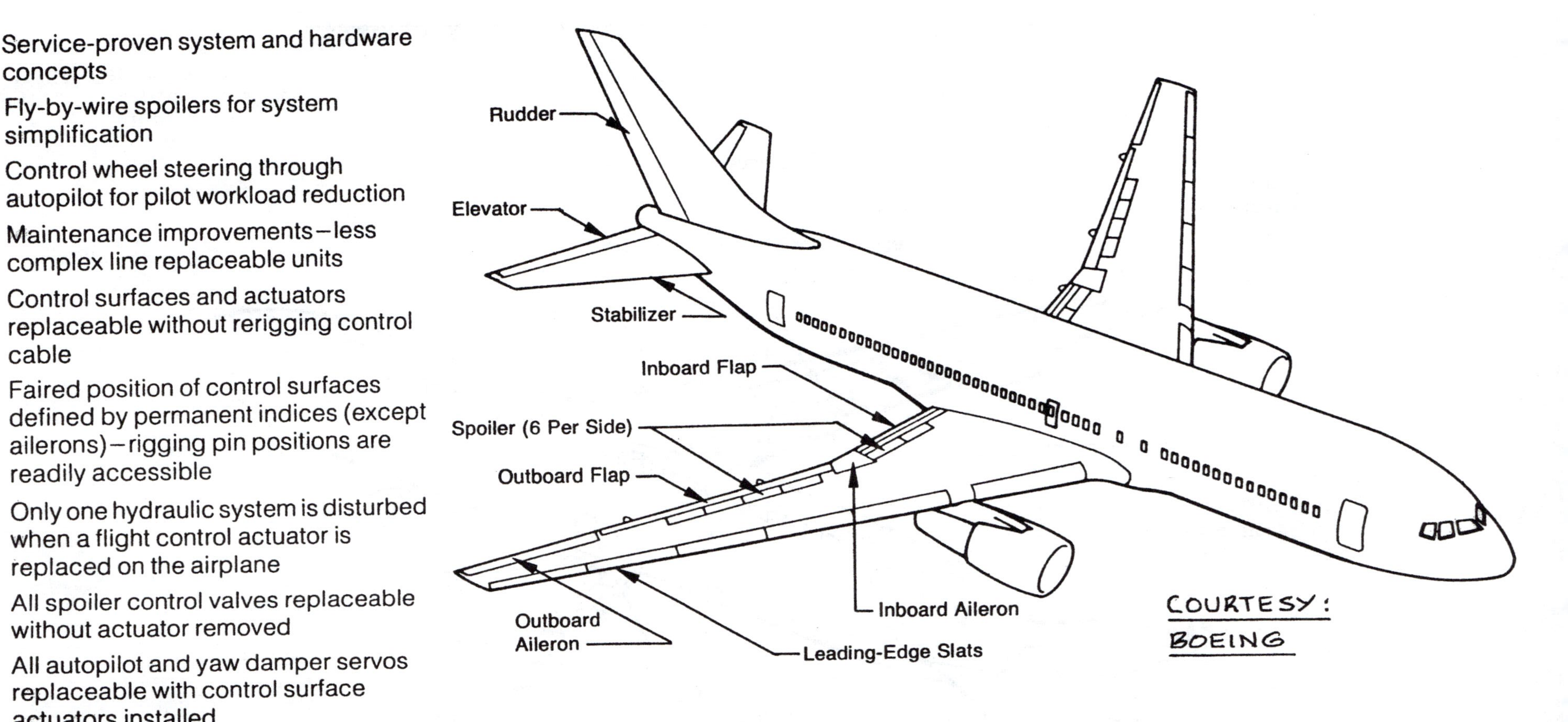

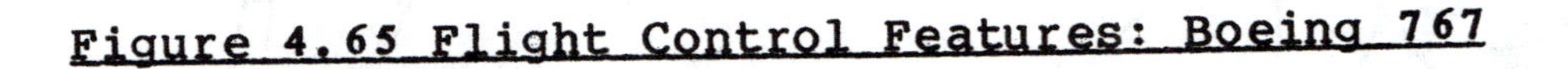

Figure 4.65 Flight Control Features: Boeing 767

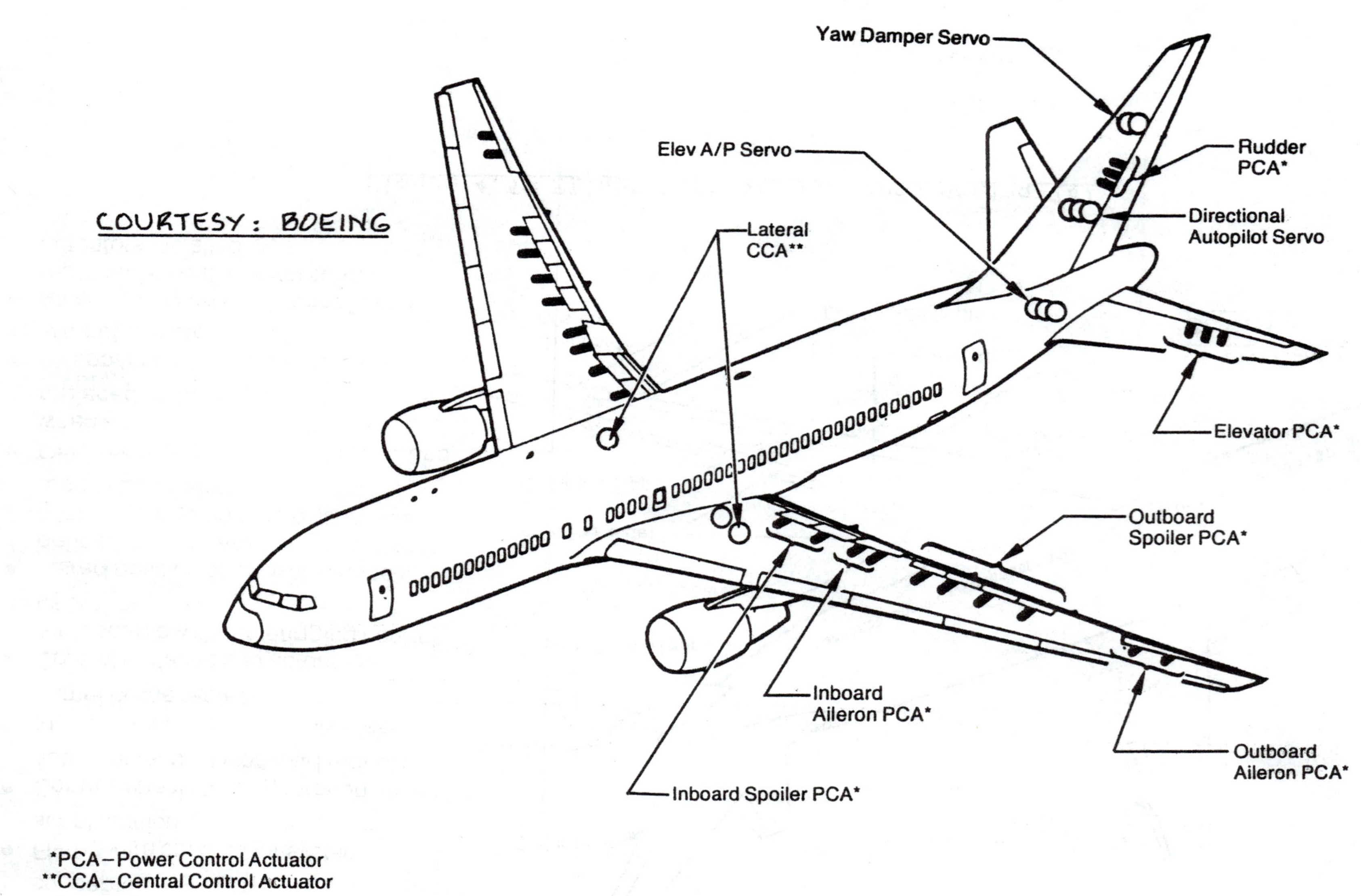

Figure 4.66 Control Surface Actuation: Boeing 767

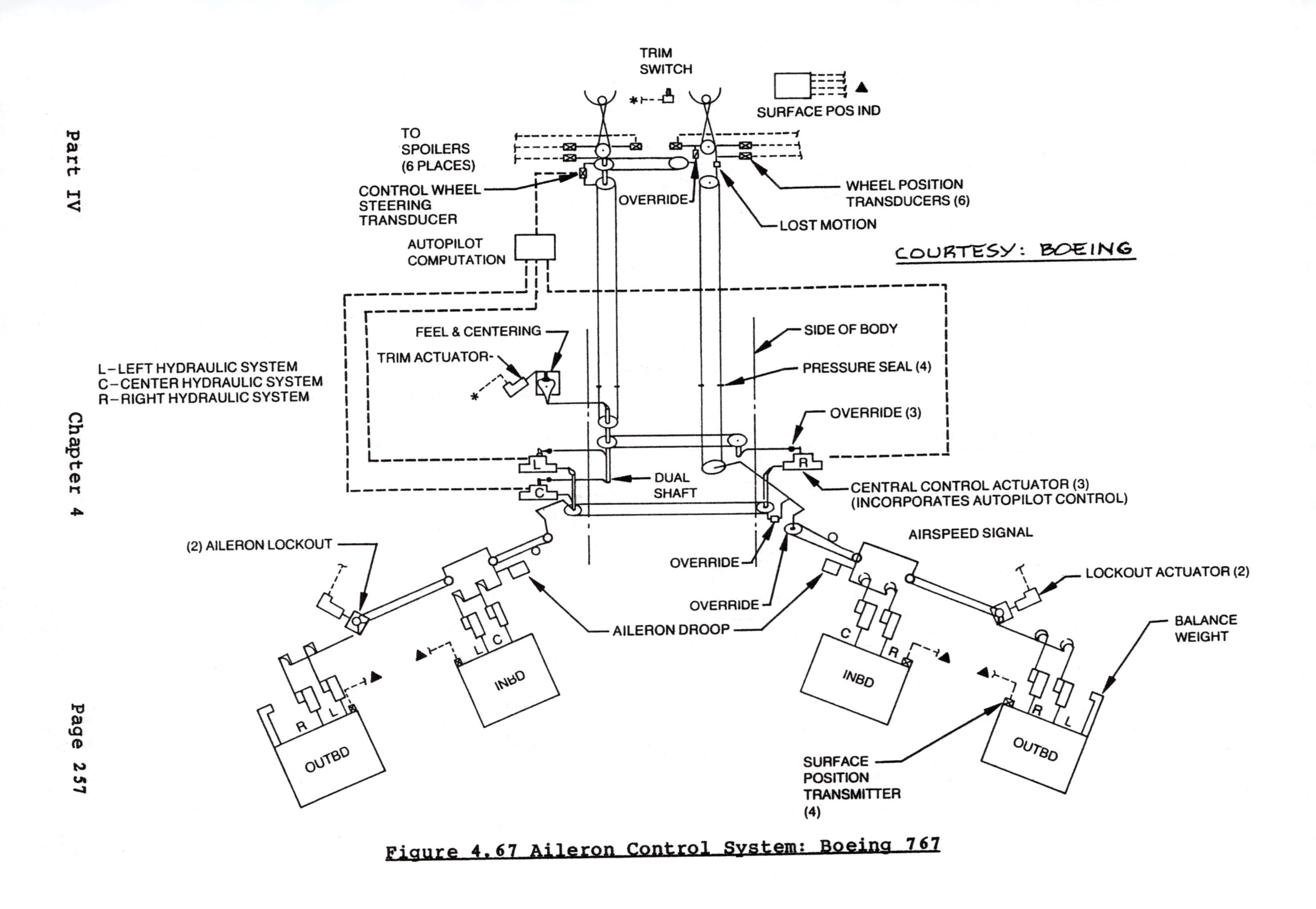

Figure 4.67 Aileron Control System: Boeing 767

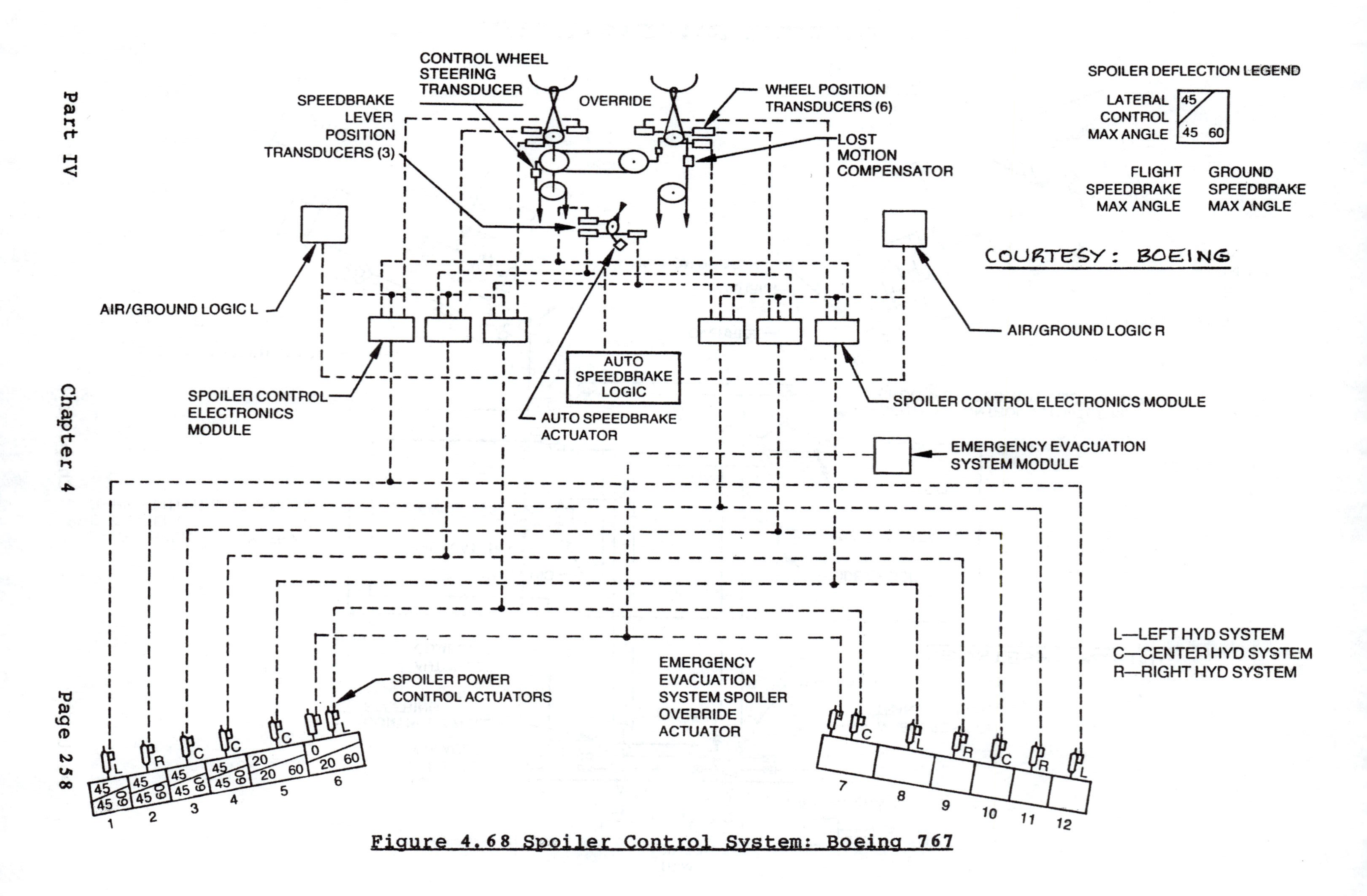

Figure 4.68 Spoiler Control System: Boeing 767

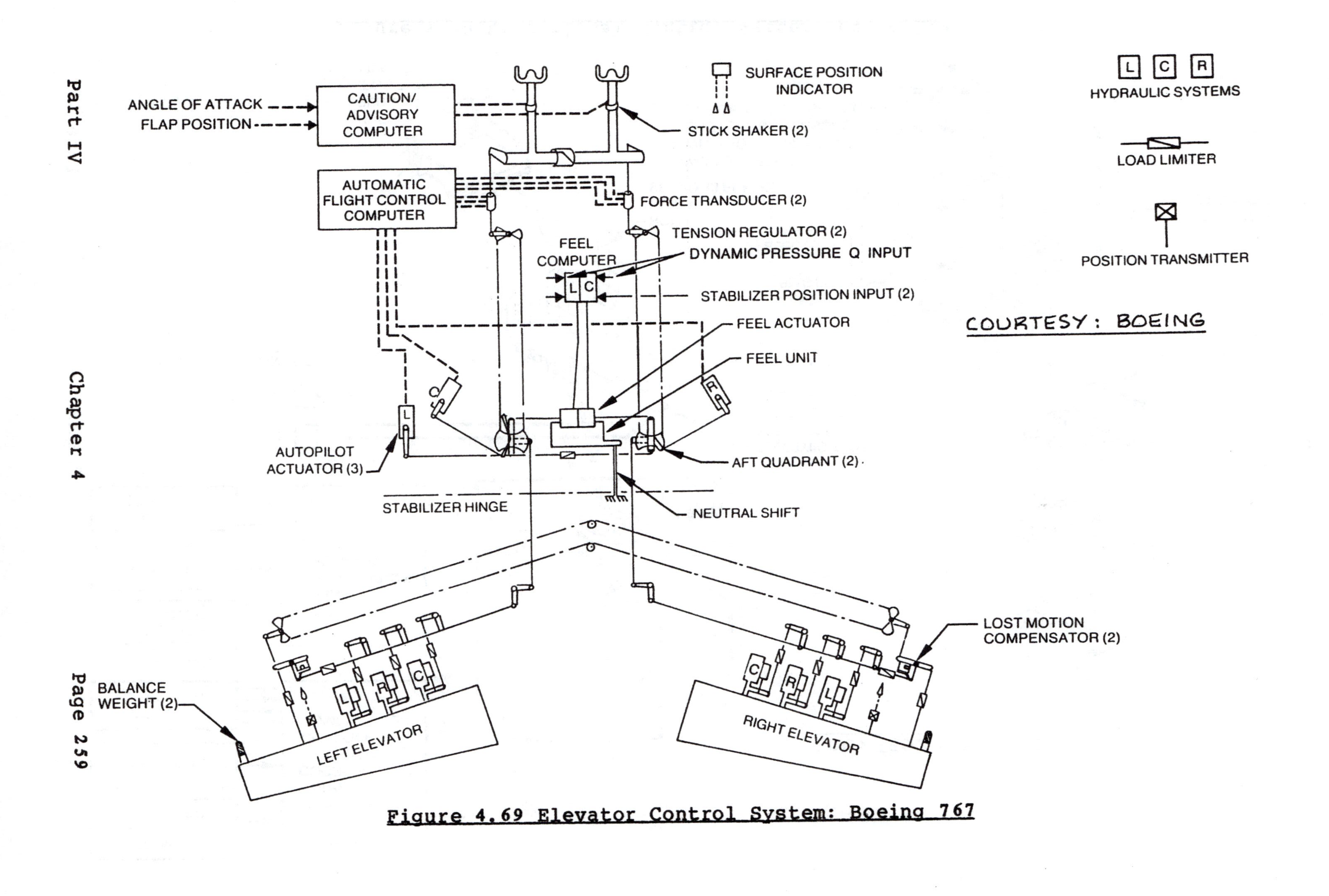

Figure 4.69 Elevator Control System: Boeing 767

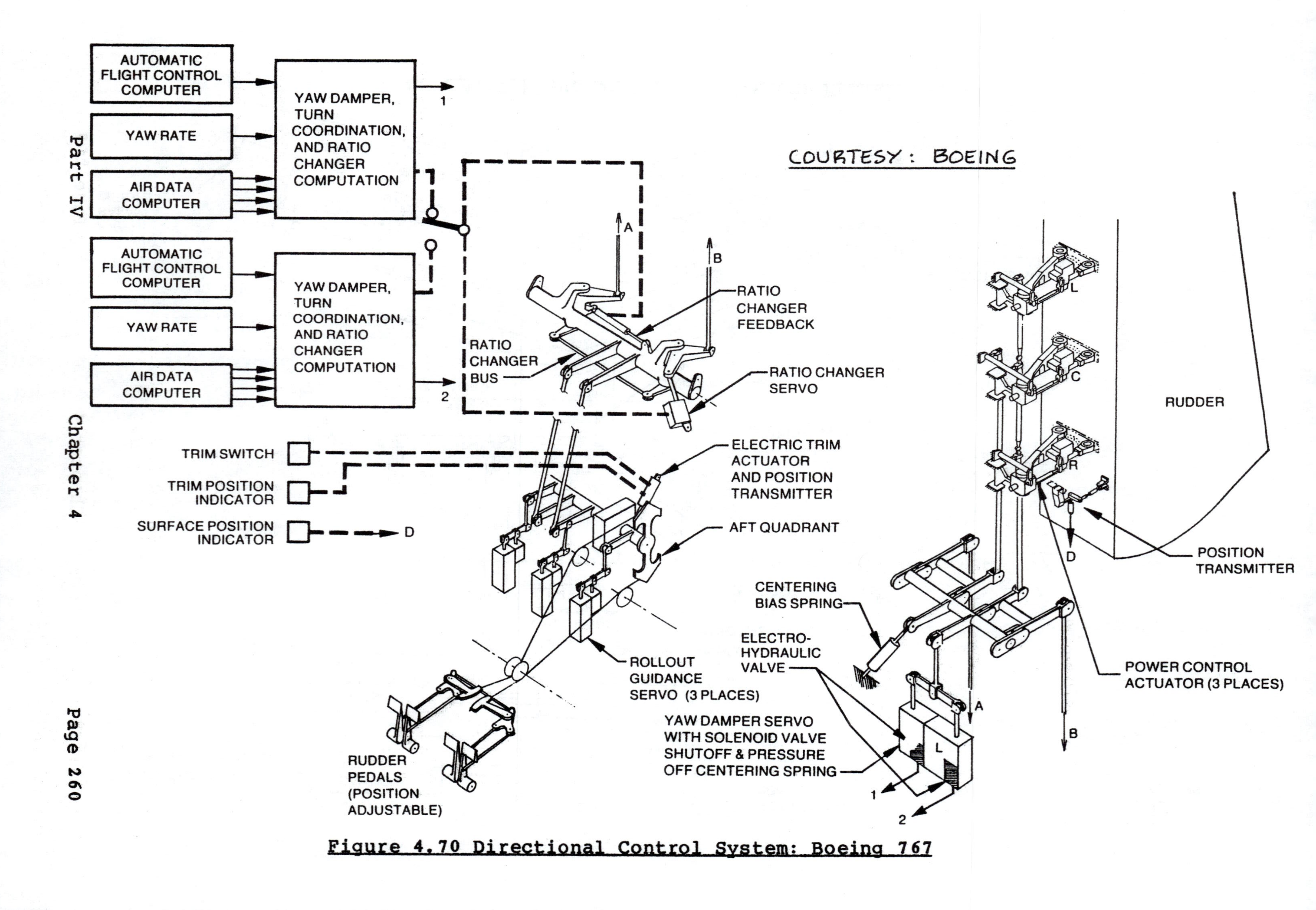

Figure 4.70 Directional Control System: Boeing 767

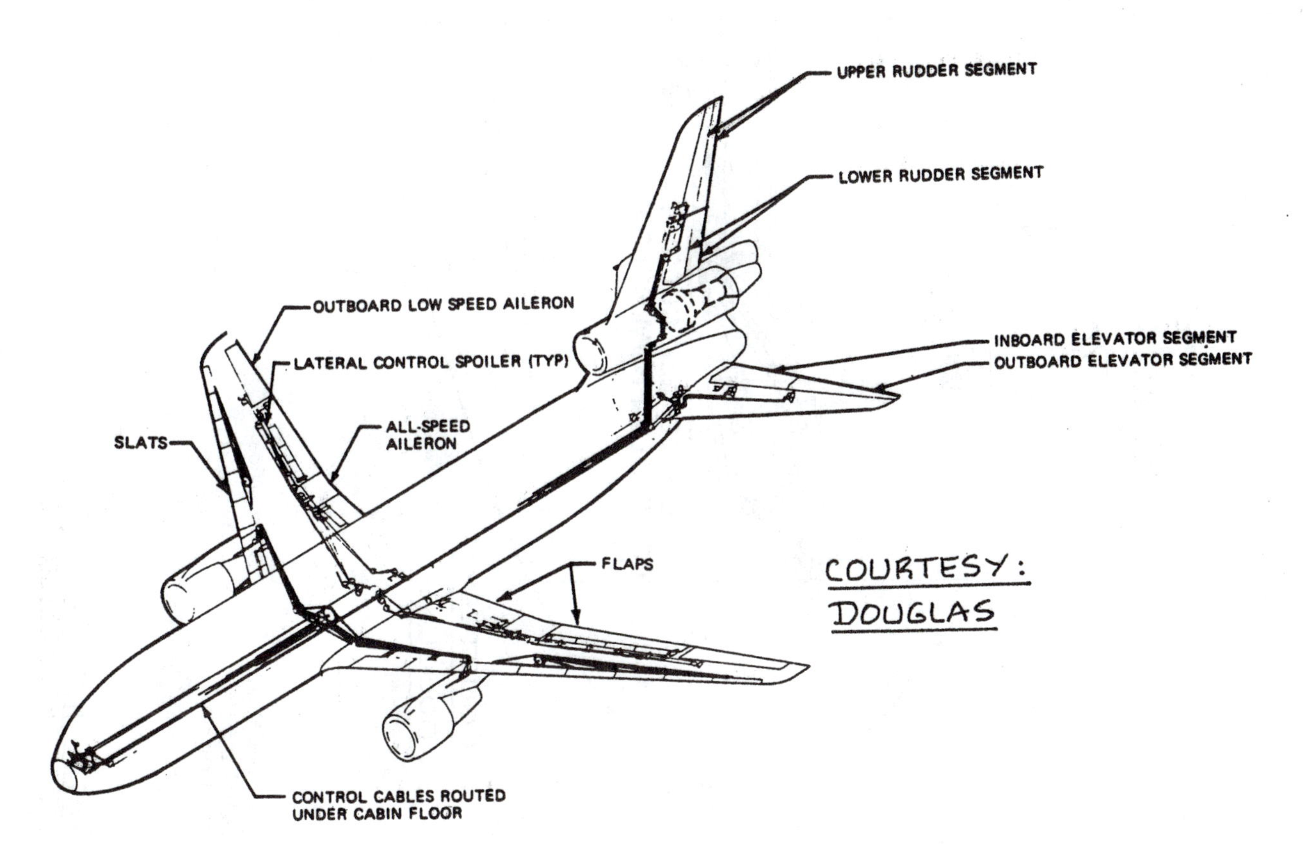

Figure 4.71 Primary Flight Control System: McDD DC-10

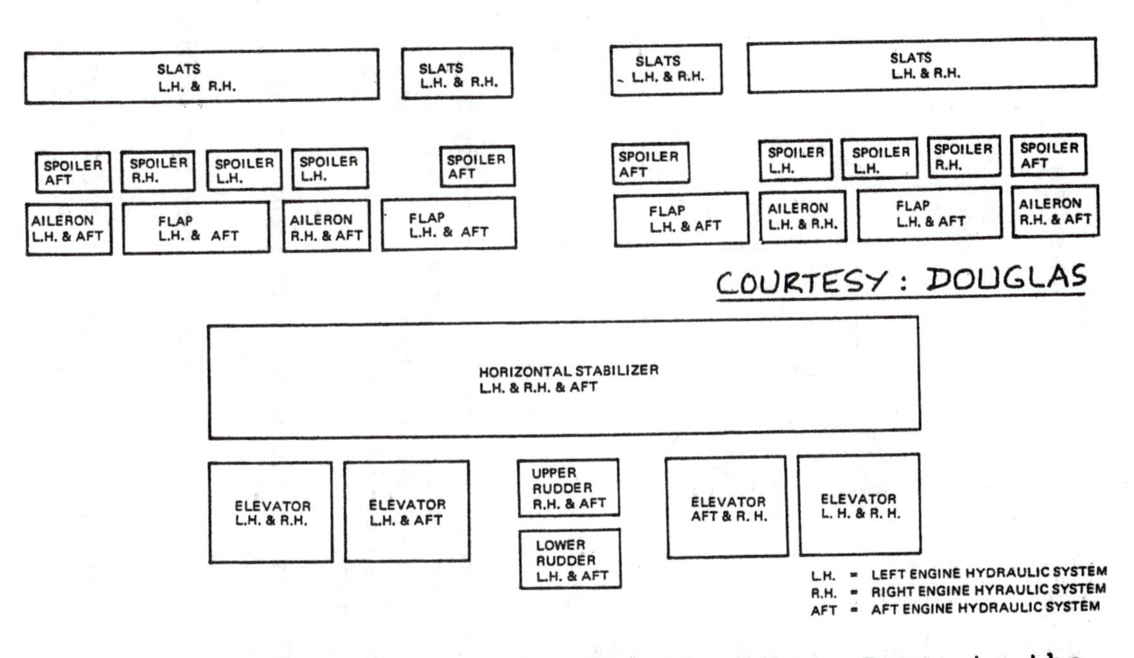

Figure 4.72 Distribution of Hydraulic System Power to the Flight Controls: McDD DC-10

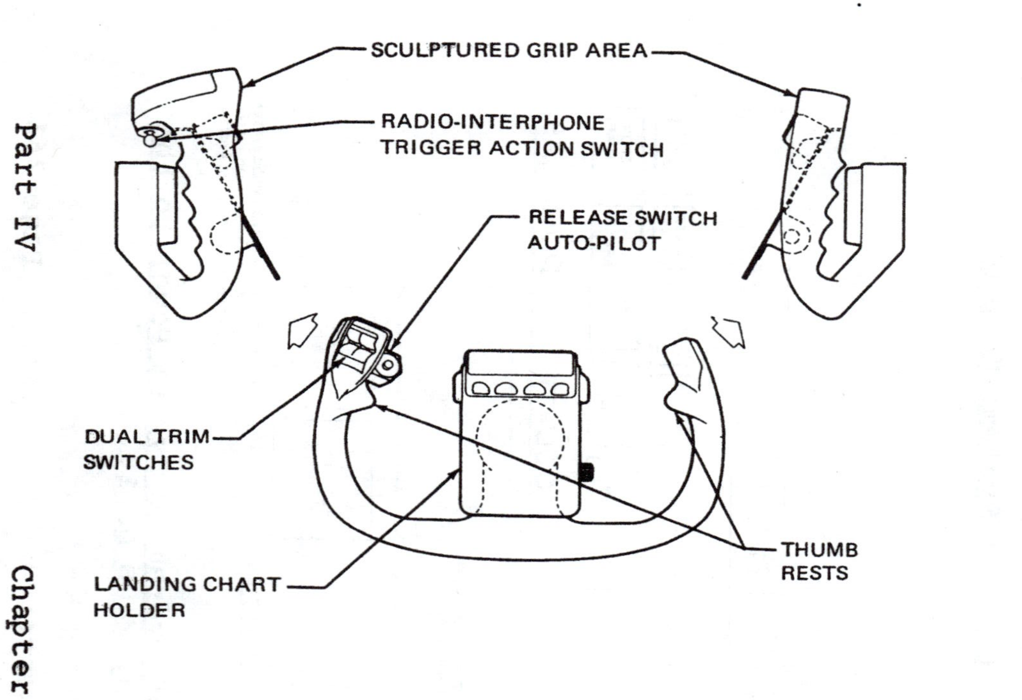

- Maximum utilization of properly designed and rigged closed-loop cable systems to provide long life, low friction, and adequate rigidity. Special cable tension regulators are not required.
- Use of carefully designed guards to prevent mechanical jamming
- Ample structural clearances on rotating or swinging parts of the control system
- Elevator, rudder, and lateral control systems provide instinctive override capability to protect against a jammed or runaway control.

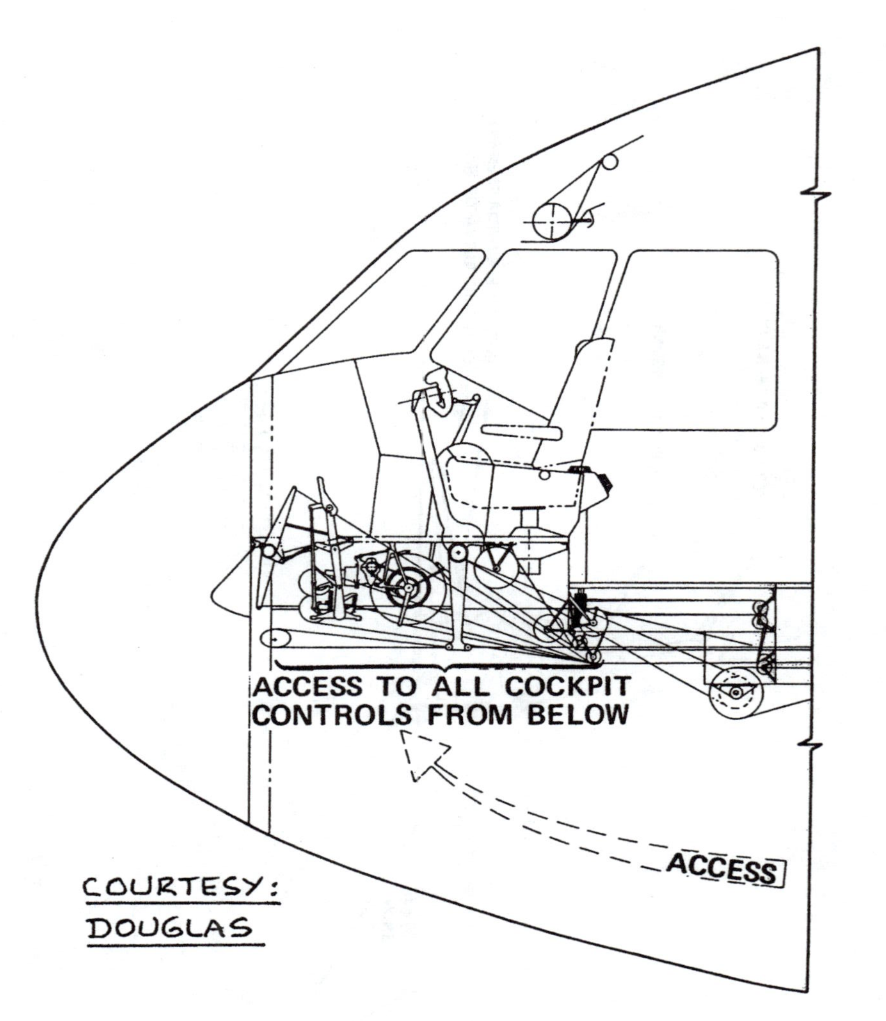

Figure 4.73 Cockpit Flight Controls: McDD DC-10

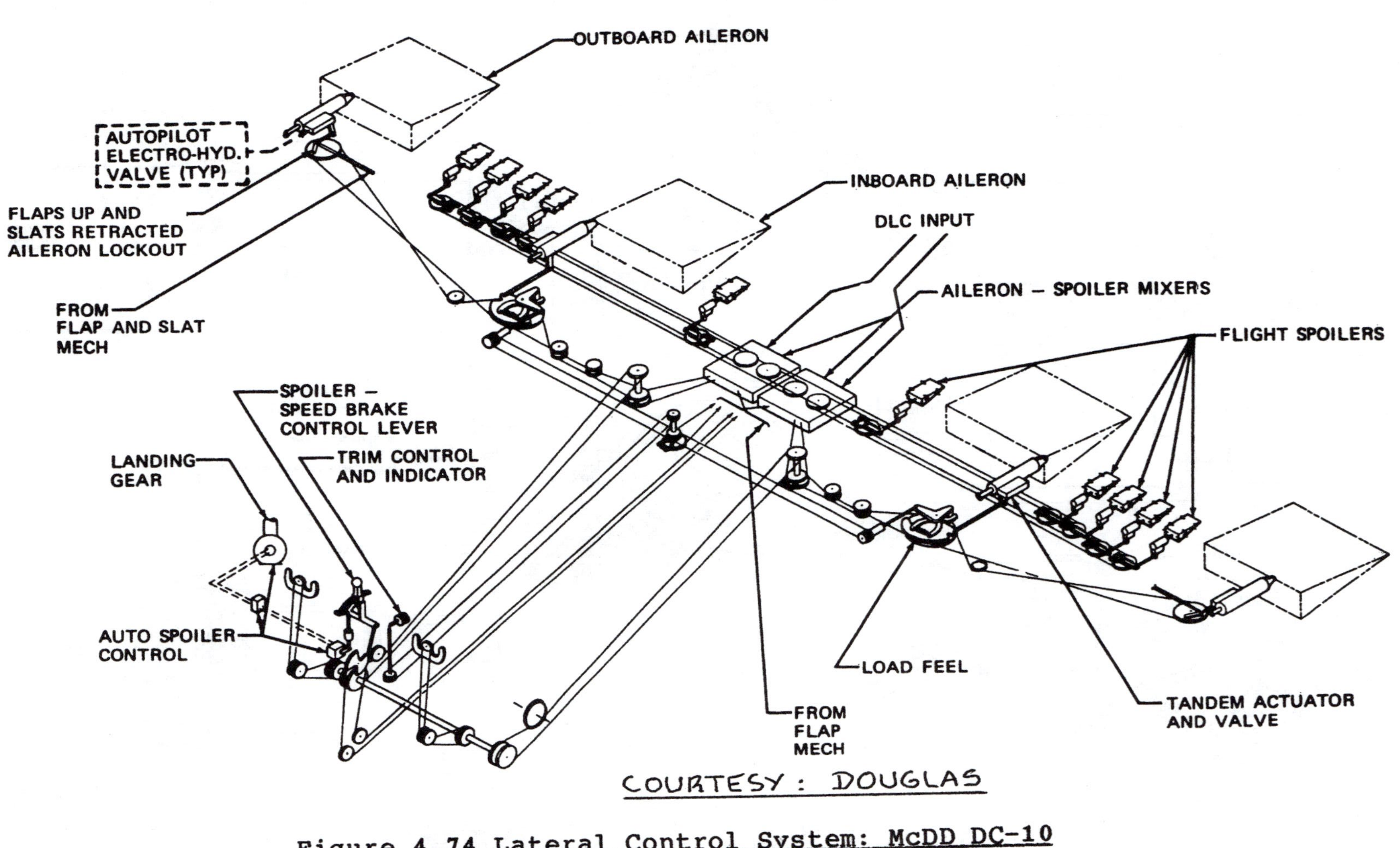

Figure 4.74 Lateral Control System: McDD DC-10

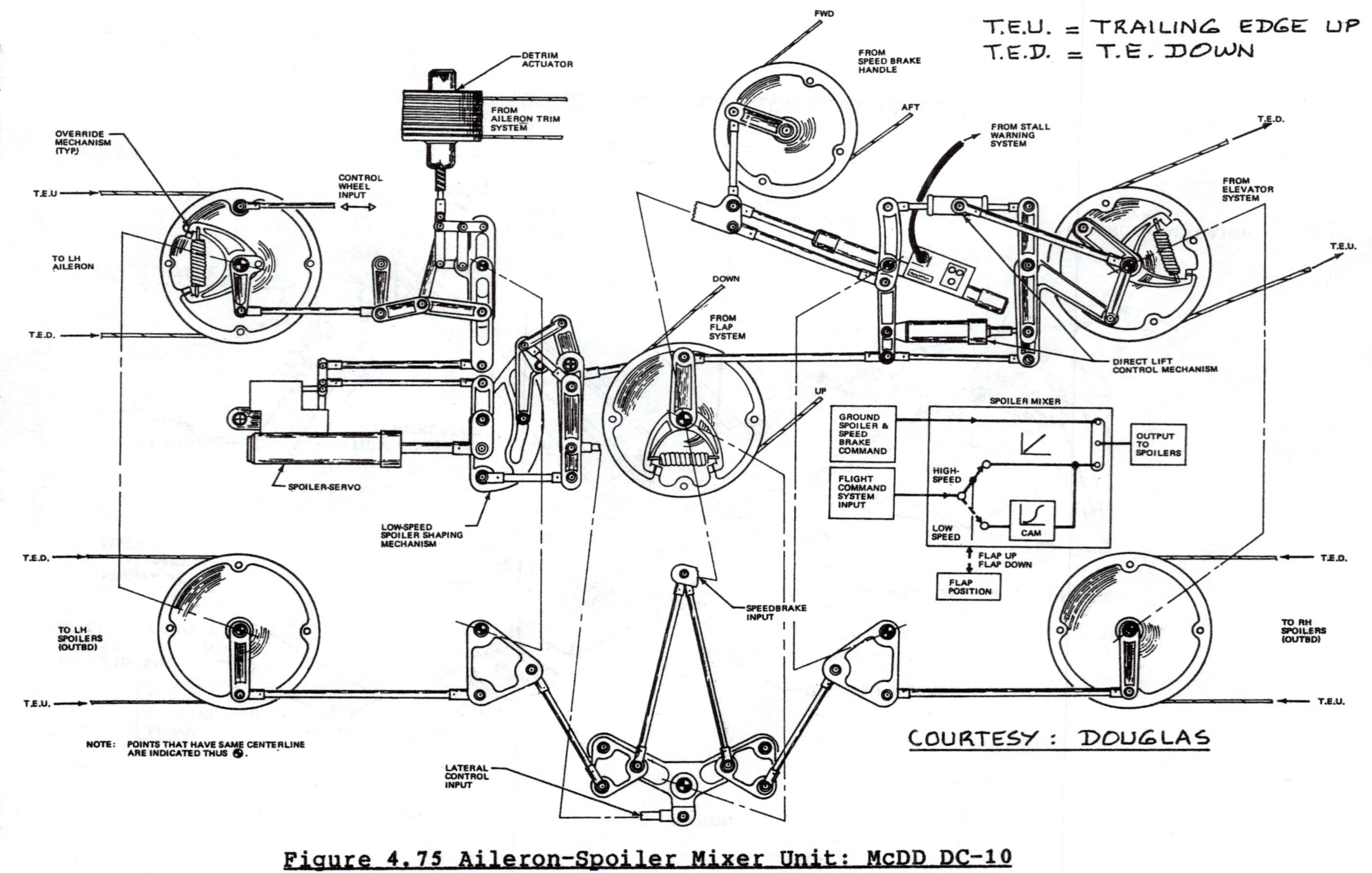

Figure 4.75 Aileron-Spoiler Mixer Unit: McDD DC-10

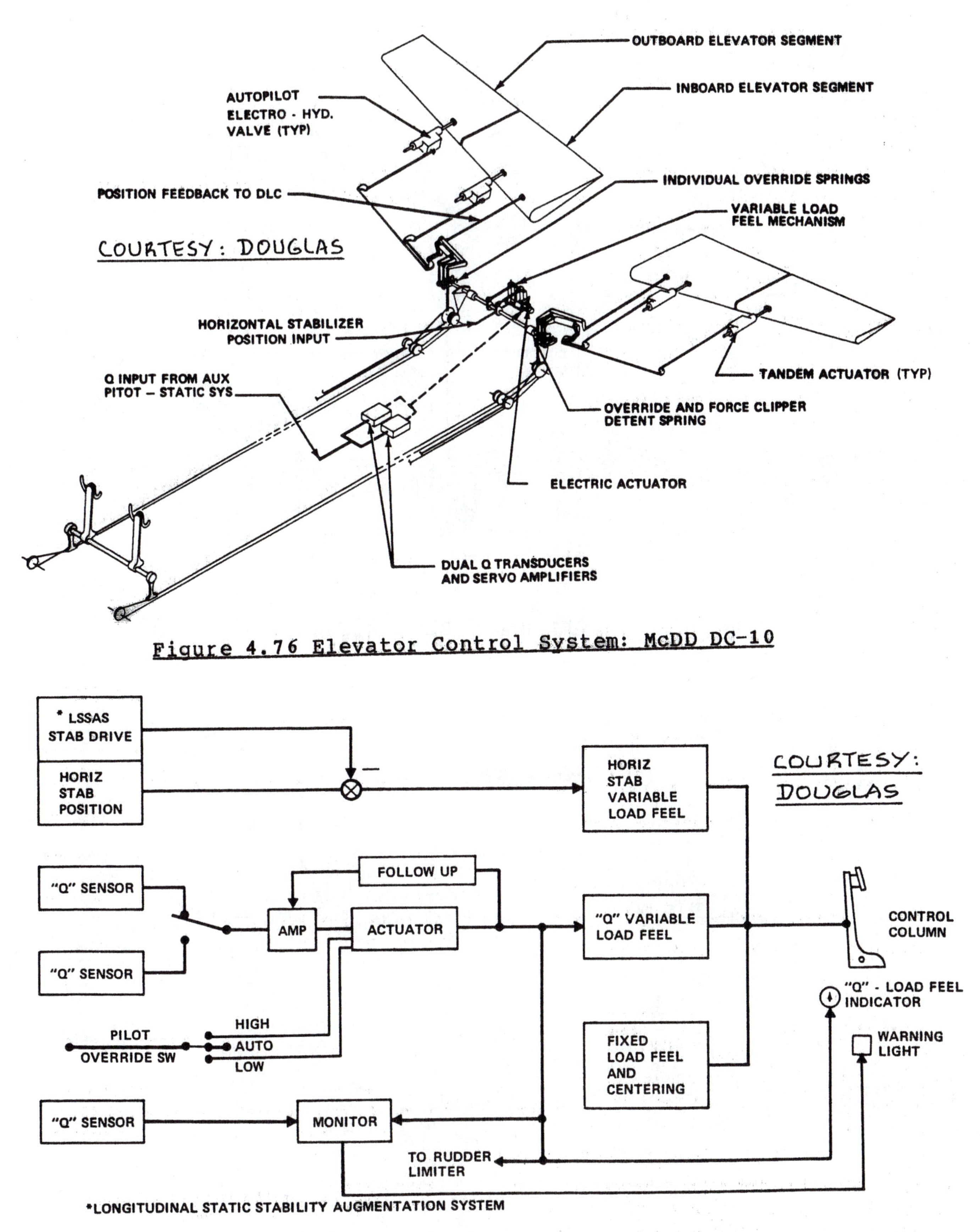

Figure 4.76 Elevator Control System: McDD DC-10

Figure 4.77 Control Feel Unit: McDD DC-10

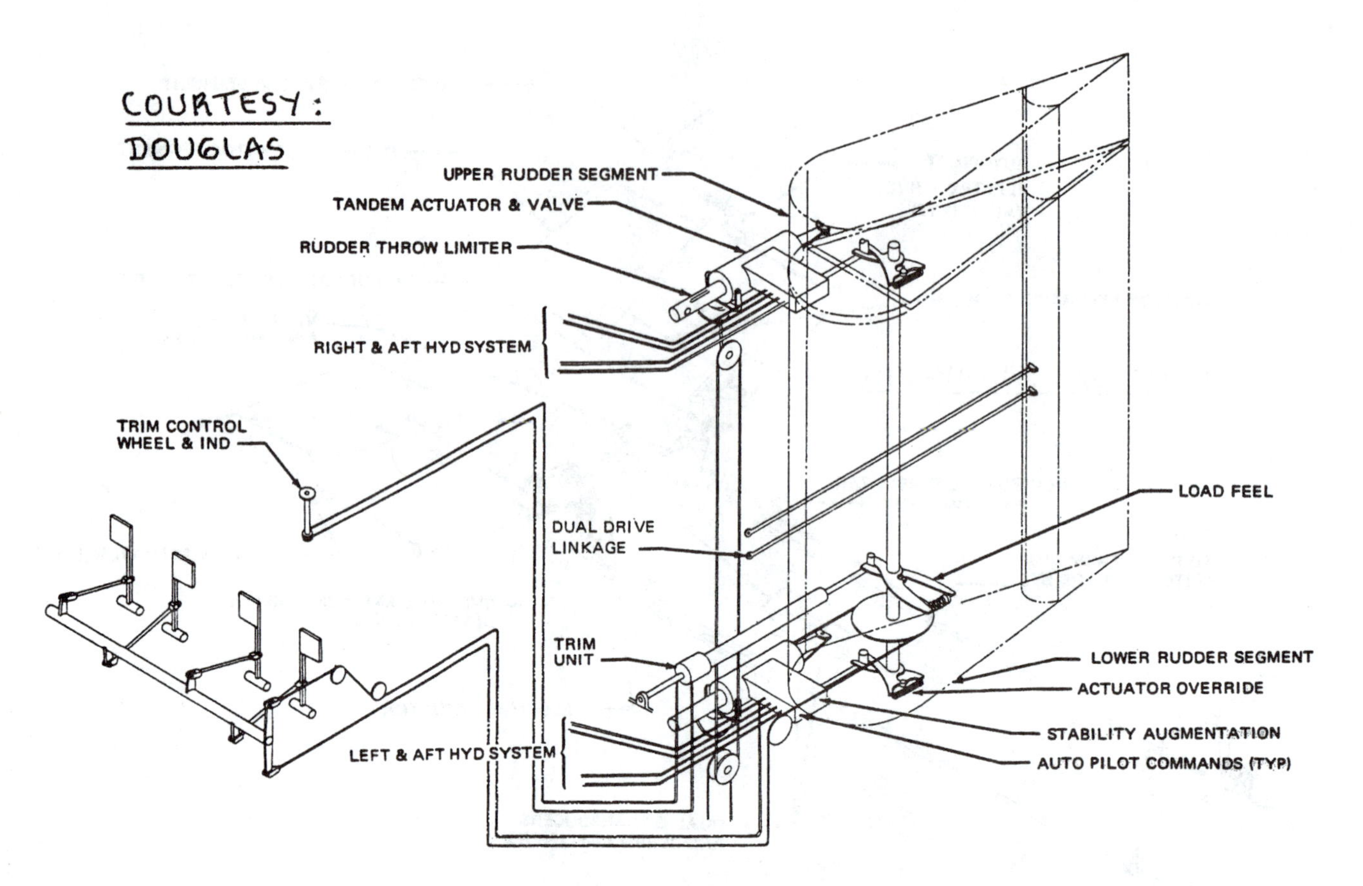

Figure 4.78 Directional Control System: McDD DC-10

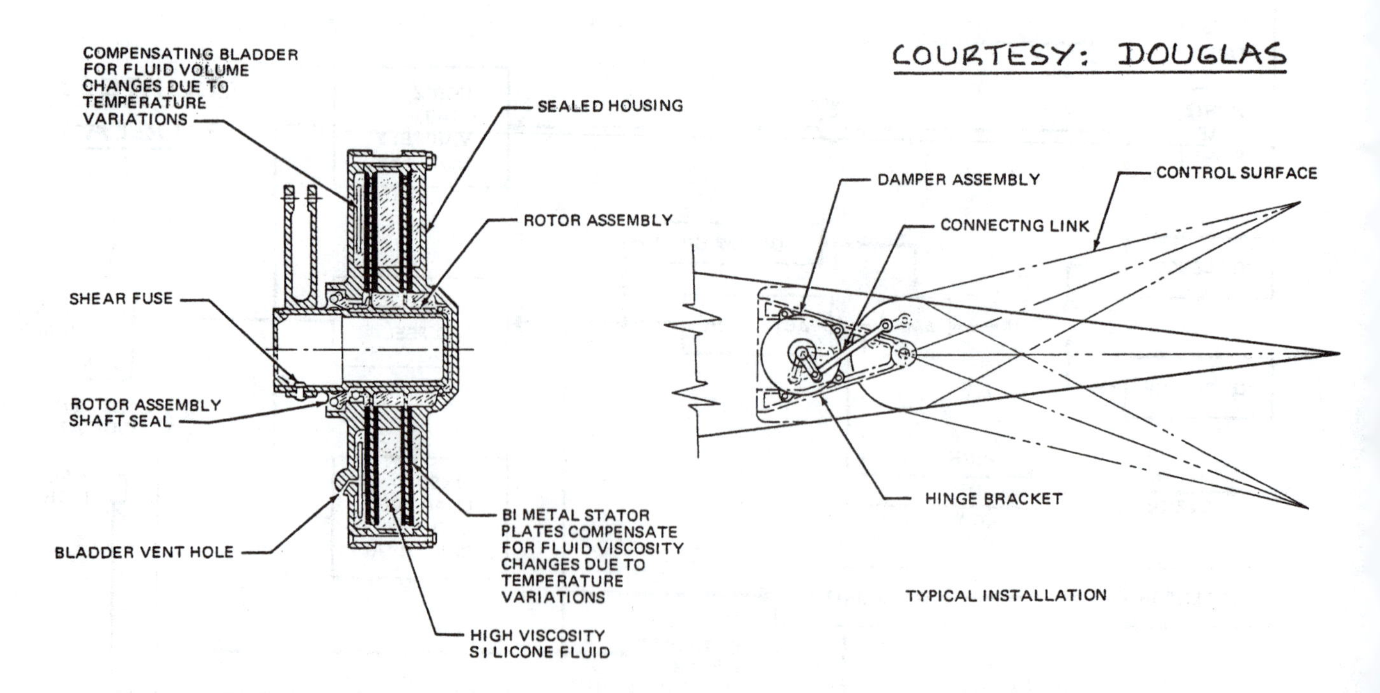

Figure 4.79 Control Surface Damper: McDD DC-10

4.5 TRIM SYSTEMS

A trim system helps the pilot maintain zero cockpit control forces in a given flight condition. In most airplanes it is desirable that cockpit control force trim can be maintained in nearly all flight conditions.

For a detailed discussion of the longitudinal and lateral-directional trim state of airplanes the reader should consult Chapter 5 of Ref.20.

Pilot force trim can be obtained as follows:

A) In reversible control systems by:

1. adding trim-tabs to the primary flight control surfaces. Figures 4.80 through 4.83 provide examples of trim tab systems.

2. the use of a variable incidence stabilizer. Figure 4.84 shows an example of a stabilizer trim system.

3. a trim-spring system as shown in Fig.4.24.

B) In an irreversible system the cockpit control forces are simulated by means of a so-called force-feel system. By nulling out the output of the force-feel system a trimmed condition can be obtained.

Figures 4.69 and 4.77 show examples of such control force feel systems.

CAUTION: In nearly all trim systems care must be taken that the airplane will not fly with the control surfaces in a so-called 'jack-knifed' condition. Figure 4.85 illustrates a stabilizer/ elevator combination in a 'jack-knifed' position. Such jack-knifing of the controls leads to extra drag. Trading of elevator angle for stabilizer angle eliminates this problem.

In most jet transports the trim system is arranged to drive the stabilizer angle to prevent jack-knifing. Figures 4.86 and 4.87 show example trim systems used in typical jet transports.

Notes: 1. Trim systems are nearly always mechanized as irreversible systems. They are either powered or unpowered.

2. If trim systems are powered (electrical or hydraulic) redundancy must be provided: run-away trim deflections can cause loss of control.

3. Figs 4.80 - 4.83 show examples of unpowered trim systems: redundancy is not needed.

4. Figs 4.84, 4.86 and 4.87 show examples of powered trim systems. Redundancy is required by Refs 21 and 22. In addition, 'trim-in-motion' indicators are required in the cockpit.

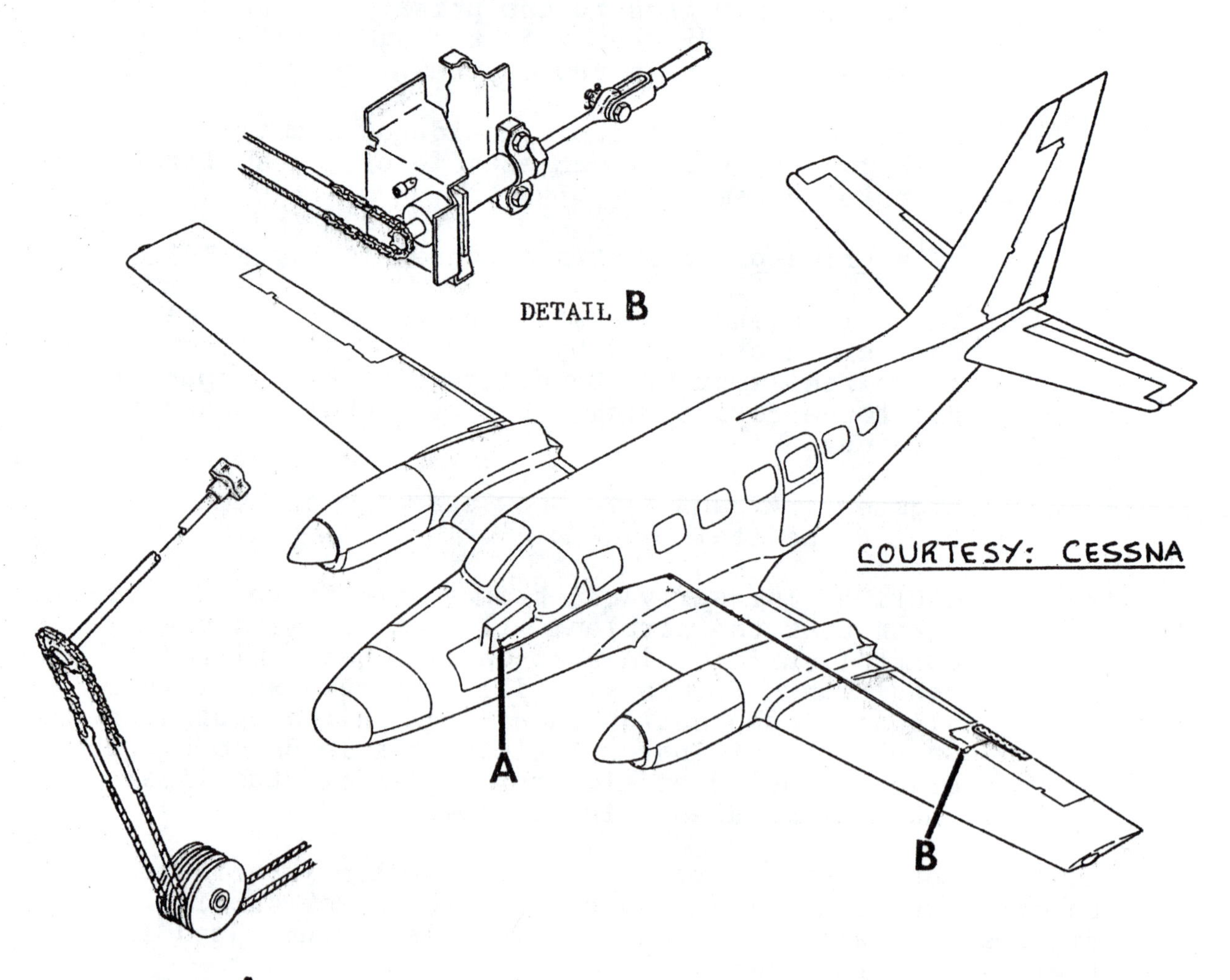

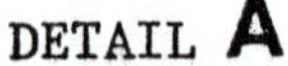

Figure 4.80 Aileron Trim System: Cessna 441

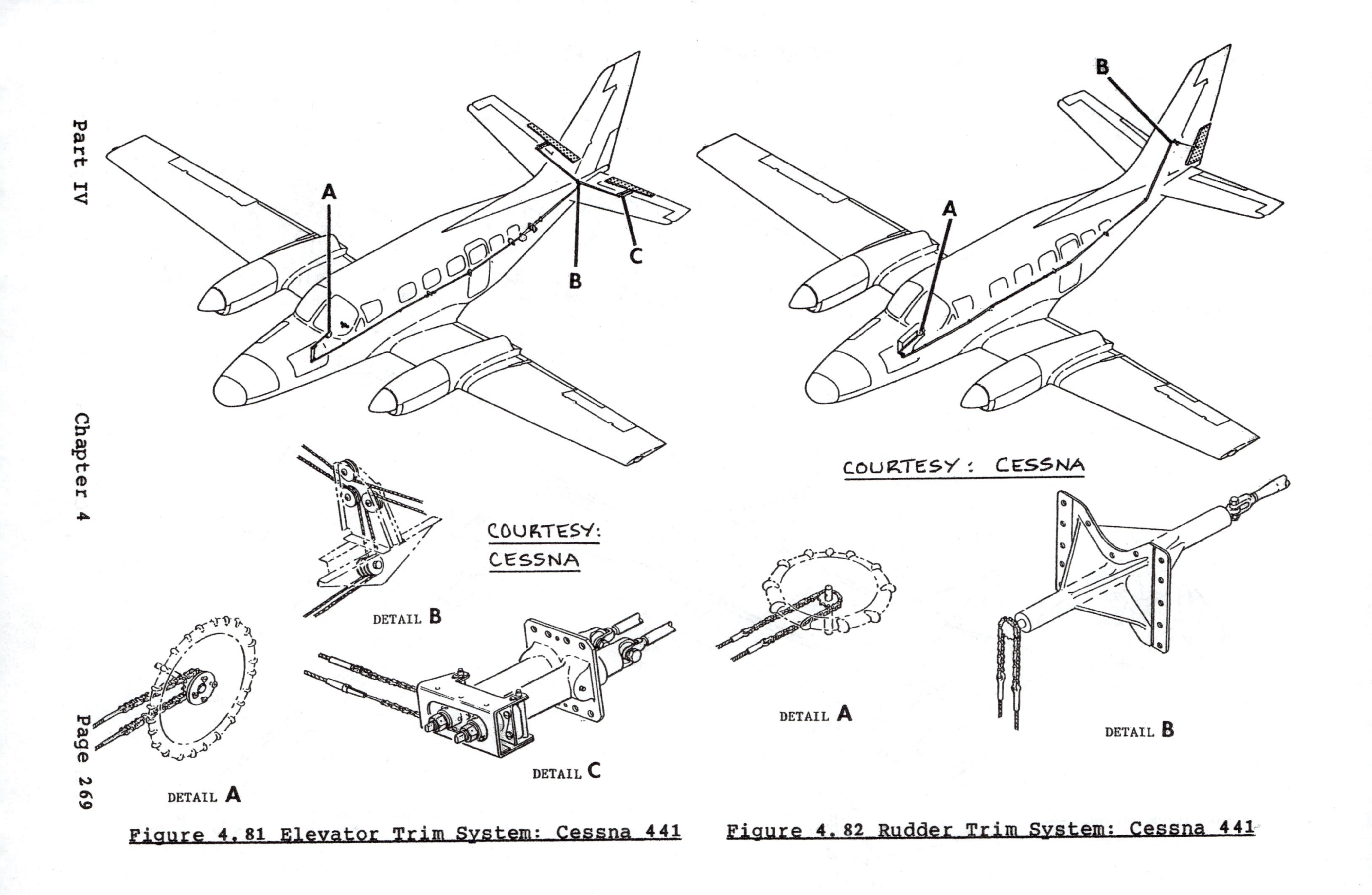

Figure 4.81 Elevator Trim System: Cessna 441

Figure 4.82 Rudder Trim System: Cessna 441

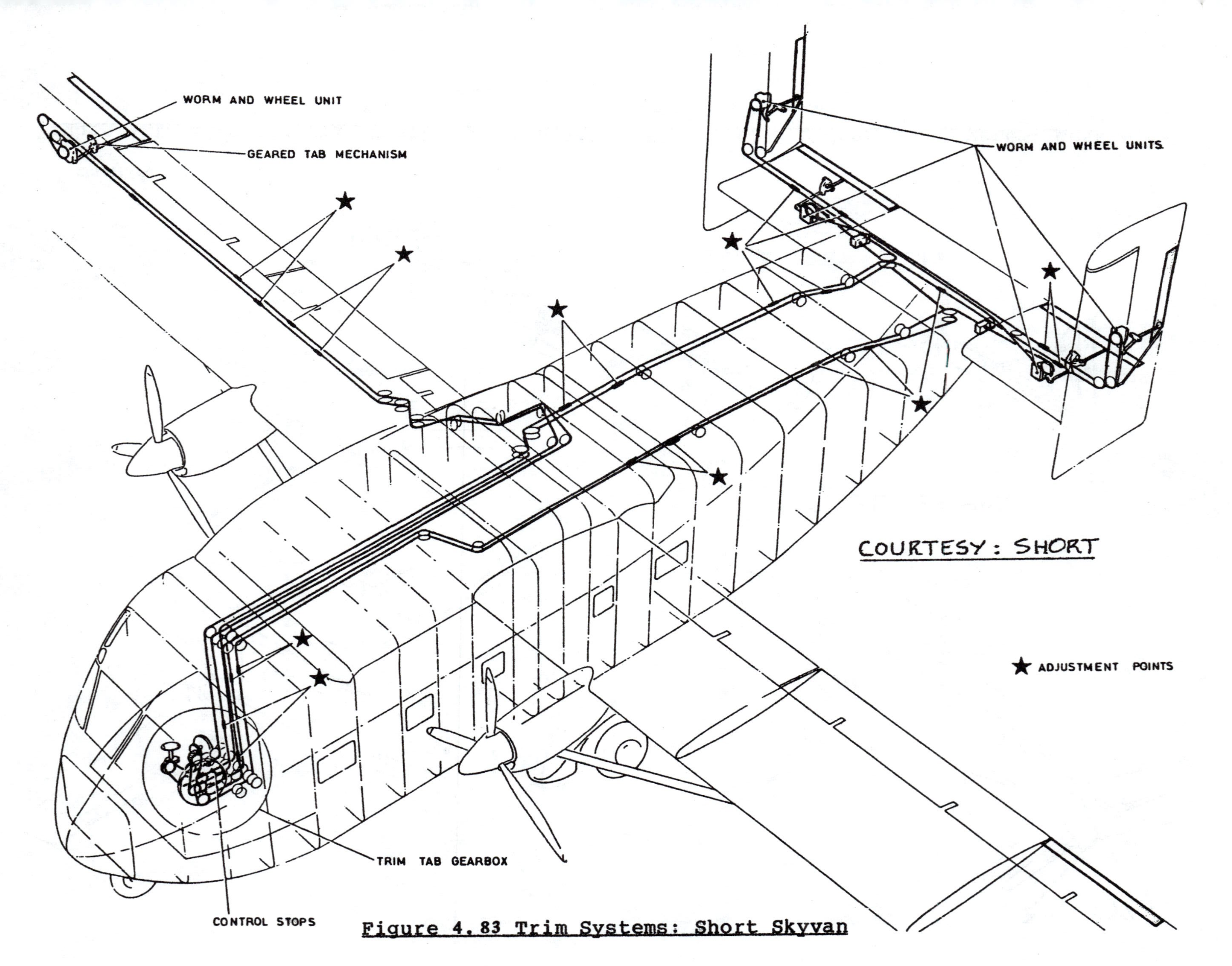

Figure 4.83 Trim Systems: Short Skyvan

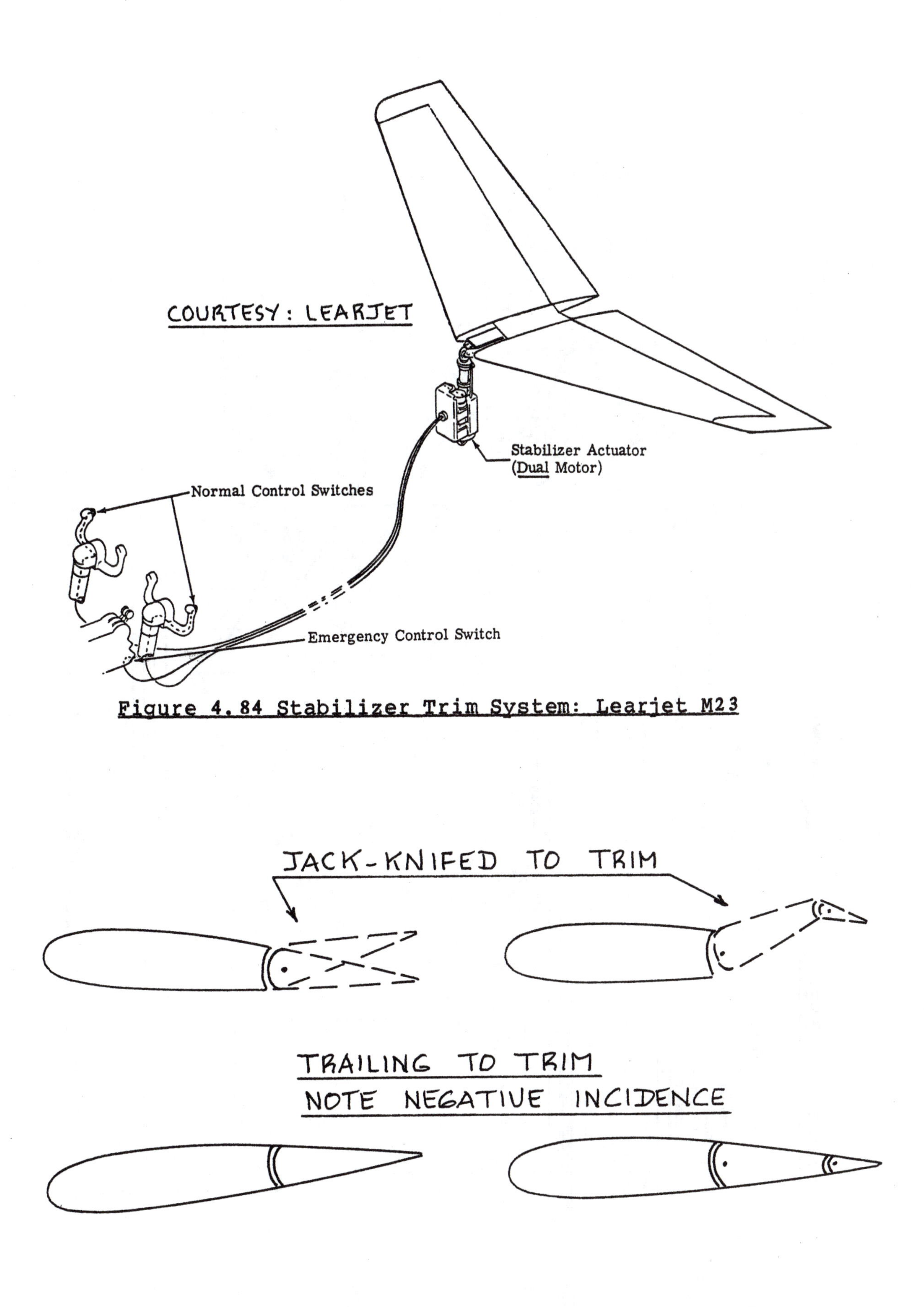

Figure 4.84 Stabilizer Trim System: Learjet M23

Figure 4.85 Examples of Jack-knifing of Controls

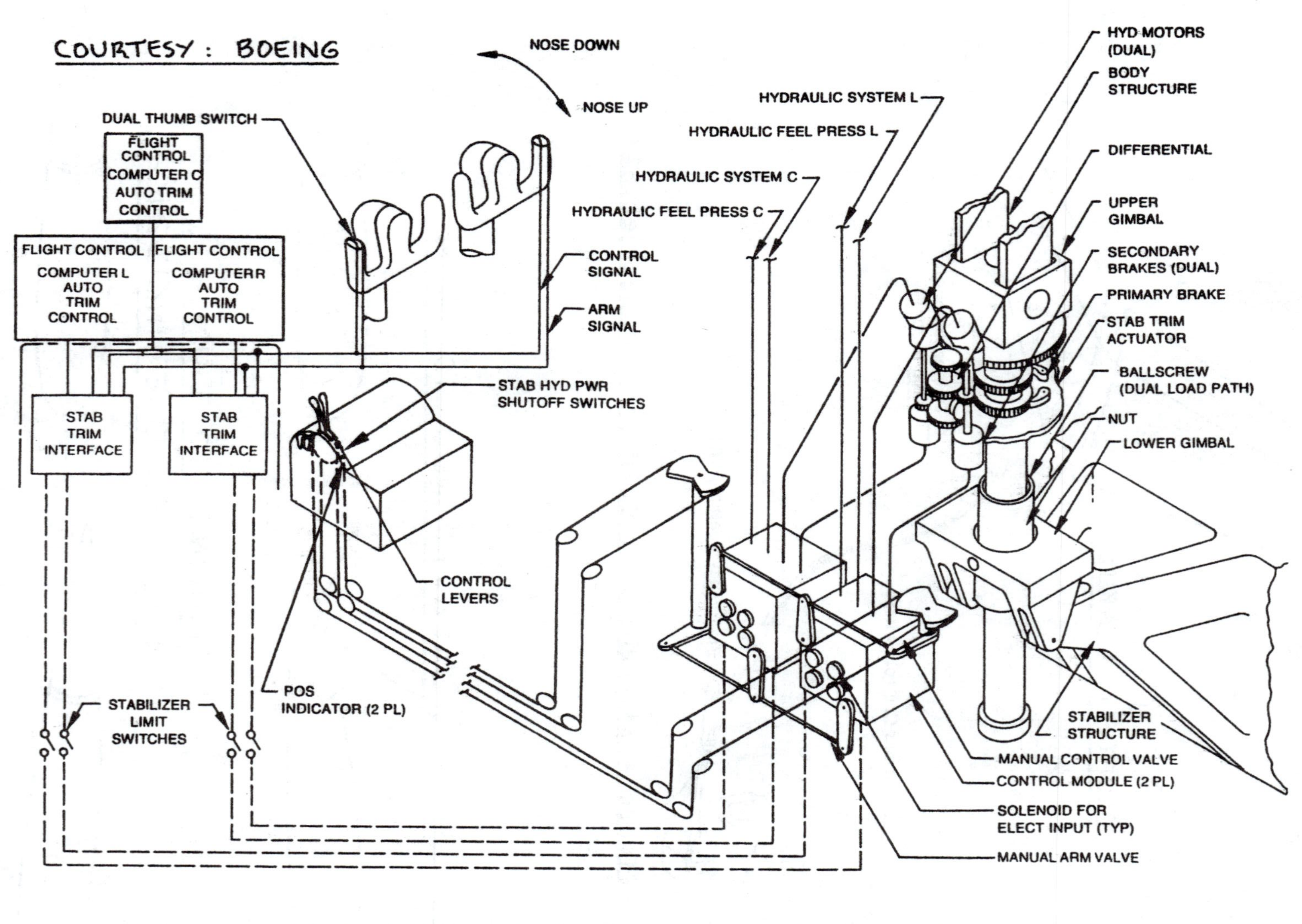

Figure 4.86 Stabilizer Trim Schematic: Boeing 767

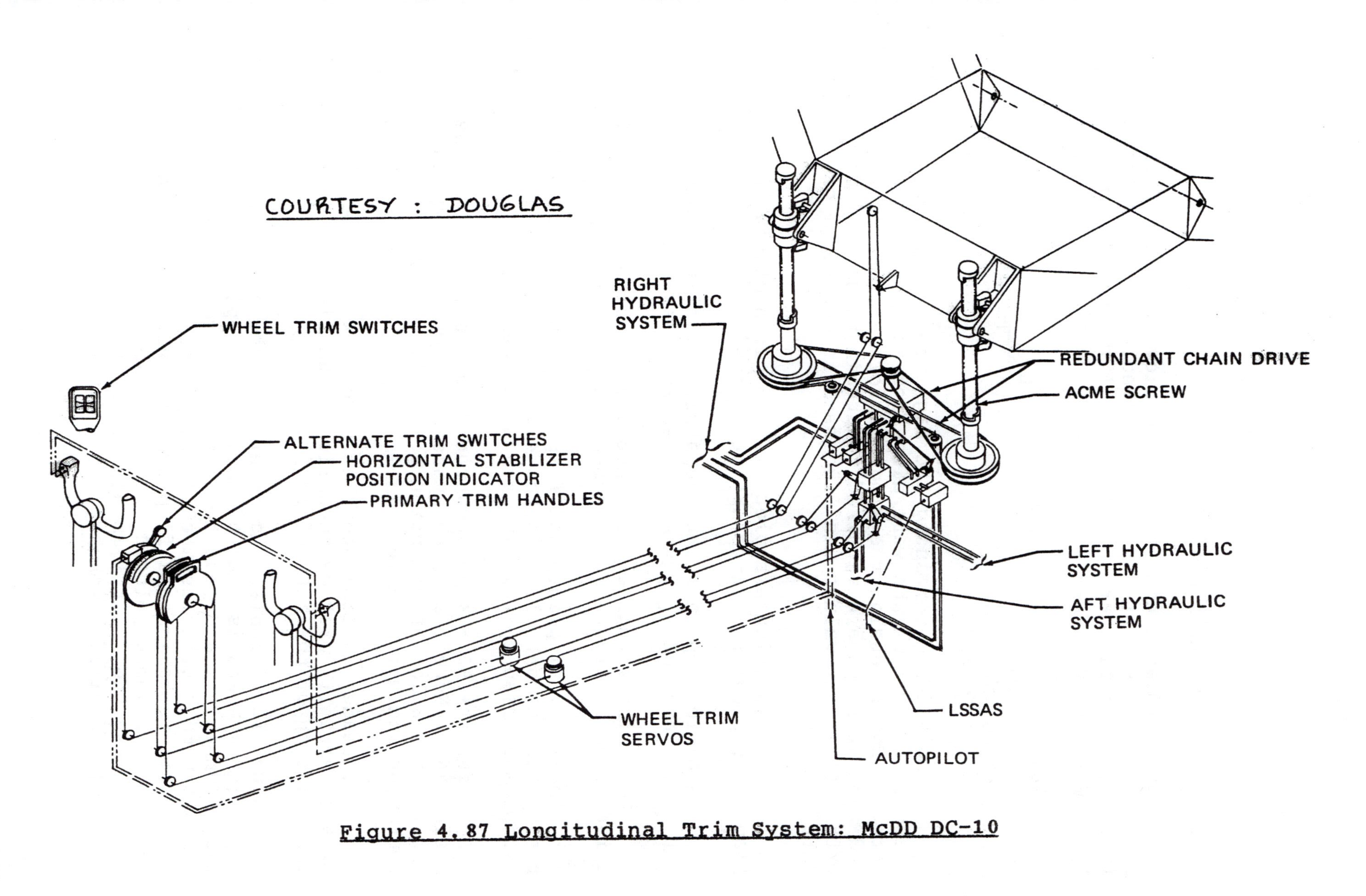

Figure 4.87 Longitudinal Trim System: McDD DC-10

4.6 HIGH LIFT CONTROL SYSTEMS

Most airplanes need flaps to achieve the lift coefficients required for take-off and landing. It was shown in Parts I and II that the following lift coefficients are important in airplane preliminary design:

Clean: $C_{L_{max}}$ Take-off: $C_{L_{max_{TO}}}$

Landing: $C_{L_{max_L}}$

Methods for determining the magnitudes required for these lift coefficients were discussed in Chapter 3 of Part I. How to design the wing planform and what size and type of flaps are required to attain these maximum lift coefficient values was discussed in Chapter 7 of Part II.

In this section examples will be given of how the flaps of typical airplanes are mechanized.

In basic airplanes the flaps are normally mechanized for manual control. Figure 4.88 shows an example.

Many single engine airplanes use electromechanically driven flaps. Figure 4.89 presents an example.

In larger airplanes the flaps are usually mechanized for hydraulic control. Figures 4.90 through 4.95 show examples of flap systems in such airplanes.

Figures 4.92 - 4.95 show the flap system used in the Boeing 767. Note that such systems become very complex and require a lot of attention to detail design, operation and maintenance.

The reader is reminded that Figures 4.71 through 4.76 in Part III provided details of the flap installation of the McDD DC-10.

A critical problem encountered in the design of most high lift systems is that of asymmetric deployment. If flap asymmetry would lead to loss of control, it is mandatory to prevent asymmetric flap deployment. One way to do this is to install flap travel sensors which in turn drive asymmetry logic circuits. These logic circuits can then halt flap deployment beyond some tolerable level of asymmetry.

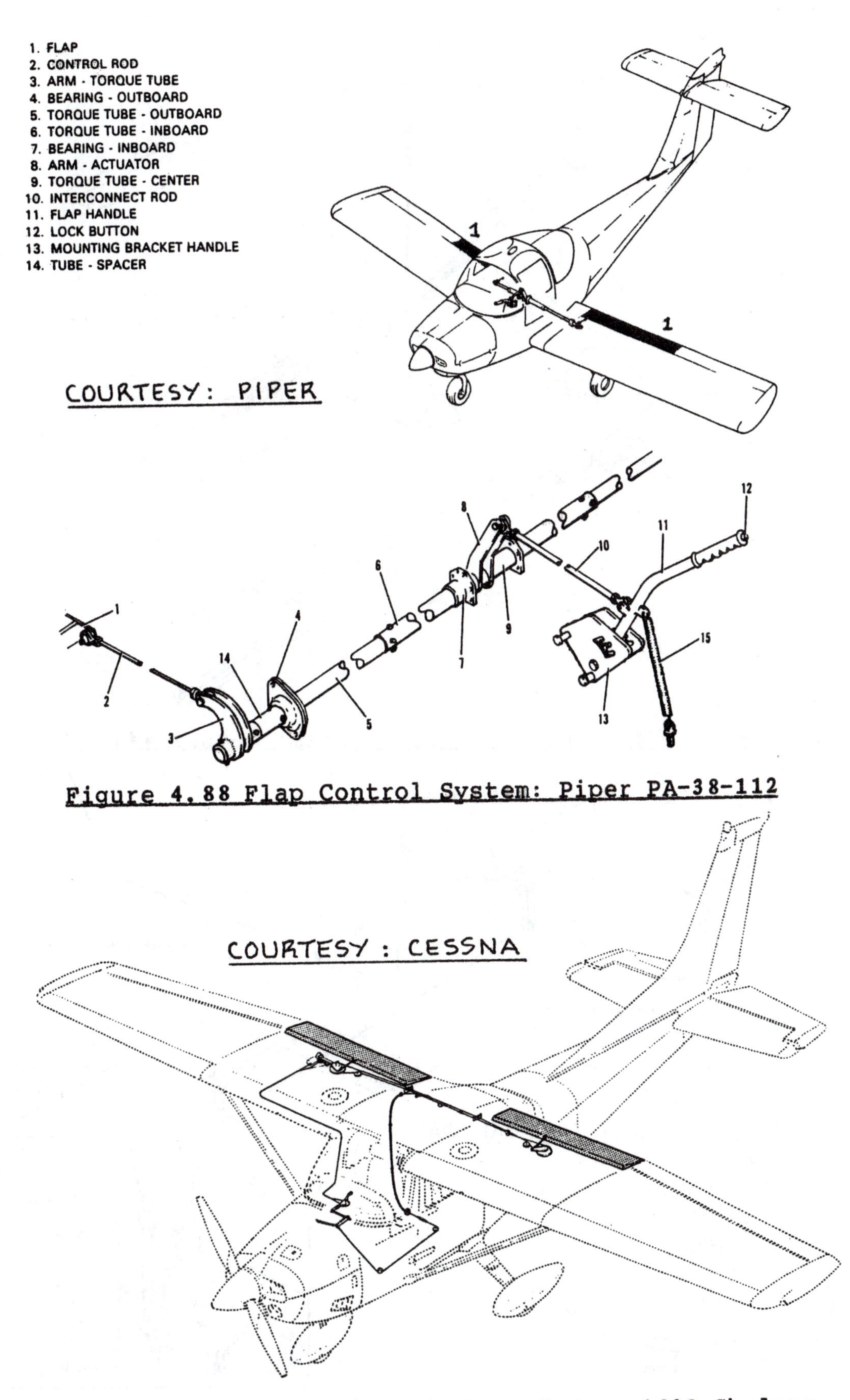

Figure 4.88 Flap Control System: Piper PA-38-112

Figure 4.89 Flap Control System: Cessna 182Q Skylane

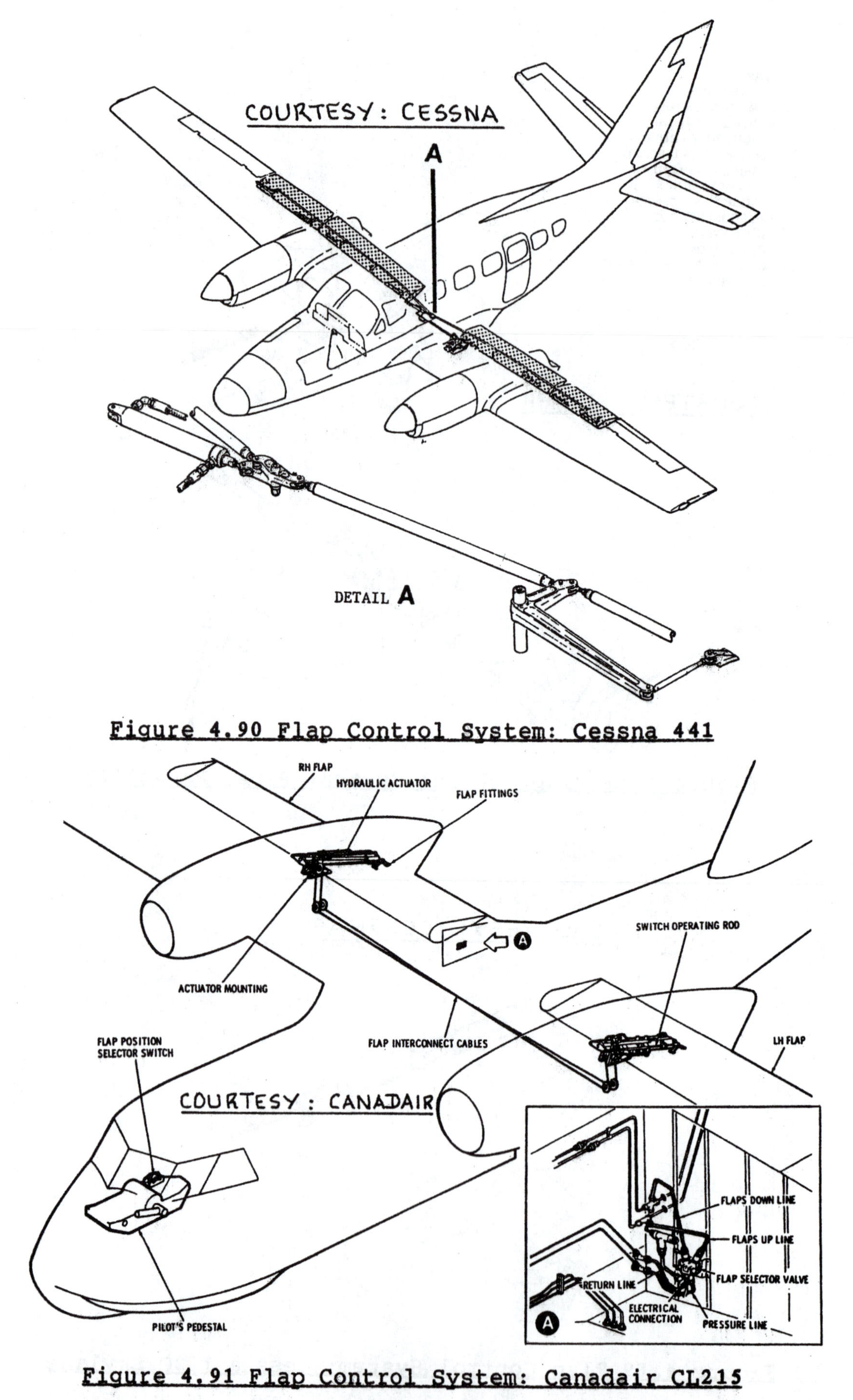

Figure 4.90 Flap Control System: Cessna 441

Figure 4.91 Flap Control System: Canadair CL215

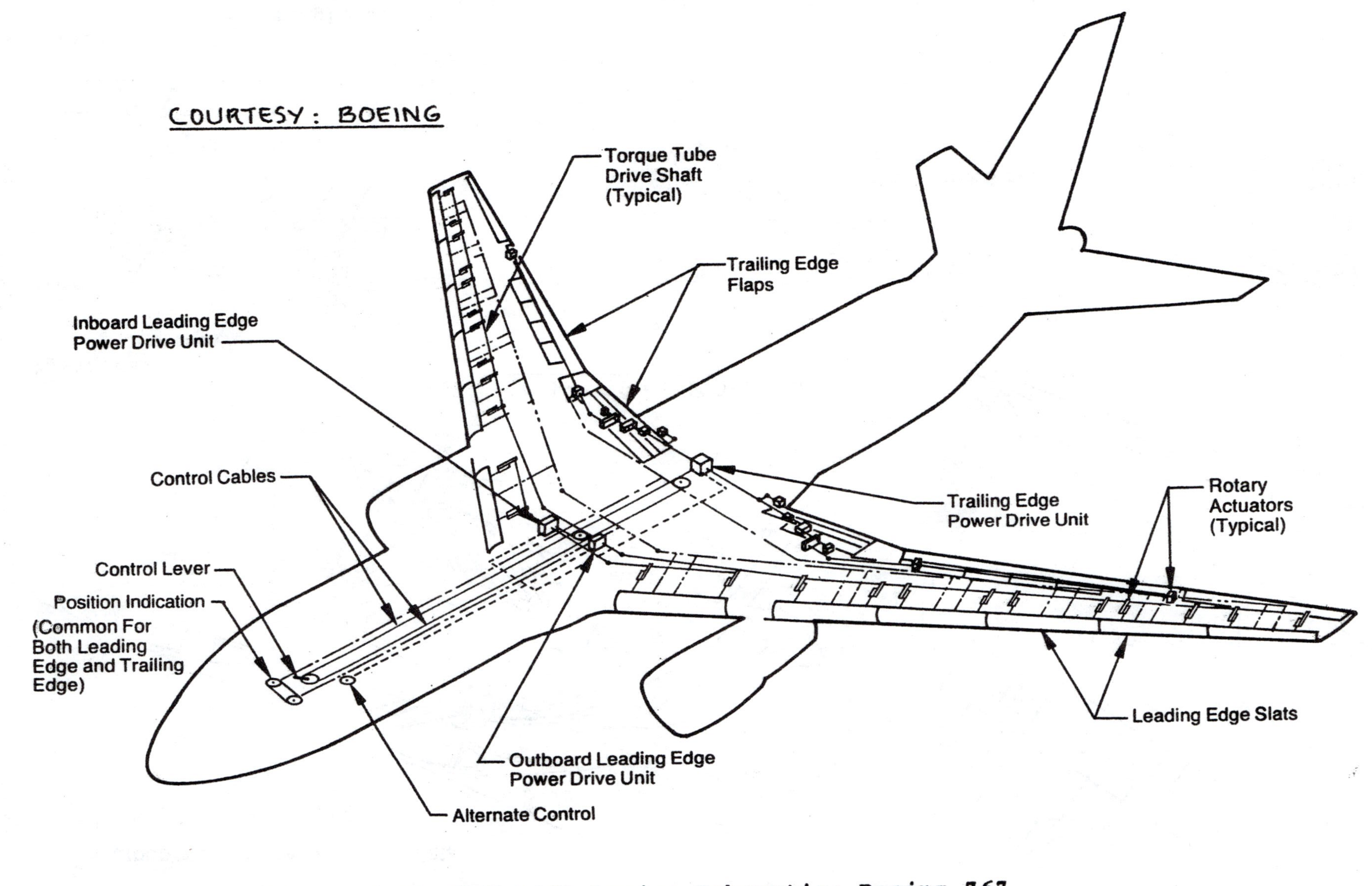

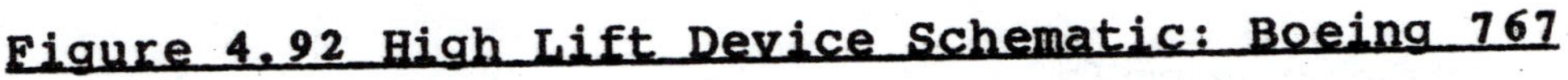

Figure 4.92 High Lift Device Schematic: Boeing 767

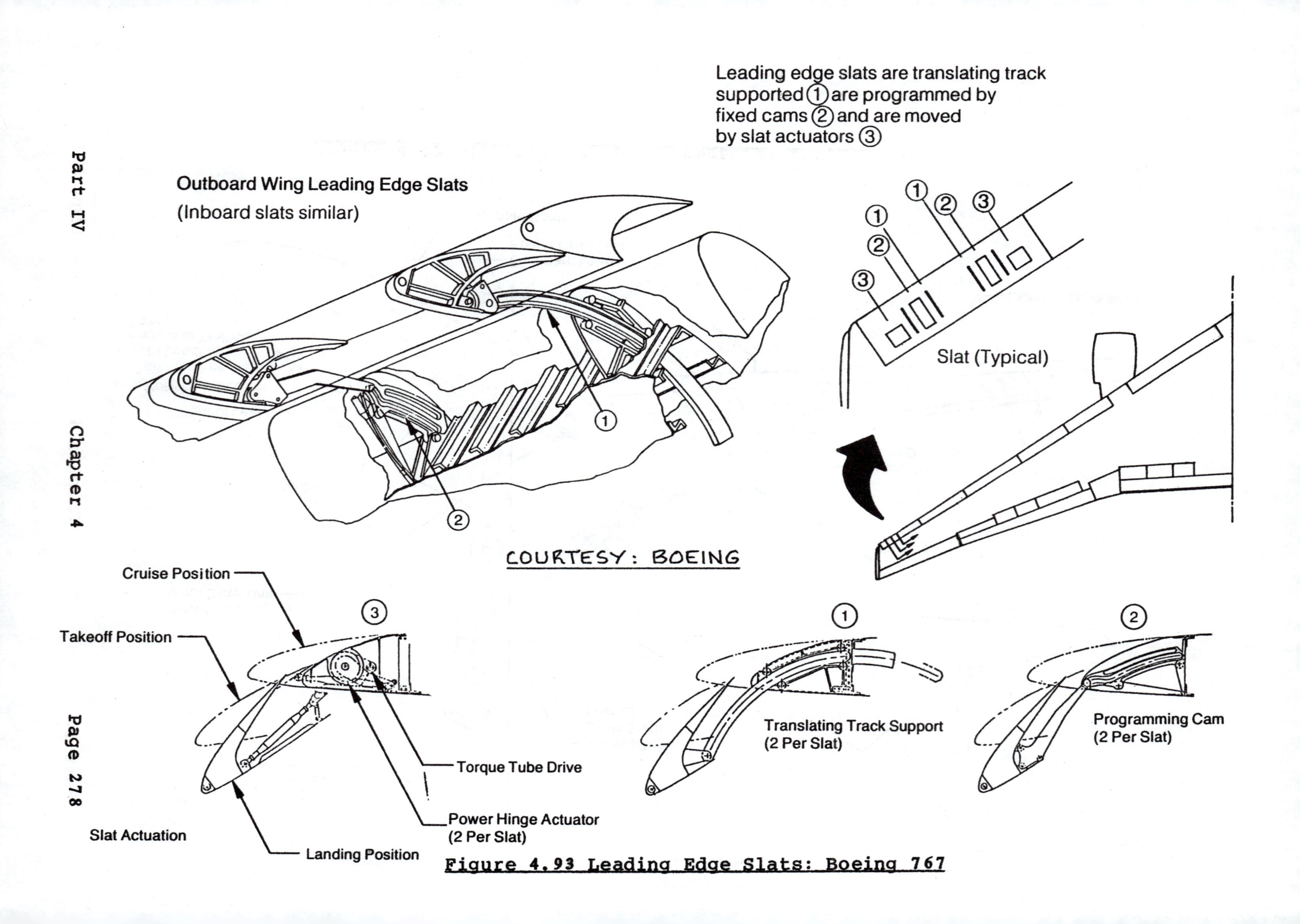

Figure 4.93 Leading Edge Slats: Boeing 767

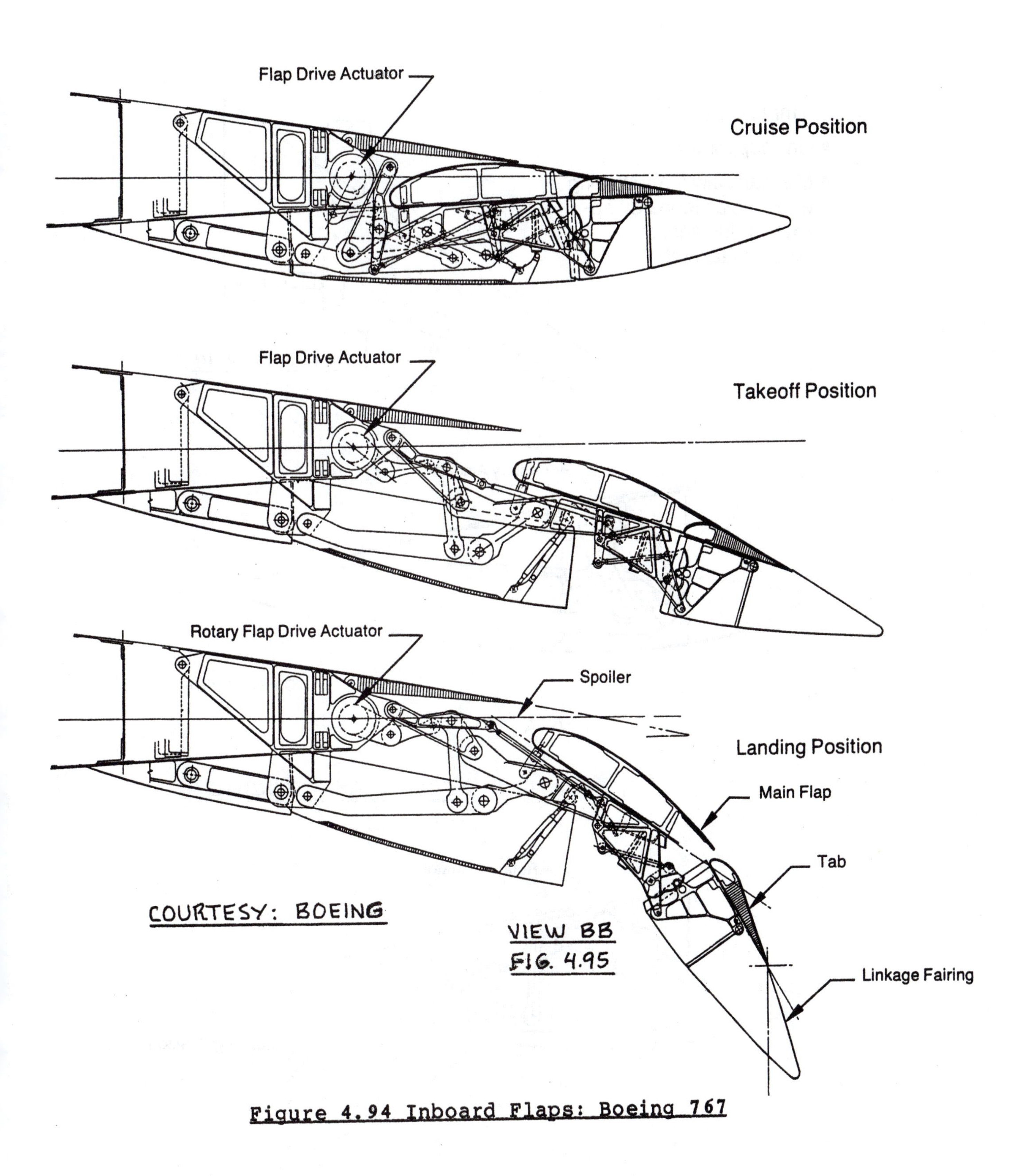

Figure 4.94 Inboard Flaps: Boeing 767

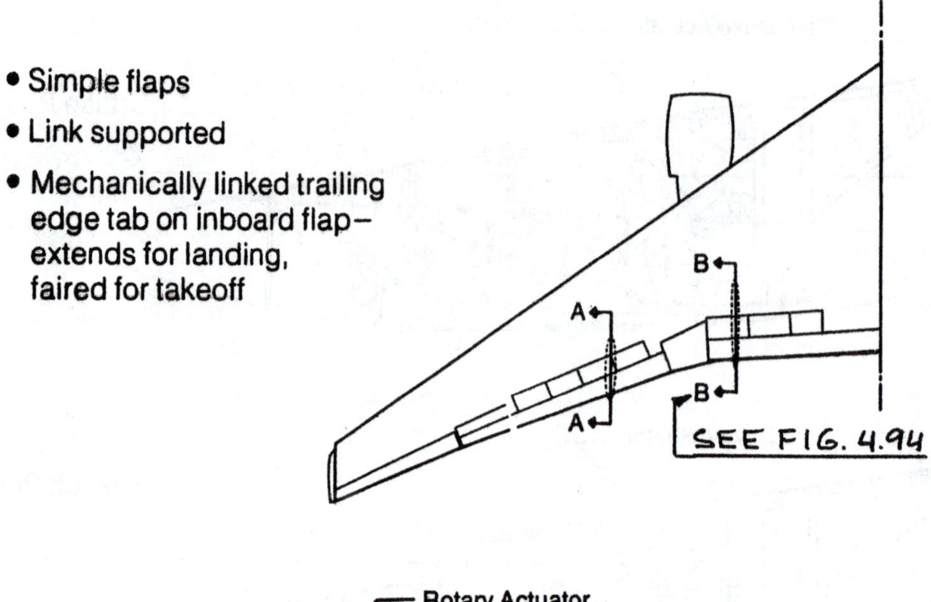

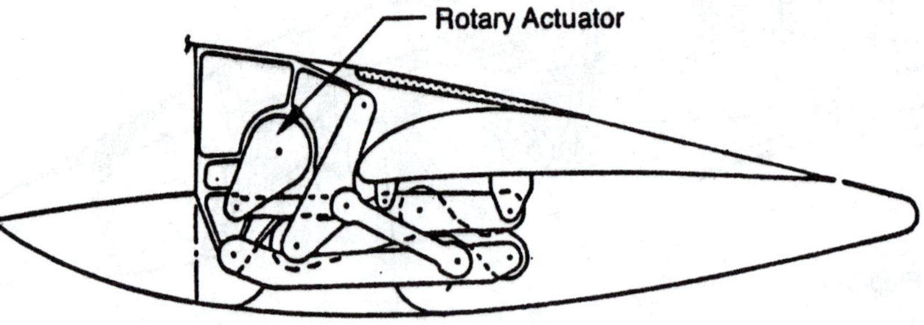

Cruise Position

A-A

COURTESY: BOEING

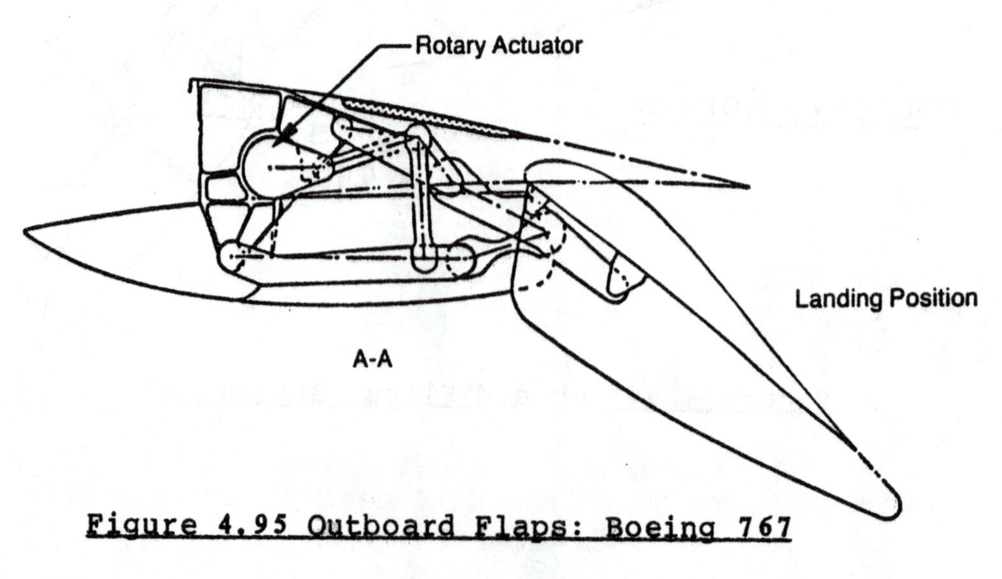

Figure 4.95 Outboard Flaps: Boeing 767

4.7 PROPULSION CONTROL SYSTEMS

Except for gliders, all other airplanes use some form of propulsion to control their flight paths. The propulsion system in turn needs to be controlled by the pilot. This is done with the help of a propulsion control system.

The reader is referred to Chapter 6 in Part III for a discussion of the integration of the propulsion system into airplanes.

Most propeller driven airplanes require six types of propulsion control systems:

1. Ignition control

2. Starter system

3. Fuel flow (=throttle) control

4. Manifold (= mixture) control

5. Propeller control

6. Cooling flap control

In most jet powered airplanes four types of propulsion control systems are needed:

1. Ignition control

2. Starter system

3. Fuel flow (= throttle) control

4. Thrust reverser control

A brief discussion of each propulsion control system follows.

Ignition Controls

Ignition controls are normally a part of the electrical system. Chapter 7 deals with the layout of electrical systems.

Starter System

In most airplanes the starter system consists of an electric starter motor which is geared to the engine. In some military jet engine installations so-called cartridge starters are used. Cartridge starters generate a burst of high velocity, hot gas which drives a small turbine geared to the engine shaft.

Fuel Flow (= Throttle) Control

The pilot controls the power or thrust output of engines with the help of throttles. In most airplanes these throttles in turn operate a cable and/or a push-rod system which is linked to the engine fuel controls.

A separate fuel system with its own controls supplies the fuel to the engines. The layout design of fuel systems is discussed in Chapter 5.

Manifold (= Mixture) Control

This type of control is found only in piston-propeller combinations. These controls are also mechanized with cable and/or push-rod systems.

Propeller Control

All propeller driven airplanes require two forms of propeller controls: propeller speed (rpm) and propeller pitch. In many airplanes the propeller pitch controls incorporate a so-called 'beta-range': in this pitch range the propeller generates reverse thrust. This is normally used only for ground (landing) operations. Propeller controls can be mechanical, electrical or hydraulic.

Cooling Flap Control

Particularly piston engines require special cooling provisions. To admit extra cooling air during take-off operations, cooling flaps are used. These cooling flaps are operated from the cockpit with a cable and/or push-rod system. In large airplanes the cooling flaps are driven by hydraulic actuators

Thrust Reverser Control

Most jet powered airplanes are equipped with thrust reversers. These are used to help slow the airplane after touchdown. Thrust reversers are normally operated with hydraulic controls.

Figure 4.96 shows the complex nature of the propulsion controls in a large piston/propeller powered airplane.

Figures 4.97 illustrates a typical thrust control system layout for a jet transport. A thrust reverser installation is shown in Figure 4.98.

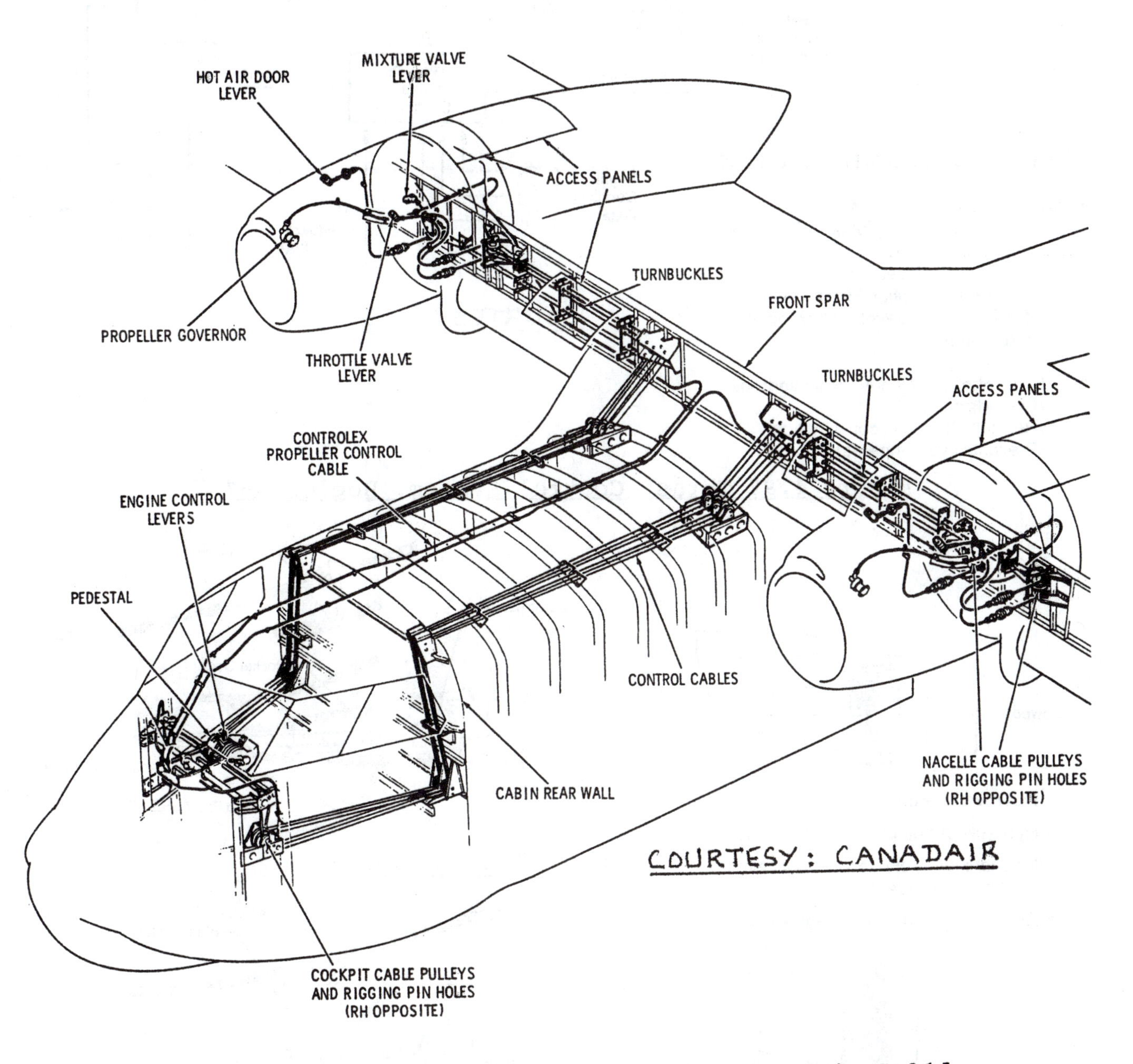

Figure 4.96 Propulsion System Controls: Canadair CL215

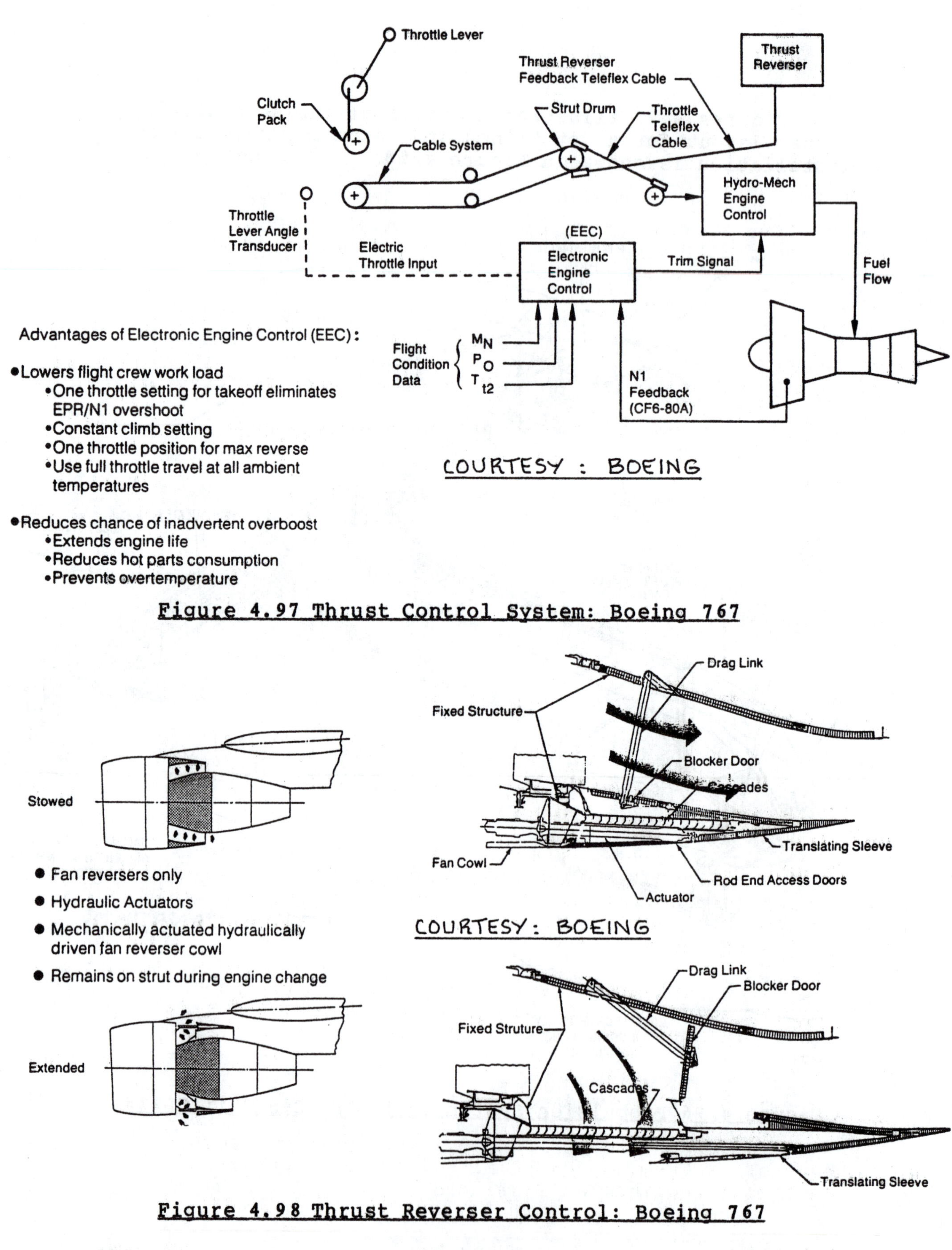

Figure 4.97 Thrust Control System: Boeing 767

Figure 4.98 Thrust Reverser Control: Boeing 767

5. FUEL SYSTEM LAYOUT DESIGN

The purpose of this chapter is to discuss the most fundamental principles of fuel system layout design for airplanes.

For a detailed discussion on airplane fuels, the operation and the layout of airplane fuel systems references 29 and 30 should be consulted.

Since airplane fuels are very combustible liquids, the design, location, operation, accessibility and maintenance aspects of the fuel system are of great importance to airplane safety as well as to airplane economy.

To operate properly, most fuel systems need the following components:

1. Fuel tanks or fuel bays with a total volume sufficient to cover the design range of the airplane plus reserves.

2. Fuel pump(s) and fuel lines to carry the fuel from the fuel tanks to the propulsion units (engines).

   The fuel pumps and fuel lines must be dimensioned so they can supply 1.5 times the maximum required fuel flow by the engines.

   In aerobatic and in fighter airplanes this fuel supply system must also be able to supply fuel under high 'g' maneuvers and/or under inverted flight conditions.

3. A fuel venting system to prevent excessive pressures from building up in the tanks (this could happen when parked in the hot sun). The fuel venting system also must provide positive pressure in the tanks during flight.

4. A fuel quantity indicating system so the crew can tell how much fuel has been used and how much fuel is left. In some systems fuel flow indicators are also needed.

5. A fuel management system (= tank selection system), to allow the crew to regulate the flow from various tanks to different engines. This includes a shut-off system so that fuel supply to inoperative engines (or egines on fire) can be stopped.

6. In airplanes where the center of gravity travel is large, automatic fuel management systems may be required.

7. An easy method for refuelling must be provided. In transport and in military airplanes single point refuelling is operationally desirable. In many military airplanes in-flight refuelling is also required.

8. If the airplane ramp weight exceeds the maximum design landing weight by more than 5 percent a fuel dumping system must be provided.

The material in this chapter is organized as follows:

5.1 Sizing of the fuel system

5.2 Guidelines for fuel system layout design

5.3 Fire extinguishing systems

5.4 In-flight refuelling systems

5.5 Examples of fuel system layouts

5.1 SIZING OF THE FUEL SYSTEM

Fuel system sizing encompasses the following design decisions:

1. Total fuel volume required

2. Size, location and number of fuel tanks needed

3. Number of fuel pumps, location of fuel pumps and required capacity of fuel pumps and fuel lines

The total amount of fuel required by an airplane depends on its mission. It was shown in Part I, Chapter 2 how a first estimate of the mission fuel required, W_F can be obtained. This estimate was later refined with the help of preliminary design Steps 14 and 28 in Chapter 2 of Part II. In any case, the mission fuel weight, W_F is assumed to be known.

In most airplanes fuel is stored in tanks located in the wing or in the empennage. Estimates of available fuel volume in lifting surfaces can be obtained with Eqn.(6.3) of page 153, Part II.

Reference 3 contains methods for determining the volume of 'irregularly' shaped fuel containers.

Fuel pumps need to be sized so that they can deliver 1.5 times the maximum required amount of fuel flow to the engines. Maximum fuel flow normally occurs during take-off. In military airplanes, maximum fuel flow normally occurs during afterburning operations.

In either case the required engine fuel flow is obtained from the engine manufacturer or, in early preliminary design by multiplying the maximum required thrust by the associated fuel consumption:

$$\text{Max. Fuel Flow} = T_{TO}(c_j) \qquad (5.1)$$

or from:

$$\text{Max. Fuel Flow} = P_{TO}(c_p) \qquad (5.2)$$

The number of fuel tanks should be kept to a minimum from a weight and cost viewpoint.

For the sizing of fuel lines and the determination of necessary fuel pump pressures the method of Ref.31 (Section 3, Flow of Liquids Through Pipes) can be used.

5.2 GUIDELINES FOR FUEL SYSTEM LAYOUT DESIGN

Figure 5.1 shows a so-called gravity fuel system in a high wing light airplane. In such a system only one fuel pump may be required.

Figure 5.2 shows a typical fuel system layout in a low wing light airplane. In this type of system at least two fuel pumps are required.

Figure 5.3 shows an example of a fuselage mounted fuel tank in a fighter airplane. This fuel tank is designed for 'inverted' operation: the inner tank is equipped with flapper valves which trap fuel around the pump during inverted flight.

Each fuel system must be equipped with a fuel vent and a fuel sump system.

The fuel vent system prevents excessive pressure from building up in the fuel tanks. It also serves to maintain ram-air pressure in the tanks while in flight. In some fighter airplanes the tanks are pressurized by

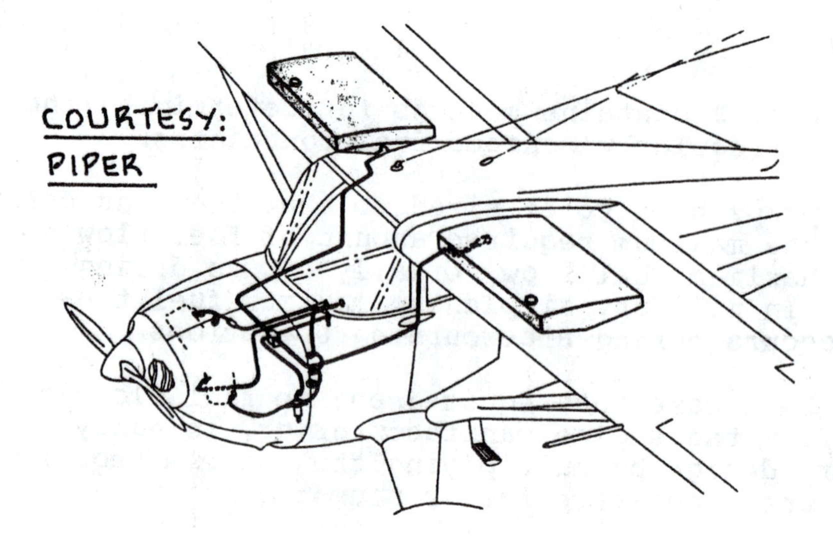

Figure 5.1 Fuel System Layout for a High Wing Airplane

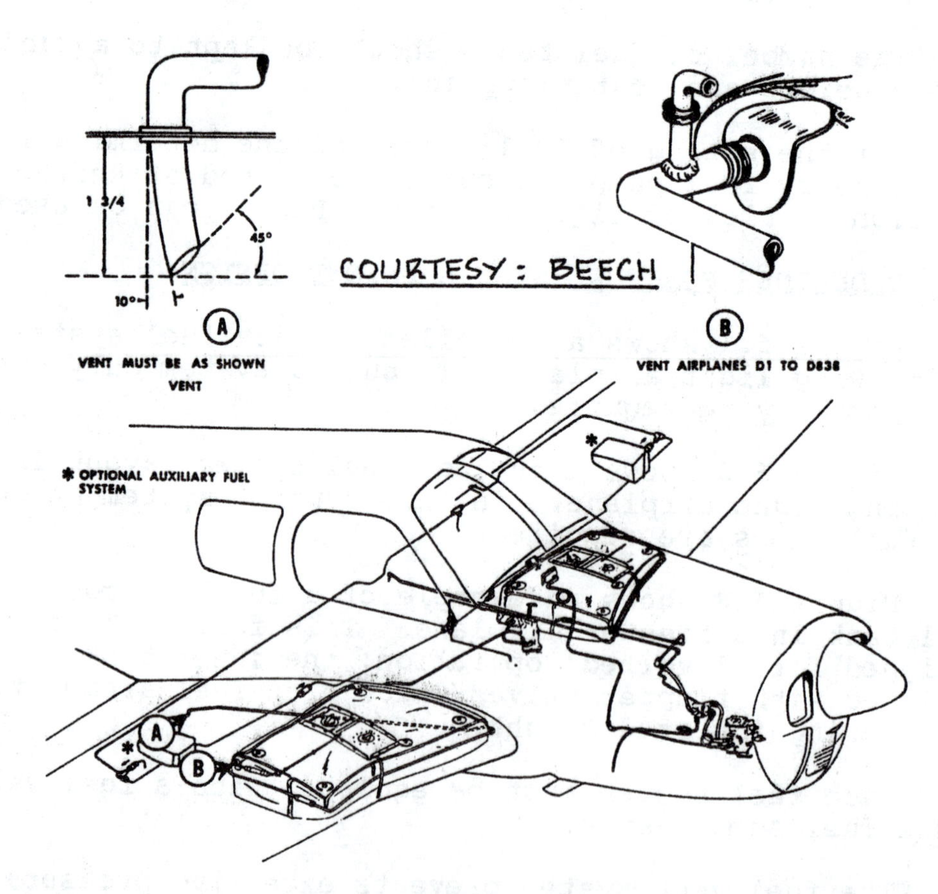

Figure 5.2 Fuel System Layout for a Low Wing Airplane

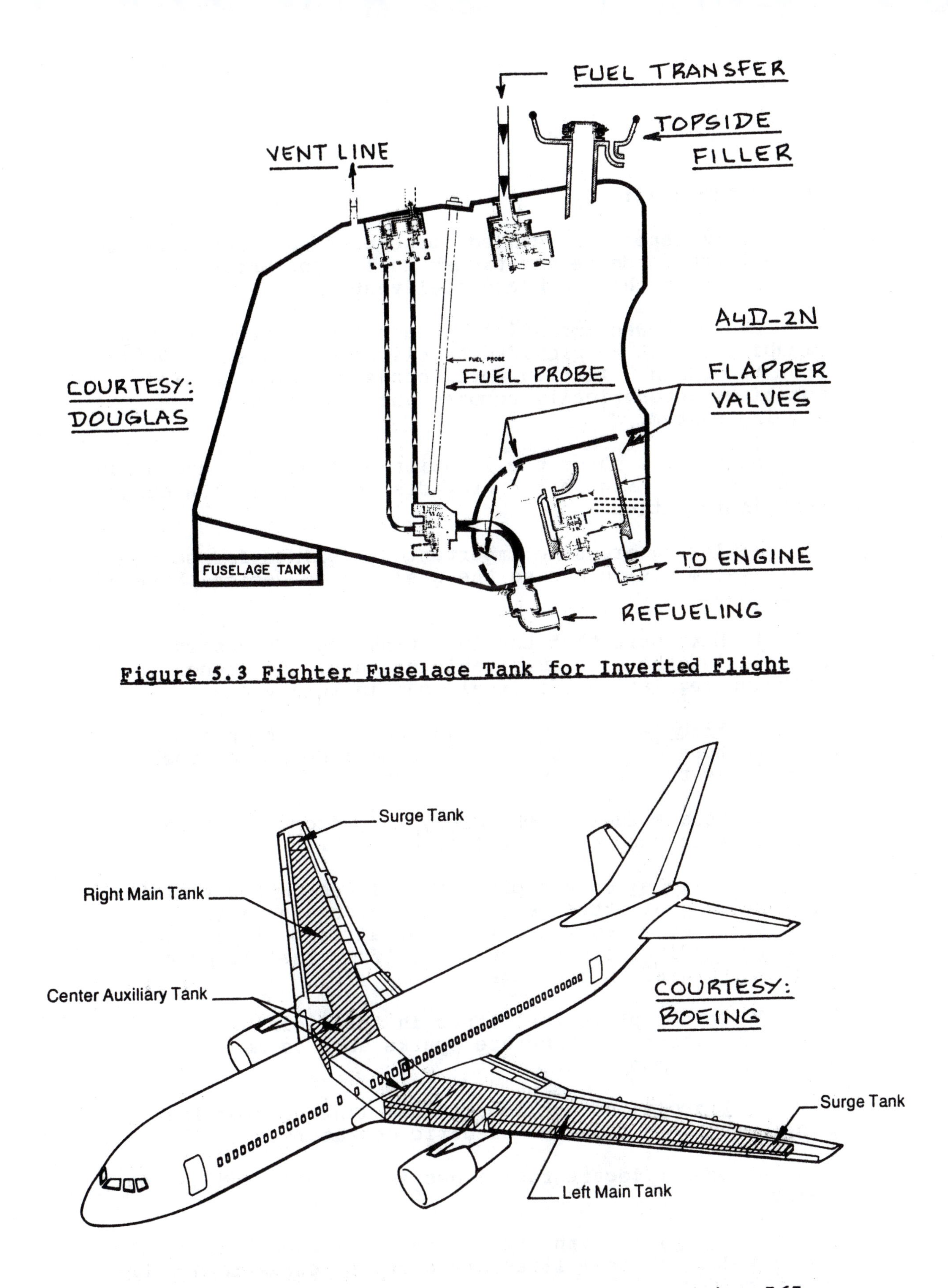

Figure 5.3 Fighter Fuselage Tank for Inverted Flight

Figure 5.4 Example of Surge Tank Location: Boeing 767

engine bleed-air.

In transports so-called surge tanks are installed to collect and condense any excess fuel vapor before it exits through the overboard fuel vents.

At the lowest point(s) in each fuel system drainage capability must be provided to eliminate condensed water from the tanks. In transports condensed liquids are sometimes automatically removed through the use of a sumping system.

Figure 5.2 shows the location of fuel vent and drain lines (sump). Figure 5.4 provides an example of a surge tank installation.

Following are some important guidelines for safe fuel system design during the preliminary design phase of airplanes.

1. Make sure that the fuel tanks cannot easily rupture in otherwise survivable crashes. Reference 21 contains regulations for fuel tank integrity.

<u>Example:</u> design the landing gear attachments so that rupture of gear struts and braces during a crash is not likely to damage any fuel tanks.

2. Locate fuel lines safely away from easily damaged structure in case of a survivable crash.

<u>Example:</u> do not place fuel lines in the vicinity of landing gear attachment points which are easily damaged or even separated from the airframe in the case of hard landings or aborted take-offs. Fuel lines are easily ruptured in such cases.

3. Do not place fuel lines in the vicinity of equipment which can generate sparks such as many electrical components.

<u>Example:</u> do not locate fuel lines and/or fuel line connectors close to electric genrators.

4. Do not locate fuel lines in or near landing gear wells.

<u>Example:</u> overheated tires can explode inside the wheel well. If fuel lines are present, catastrophic fire may result.

5. Do not locate the fuel tanks close to engines if at all possible. Use dry bays to separate engines from fuel tanks. Fire walls (stainless steel) must separate all engines and all passenger and crew compartments from fuel tanks.

<u>Example:</u> Figure 5.4 shows small dry bays directly behind the engines.

6. Locate the fuel vent lines and the fuel dump lines so that positive separation of fuel and fuel vapor from the airframe is assured.

<u>Example:</u> determine the local pressures which surround fuel vents. It has occurred that due to adverse pressure gradients vented fuel was drawn into a cabin heating system leading to a fatal accident.

7. Locate ram-air inlets for vent systems so that large asymmetric pressures are unlikely to occur.

<u>Example:</u> when asymmetric ram pressures occur in the vent system it is possible that fuel will be transferred from one wing into the other. This can lead to serious lateral control problems.

8. Locate fuel quantity sensors in fuel tanks such that in extreme airplane attitudes (climb, dive or glide) the correct level of fuel remaining is indicated to the crew.

<u>Example:</u> in wing tip mounted fuel tanks, placing the fuel quantity sensors at one end of the tank may cause it to indicate wrongly during prolonged climbs or glides.

9. Locate fuel pumps in fuel systems so that in extreme airplane attitudes fuel is still delivered to the engine(s).

<u>Example:</u> locating fuel pumps at one end of a long fuel container may cause it to run dry (which can cause pump failure and/or engine flame-out) during prolonged climbs or glides.

10. Avoid the use of tiptanks.

<u>Example:</u> tiptanks may cause large changes in the airplane rolling moment of inertia between their full and empty condition. This in turn may cause detrimental effects on the lateral control of the airplane.

11. Most fuel tanks require access panels and fuel caps. If improperly designed, these can cause explosions due to arcing inside the fuel tank when struck by lightning. Figure 5.5 shows airplane regions where lightning strikes occur most frequently.

Example: Figure 5.6 shows a properly and improperly designed fuel cap. Reference 32 presents design procedures which minimize adverse effects of lightning strikes on airplanes.

5.3 FIRE EXTINGUISHING SYSTEM

Reference 21 contains regulations regarding the need for and installation of fire extinguishing systems. All FAR 25 certified airplanes must have such systems aboard.

Figures 5.7 and 5.8 show example installations for fire extinguishing systems.

5.4 IN-FLIGHT REFUELING SYSTEMS

In-flight refueling is required in many military airplanes. Two types of systems are in use:

1. Refueling Boom System

2. Probe and Drogue System

An example of a refueling boom installation is shown in Figure 5.9. This system requires a boom operator stationed in the tanker airplane.

An example of a probe and drogue system is shown in Figure 5.10.

Figure 5.11 presents an example of an airplane modification proposal which includes both refueling systems.

The receiving airplane must be equipped with a refueling probe. Figure 5.12 shows an example.

5.5 EXAMPLES OF FUEL SYSTEM LAYOUTS

Figures 5.13 through 5.20 contain examples of fuel system schematics and/or layouts. Commentary on these drawings is given next.

Figure 5.13 Fuel System Installation: Piper PA-38-112

This is a low wing airplane: two pumps are required to operate this system. Note the location of the fuel vents below each wing. Also note that the fuel tanks are well away from the engine compartment.

==

Figure 5.14 Fuel System Schematic: Canadair CL215

Only the right side of the system is shown. The left side is mirror symmetric. Note the ram air scoop associated with the vent system. In this system fuel is pumped from the main fuel tanks (numbers 1-8) to an engine collector tank, one for each engine. The engines are fed only from the collector tanks.

==

Figure 5.15 Fuel System Schematic: Short Skyvan

Note again the presence of the collector tanks. The fuel vent lines in this case protrude into the airstream adjacent to the fuselage.

==

Figure 5.16 Fuel System Schematic: Boeing 767

Observe the dry bays behind the engines. Note also the surge tanks near the wing tips. These surge tanks are not normally filled with fuel. The system can be filled from over the wing if the airplane is landed at an airport where pressure refueling is not available.

Figure 5.17 Fueling/Defueling System: Boeing 767

Note the single point refueling in the left wing only. The overfill sensor installed in the surge tanks prevents overfilling and overpressuring of the fuel tanks. Defueling is done close to the center of the airplane: low point in the fuel system.

Figure 5.18 Engine Fuel Feed System: Boeing 767

Observe the tie-ins with the airplane APU system located in the fuselage tailcone.

Figure 5.19 Fuel Tank Vent System: Boeing 767

Note that the vent lines are equipped with flame arrestors. The center wing drainmast allows fuel dumping if required.

Figure 5.20 Automatic Sumping System: Boeing 767

In airline operation it is desirable not to have to inspect the fuel system all the time for condensates which may collect at the bottom of the tanks. The automatic sumping system clears condensates automatically.

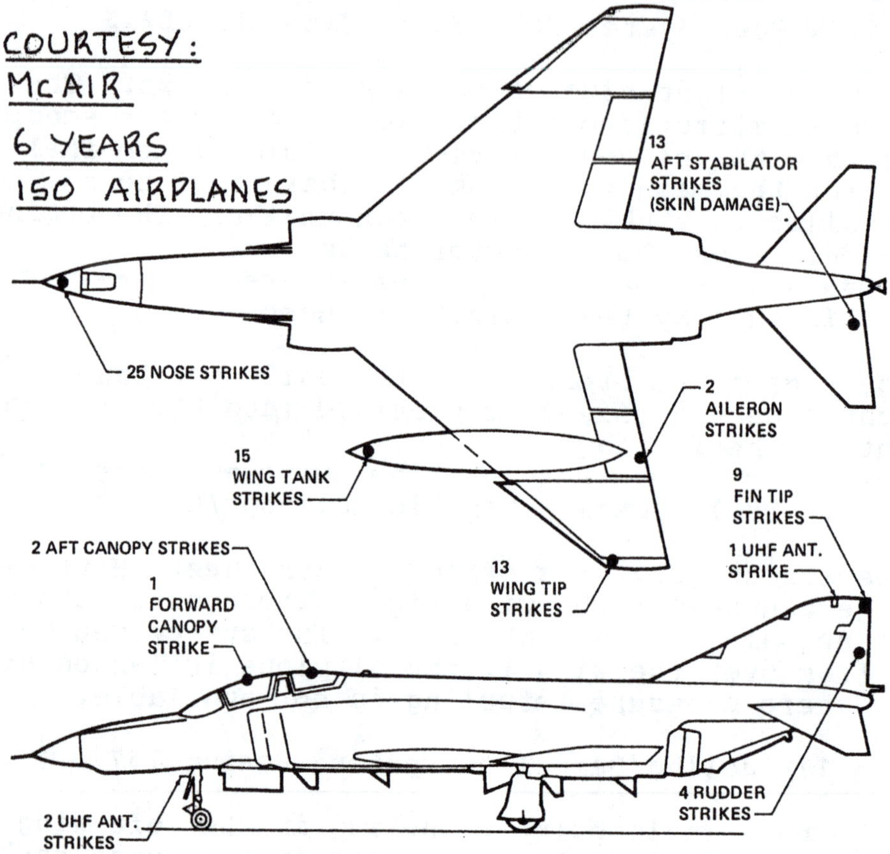

COURTESY:
McAIR
6 YEARS
150 AIRPLANES
13 AFT STABILATOR STRIKES (SKIN DAMAGE)
25 NOSE STRIKES
2 AILERON STRIKES
15 WING TANK STRIKES
9 FIN TIP STRIKES
2 AFT CANOPY STRIKES
1 UHF ANT. STRIKE
13 WING TIP STRIKES
1 FORWARD CANOPY STRIKE
4 RUDDER STRIKES
2 UHF ANT. STRIKES

Figure 5.5 Likely Locations for Lightning Strikes

COURTESY:
McAIR
PLASTIC
SPRING TINES
B
GOOD
BAD
LIGHTNING PROTECTED FUEL CAP ARCING OUTSIDE TANK IN SAFE AREA
USUALLY SEMI-INSULATED FUEL CAP ARCING INSIDE TANK DUE TO DIRECT STROKE

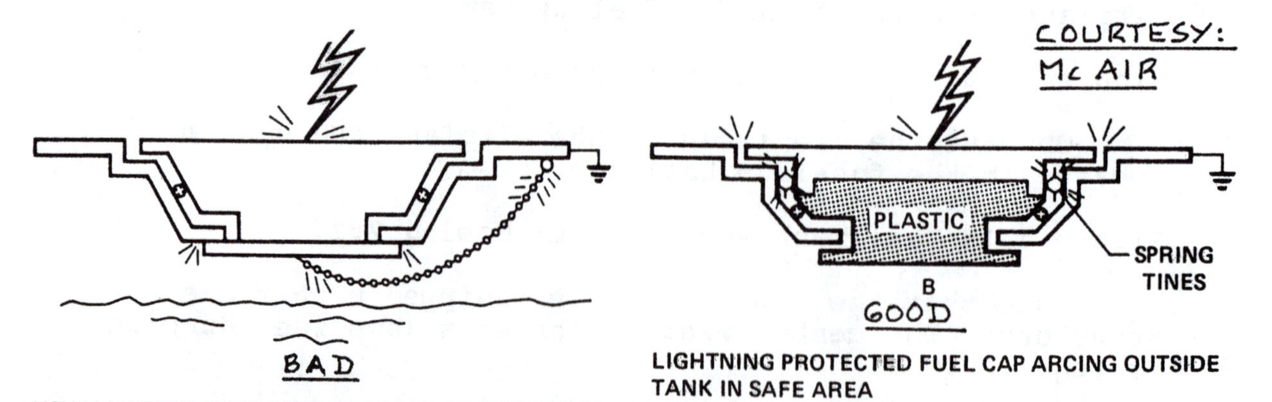

Figure 5.6 Good and Bad Design Of Fuel Caps

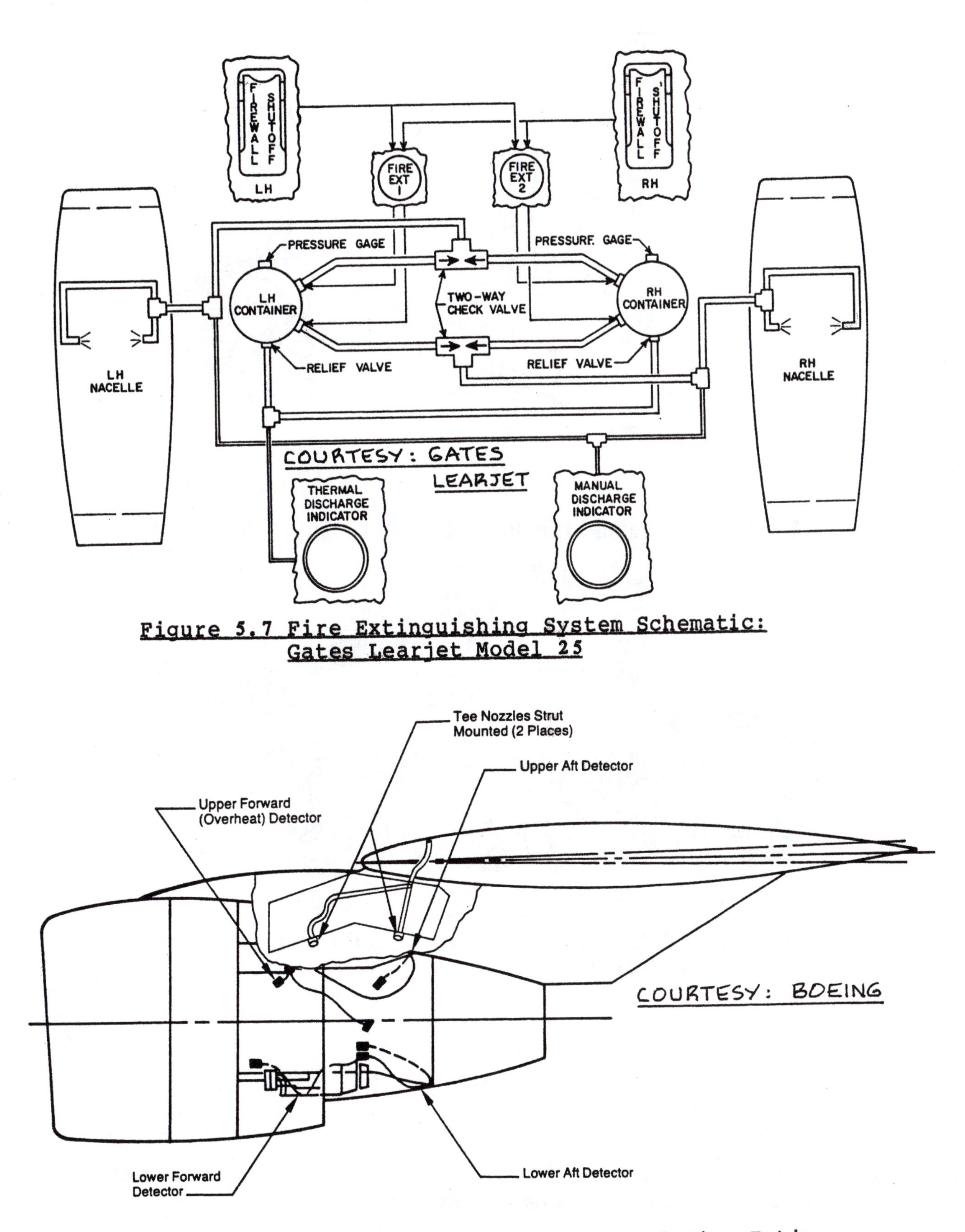

Figure 5.7 Fire Extinguishing System Schematic: Gates Learjet Model 25

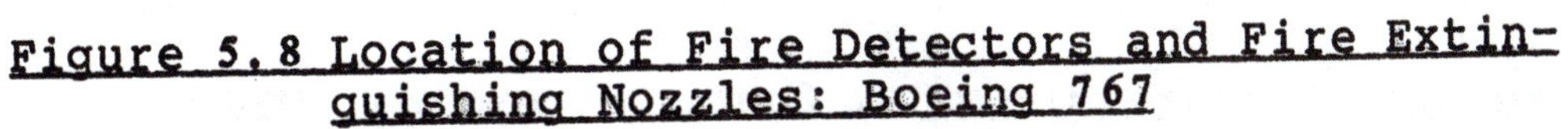

Figure 5.8 Location of Fire Detectors and Fire Extinguishing Nozzles: Boeing 767

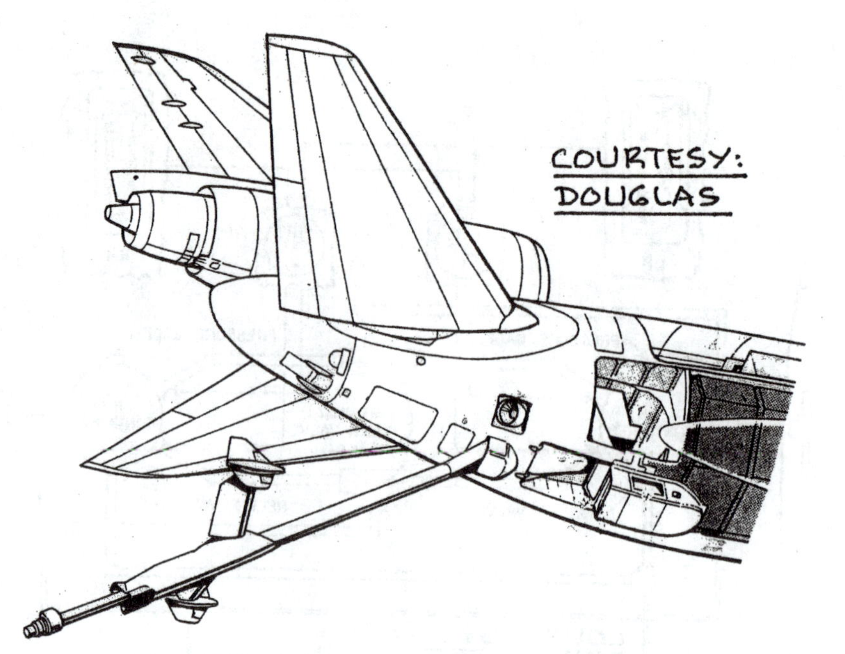

Figure 5.9 Refueling Boom Installation McDD KC-10A

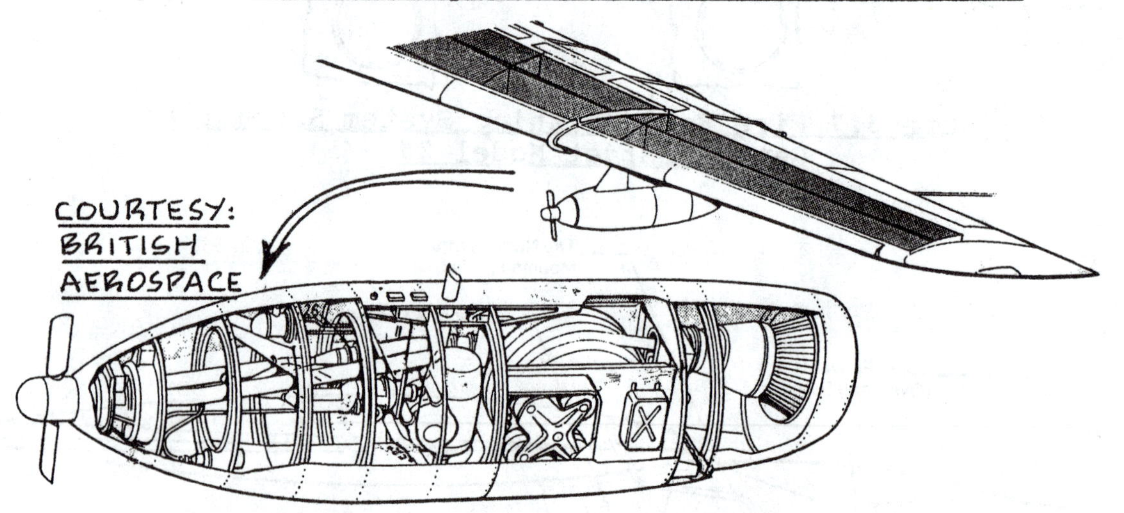

Figure 5.10 Probe and Drogue Refueling Installation: Vickers VC10 K.2

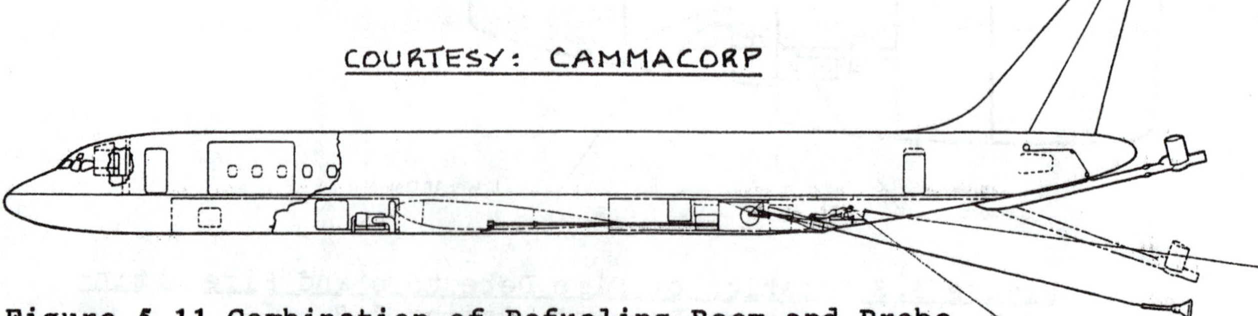

Figure 5.11 Combination of Refueling Boom and Probe and Drogue Unit: DC-8

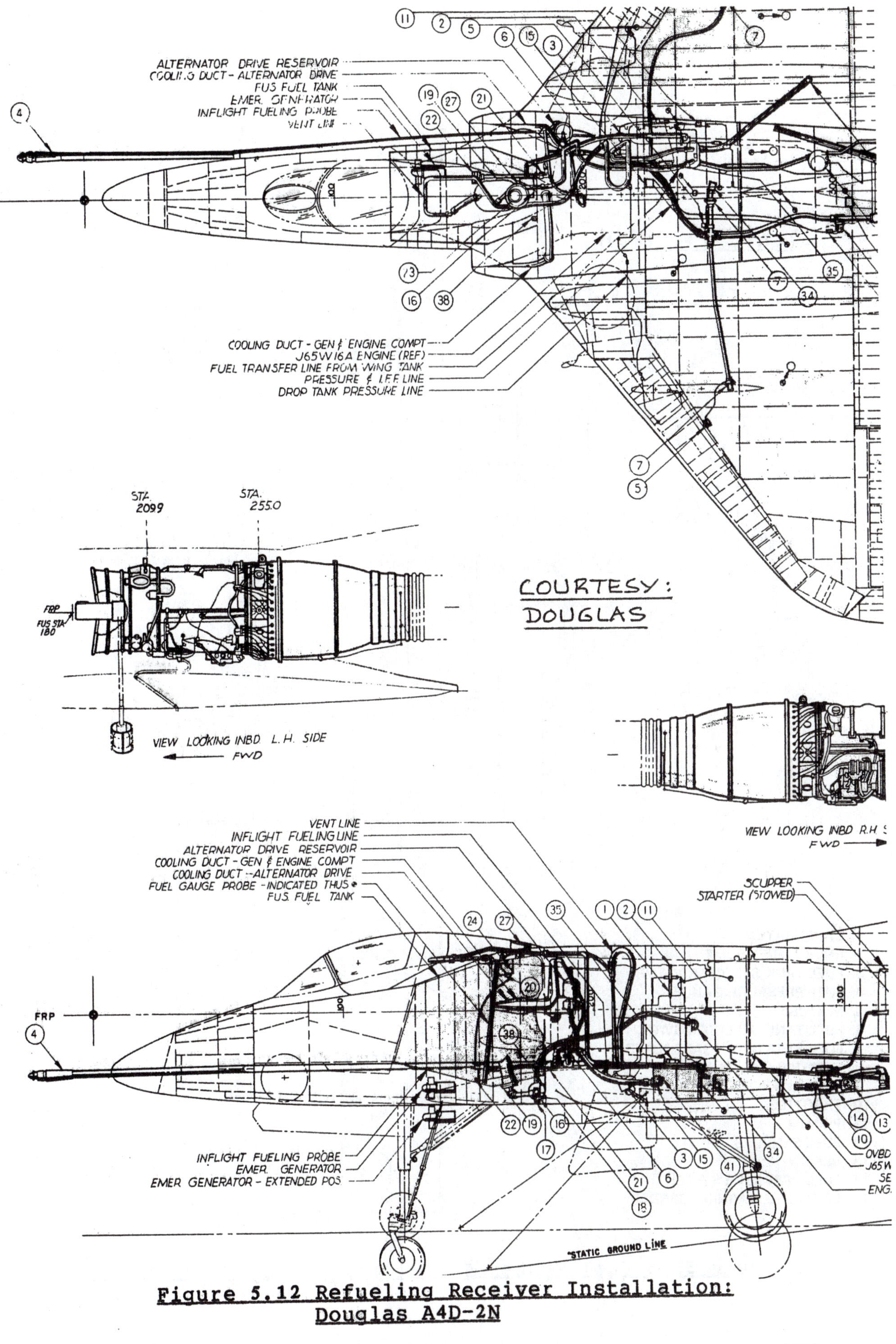

Figure 5.12 Refueling Receiver Installation: Douglas A4D-2N

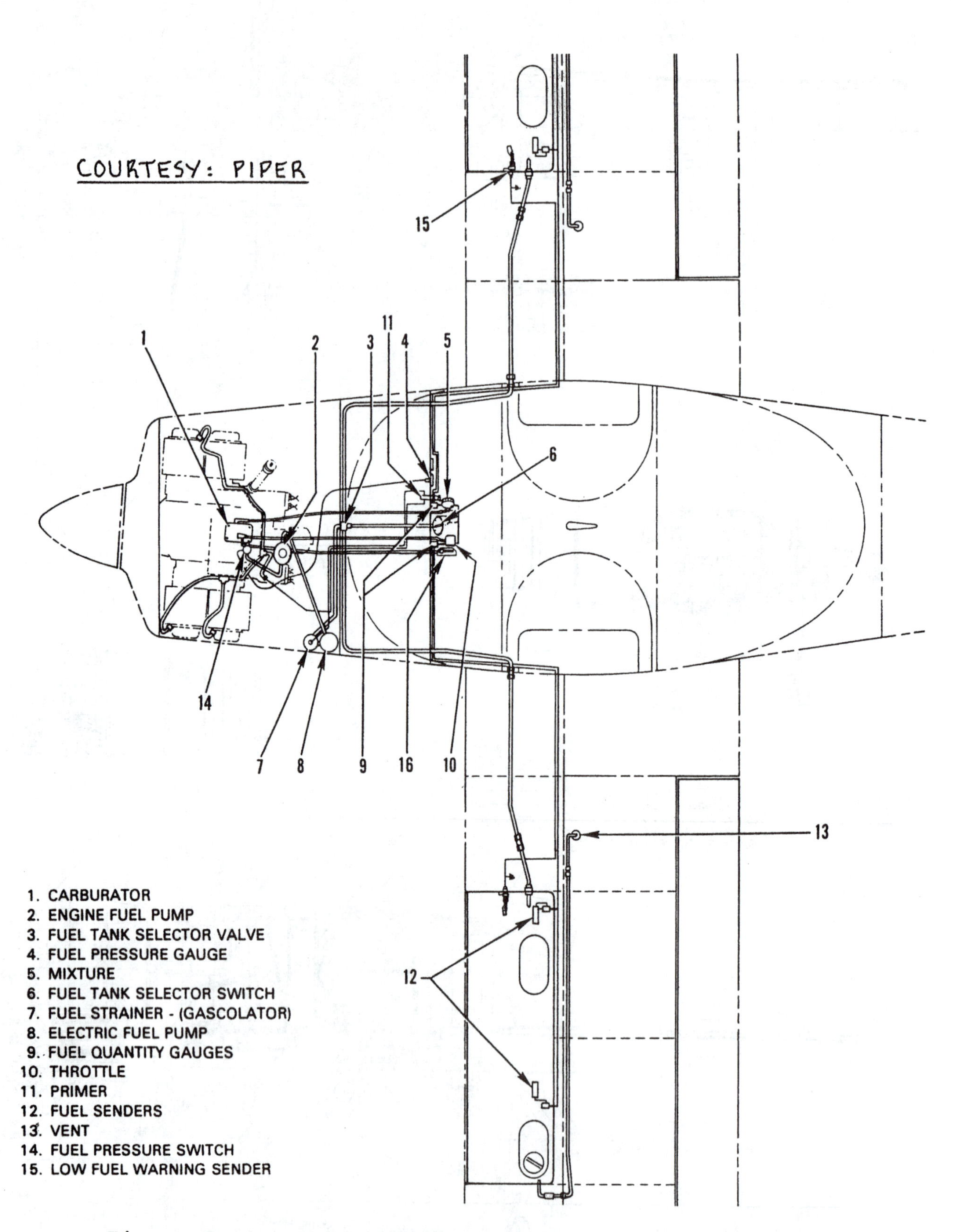

Figure 5.13 Fuel System Installation: Piper PA-38-112

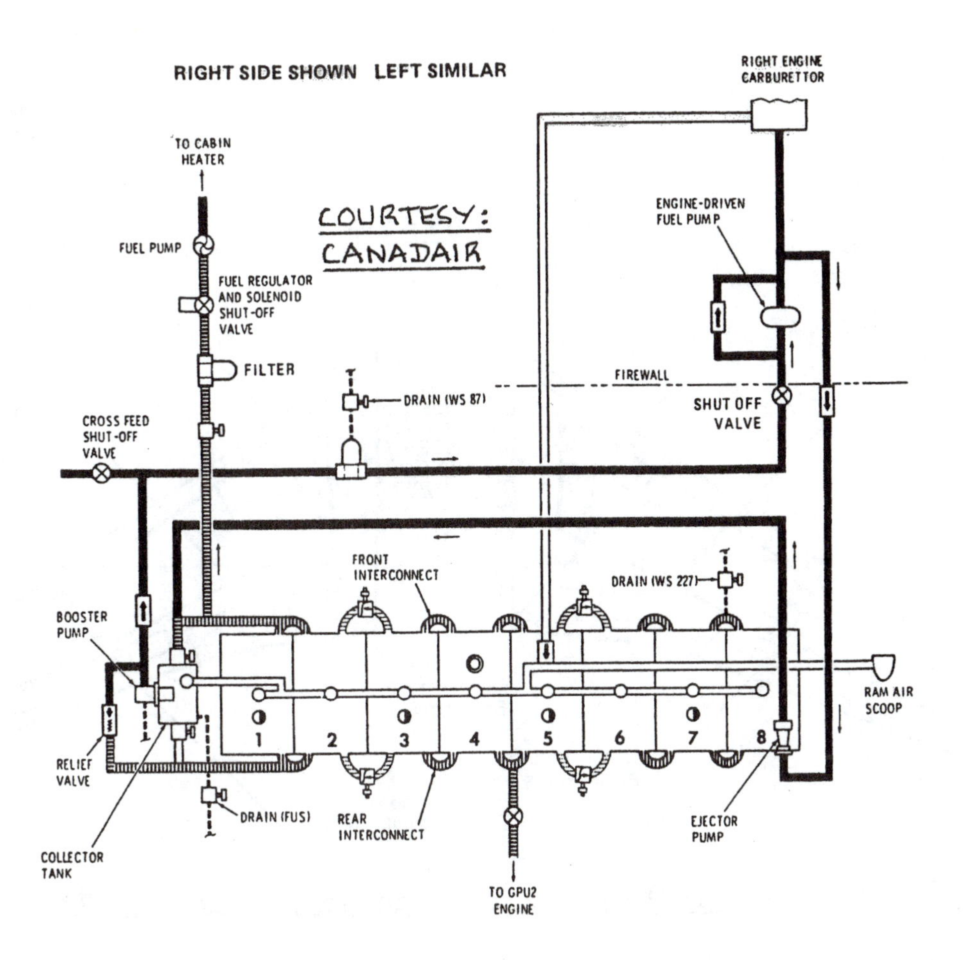

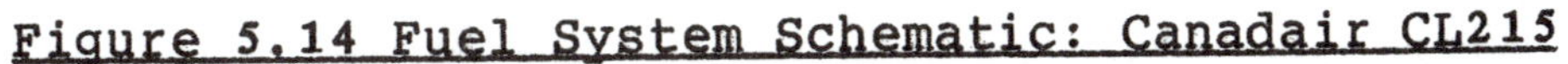

Figure 5.14 Fuel System Schematic: Canadair CL215

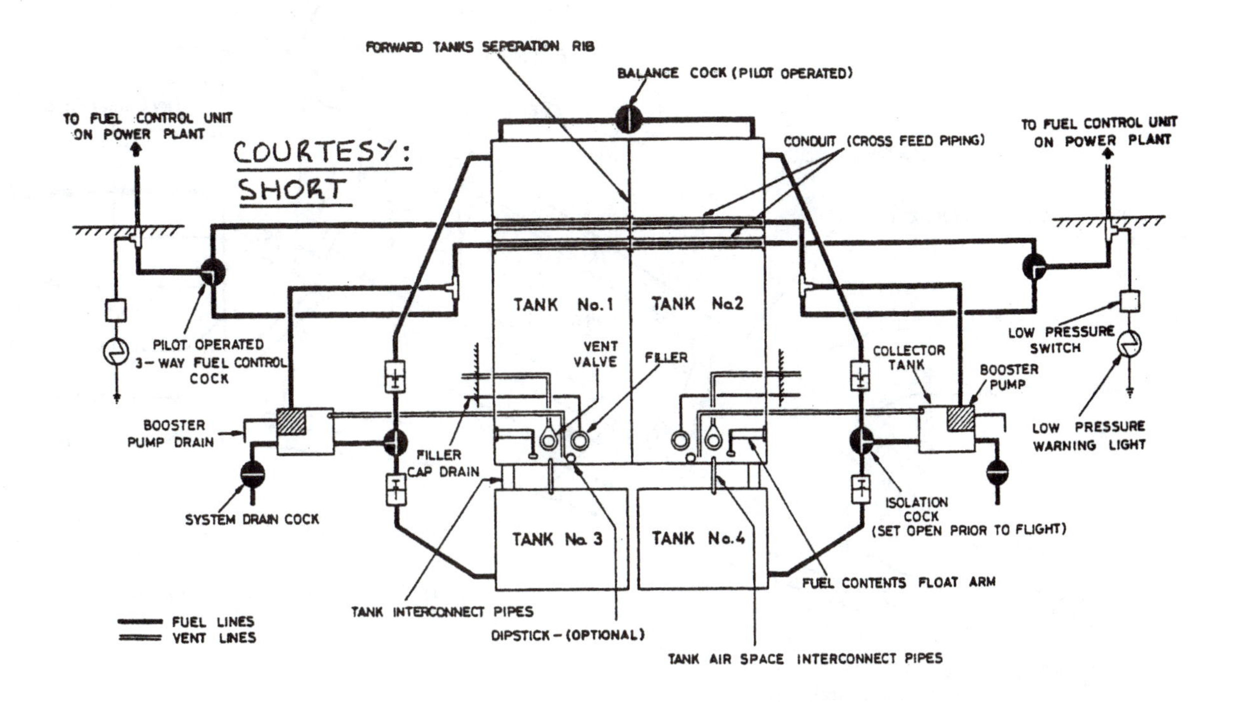

Figure 5.15 Fuel System Schematic: Short Skyvan

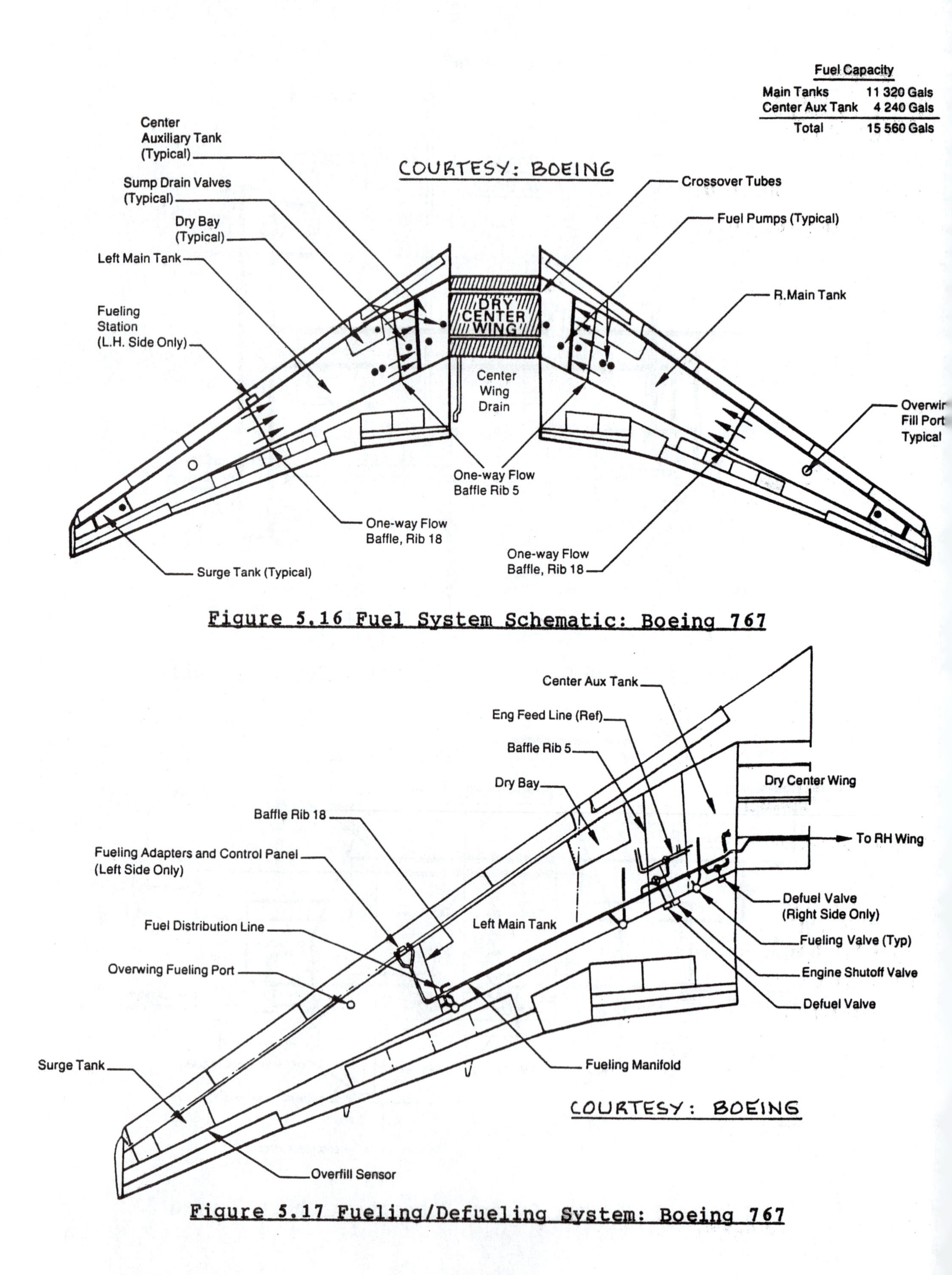

Figure 5.16 Fuel System Schematic: Boeing 767

Figure 5.17 Fueling/Defueling System: Boeing 767

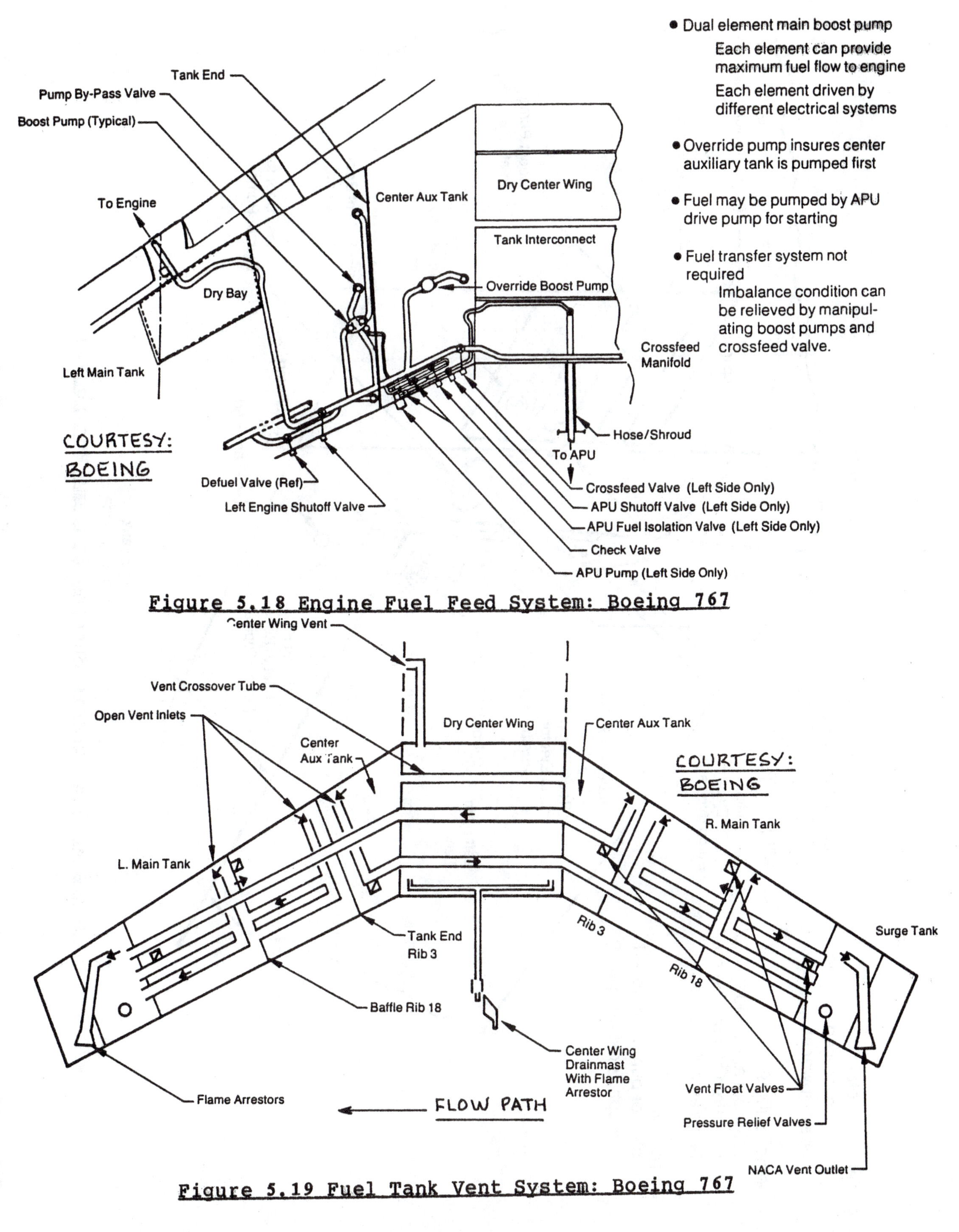

Figure 5.18 Engine Fuel Feed System: Boeing 767

Figure 5.19 Fuel Tank Vent System: Boeing 767

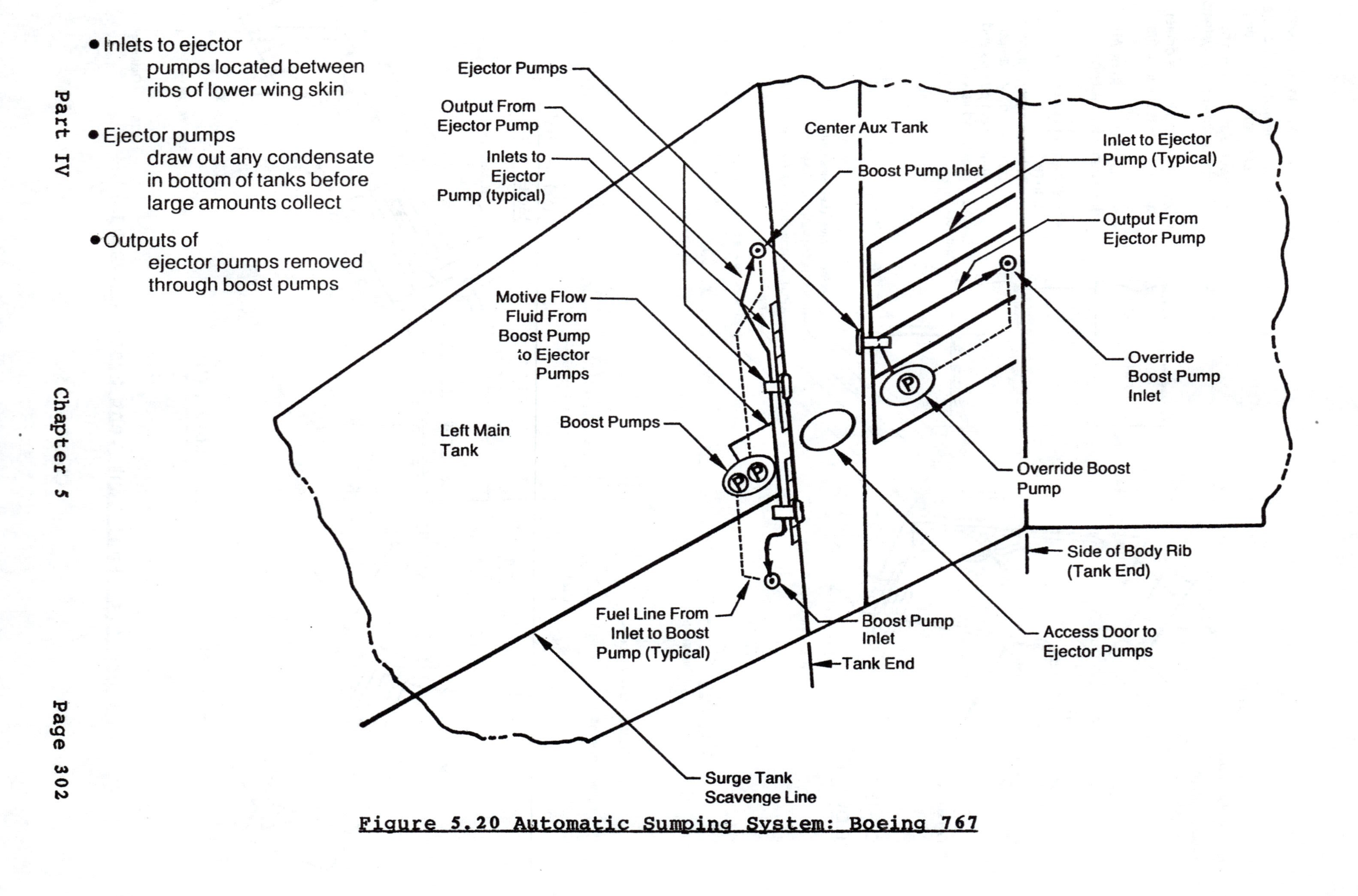

Figure 5.20 Automatic Sumping System: Boeing 767

6. HYDRAULIC SYSTEM LAYOUT DESIGN

In this chapter some fundamental design layout requirements for hydraulic systems will be discussed. The material is organized as follows:

6.1 Functions of the hydraulic system

6.2 Sizing of the hydraulic system

6.3 Guidelines for hydraulic system design

6.4 Hydraulic system layout examples

6.1 FUNCTIONS OF HYDRAULIC SYSTEMS

The functions of hydraulic systems vary from one airplane to the other. Typical functions are to provide hydraulic power to actuators with the following tasks:

1. Moving primary flight controls: ailerons, elevator, stabilizer, rudder and spoilers.

2. Moving secondary flight controls:

   a) high lift devices (flaps)
   b) trim controls
   c) speed brakes

3. Extending and retracting of the landing gear

4. Controlling wheel brakes

5. Landing gear steering

6. Operating thrust reversers

Hydraulic systems usually consist of the following components:

1. Hydraulic fluid reservoir: Figure 6.1 shows an example.

2. Hydraulic pumps (engine driven, air-driven, electric driven or RAT driven): Figure 6.2 shows an example. RAT = Ram Air Turbine.

   Note: in some light airplanes a hand driven pump is provided for emergencies.

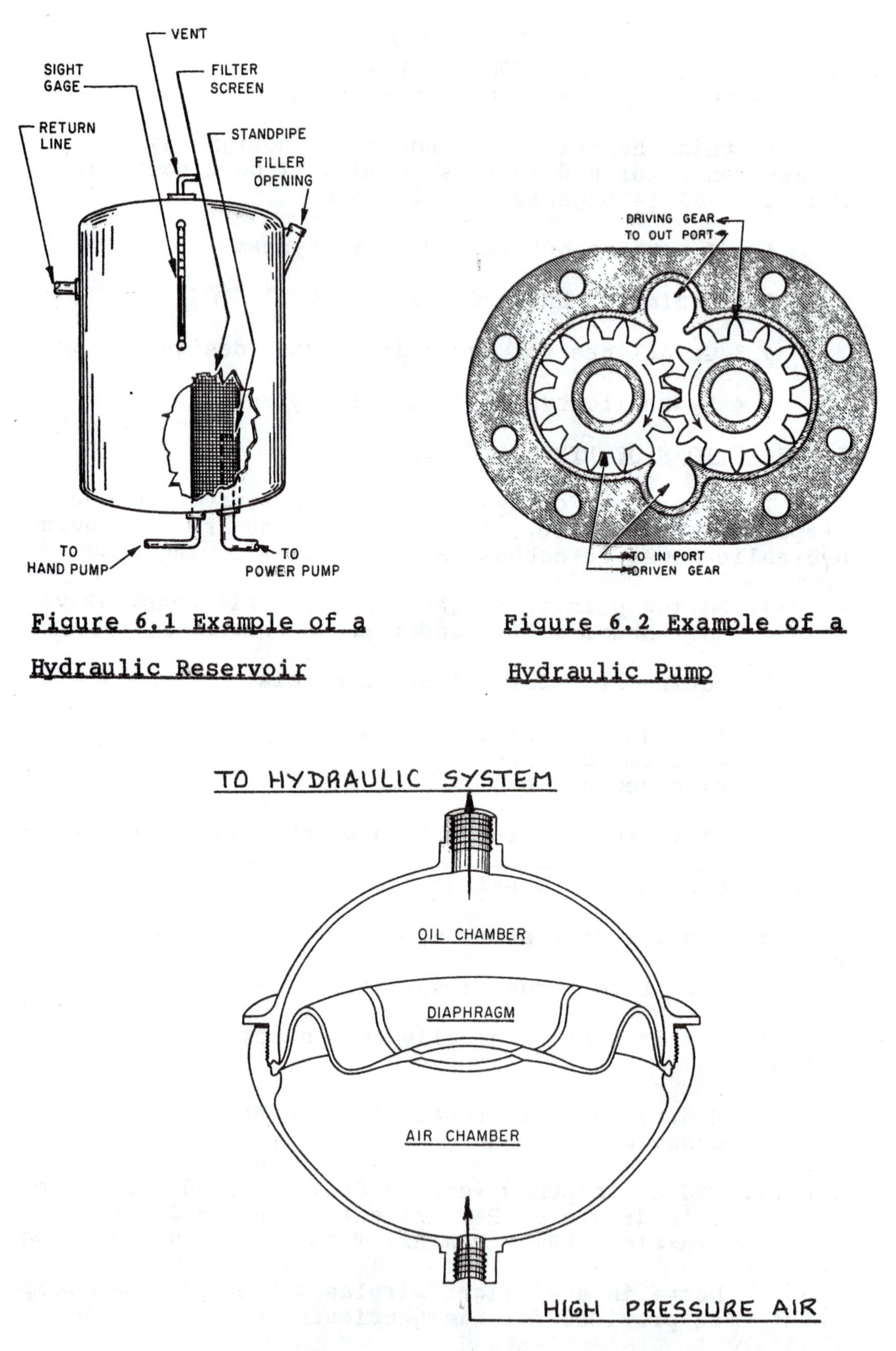

Figure 6.1 Example of a Hydraulic Reservoir

Figure 6.2 Example of a Hydraulic Pump

Figure 6.3 Example of a Hydraulic Accumulator

3. Accumulators (mostly used for emergencies): Figure 6.3 shows an example.

4. a) Lines and valves for fluid distribution to all operating points.

   b) Cockpit controls to operate the functions served by the hydraulic system.

   Figure 6.4 shows an example schematic of a hydraulic system used in a light twin. Note the hand driven pump.

The number of hydraulic pumps required depends on the criticality of the hydraulic system to safe flight operations. If the hydraulic system is essential for safe flight operations a large number of pumps and 3-4 independent hydraulic systems may be required. If the hydraulic system is not critical for safe flight operations, a hydraulic accumulator is often used to provide temporary hydraulic pressure if the pumps have failed.

Most hydraulic systems today operate at a pressure level of 3,000 psi and use some version of Skydrol 500 as the operating fluid.

Reference 28 contains a discussion of the design requirements for future 8,000 psi systems. The major advantages of such high pressures are much reduced weight and installed volume.

References 29 and 33 contain excellent detailed discussions of the functions of hydraulic system components.

In Chapter 4 the possible use of electrohydrostatic actuators was mentioned. With such actuators it is in principle possible to eliminate the hydraulic system entirely from future airplane.

6.2 SIZING OF THE HYDRAULIC SYSTEM

6.2.1 Normal Operation

The maximum amount of hydraulic fluid flow required for the operation of an airplane normally occurs in the landing phase: the primary and secondary flight controls, the landing gear and speed brakes may all have to be operated simultaneously.

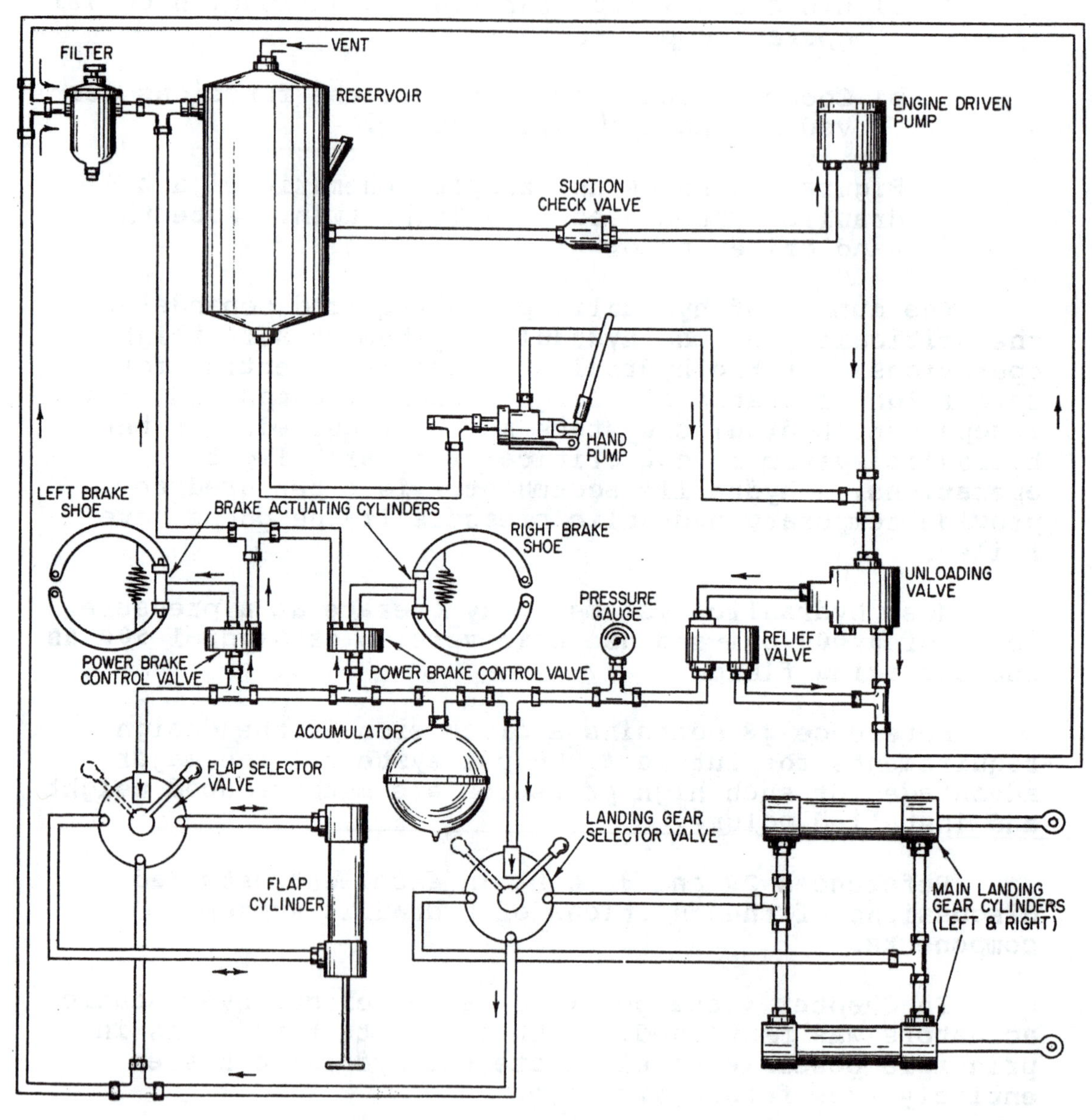

COURTESY : NORTHROP UNIVERSITY

Figure 6.4 Hydraulic System Schematic for a Light Twin

A convenient way to analyze the total fluid flow requirements is to make a list of the actuator rate and force requirements from which the 'gallons-per-minute' flow requirement is obtainded by summation. Such a hydraulic system load analysis will be different for each type of airplane.

Table 6.1 provides some data on hydraulic system capabilities of several airplanes.

Power requirements for hydraulic systems range from 1-5 hp in light airplanes to 200-300 hp in transports and up to 700 hp in large tactical fighters.

6.2.2 Emergency Operation

Hydraulic pumps or their power source (such as the engines or electric power sources) can fail. It depends on the type of airplane and the criticality of functions served by the hydraulic system whether or not back-up systems need to be provided.

Typical back-up systems used in many airplanes include:

1. Accumulators to provide short duration hydraulic pressure for lowering the landing gear. Figure 6.4 gives an example.

2. APU and/or RAT to provide long duration emergency stand-by popwer to operate the flight controls. APU installations are discussed in Chapter 7. Fig.6.5 gives an example of a RAT installation. APU = Auxiliary Power Unit, RAT = Ram Air Turbine.

6.3 GUIDELINES FOR HYDRAULIC SYSTEM DESIGN

1. It is essential to make a list of functions to be served by the hydraulic system under normal and under emergency operating conditions. Table 6.2 shows such a list for the Boeing 737-200.

2. Many hydraulic system components require service and maintenance. Make sure that these components can be accessed.

3. Do not route the hydraulic supply lines to flight critical controls such that they all fail together in case of:

a) engine component disintegration

Table 6.1 Airplane Hydraulic System Capabilities

Airplane Type	Hydraulic Pressure	Systems Serviced	System Flow Capacity	System Pumps
Canadair CL215	3,000 psi	Flaps, landing gear, wheel brakes, water doors	4 gpm 1.1 gpm	2 engine driven 1 auxiliary
Short Skyvan	2,500 psi	Flaps, nosewheel steering, wheel brakes		1 electric 1 accumulator
Gates Learjet M25	1,500 psi	Flaps, spoilers, landing gear, wheel brakes	4 gpm 0.3 gpm	2 engine driven 1 auxiliary
Boeing 757	3,000 psi	Flight controls, landing gear, wheel brakes, nose-wheel steering, thrust reversers	74 gpm 27 gpm 11 gpm	2 engine driven 3 electric RAT*
Boeing 767	3,000 psi	Flight controls, landing gear, nose-wheel steering wheel brakes, thrust reversers	74 gpm 37 gpm 16 gpm 9.6 gpm	2 engine driven 1 air driven by APU* 2 electric 1 RAT*

* RAT = Ram Air Turbine   APU = Auxiliary Power Unit

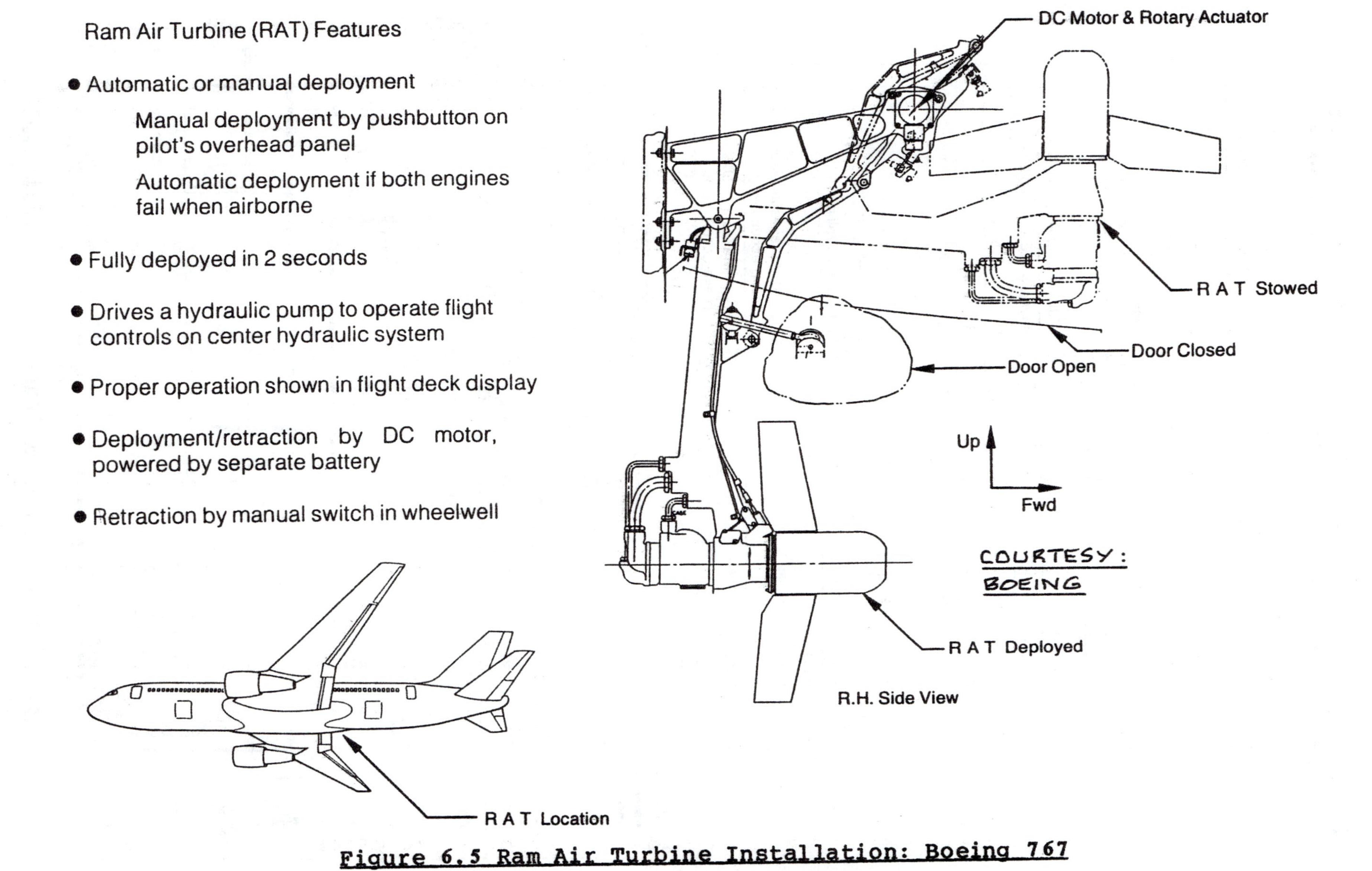

Figure 6.5 Ram Air Turbine Installation: Boeing 767

Table 6.2 Hydraulic System Function Distributions for the Boeing 737-200

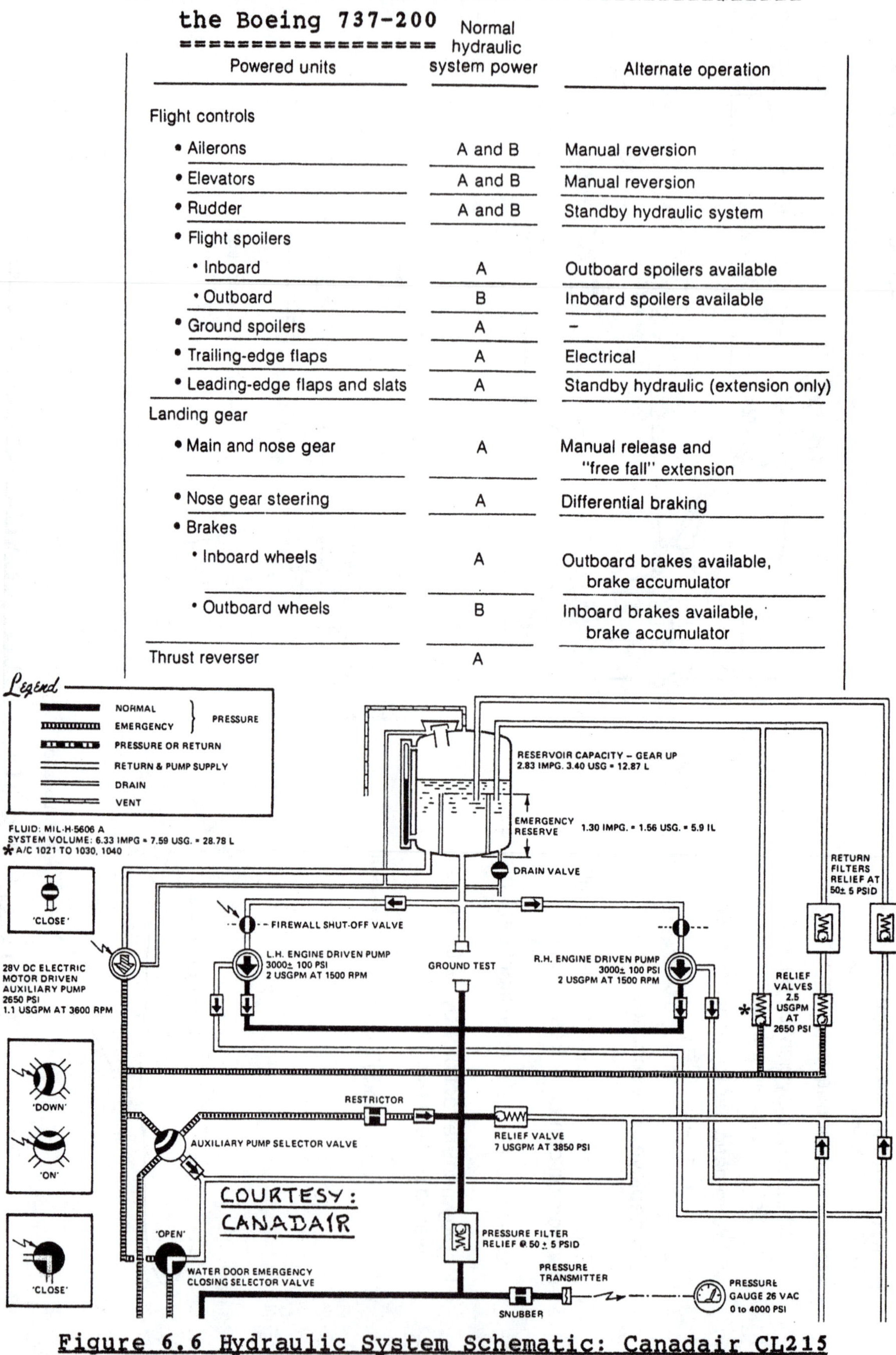

Powered units	Normal hydraulic system power	Alternate operation
Flight controls		
• Ailerons	A and B	Manual reversion
• Elevators	A and B	Manual reversion
• Rudder	A and B	Standby hydraulic system
• Flight spoilers		
• Inboard	A	Outboard spoilers available
• Outboard	B	Inboard spoilers available
• Ground spoilers	A	–
• Trailing-edge flaps	A	Electrical
• Leading-edge flaps and slats	A	Standby hydraulic (extension only)
Landing gear		
• Main and nose gear	A	Manual release and "free fall" extension
• Nose gear steering	A	Differential braking
• Brakes		
• Inboard wheels	A	Outboard brakes available, brake accumulator
• Outboard wheels	B	Inboard brakes available, brake accumulator
Thrust reverser	A	

Figure 6.6 Hydraulic System Schematic: Canadair CL215

b) terrorist action
c) failure of adjacent structure

In other words: do not route critical hydraulic lines close together.

4. Make sure that 'independent' systems are truly independent.

6.4 HYDRAULIC SYSTEM LAYOUT EXAMPLES

Figures 6.6 through 6.13 provide examples of hydraulic system layouts. Comments on these figures are given next.

==

Figure 6.6 Hydraulic System Schematic: Canadair CL215

Note that this system has two engine driven pumps and one electric auxiliary pump.

==

Figure 6.7 Hydraulic System Schematic: Short Skyvan

The hydraulic power pack employs one electric pump and a nitrogen charged accumulator for emergency use.

==

Figure 6.8 Hydraulic System Schematic: Gates Learjet M25

This system has two engine driven pumps and an accumulator for emergency services.

==

Figure 6.9 Hydraulic System Functions: Boeing 767

Because of the flight criticality of this hydraulic system it uses three independent systems each with their own power supply. Note that there are three different types of hydraulic pump: engine driven, electric driven and air driven. On top of these there is a RAT driven pump, if all else fails.

==

Figure 6.10 Hydraulic System Schematic: McDD DC-10

Because of the flight criticality of this hydraulic system it uses three independent systems. Each system is driven by two engine driven pumps. One system may also be driven by an electric pump.

Note that the landing gear has a mechanical free-fall capability.

Figures 6.11 - 6.13 illustrate the redundant aspects of critical flight control hydraulics in more detail.

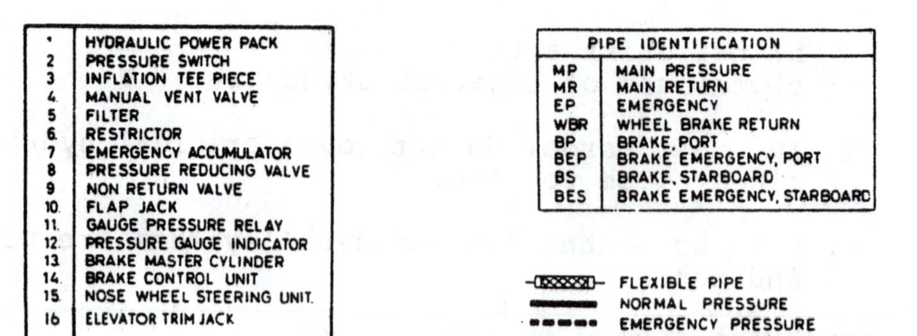

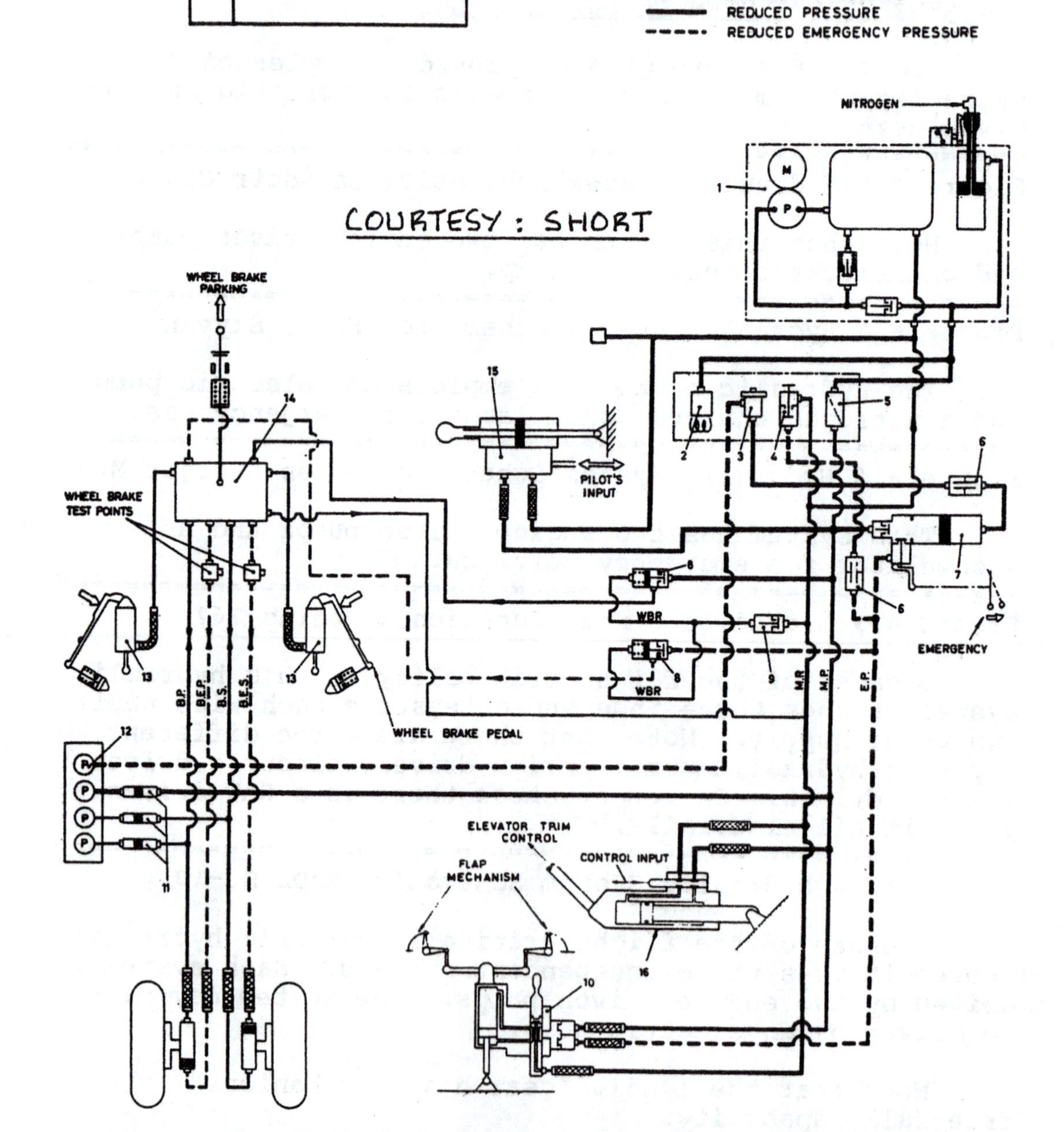

Figure 6.7 Hydraulic System Schematic: Short Skyvan

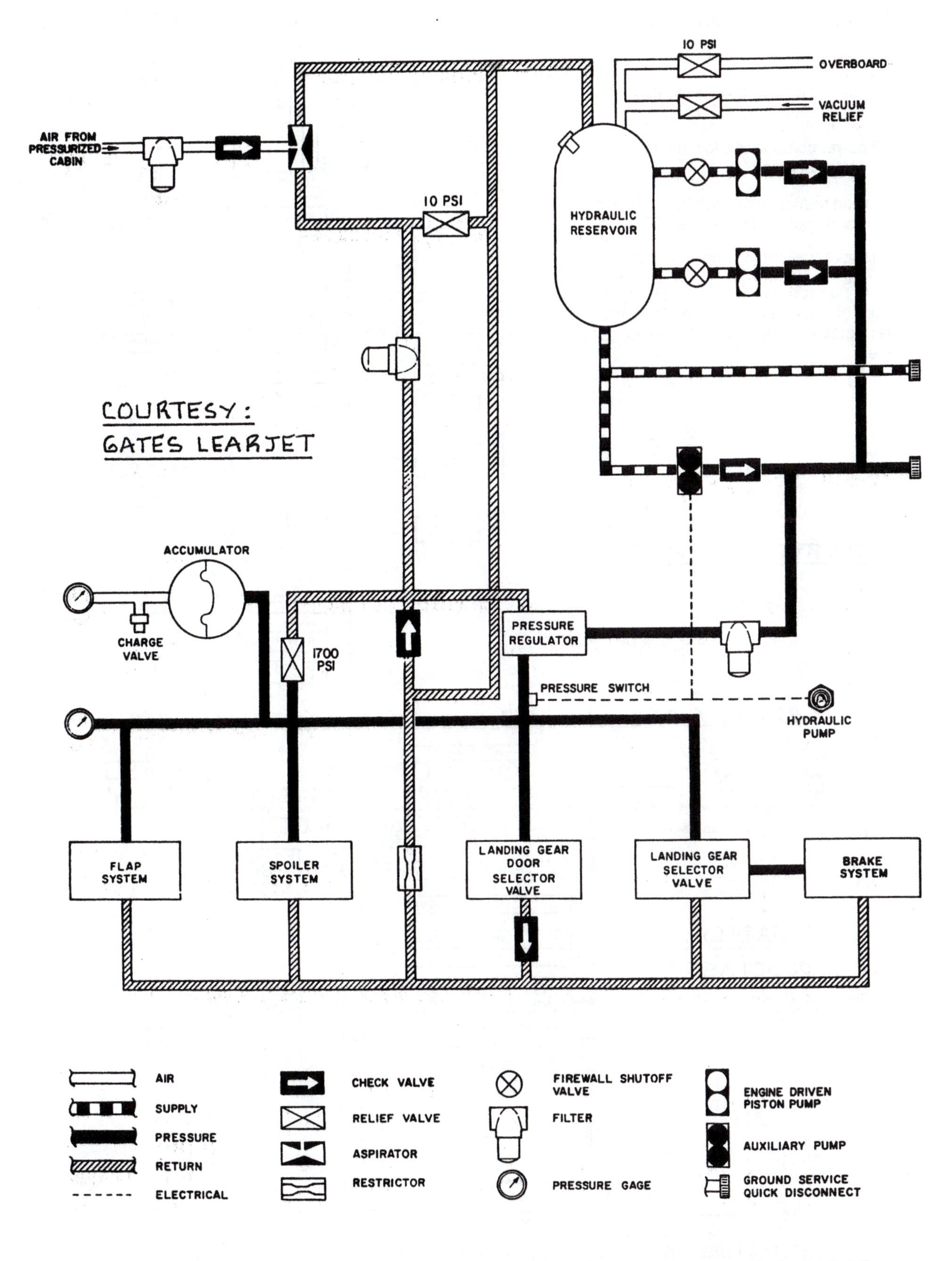

Figure 6.8 Hydraulic System Schematic: Gates Learjet M25

- Three independent 3000 psi systems
- 747 type engine-driven pumps
- Electric motor pumps on all systems available for ground operations
- All power components located to facilitate servicing
- Erosion reduction through use of type IV fluid and improved valve design
- Titanium pressure tubing (3Al-2.5V)
- Ram air turbine for emergency flight control hydraulic power source—retractable, located aft of landing gear wells

COURTESY: BOEING

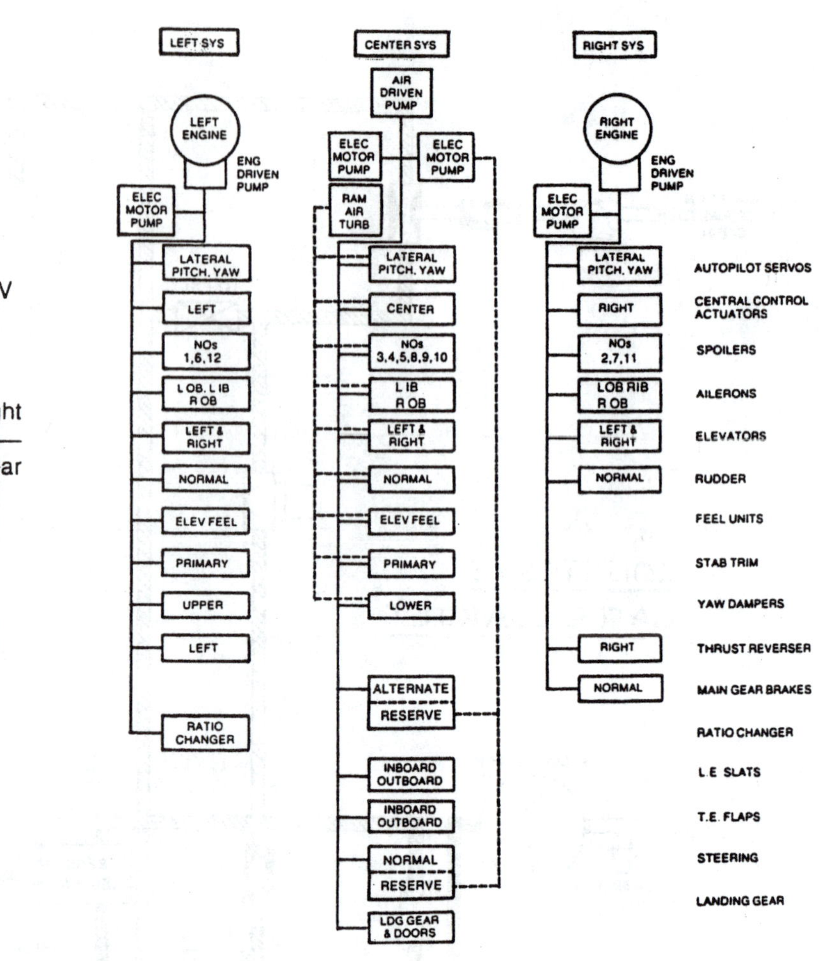

Figure 6.9 Hydraulic System Functions: Boeing 767

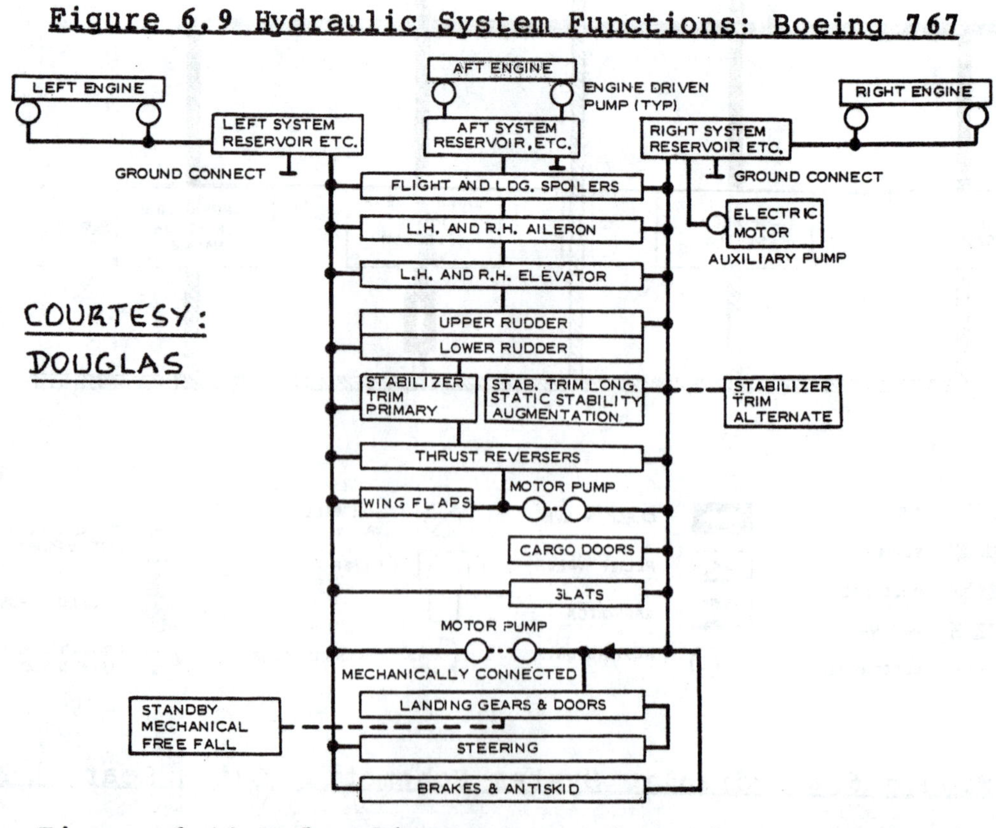

Figure 6.10 Hydraulic System Schematic: McDD DC-10

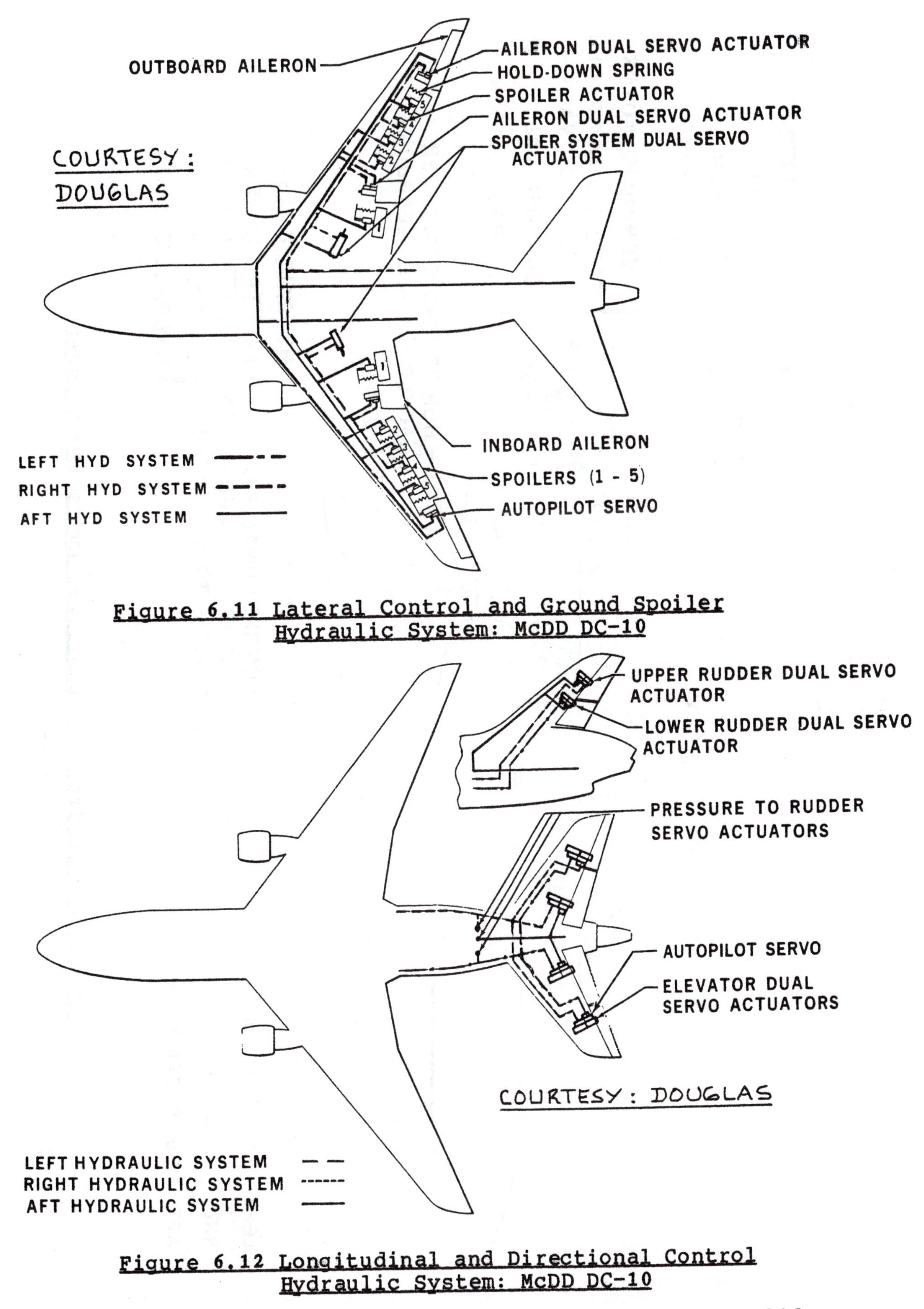

Figure 6.11 Lateral Control and Ground Spoiler Hydraulic System: McDD DC-10

Figure 6.12 Longitudinal and Directional Control Hydraulic System: McDD DC-10

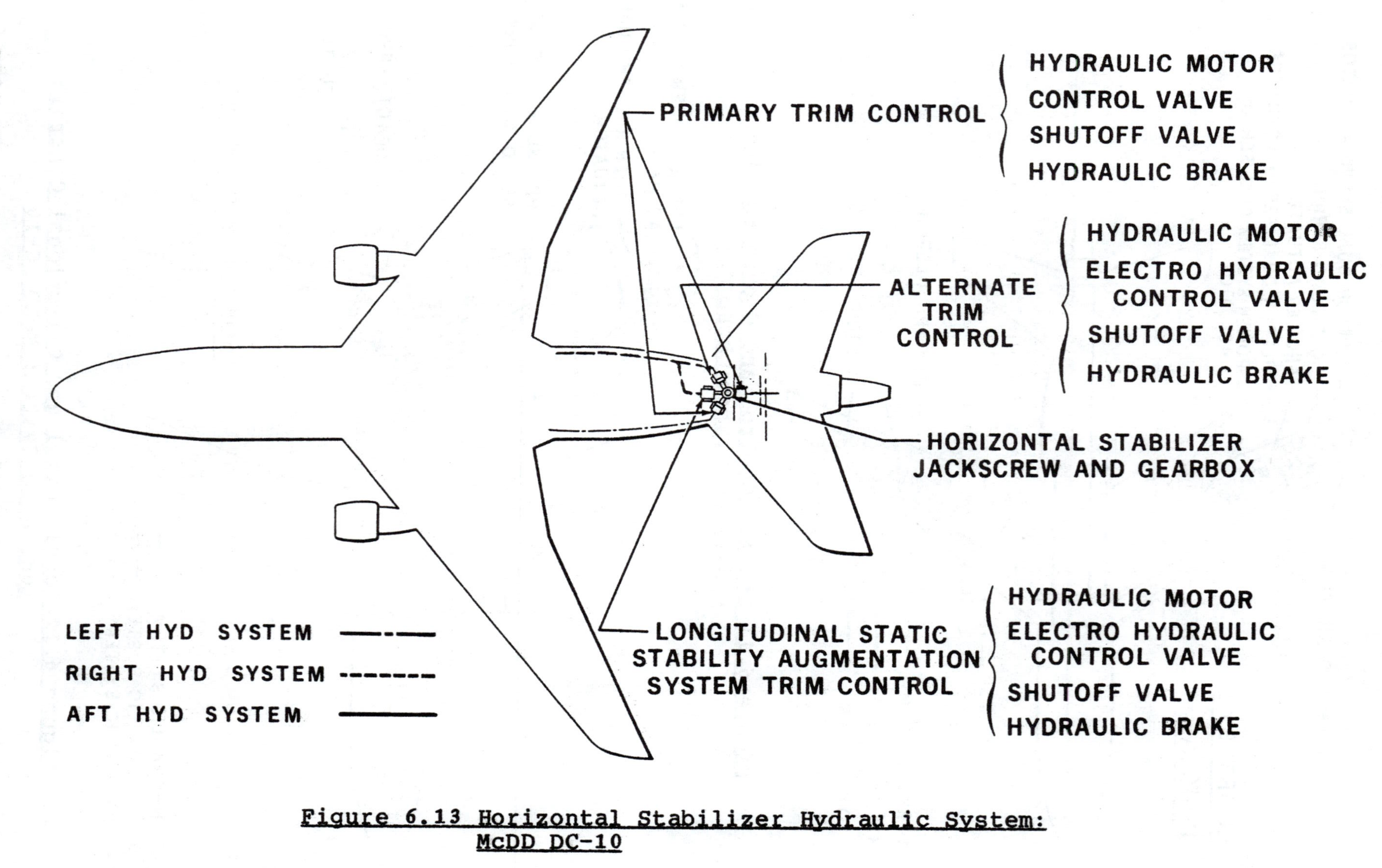

Figure 6.13 Horizontal Stabilizer Hydraulic System: McDD DC-10

7. ELECTRICAL SYSTEM LAYOUT DESIGN

Virtually all airplanes require electrical power for the operation of a large number of systems. Examples of systems requiring electrical power are:

1. Internal and external lighting

   Figure 7.1 shows an example of the external light installation in a jet transport. The angles from which landing lights and navigation lights must be visible are important. Figure 7.2 illustrates typical visibility angles for exterior lights.

2. Flight instruments and avionics systems

   Chapter 9 addresses the layout of these systems.

3. Food and beverages heating systems

   These systems are not addressed in this text: the reader should consult manufacturers brochures which cover passenger amenities.

4. Engine starting systems

   These systems are not addressed in this text.

5. Flight control systems (primary and secondary)

   Chapter 4 contains detailed descriptions of flight control systems.

Electric power is normally provided by two systems:

1.) Primary power generating system

   Primary power is usually delivered by engine driven generators.

2.) Secondary (stand-by) power generating system(s)

   Secondary power systems supply electrical power in case of failure of the primary system. These secondary power systems may consist of:

   a) Battery system
   b) Auxiliary power unit (APU)
   c) Ram-air turbine (RAT)

- Emergency lights battery powered
  - Illuminated automatically with electric power loss
  - May be illuminated upon command from flight deck panel or forward attendant panel
- Landing lights in each wing root (1) and on nose gear (2)
  - Each wing root light and nose gear light may be illuminated separately
  - Nose gear lights aimed along glide path
  - Wing root lights aimed horizontally
- Anticollision strobe lights on fuselage (red) and wingtips (white)
- Two forward-facing position lights and two aft-facing position lights on each wingtip
- Taxi light, logo lights, and external cargo handling lights optional

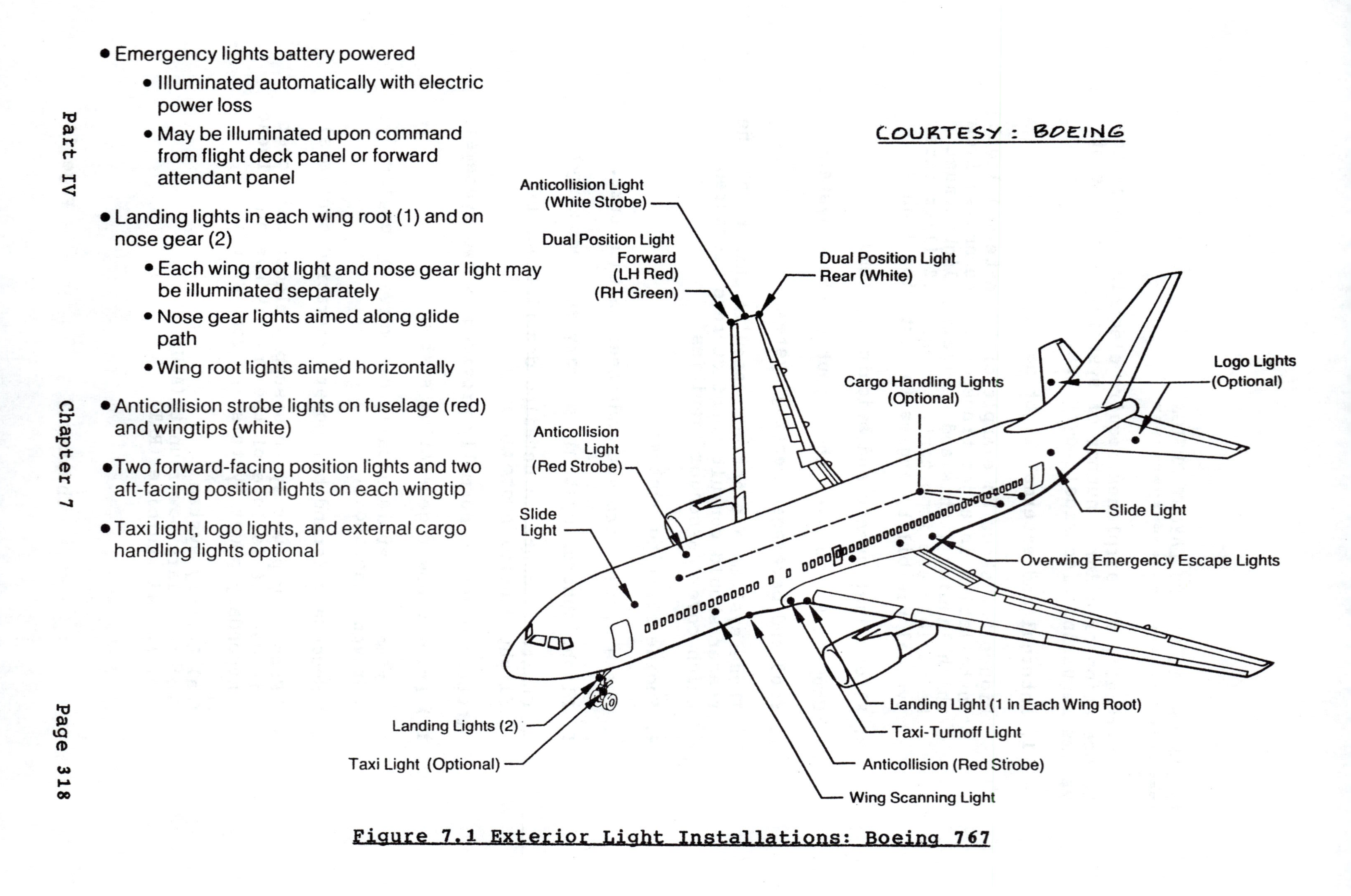

Figure 7.1 Exterior Light Installations: Boeing 767

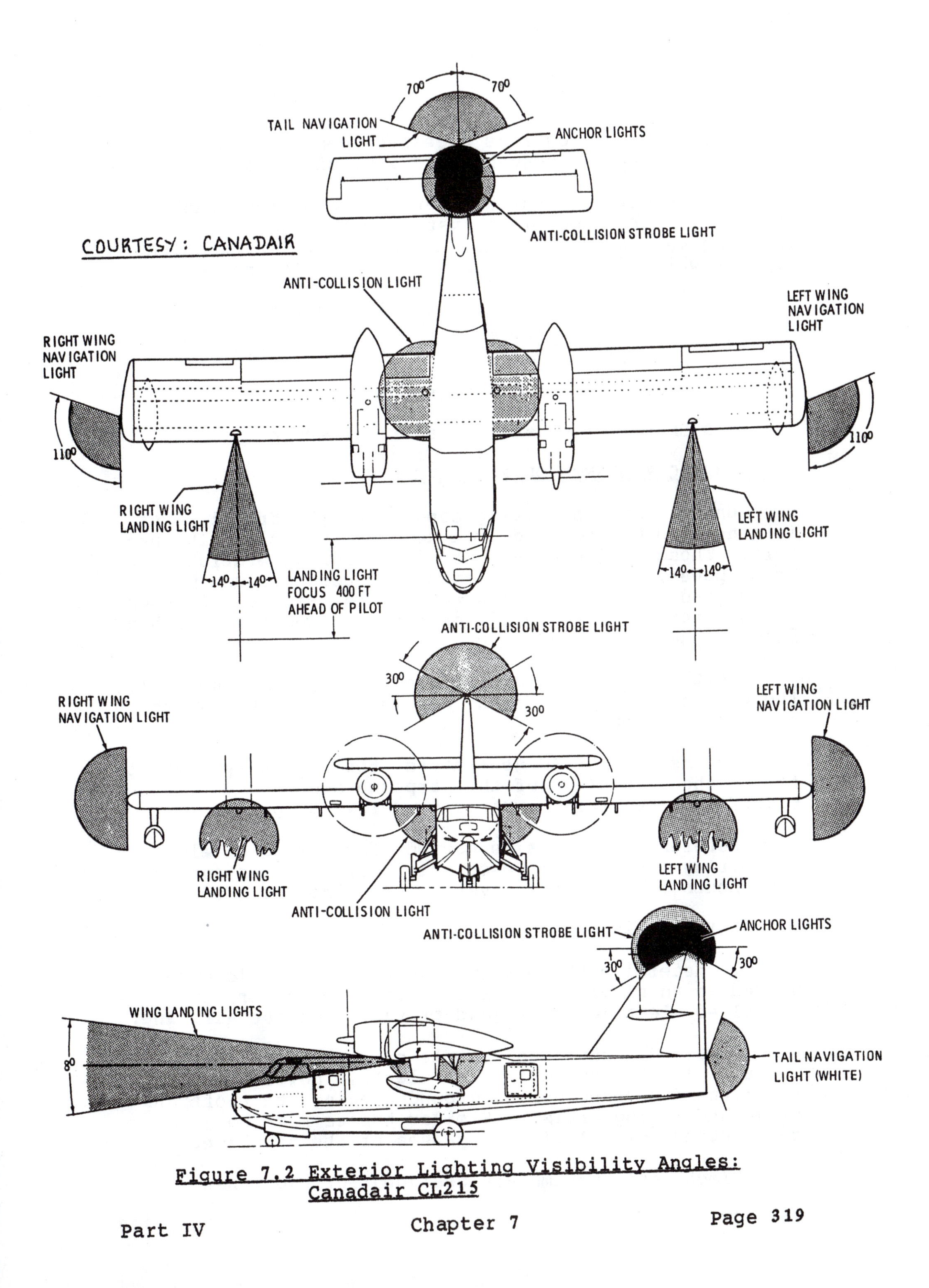

Figure 7.2 Exterior Lighting Visibility Angles: Canadair CL215

Reference 34 contains excellent descriptions of airplane electrical system operation, sizing and design.

Many experts in the airplane field believe that in the near future electrical systems will also replace existing pneumatic and hydraulic systems. An excellent article which foreshadows this change is Reference 35.

The material in this chapter is organized as:

7.1 Major components of electrical systems

7.2 Sizing of electrical systems

7.3 Example layouts of electrical systems

7.1 MAJOR COMPONENTS OF ELECTRICAL SYSTEMS

Under normal operating conditions electrical power is generated by engine driven generators and/or alternators. These devices may be designed for generation of DC or AC power. In some airplanes DC generators can be reversed and also used as starter motors. Figure 7.3 shows a cross section through a modern 90 KVA generator such as used in the Boeing 767.

In the case of DC generators their primary power is fed to the DC bus(es) of the airplane and/or to inverters to derive AC power.

In the case of AC generators, their primary power is fed to the AC bus(es) of the airplane and to transformer/rectifier systems to derive DC power.

Reference 34 contains methods for sizing of electrical wiring and buses. Descriptions of other essential components of electrical systems are also included.

7.2 SIZING OF ELECTRICAL SYSTEMS

To determine the maximum amount of electric power needed in an airplane it is necessary to construct so-called electric power load profiles for the airplane.

Figure 7.4 shows an example of such a profile. Tables 7.1 - 7.3 summarize the electric power requirements needed during ground loading, take-off and climb and during cruise. Similar tables must be constructed for all mission phases of the airplane.

Electrical systems are designed for two types of

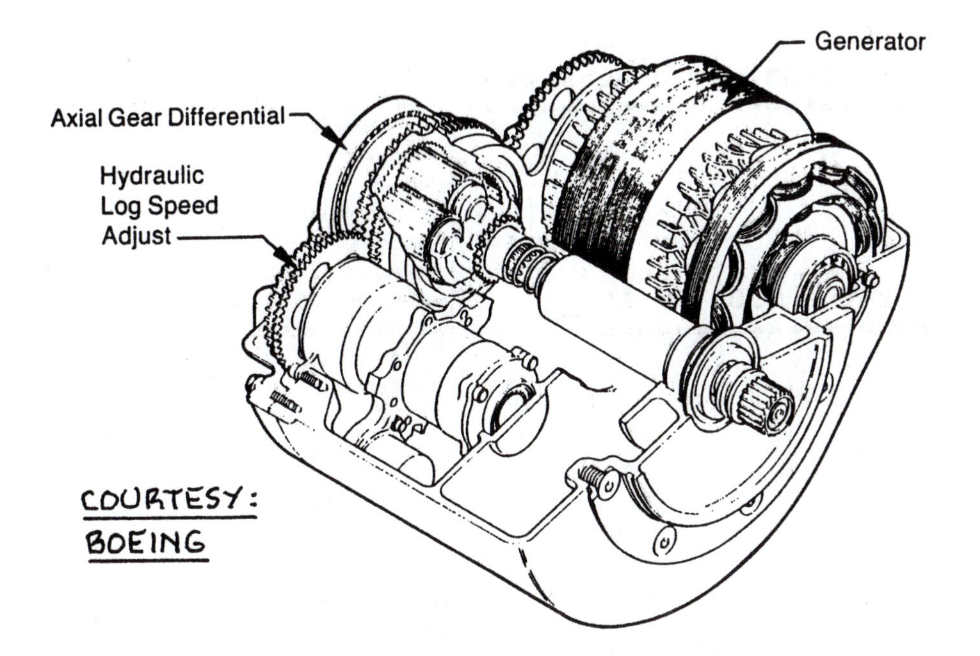

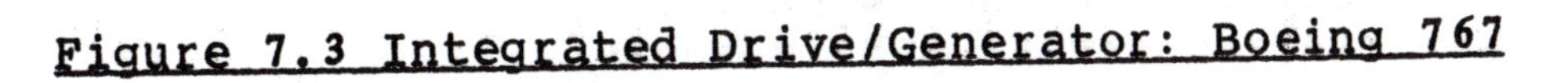
Figure 7.3 Integrated Drive/Generator: Boeing 767

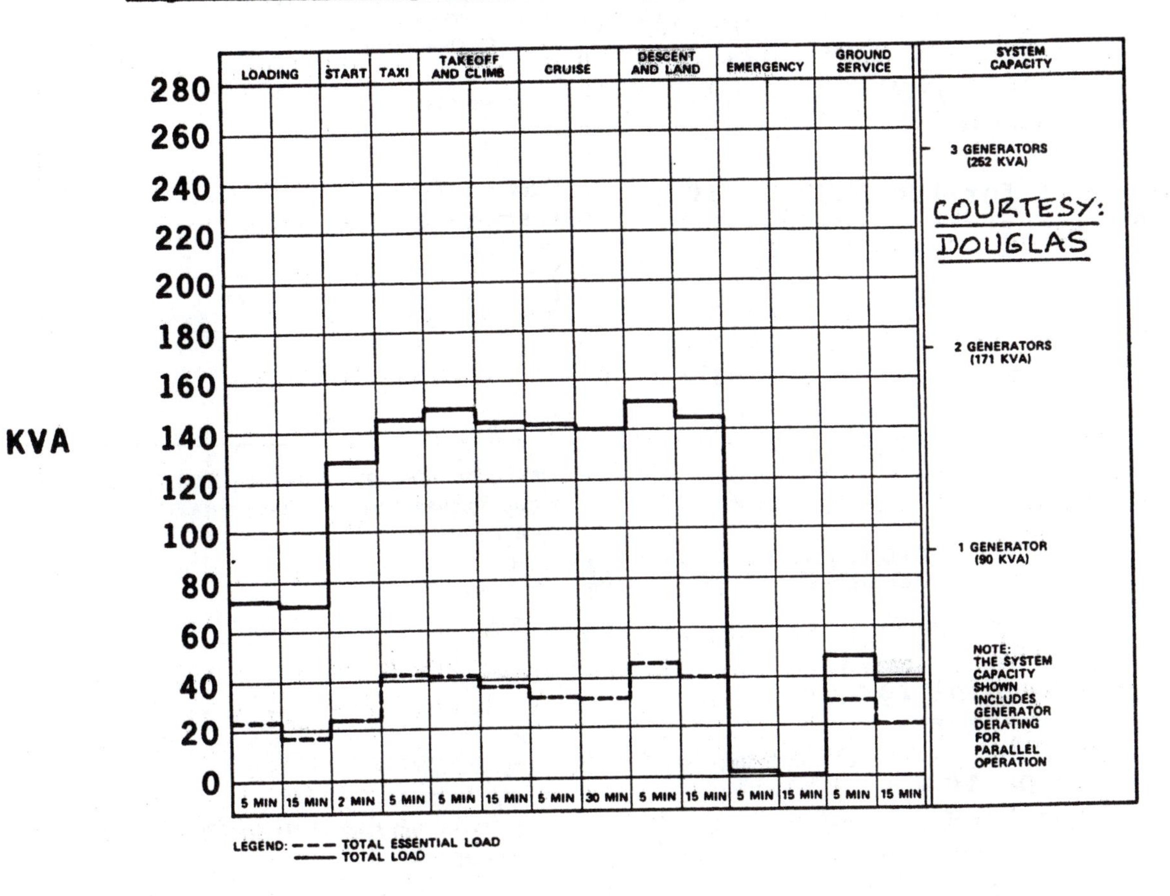

Figure 7.4 Electrical Load Profile: McDD DC-10

COURTESY : DOUGLAS

Table 7.1 Electrical Load Summary:

Ground Loading

(15 Min.) for the McDD DC-10

EXTERIOR LIGHTING	2,600
FLIGHT COMPARTMENT LIGHTING	650
PASSENGER CABIN LIGHTING	13,400
GALLEY	30,000
TOILETS	3,100
ENTERTAINMENT	50
WINDSHIELD HEATING	1,000
AVIONICS	5,300
AIR CONDITIONING	1,600
FUEL	0
HYDRAULIC	0
FLIGHT CONTROL	1,250
ELECTRICAL POWER (CONVERTED DC)	7,900
CARGO HANDLING	2,750
MISCELLANEOUS	200
TOTAL	69,800 VA

Table 7.2 Electrical Load Summary:

Take-off and Climb

(15 Min.) for the McDD DC-10

EXTERIOR LIGHTING	3,850
FLIGHT COMPARTMENT LIGHTING	1,200
PASSENGER CABIN LIGHTING	14,200
GALLEY	84,000
TOILETS	0
ENTERTAINMENT	0
WINDSHIELD HEATING	6,000
AVIONICS	7,400
AIR CONDITIONING	1,600
FUEL	6,500
HYDRAULICS	8,800
FLIGHT CONTROL	2,000
ELECTRICAL POWER (CONVERTED TO DC)	7,900
MISCELLANEOUS	250
TOTAL	143,700 VA

Table 7.3 Electrical Load Summary:

Cruise (30 min) for

the McDD DC-10

EXTERIOR LIGHTING	200
FLIGHT COMPARTMENT LIGHTING	1,200
PASSENGER CABIN LIGHTING	14,700
GALLEY	84,000
TOILETS	6,100
ENTERTAINMENT	3,000
WINDSHIELD HEATING	7,200
AVIONICS	7,250
AIR CONDITIONING	1,600
FUEL	6,500
HYDRAULICS	0
FLIGHT CONTROL	2,000
ELECTRICAL POWER (CONVERTED TO DC)	6,000
MISCELLANEOUS	250
TOTAL	140,000 VA

load requirements:

1. Essential load requirements

Essential load requirements are determined by the sum of all electric loads required by systems which are essential to the safe operation of the airplane.

2. Normal operating load requirements

Normal operating load requirements are determined by the maximum sum of all electric power requirements during certain mission phases.

The probability that the essential electric power requirements cannot be met during any flight must be extremely remote. This is why in passenger transports usually a minimum of three main electrical systems is employed with one or more back-up systems just in case.

In the case of the DC-10 in Figure 7.4 it is seen that two generators can supply the maximum electric power load needed during a typical mission. One generator can handle the essential load. That is why three generators were selected for the electrical system. Each generator is driven from an engine mounted accessory pad. A fourth generator of the same capacity is driven by the APU. Figure 7.5 shows the APU system integration with its 90 KVA generator attached.

Figure 7.6 shows an example of one possible DC-10 APU installation. Figure 7.7 shows a similar installation in the Boeing 767.

7.3 GUIDELINES FOR LAYOUT OF ELECTRICAL SYSTEMS

Following are some basic guidelines which must be followed when laying out electrical systems:

1. Electrical systems must be carefully shielded from the effects of lightning strike.

2. Electrical systems must be designed so that they are shielded from each other. Electromagnetic interference is one of the major development headaches during the certification of airplanes.

3. Electrical systems must be designed so that airplane dispatch is possible with certain system components failed. For example: in most passenger transports it is highly desirable to be able to dispatch

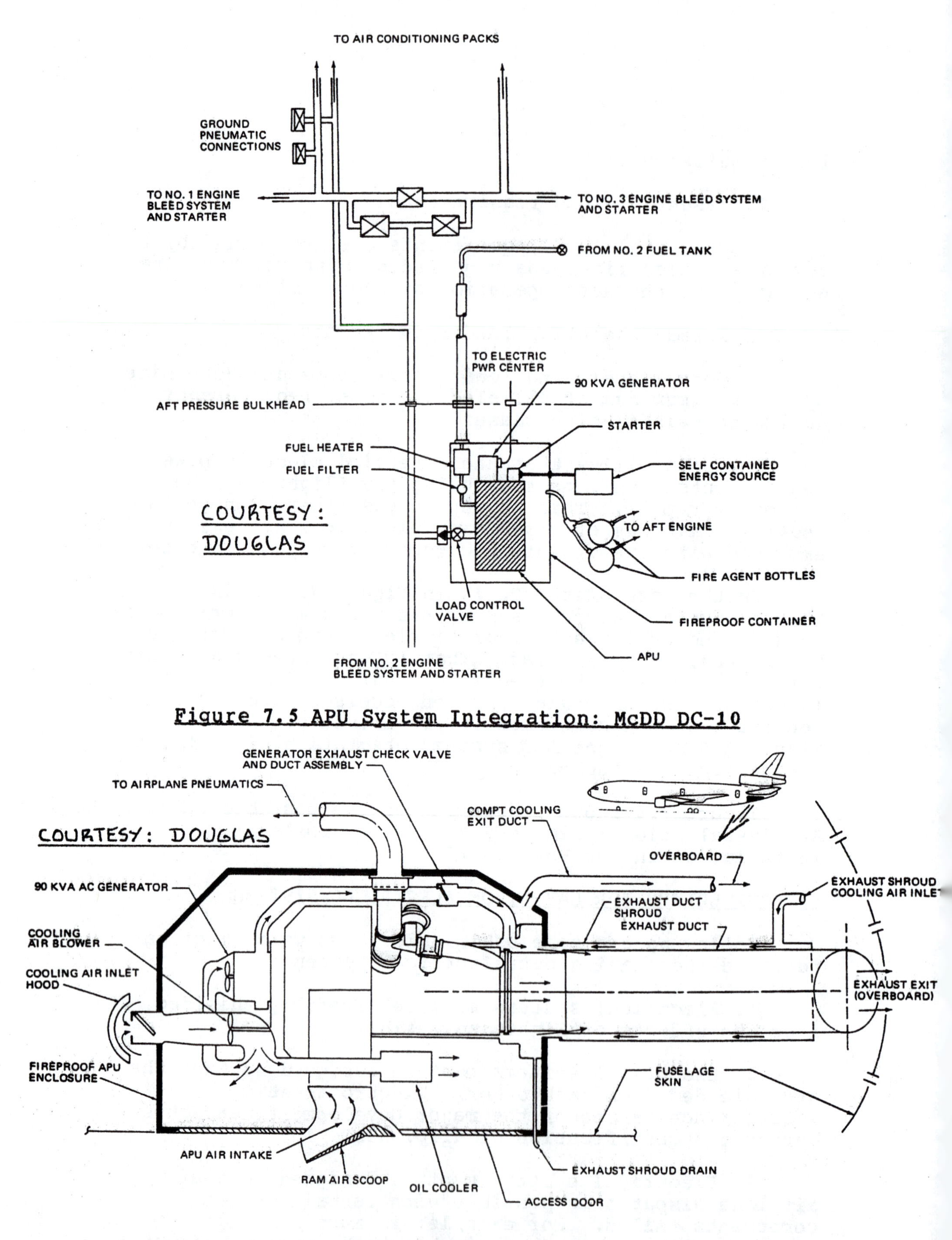

Figure 7.5 APU System Integration: McDD DC-10

Figure 7.6 APU Installation Schematic: McDD DC-10

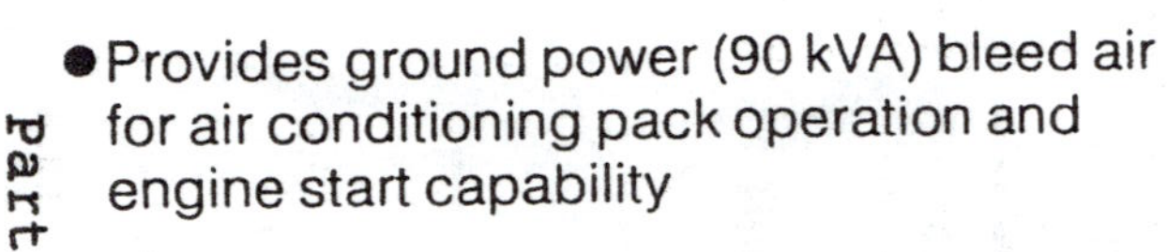

- Provides ground power (90 kVA) bleed air for air conditioning pack operation and engine start capability
- Inflight power capability to 35 000 feet, air start to 25 000 feet
- Supplies air-driven pump for hydraulic system (if one engine air source unavailable)
- Electronic APU control unit installed in aft cargo compartment
- Low noise level at ramp work-stations
  - APU installed in aft fuselage
  - Inlet high above ramp
  - Inlet muffler
  - APU engine acoustic treatment incorporated

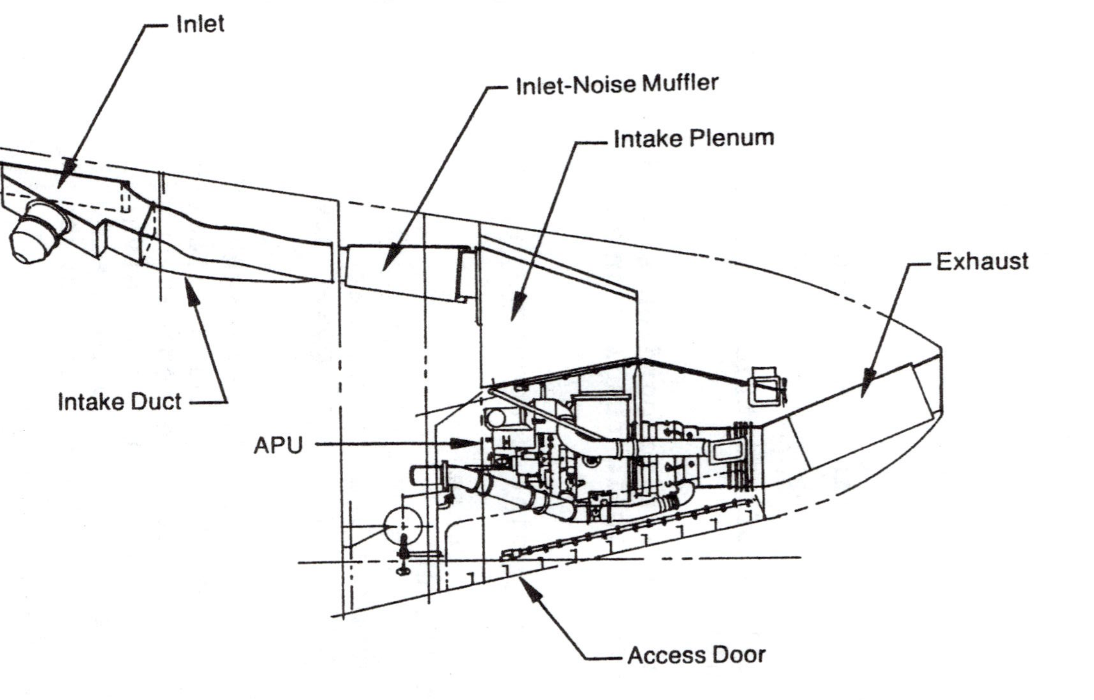

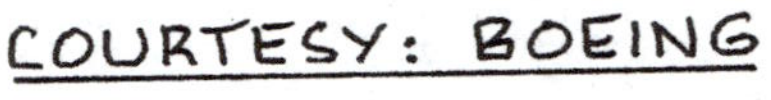

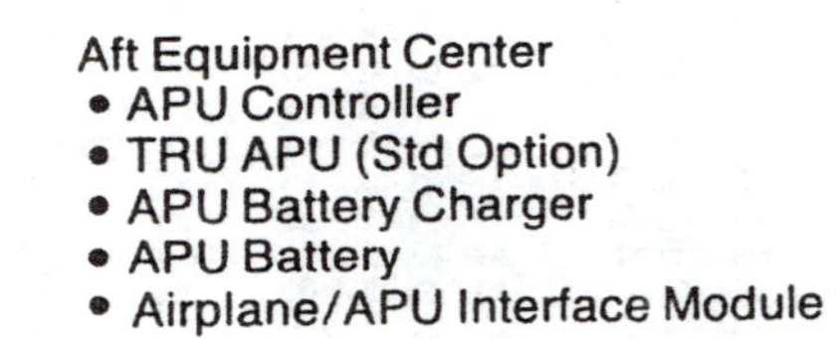

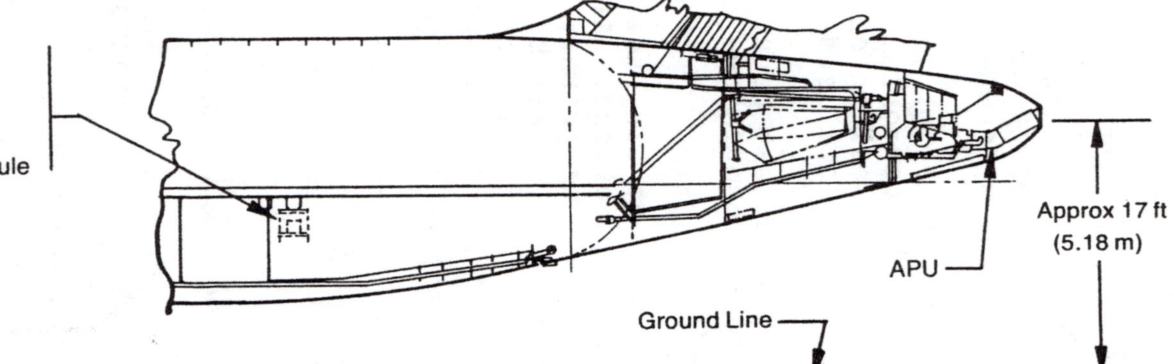

Figure 7.7 APU Installation Schematic: Boeing 767

with one failed system. That means a minimum of four independent systems are needed.

4. Servicing and accessibility of electrical system components must be easy and safe.

5. Flight crucial buses and/or wiring bundles should be widely separated to avoid catastrophic results under the following scenarios:

a) uncontained failure of engine components
b) terrorist action
c) failure of adjacent structure
d) localized in-flight fires

6. In most passenger transports and in military airplanes provisions for hooking up to ground power are required.

7. Most airplanes require batteries for various standby functions. Airplane batteries must be physically shielded from primary structure and from essential airplane services: batteries can leak highly corrosive fluids. Batteries can als pose a fire hazard following certain failure scenarios. They must therefore also be shielded from flammable materials.

For these reasons many batteries are installed in fire-proof and corrosion proof containers. These containers may have to be vented to the atmosphere to prevent excessive pressures from building up.

7.4 EXAMPLE LAYOUTS OF ELECTRICAL SYSTEMS

Figures 7.8 - 7.11 show examples of electrical system layouts. Comments on these figures are given next.

==

Figure 7.8 Electrical System Schematic: Canadair CL215

This airplane is equipped with two engine driven 28-volt, 200 amp., DC generators. AC power is provided by monophase, 400 cycle, 115 AC static inverters.

A 36 ampere-hour lead-acid battery is capable of carrying essential electric power loads.

==

Figure 7.9 Electrical System Schematic: Short Skyvan

This airplane uses two engine driven, 28-volt, 200 amp., DC starter generators. For some systems the DC power is inverted to 115 and 26 volt, 400 cycle, AC power.

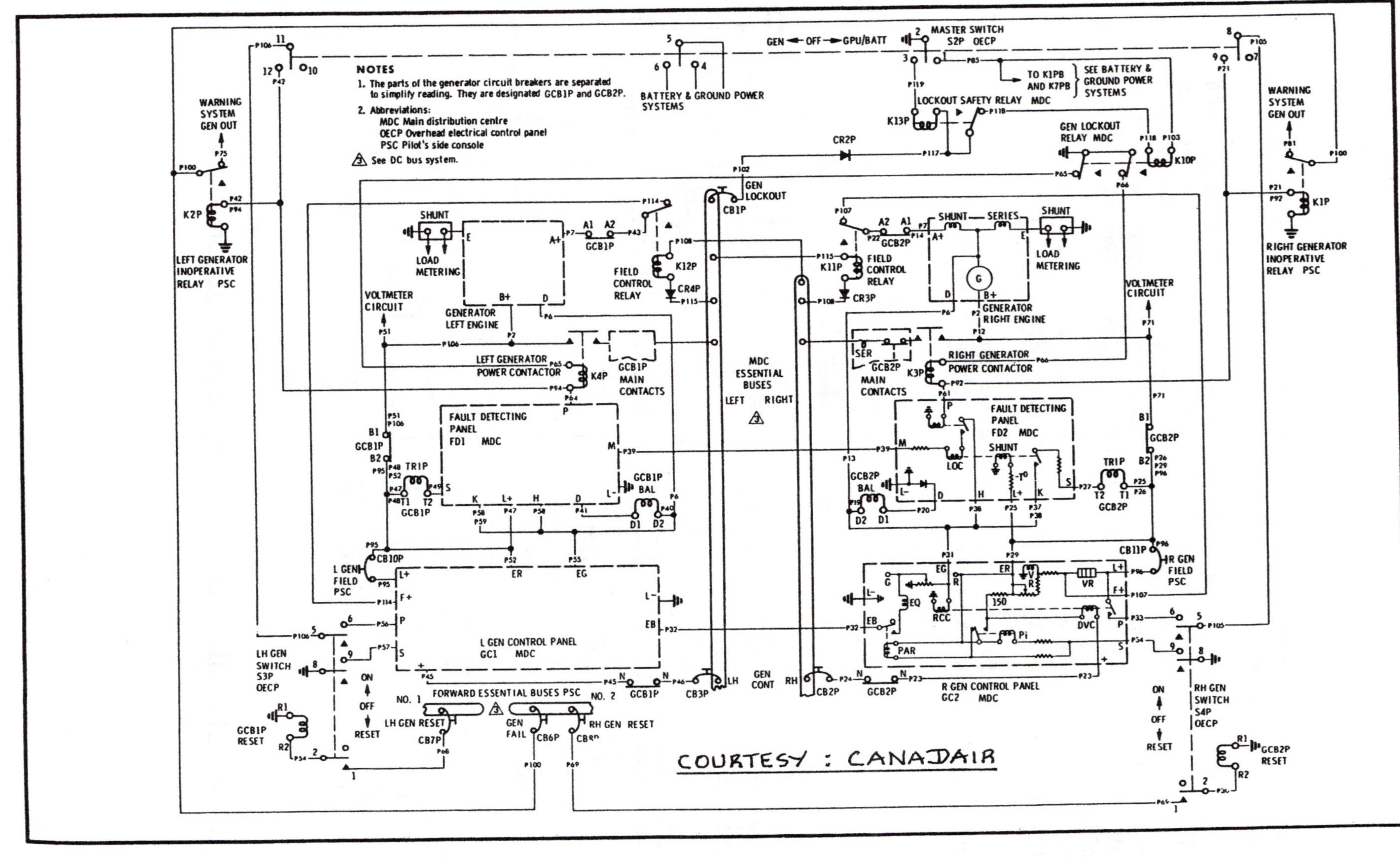

Figure 7.8 Electrical System Schematic: Canadair CL 215

Figure 7.10 AC And DC Power Distribution Schematics: Gates Learjet M25

This airplane uses two engine driven, 28-volt, 400 amp., DC starter generators. The schematics show which types of power are required for the various functions. Note that the DC power is inverted to 115 and 26 volt, 400 cycle, AC power. The inverters are rated at 400 volt-amperes.

==

Figure 7.11 Electrical Power System McDD DC-10

Note the four generators used in this system. Section 7.3 gave the reasons for this selection.

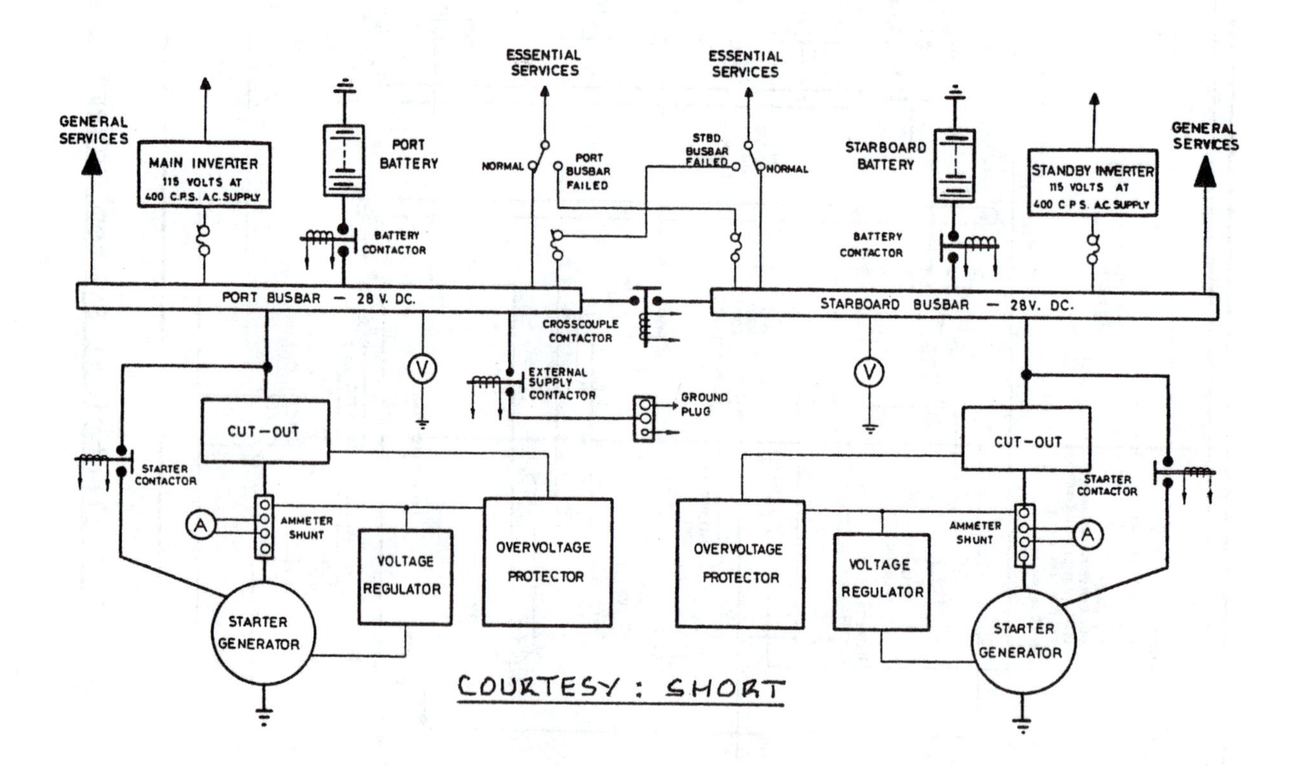

Figure 7.9 Electrical System Schematic: Short Skyvan

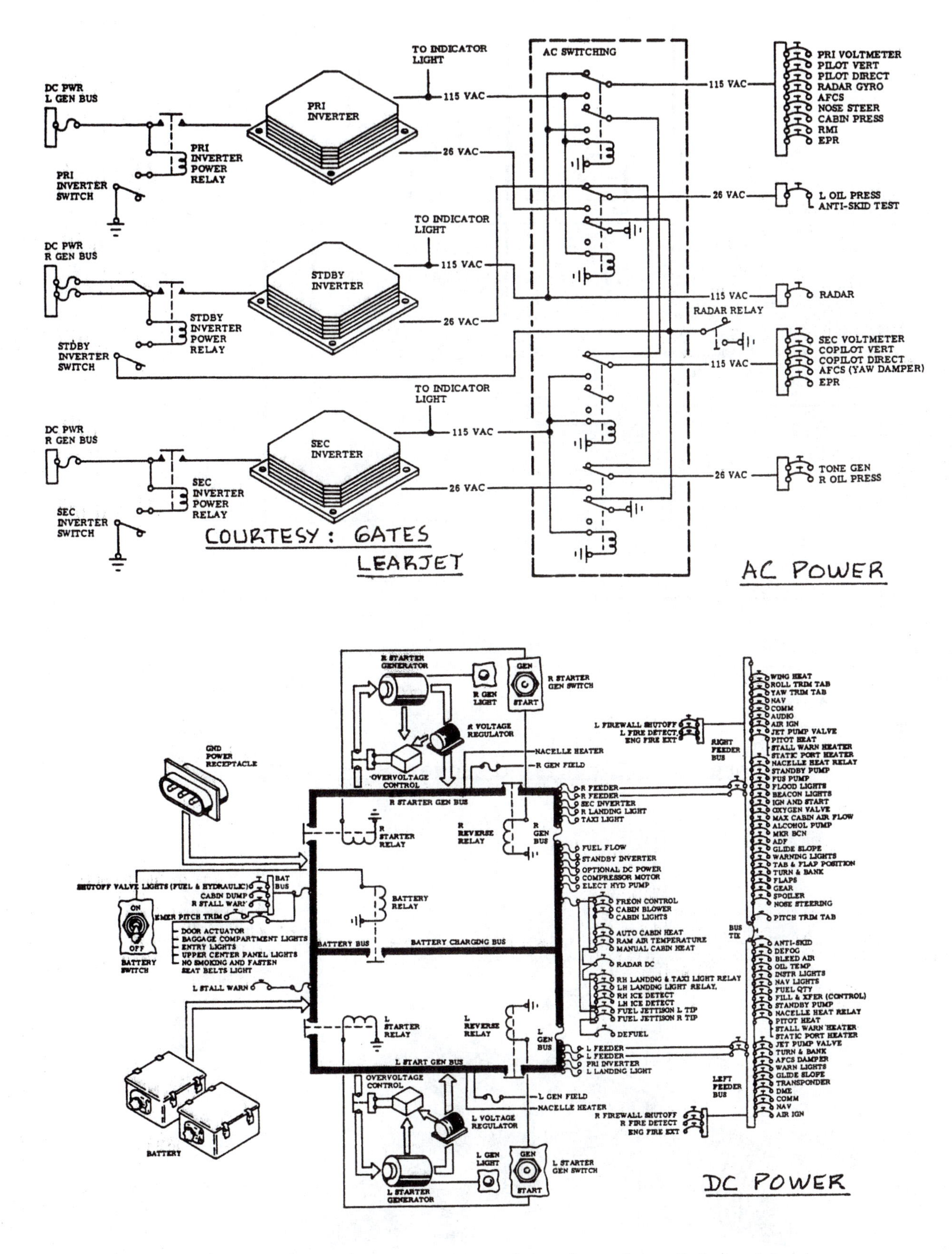

Figure 7.10 AC and DC Power Distribution Schematics: Gates Learjet M25

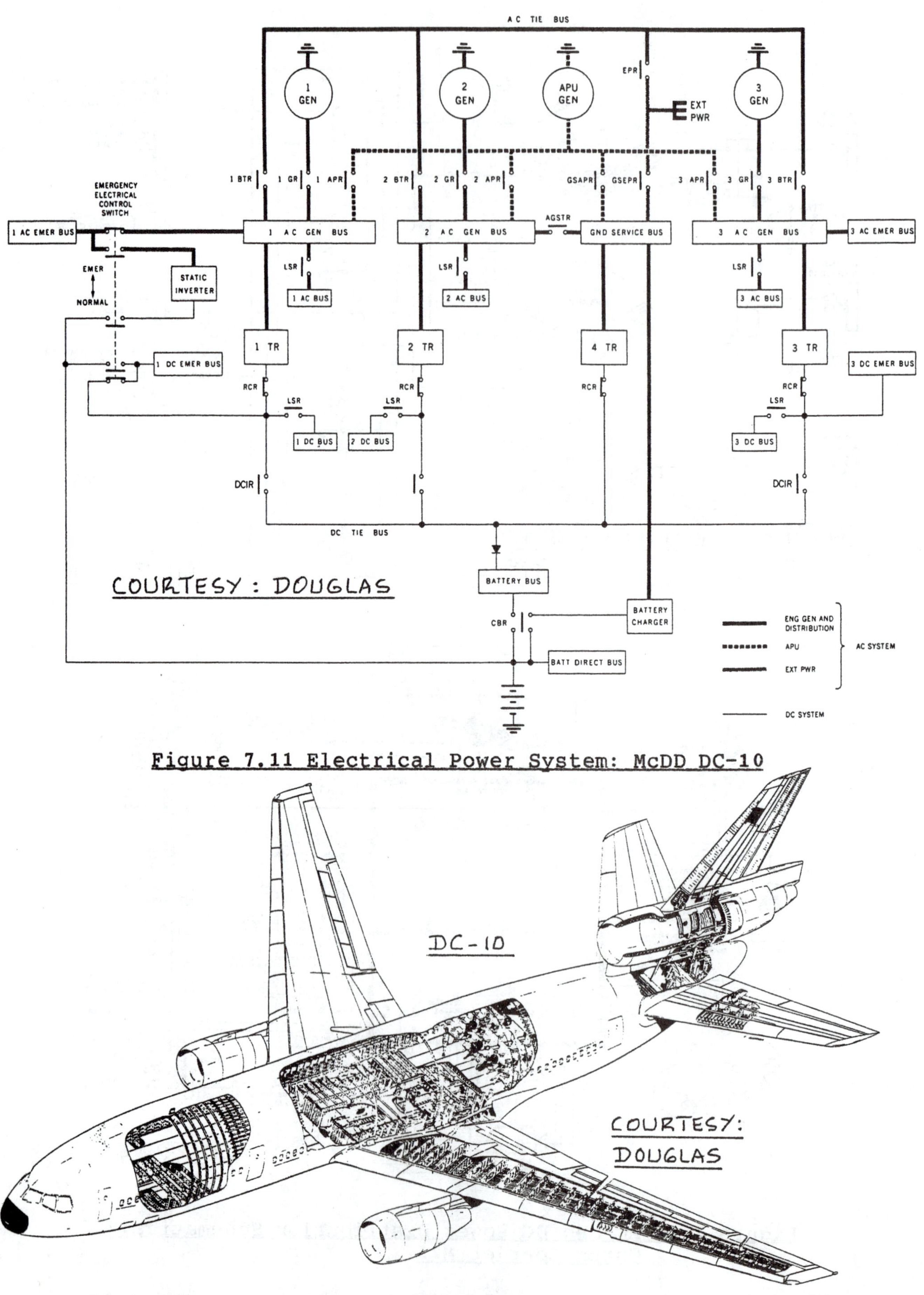

Figure 7.11 Electrical Power System: McDD DC-10

8. ENVIRONMENTAL CONTROL SYSTEMS LAYOUT DESIGN

In this chapter some fundamental design layout requirements for environmental control systems will be discussed. The material is organized as follows:

8.1 Pressurization system

8.2 Pneumatic system

8.3 Airconditioning system

8.4 Oxygen system

8.1 PRESSURIZATION SYSTEM

The purpose of this system is to maintain sufficient cabin air pressure during flight at high altitudes so that passengers remain comfortable. Typical differential pressure capabilities in jet transports are designed to maintain a cabin altitude from 1000 ft below sealevel to 10,000 ft above sealevel.

Figure 8.1 shows a typical pressurization schedule used in a transport. To reduce crew workload the pressurization schedule with altitude should be automatic.

Cabin pressurization systems need the following components:

1. A source of high pressure air (high relative to the outside air pressure). The air source is normally the airplane pneumatic system. The pneumatic system is discussed in Section 8.2.

2. A control and metering system to:

    a) provide positive pressure relief to protect the structure. This pressure relief is typically set for a pressure differential larger than 9-10 psi.

    b) provide negative pressure relief. This lets air into the cabin when outside air pressure exceeds the inside pressure. This system is normally set for a pressure differential corresponding to about 10 inches of water.

The reliability of the cabin pressurization system is of great importance to the safety of the passengers.

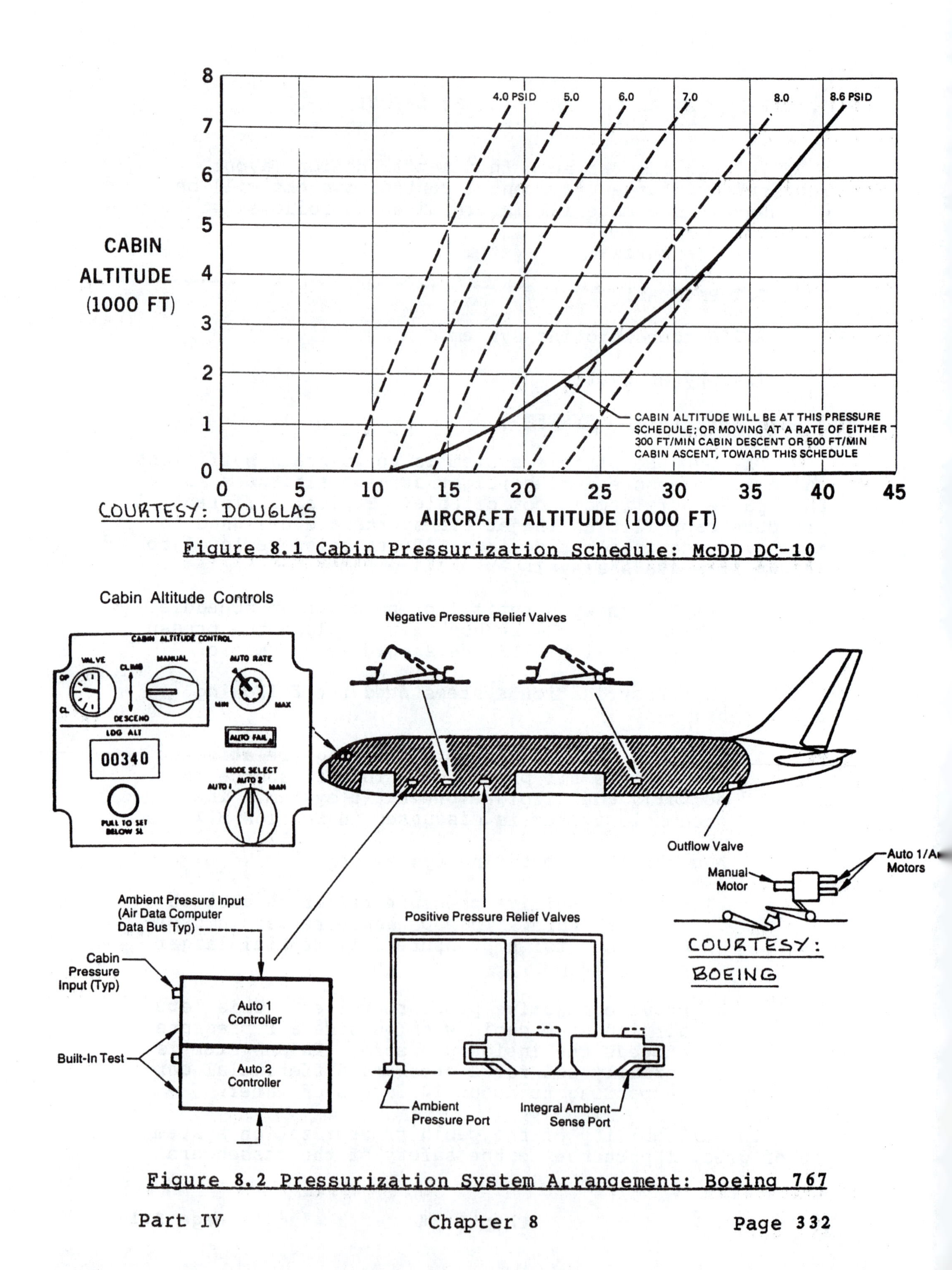

Figure 8.1 Cabin Pressurization Schedule: McDD DC-10

Figure 8.2 Pressurization System Arrangement: Boeing 767

If this system fails at altitude, serious breathing problems would arise. That is the reason why an emergency oxygen system must be installed in all jet transports. Section 8.4 covers oxygen system layout design.

If the pressurization system fails at landing it would be impossible to open the cabin doors. For that reason a cabin depressurization system is also included.

A major problem can arise if cargo doors located below the cabin floor accidentally blow out. In that case the pressure in the cargo hold bleeds off very rapidly. The pressure differential which then exists between the passenger cabin and the cargo hold may cause the cabin floor to fail. If flight crucial systems (such as flight control cables or wire bundles) are located in that area, loss of control can occur. These problems can be avoided only by a combination of:

a) proper location of essential controls
b) providing pressure relief for the cabin if a cargo door fails
c) fail-safe design of the cargo door latch mechanism

Figure 8.2 shows a typical arrangement for a cabin pressurization system. For detailed discussions of the associated control systems the reader should consult Ref.33 and manufacturers brochures describing such systems for a particular airplane.

8.2 PNEUMATIC SYSTEM

The purpose of the pneumatic system is to supply air for the following functions:

1. Cabin pressurization and airconditioning
2. Ice protection system (See Chapter 10)
3. Cross engine starting (not in all airplanes)

Figure 8.3 shows a typical schematic for a pneumatic system used in jet transports. Note that the primary source of air is engine compressor bleed air. A secondary source is air from the APU.

8.3 AIRCONDITIONING SYSTEM

The purpose of the airconditioning system is to 'condition' cabin air in terms of temperature (cooling or heating) and humidity.

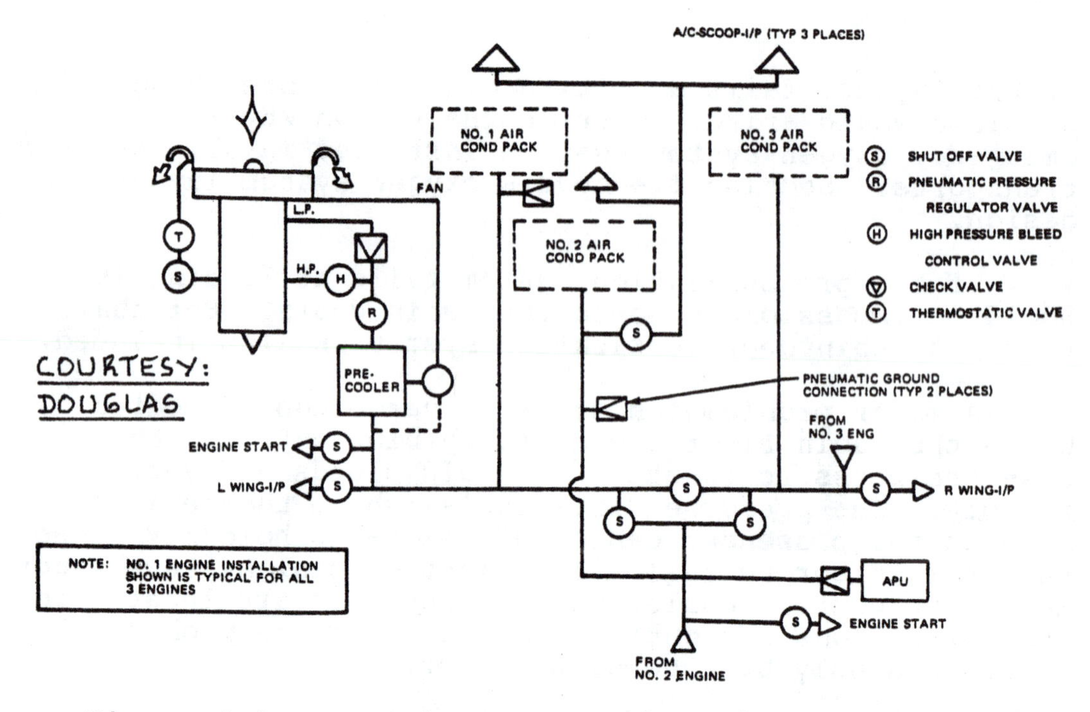

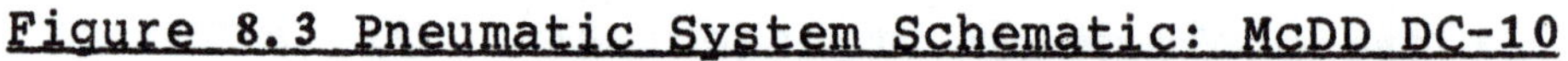

Figure 8.3 Pneumatic System Schematic: McDD DC-10

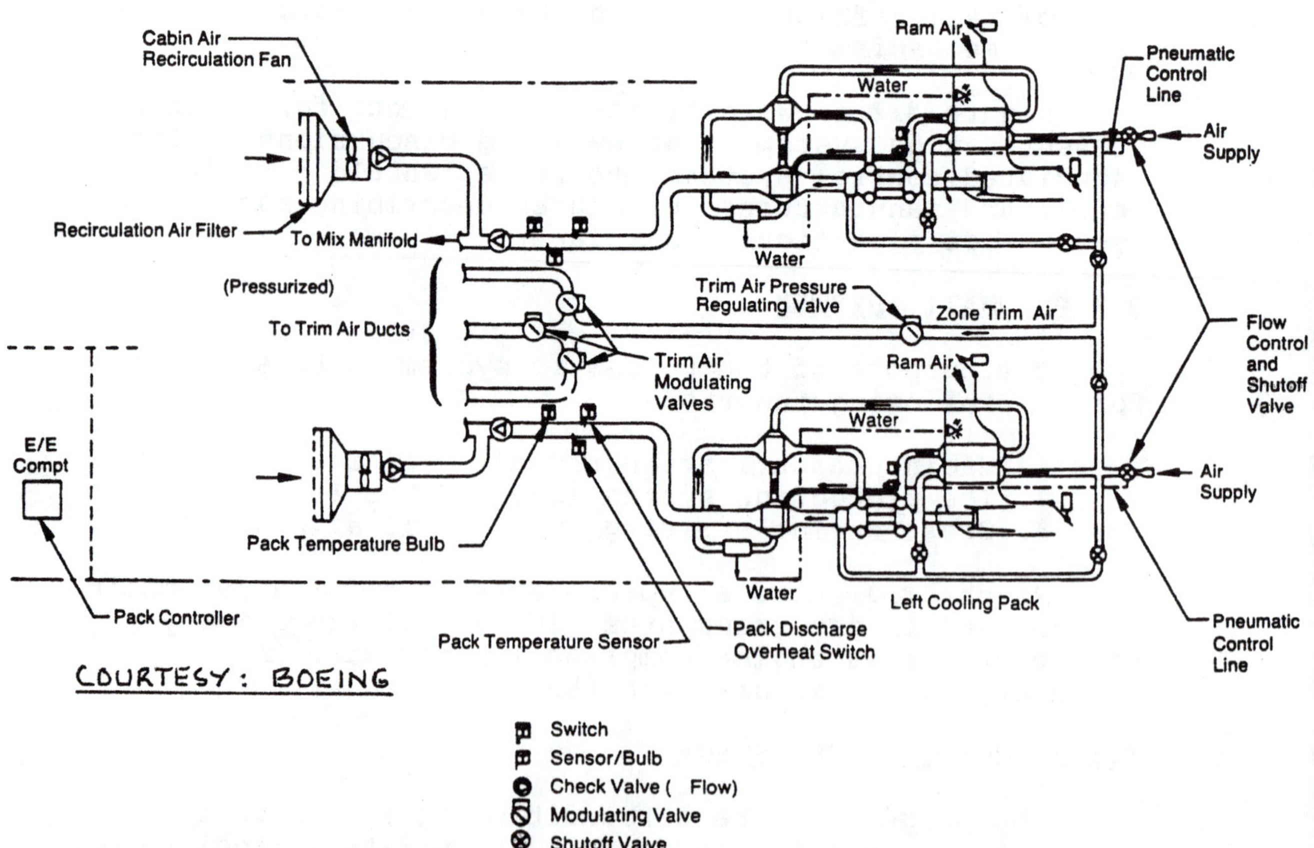

Figure 8.4 Airconditioning System: Boeing 767

Figure 8.4 provides a schematic of a typical airconditioning system used in jet transports. Since the engine bleedair is always hot, cabin heating requires less air cooling than cabin cooling. In the design of these systems a critical condition is normally a descent at flight idle power. This flight condition leads to minimal bleedair availability.

The overall efficiency of the cabin airconditioning system depends a great deal on the thermal insulation of the cabin wall. Figure 8.5 shows a cross section through a cabin wall with the associated temperature gradients.

The air coming from the airconditioning system must be distributed into the cabin. In passenger transports this requires a complex network of airducts. In large transports the cabin temperature is controlled in individual cabin zones. This temperature control is achieved by a system which mixes hot air with cooled air in the desired proportion.

The amount of cabin air required in jet transports is typically 20 cubic feet per minute per passenger.

Figure 8.6 shows the location of airconditioning units in a jet transport with the primary air ducts indicated. Note the separate line leading to the electrical/avionics bay: these systems require a significant amount of cooling or they will malfunction.

Figure 8.7 shows an air distribution network in more detail. A cross section through the passenger cabin with the various air distribution ducts indicated is given in Figure 8.8.

In the detail design of the air distribution system careful attention must be paid to duct acoustics: improperly designed air distribution systems can be extremely noisy.

In many unpressurized airplanes cabin heating systems are used. Figures 8.9 and 8.10 show schematics of such systems. Note that these systems use fuel and a combustor (heat exchanger) as a source of heat. Great care must be taken in the design and location of these systems: they are a potential source of hazards: carbon-monoxide poisoning and fires are two of these.

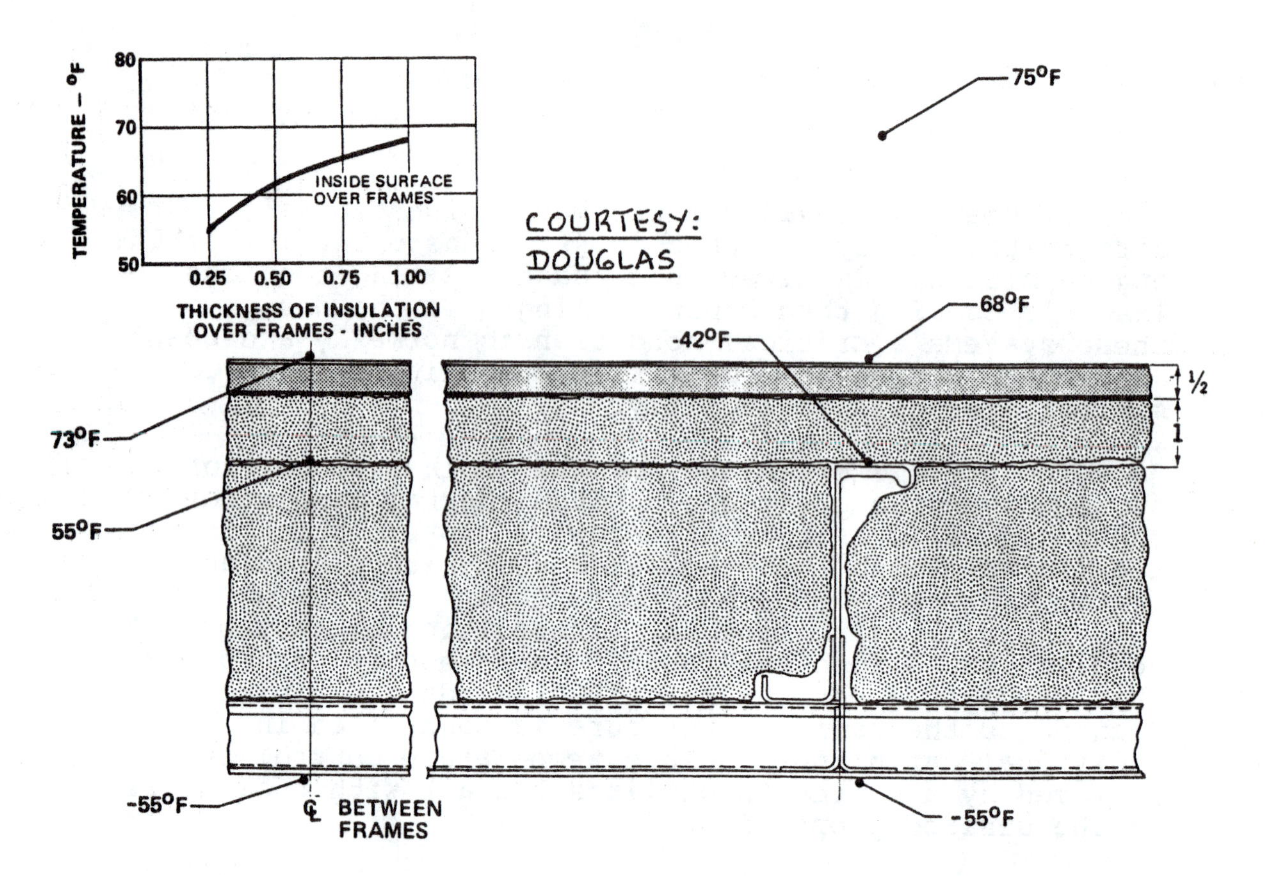

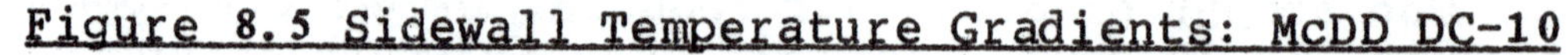

Figure 8.5 Sidewall Temperature Gradients: McDD DC-10

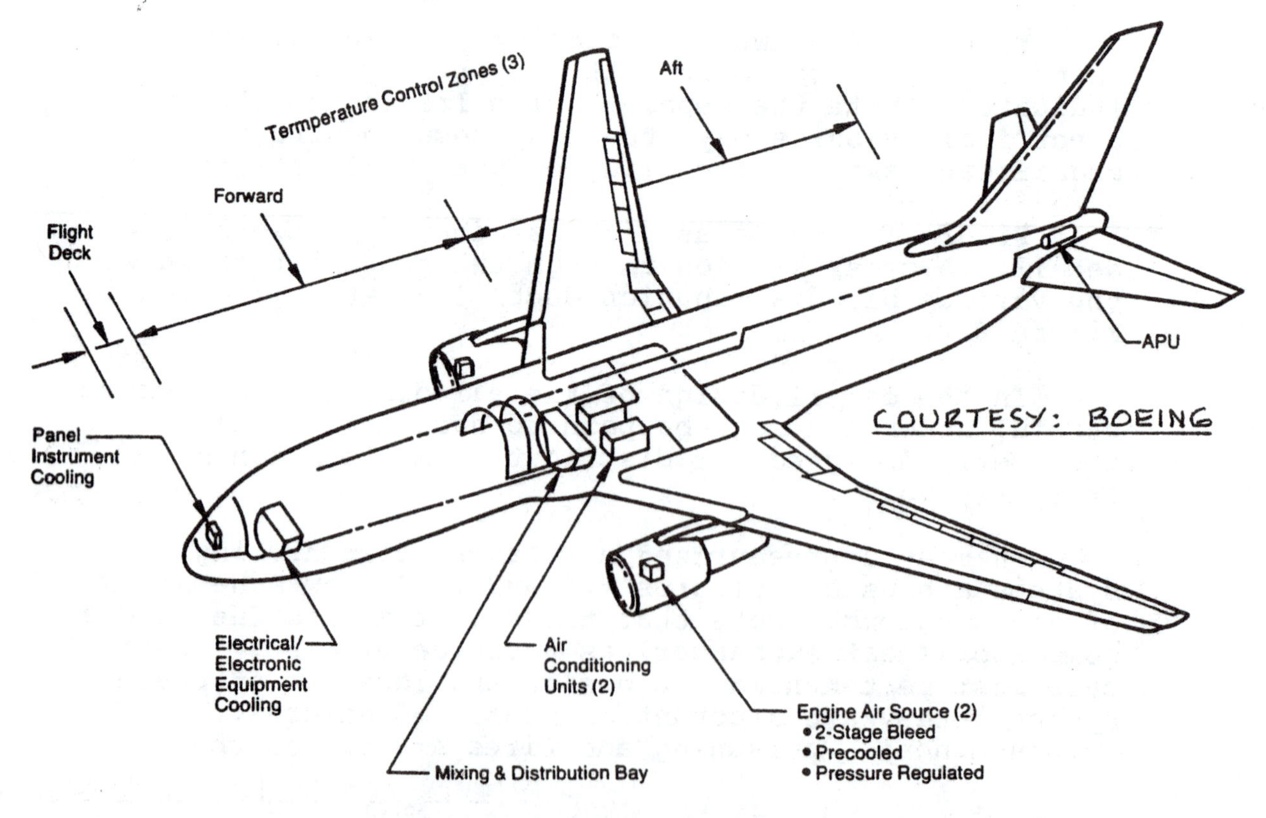

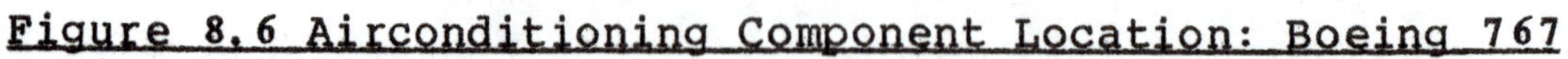

Figure 8.6 Airconditioning Component Location: Boeing 767

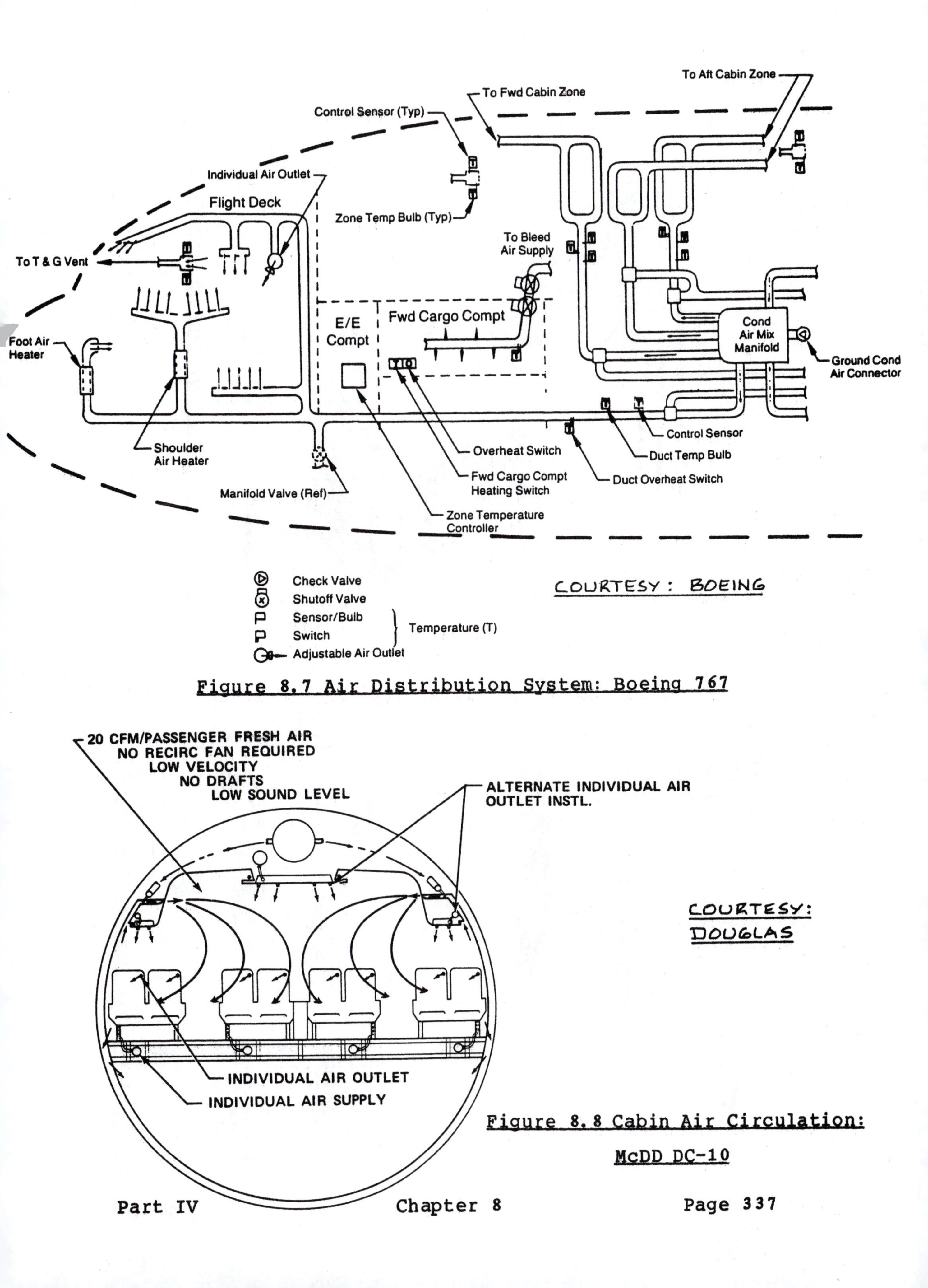

Figure 8.7 Air Distribution System: Boeing 767

Figure 8.8 Cabin Air Circulation:
McDD DC-10

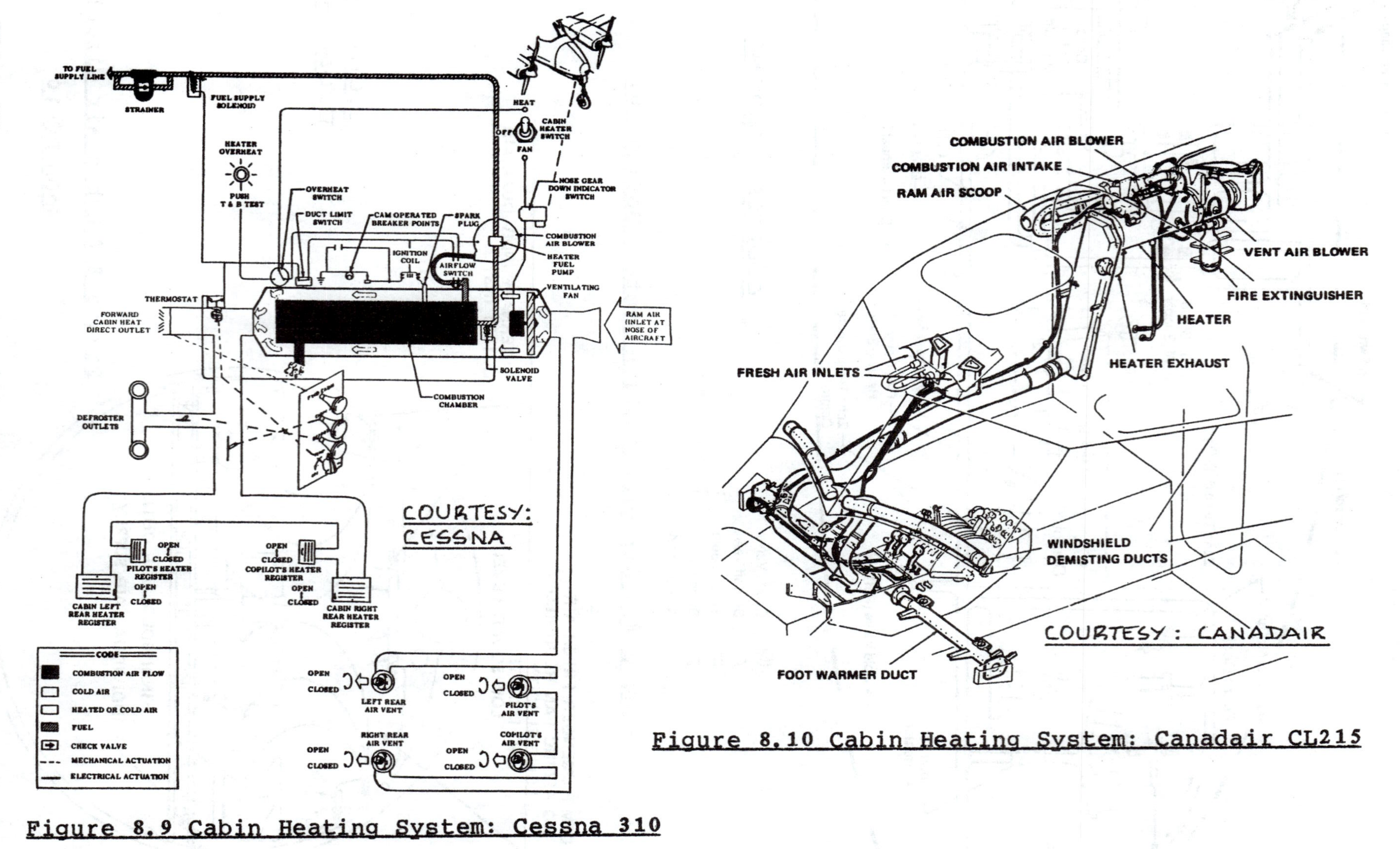

Figure 8.9 Cabin Heating System: Cessna 310

Figure 8.10 Cabin Heating System: Canadair CL215

8.4 OXYGEN SYSTEM

For flight at high altitudes oxygen is required after failure of the cabin pressurization system.

Oxygen systems use either gaseous oxygen or chemically obtained oxygen. Crew oxygen systems are normally supplied from a gaseous source. Passenger oxygen is usually supplied from a chemical source.

The main disadvantage associated with gaseous oxygen is that it presents a fire hazard during servicing and during cylinder replacement. The main disadvantage of chemical oxygen systems is their larger weight.

Figure 8.11 shows a crew oxygen delivery system in detail. Figure 8.12 shows a schematic of the entire oxygen system in a typical jet transport.

Reference 33 is an excellent source for more information on oxygen systems.

In military airplane liquid oxygen systems are often used. Although more hazardous, these systems save weight and volume. Figure 8.13 shows a typical oxygen system layout used in a fighter airplane.

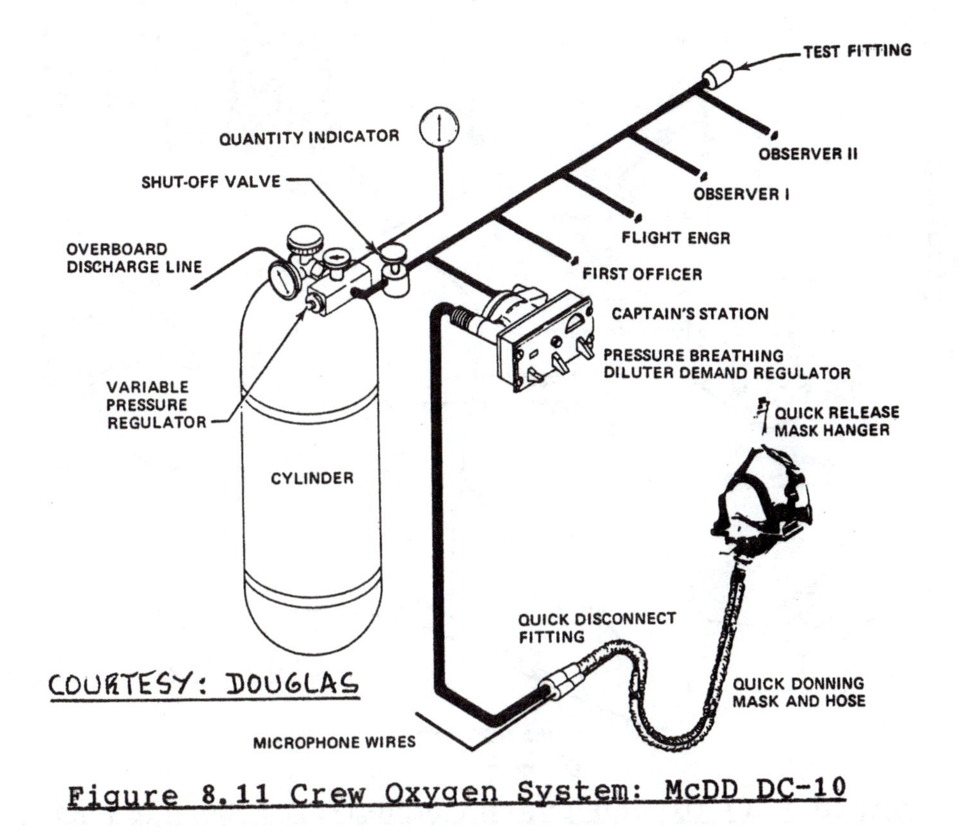

Figure 8.11 Crew Oxygen System: McDD DC-10

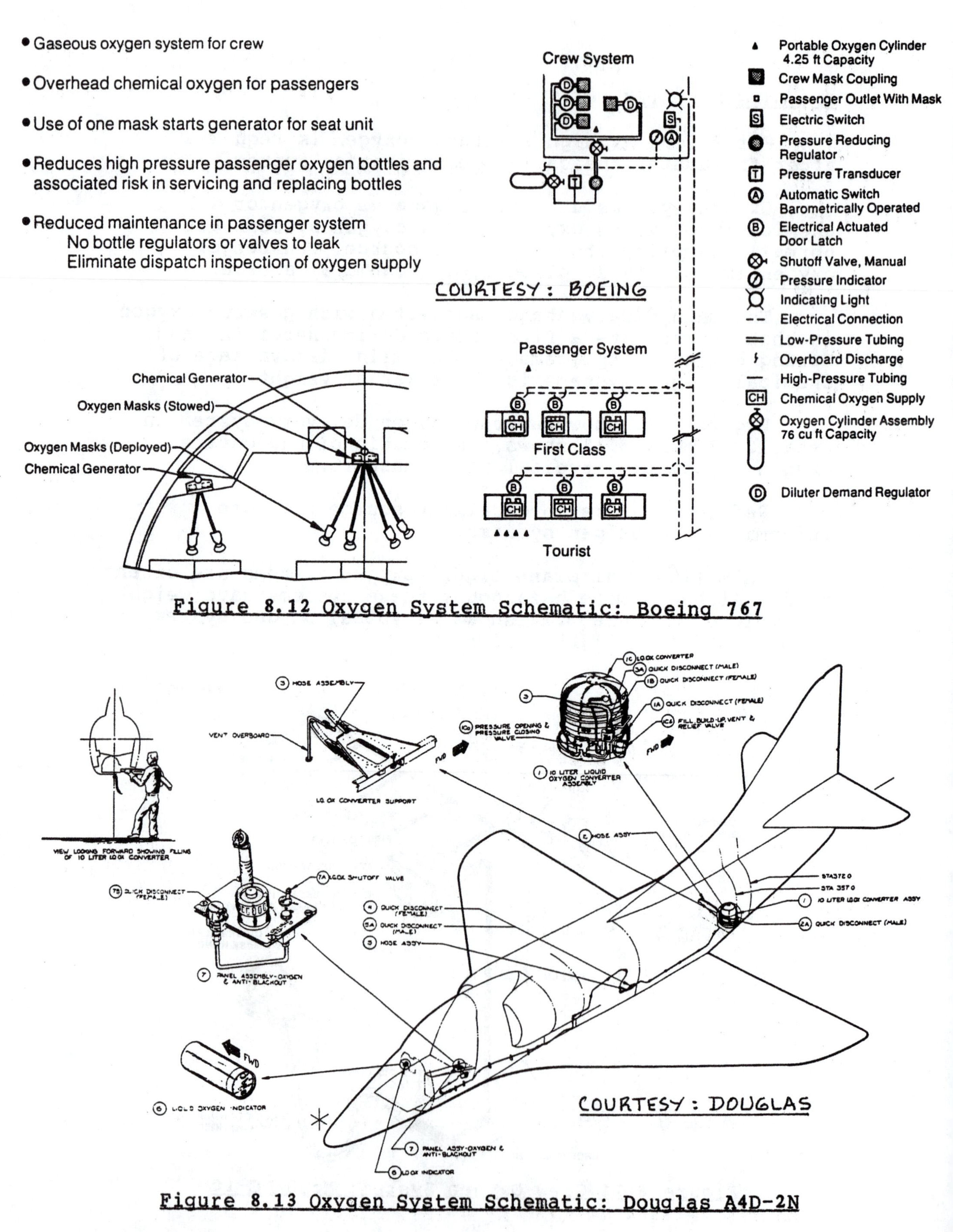

Figure 8.12 Oxygen System Schematic: Boeing 767

Figure 8.13 Oxygen System Schematic: Douglas A4D-2N

9. COCKPIT INSTRUMENTATION, FLIGHT MANAGEMENT AND AVIONICS SYSTEM LAYOUT DESIGN

In this chapter some fundamental design layout requirements for cockpit instrumentation, flight management systems and avionics systems will be discussed.

The material is organized as follows:

9.1 Cockpit instrumentation system

9.2 Flight management and avionics system layout

9.3 Antenna system layout

9.4 Installation, maintenance and servicing considerations

9.1 COCKPIT INSTRUMENTATION LAYOUT

The cockpit instrumentation layout should be uncluttered and functional. The crew must be able to see all flight crucial instruments, controls and warning devices.

Figure 9.1 shows a typical cockpit instrumentation layout used in light airplanes.

In larger airplanes the cockpit instrumentation is organized into panels. Figure 9.2 shows an example of a panel organization in a small turboprop twin.

Figure 9.3 gives a general flight compartment view for a piston/prop twin engine amphibian.

A cockpit impression of an older technology twin jet transport is given in Figure 9.4. This is contrasted with the more recent CRT (Cathode Ray Tube) type of instrumentation in Figure 9.5.

Since cockpit instrumentation and airplane avionics are evolving rapidly, changes occur almost each year. Figure 9.6 speculates on a near future cockpit arrangement consisting almost exclusively of advanced flat panel displays.

Because of classification problems no cockpit arrangements of recent operational fighters are included in this text.

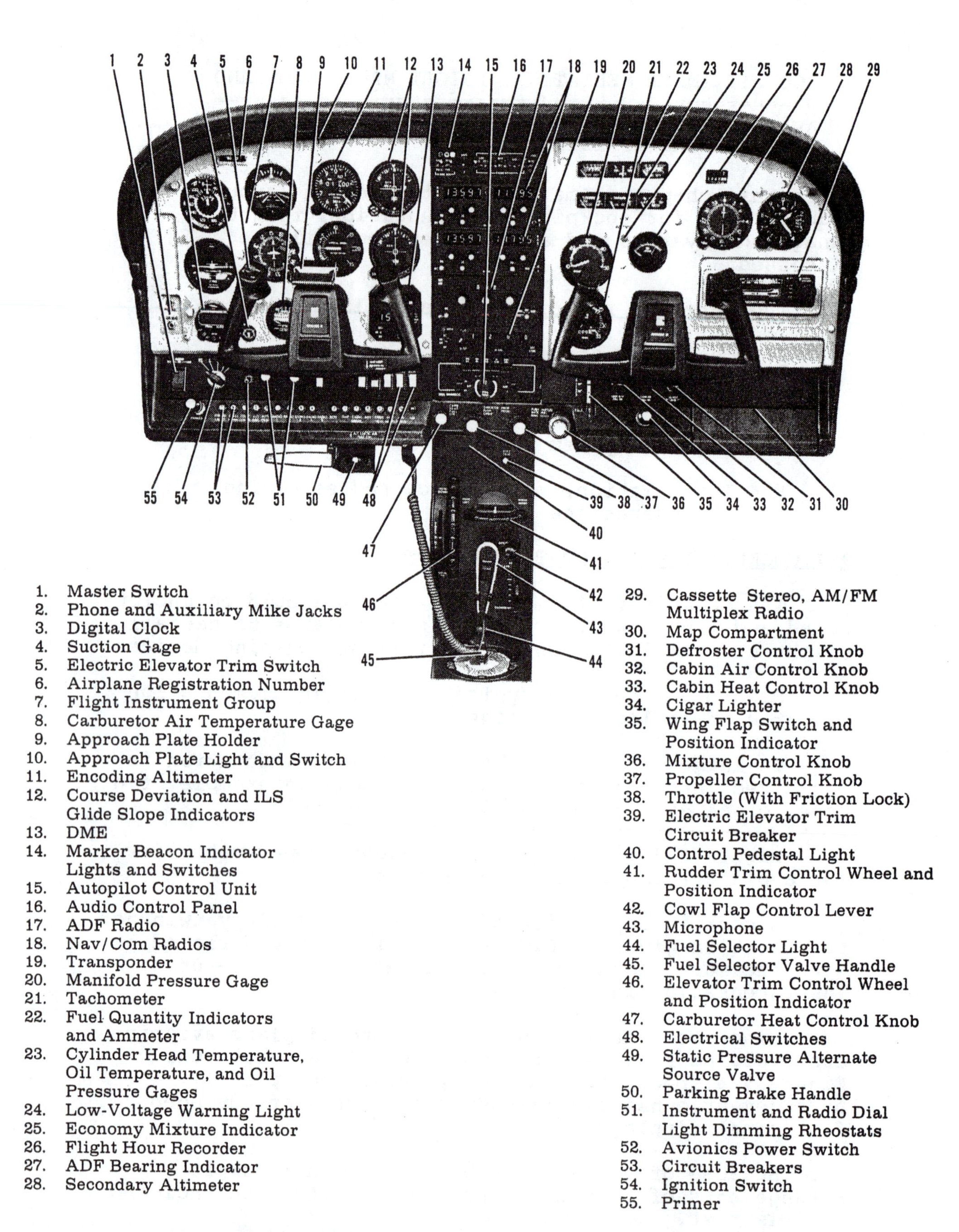

1. Master Switch
2. Phone and Auxiliary Mike Jacks
3. Digital Clock
4. Suction Gage
5. Electric Elevator Trim Switch
6. Airplane Registration Number
7. Flight Instrument Group
8. Carburetor Air Temperature Gage
9. Approach Plate Holder
10. Approach Plate Light and Switch
11. Encoding Altimeter
12. Course Deviation and ILS Glide Slope Indicators
13. DME
14. Marker Beacon Indicator Lights and Switches
15. Autopilot Control Unit
16. Audio Control Panel
17. ADF Radio
18. Nav/Com Radios
19. Transponder
20. Manifold Pressure Gage
21. Tachometer
22. Fuel Quantity Indicators and Ammeter
23. Cylinder Head Temperature, Oil Temperature, and Oil Pressure Gages
24. Low-Voltage Warning Light
25. Economy Mixture Indicator
26. Flight Hour Recorder
27. ADF Bearing Indicator
28. Secondary Altimeter
29. Cassette Stereo, AM/FM Multiplex Radio
30. Map Compartment
31. Defroster Control Knob
32. Cabin Air Control Knob
33. Cabin Heat Control Knob
34. Cigar Lighter
35. Wing Flap Switch and Position Indicator
36. Mixture Control Knob
37. Propeller Control Knob
38. Throttle (With Friction Lock)
39. Electric Elevator Trim Circuit Breaker
40. Control Pedestal Light
41. Rudder Trim Control Wheel and Position Indicator
42. Cowl Flap Control Lever
43. Microphone
44. Fuel Selector Light
45. Fuel Selector Valve Handle
46. Elevator Trim Control Wheel and Position Indicator
47. Carburetor Heat Control Knob
48. Electrical Switches
49. Static Pressure Alternate Source Valve
50. Parking Brake Handle
51. Instrument and Radio Dial Light Dimming Rheostats
52. Avionics Power Switch
53. Circuit Breakers
54. Ignition Switch
55. Primer

Figure 9.1 Cockpit Instrumentation: Cessna 182Q

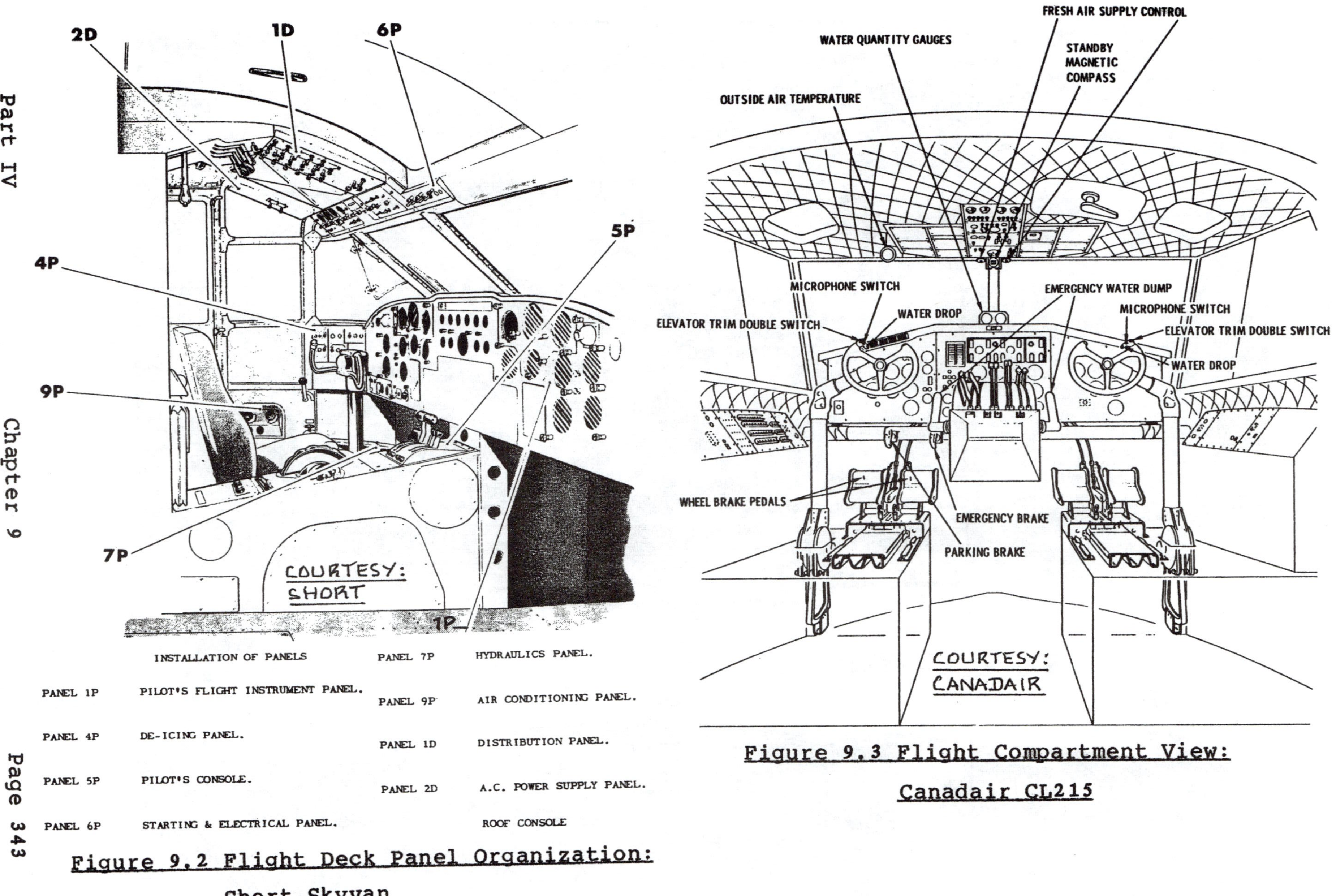

Figure 9.2 Flight Deck Panel Organization: Short Skyvan

Figure 9.3 Flight Compartment View: Canadair CL215

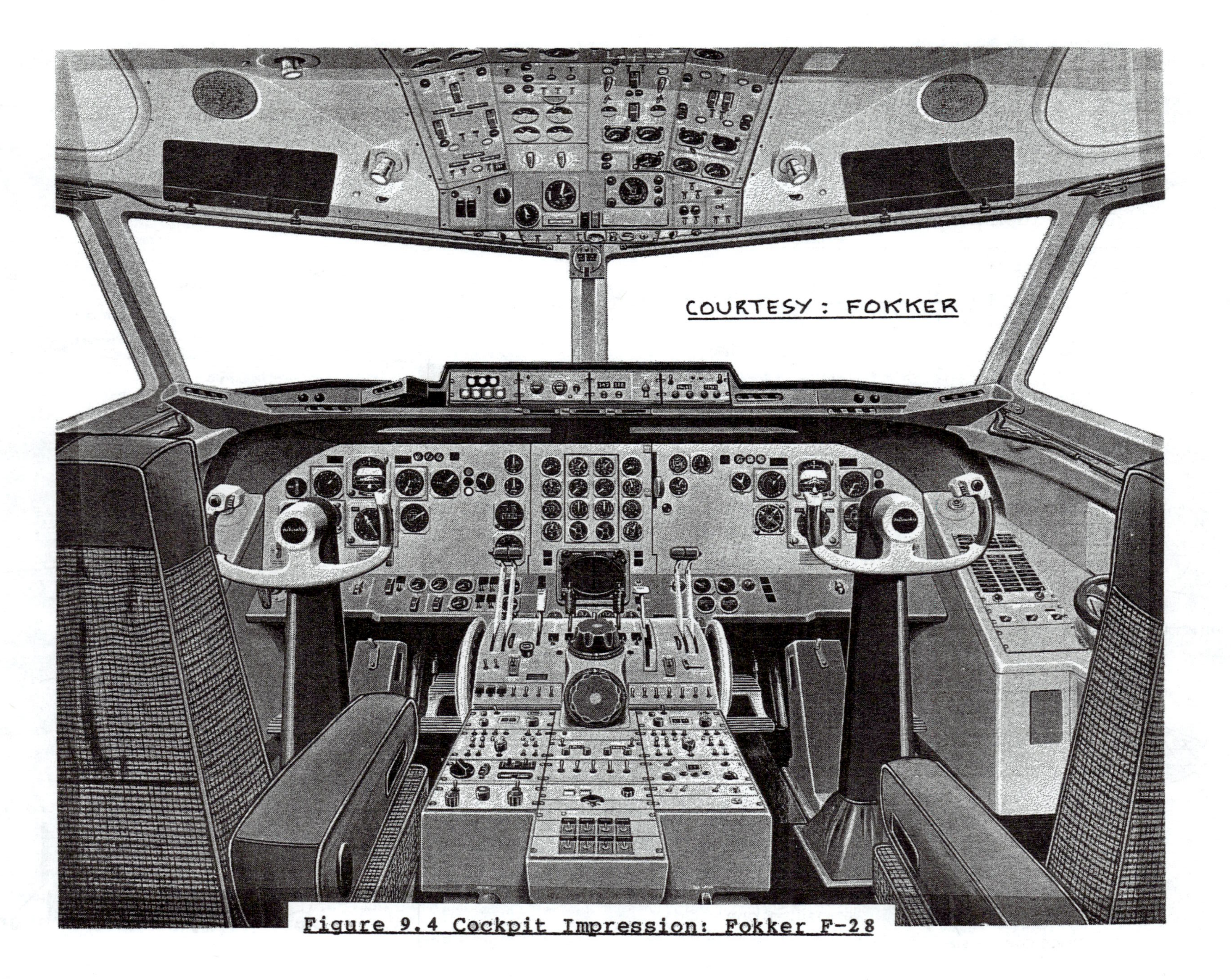

Figure 9.4 Cockpit Impression: Fokker F-28

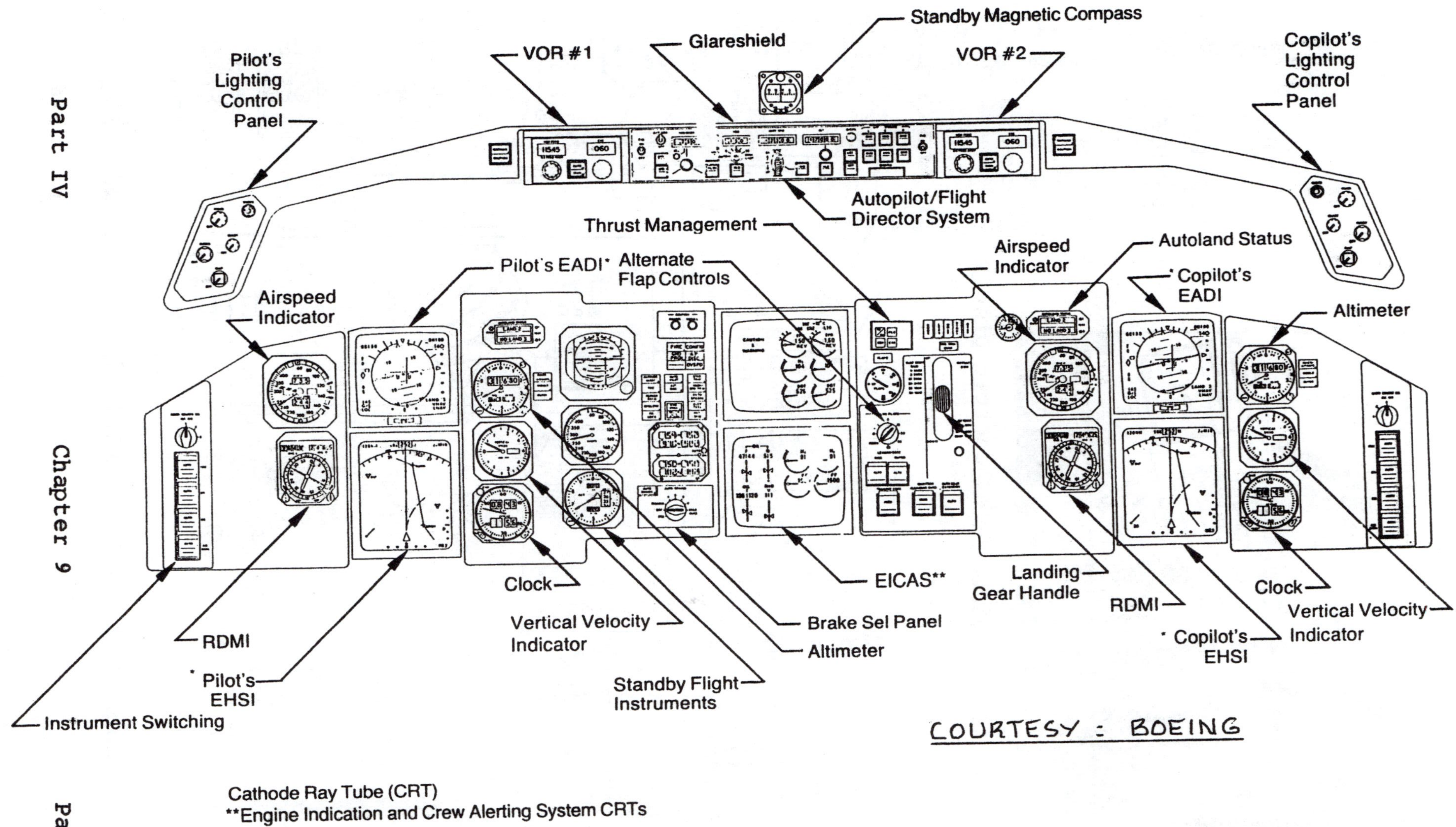

* Cathode Ray Tube (CRT)
** Engine Indication and Crew Alerting System CRTs

Figure 9.5 Cockpit Instrumentation: Boeing 767

COURTESY:
LOCKHEED

Figure 9.6 Proposed Flat Panel Cockpit Display

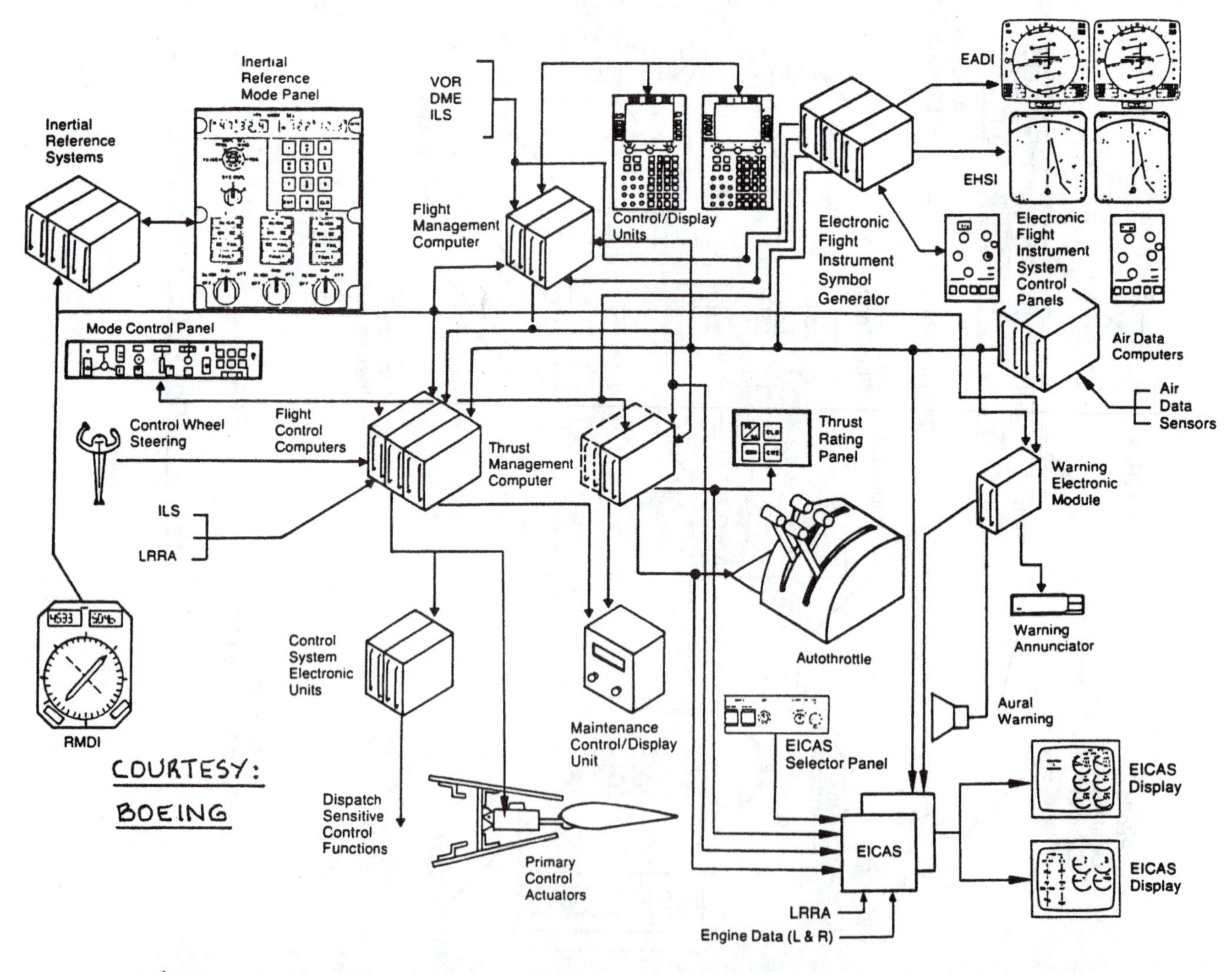

Figure 9.7 Flight Management System Schematic Boeing 767

9.2 FLIGHT MANAGEMENT AND AVIONICS SYSTEM LAYOUT

Flight management and avionics systems are undergoing very rapid development. Also, the number of different systems available to the user is so large that a concise summary in a textbook on airplane design cannot be given in a few pages. For older avionics and flight management systems the reader should refer to Ref.34. For an annual summary of available avionics systems the reader should consult the magazine: 'Business and Commercial Aviation' which each year publishes an up-to-date list of available equipment including data on weight, power requirements and pricing. Reference 36 contains a 1986 state of the art description of what advanced avionics can do for an airplane.

In nearly all recently built transports the pilot interfaces with the airplane controls through the so-called flight management system. These systems have now progressed to the point of integrating propulsion controls, flight controls and autopilot functions. Figure 9.7 shows a flight management system schematic as used in a modern jet transport.

A flight management system requires a number of sub-systems. In the following the sub-system breakdown employed in the Boeing 767 will be illustrated.

1. Flight control computer

Figure 9.8 shows a diagram indicating the flight control computer functions.

2. Autopilot/autothrottle controls

Figure 9.9 shows the autopilot/autothrottle control panel with its various function selectors indicated.

3. Thrust management computer

Figure 9.10 presents a schematic of the thrust management computer functions.

Figure 9.11 summarizes the flight control avionics functions of the 767 flight mangement system.

4. Inertial reference system

In long range airplanes in particular, inertial re-

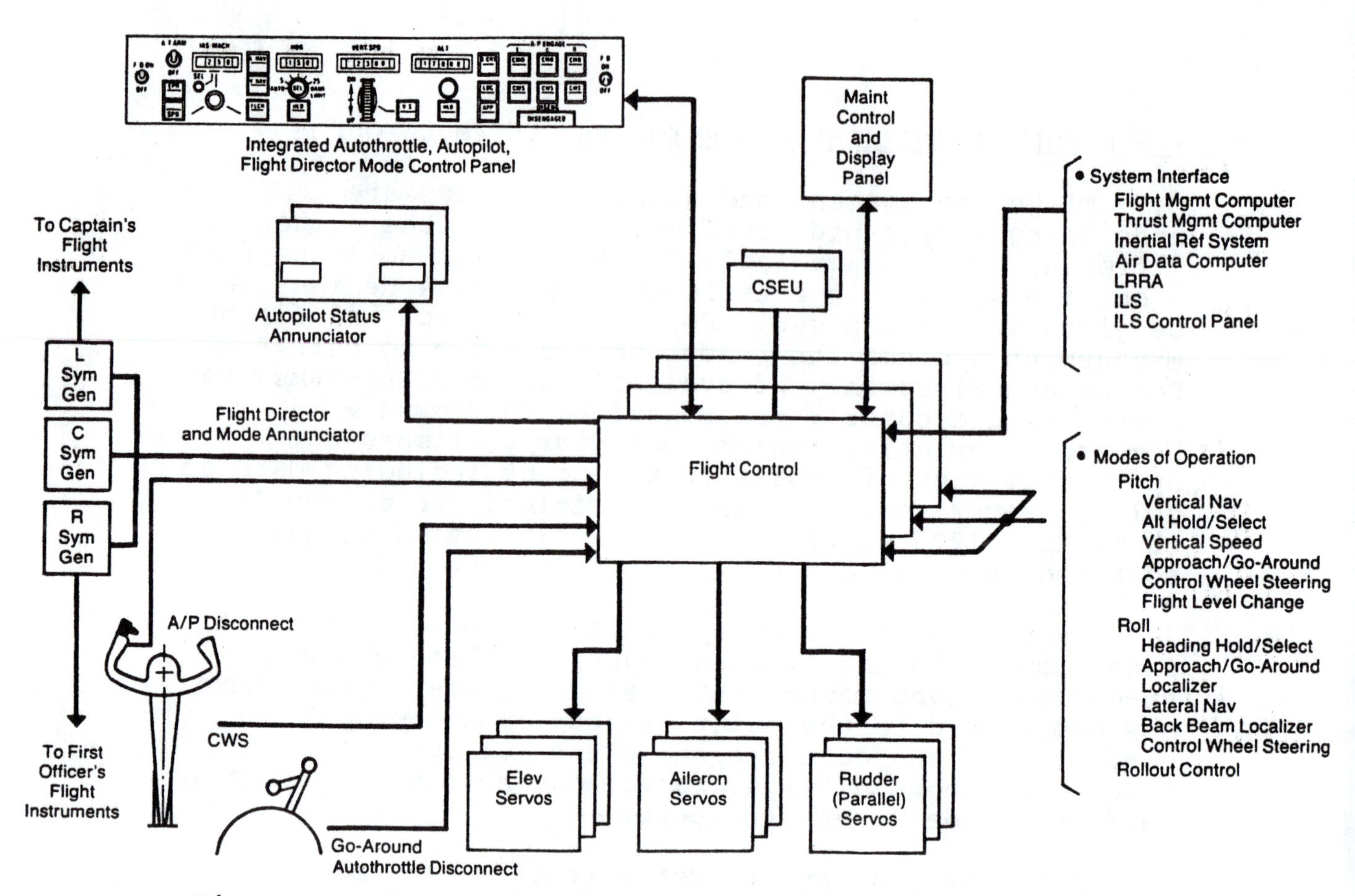

Figure 9.8 Flight Control Computer Functions: Boeing 767

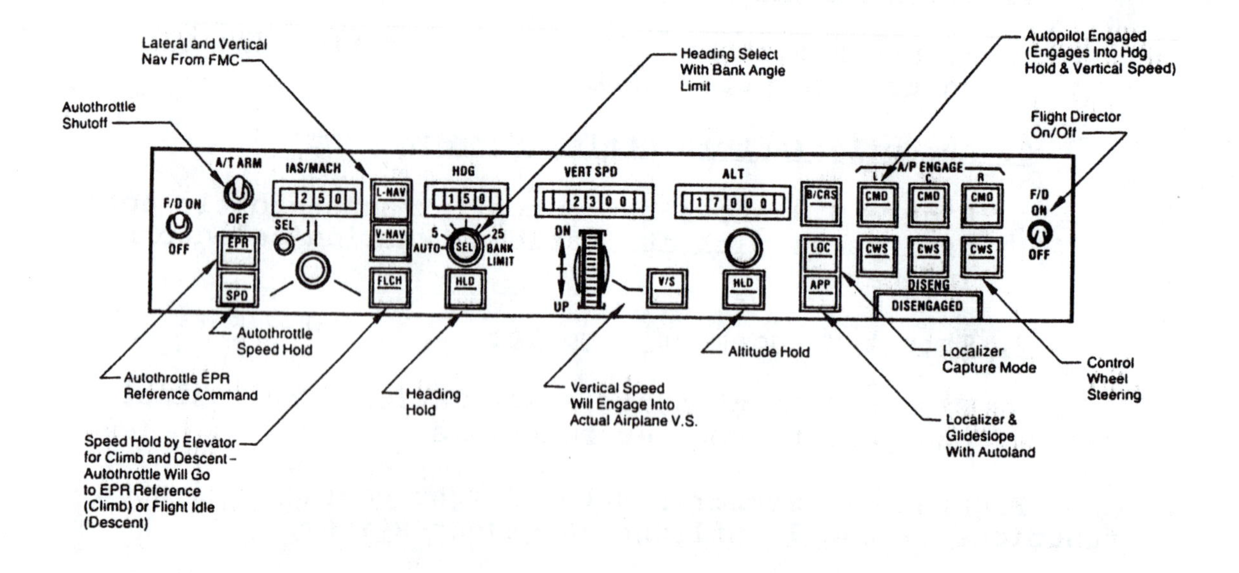

Figure 9.9 Autopilot/Autothrottle Control Panel: Boeing 767

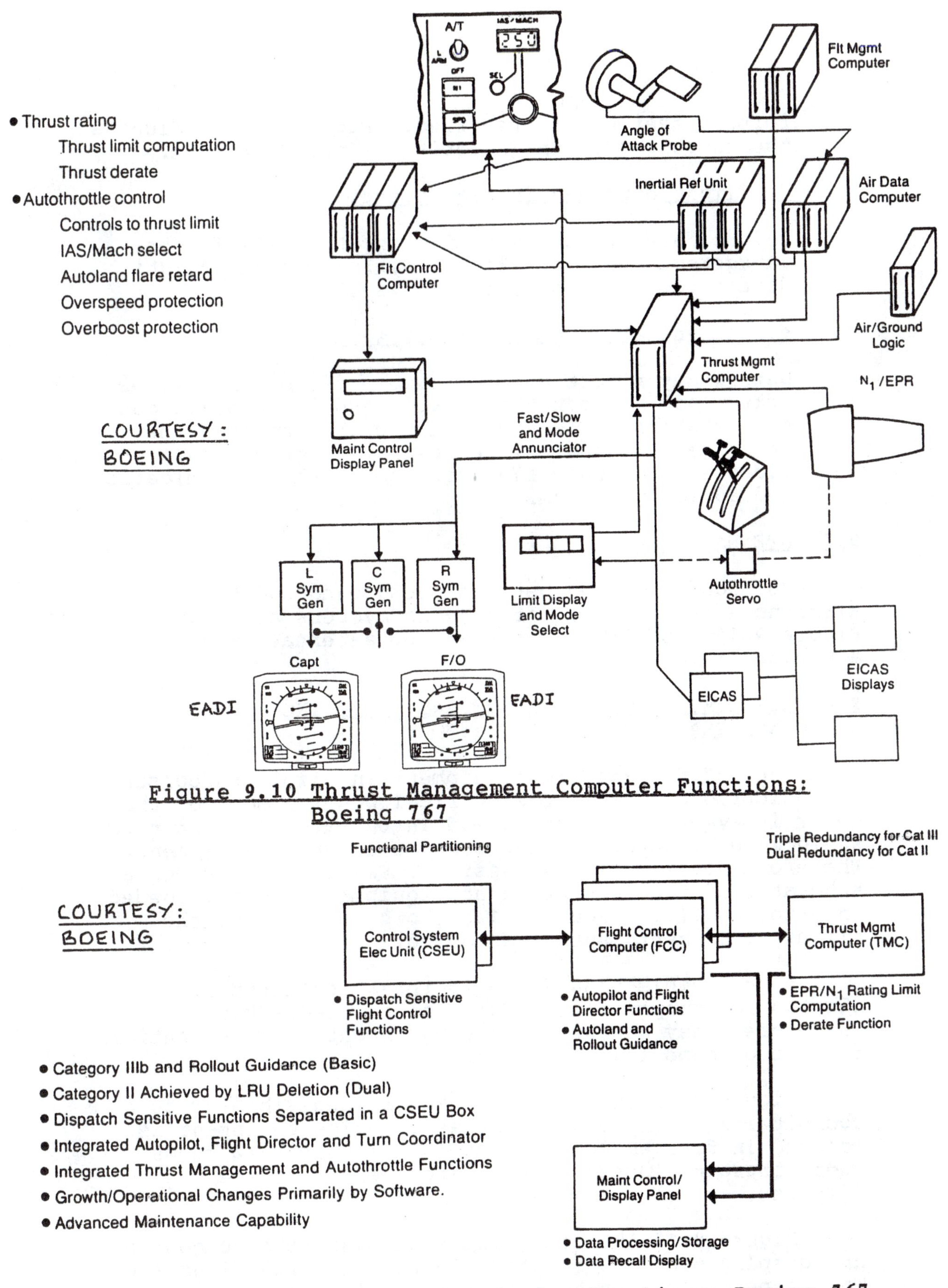

Figure 9.10 Thrust Management Computer Functions: Boeing 767

Figure 9.11 Flight Control Avionics Functions: Boeing 767

ference systems are required. Figure 9.12 indicates the functions of the inertial reference system.

5. Flight data acquisition system

Flight data are an essential input to all flight management systems. Figure 9.13 shows the flight data inputs needed by the 767 system.

6. Communication and advisory system

During any flight a large number of communications take place between the cockpit and air traffic control stations. In addition the crew advises the passengers of the flight status. Figure 9.14 provides a functional flow diagram of the communication and advisory system.

9.3 ANTENNA SYSTEM LAYOUT

For communication between the ground and the airplane a large number of antenna systems are required. Figure 9.15 shows an example of the antennae installed in a Boeing 767.

9.4 INSTALLATION, MAINTENANCE AND SERVICING CONSIDERATIONS

Much of the avionics equipment in airplanes consumes a considerable amount of electrical power. Most of that power is eventually transformed into heat. In turn this would lead to major malfunctions in avionics equipment. Therefore cooling is a necessity. Figure 9.16a shows a schematic of the instrument and equipment cooling needed in a Boeing 767. Figure 9.16b shows how some of the flight deck instruments are cooled.

A reasonable assumption is that most electrical and electronic equipment will fail rather frequently. Therefore, easy access to this equipment is an essential feature of good layout design.

Figure 9.17 shows an example of accessibility to a cockpit instrument panel. Other avionics equipment is located in five equipment centers. Figure 9.18 shows where these equipment centers are located in the airplane.

Figure 9.19 indicates where the maintenance control and display panel is located and what its functions are. This type of maintenance control is also being used in

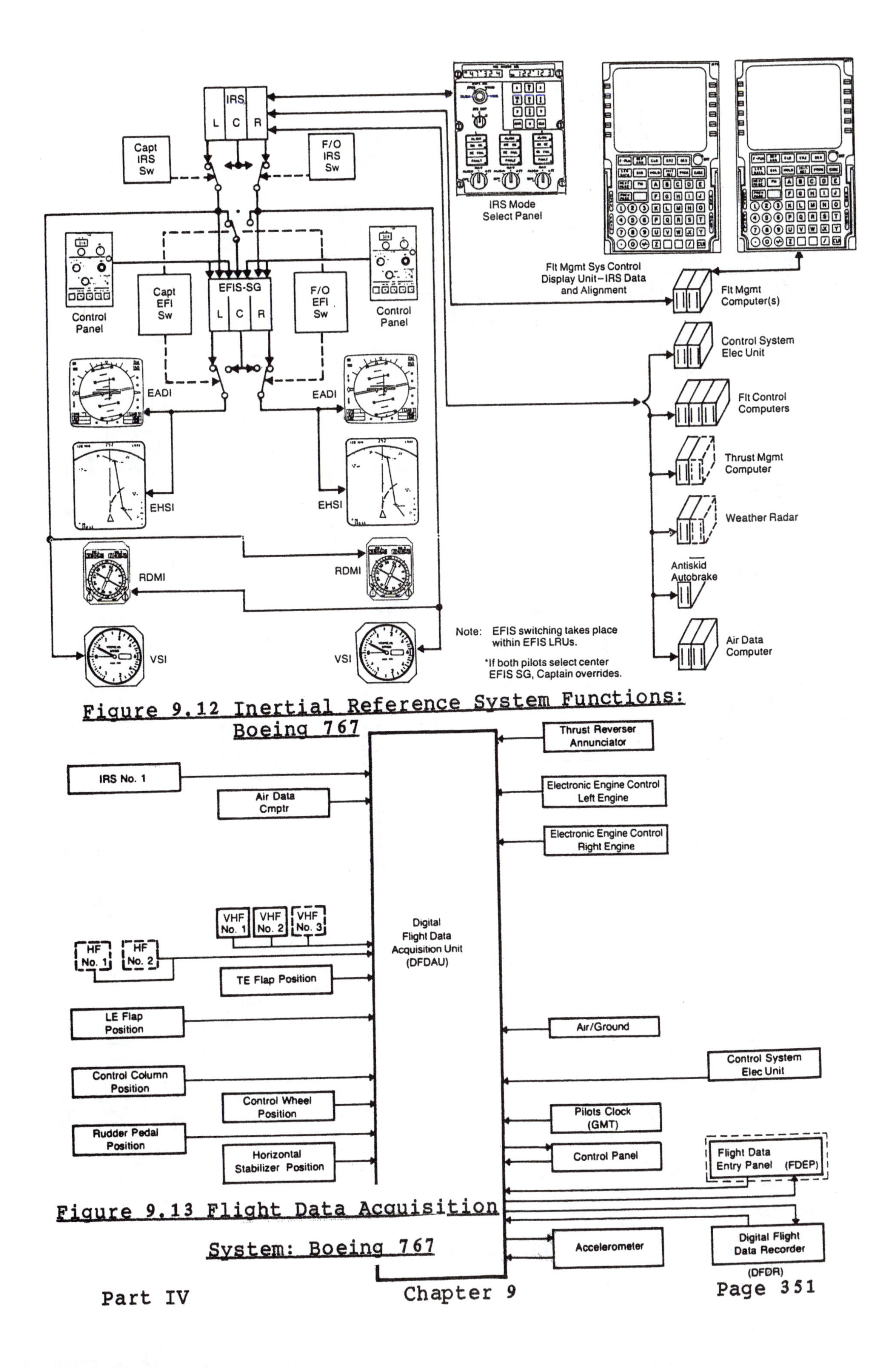

Figure 9.12 Inertial Reference System Functions: Boeing 767

Figure 9.13 Flight Data Acquisition System: Boeing 767

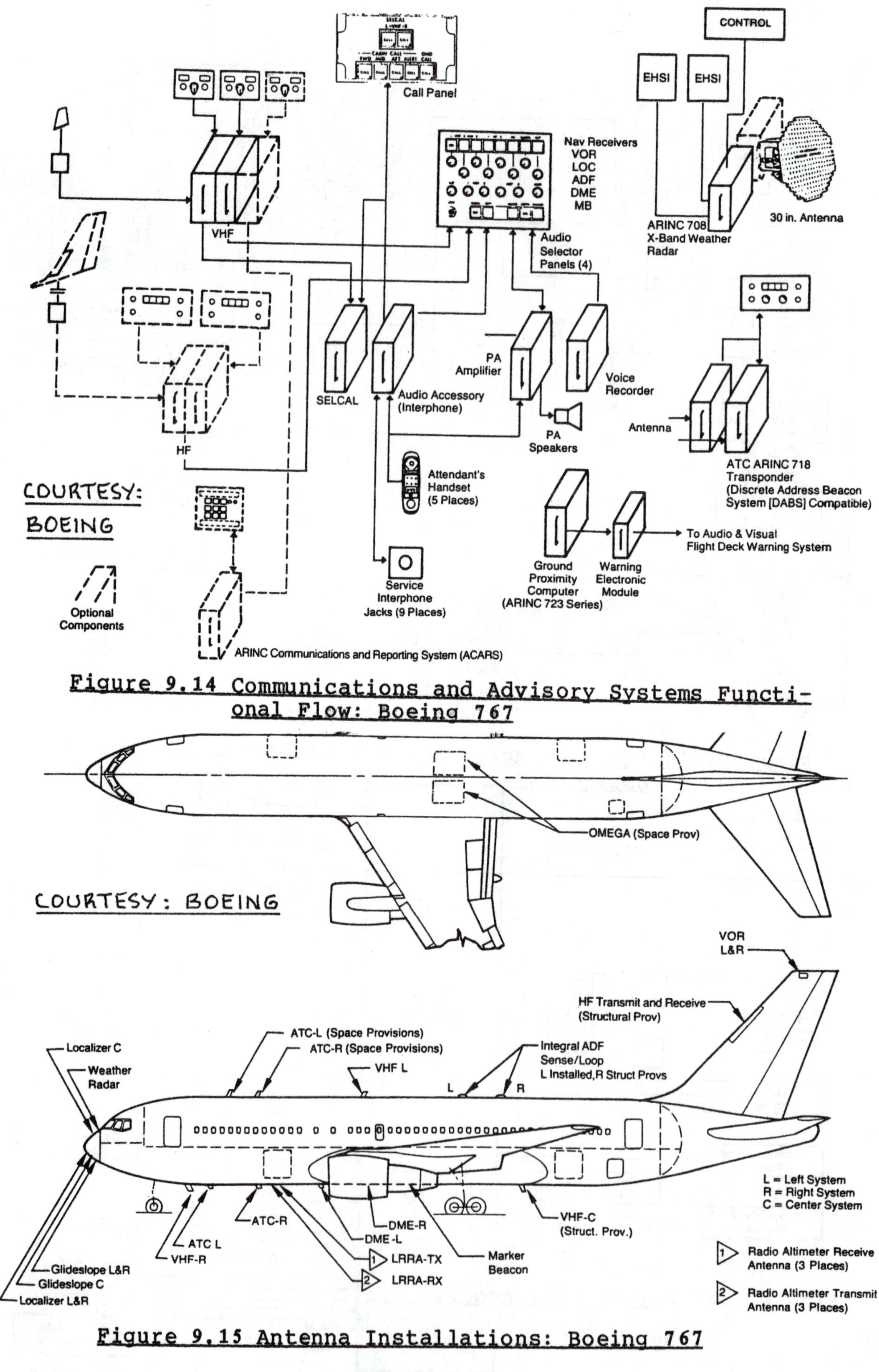

Figure 9.14 Communications and Advisory Systems Functional Flow: Boeing 767

Figure 9.15 Antenna Installations: Boeing 767

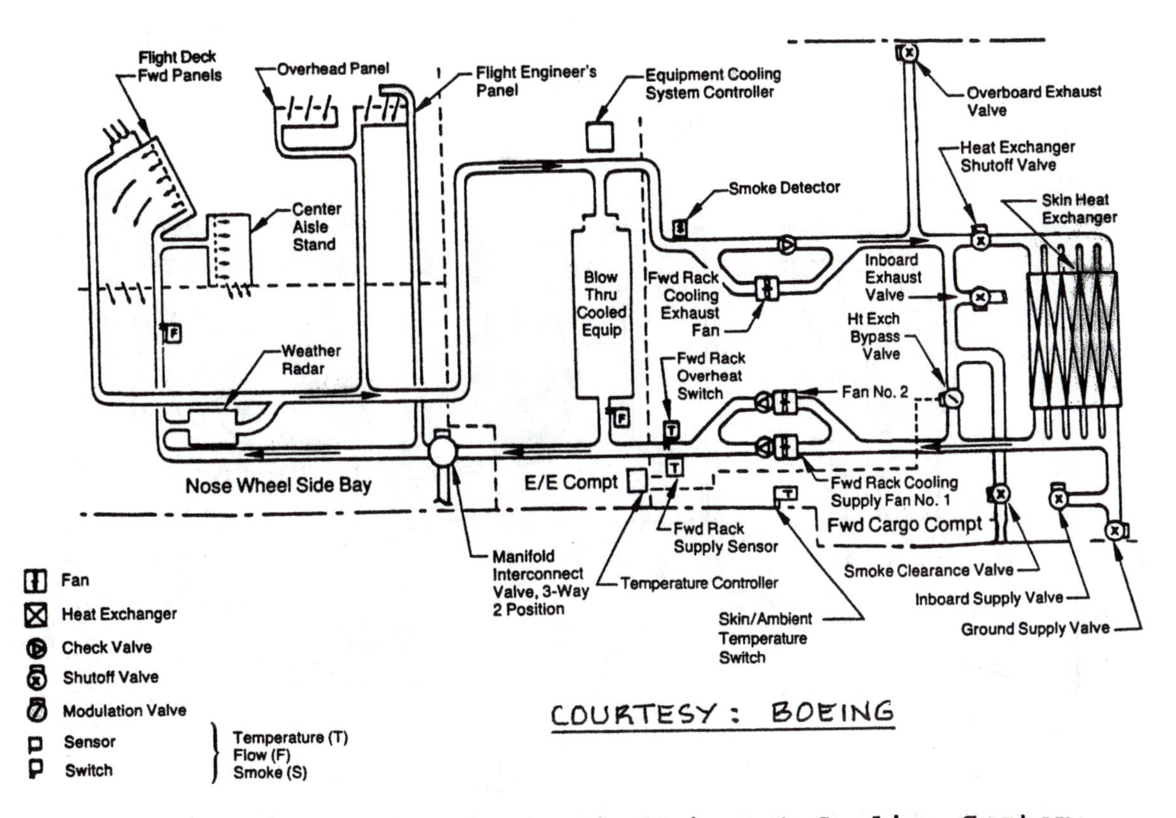

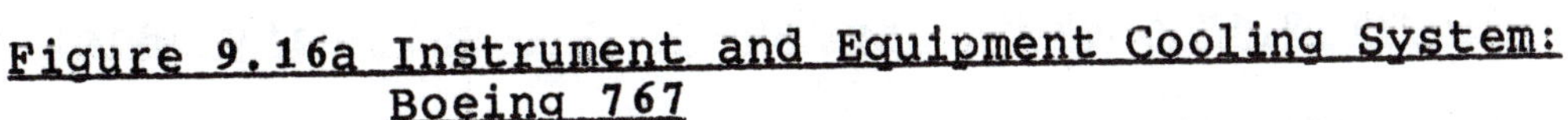
Figure 9.16a Instrument and Equipment Cooling System: Boeing 767

Instrument Cooling

- Positive cooling for flight deck instruments
- Improved ground cooling
- Closed loop in-flight cooling
    - Reduced equipment contamination without using filters
    - Reduces system impact on main cabin air distribution
    - Permits control of cooling air temperature
    - Skin heat exchanger

COURTESY: BOEING

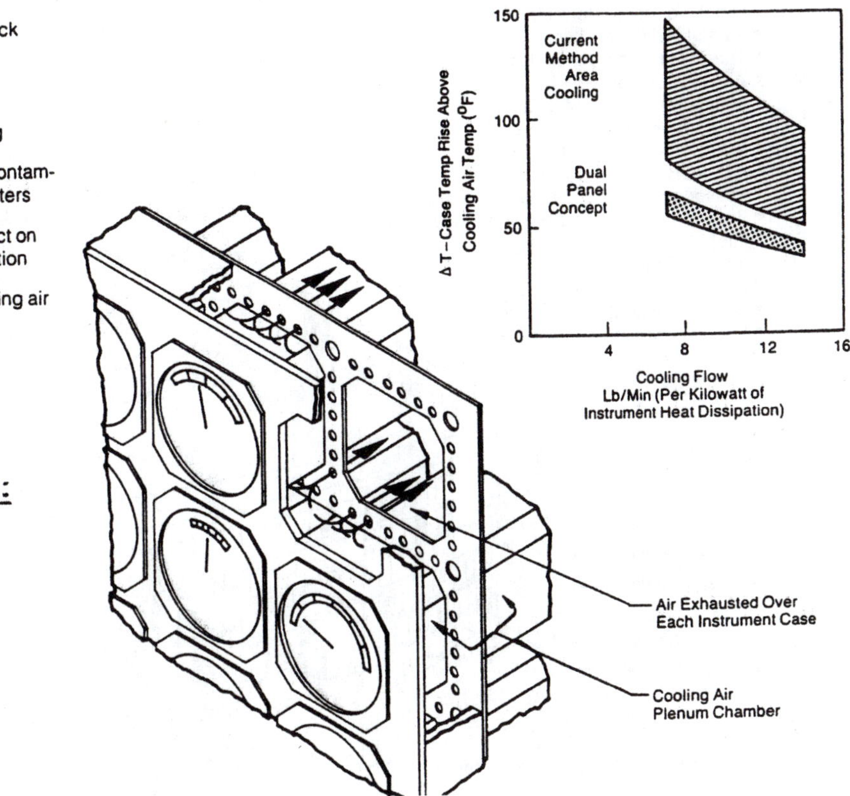

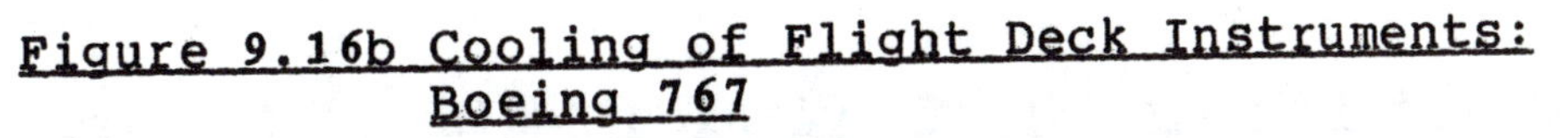
Figure 9.16b Cooling of Flight Deck Instruments: Boeing 767

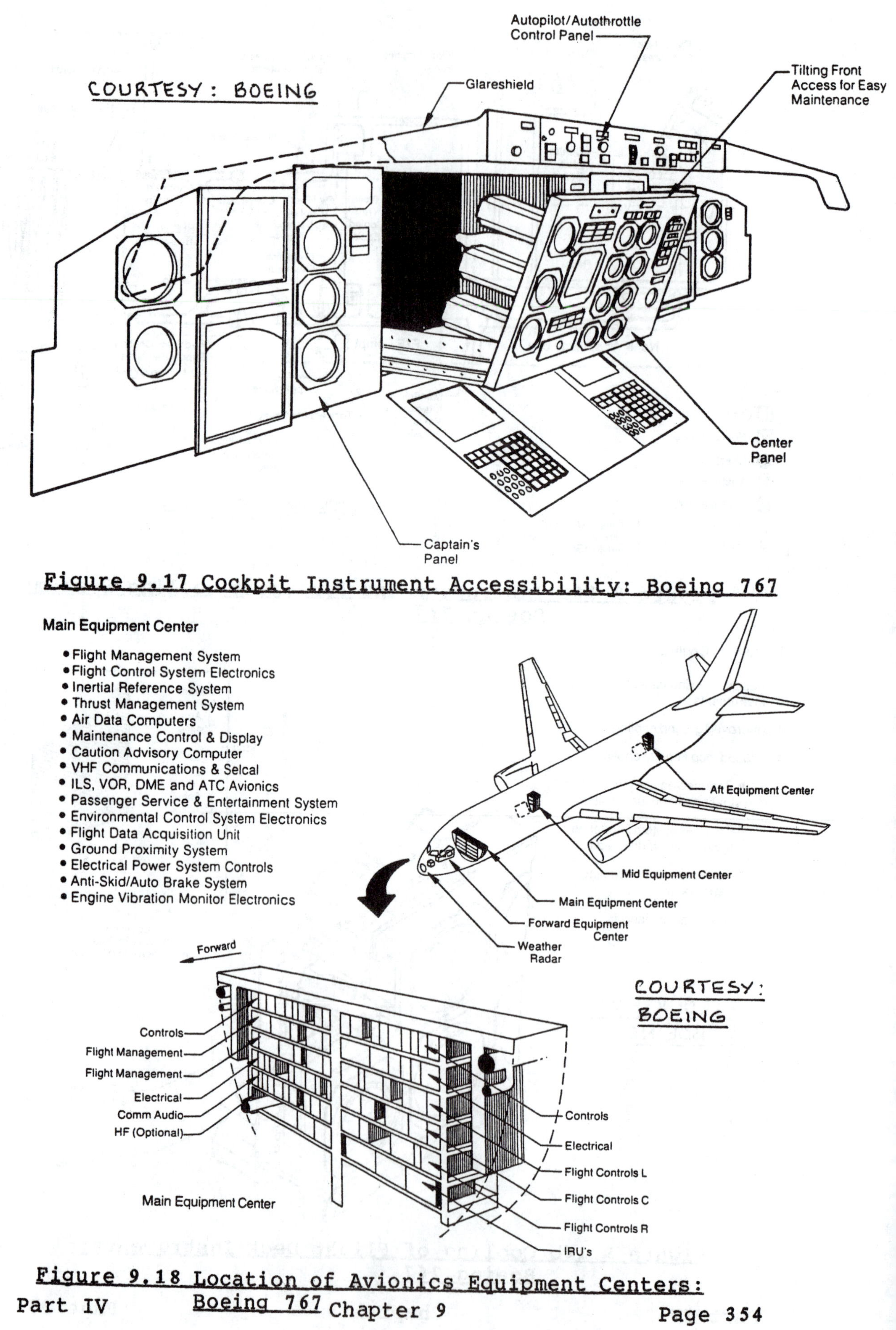

Figure 9.17 Cockpit Instrument Accessibility: Boeing 767

Figure 9.18 Location of Avionics Equipment Centers: Boeing 767

many fighter airplanes.

Figure 9.20 shows how the mid and aft equipment centers can be accessed.

The radar system and flight control antennae are accessed through removal of the radome. Figure 9.21 shows how the forward equipment center can be accessed.

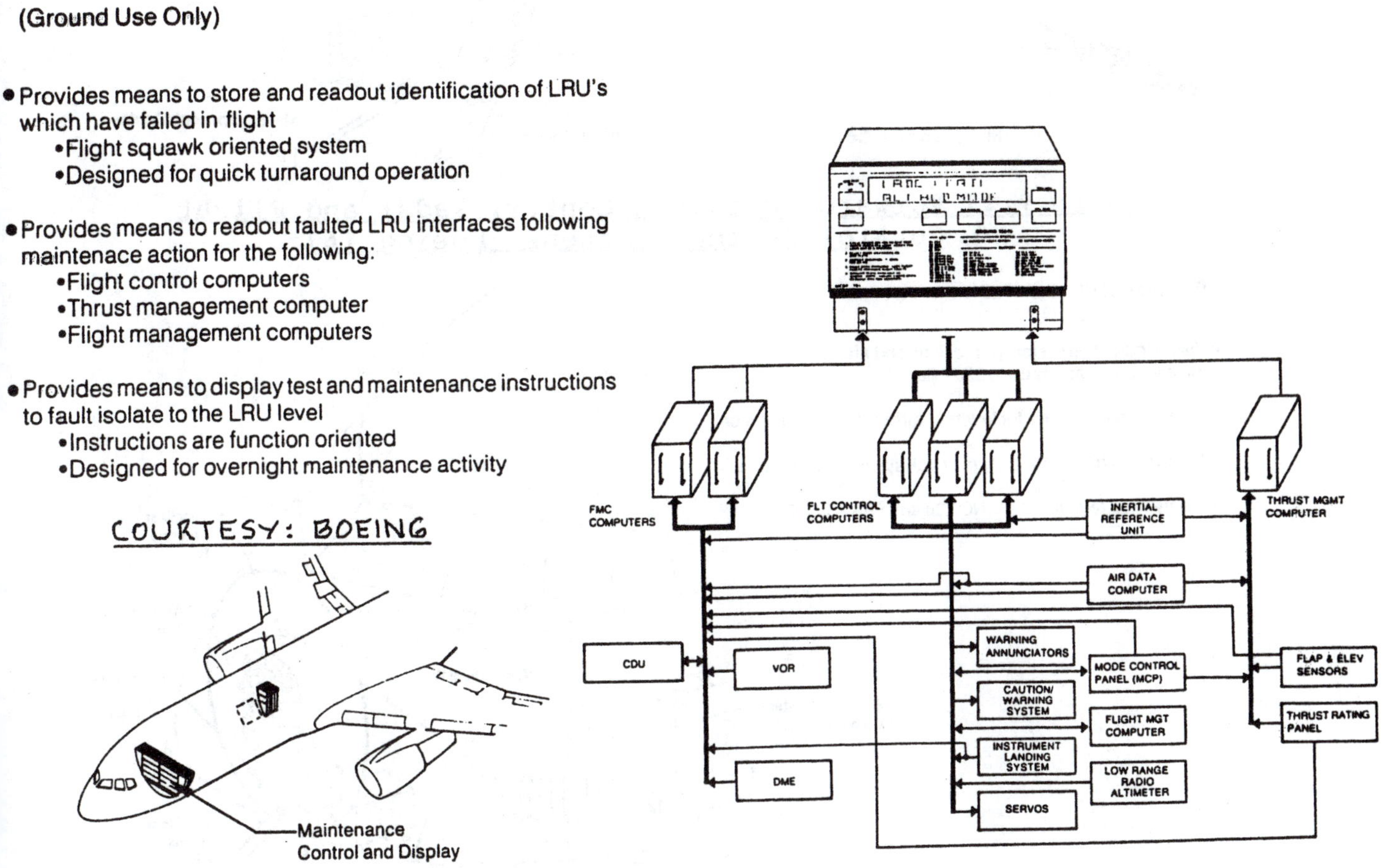
(Ground Use Only)
• Provides means to store and readout identification of LRU's which have failed in flight
•Flight squawk oriented system
•Designed for quick turnaround operation
• Provides means to readout faulted LRU interfaces following maintenace action for the following:
•Flight control computers
•Thrust management computer
•Flight management computers
• Provides means to display test and maintenance instructions to fault isolate to the LRU level
•Instructions are function oriented
•Designed for overnight maintenance activity
COURTESY: BOEING
Maintenance Control and Display
FMC COMPUTERS
FLT CONTROL COMPUTERS
THRUST MGMT COMPUTER
INERTIAL REFERENCE UNIT
AIR DATA COMPUTER
CDU
VOR
DME
WARNING ANNUNCIATORS
CAUTION/ WARNING SYSTEM
INSTRUMENT LANDING SYSTEM
SERVOS
MODE CONTROL PANEL (MCP)
FLIGHT MGT COMPUTER
LOW RANGE RADIO ALTIMETER
FLAP & ELEV SENSORS
THRUST RATING PANEL

Figure 9.19 Maintenance Control and Display Panel: Boeing 767

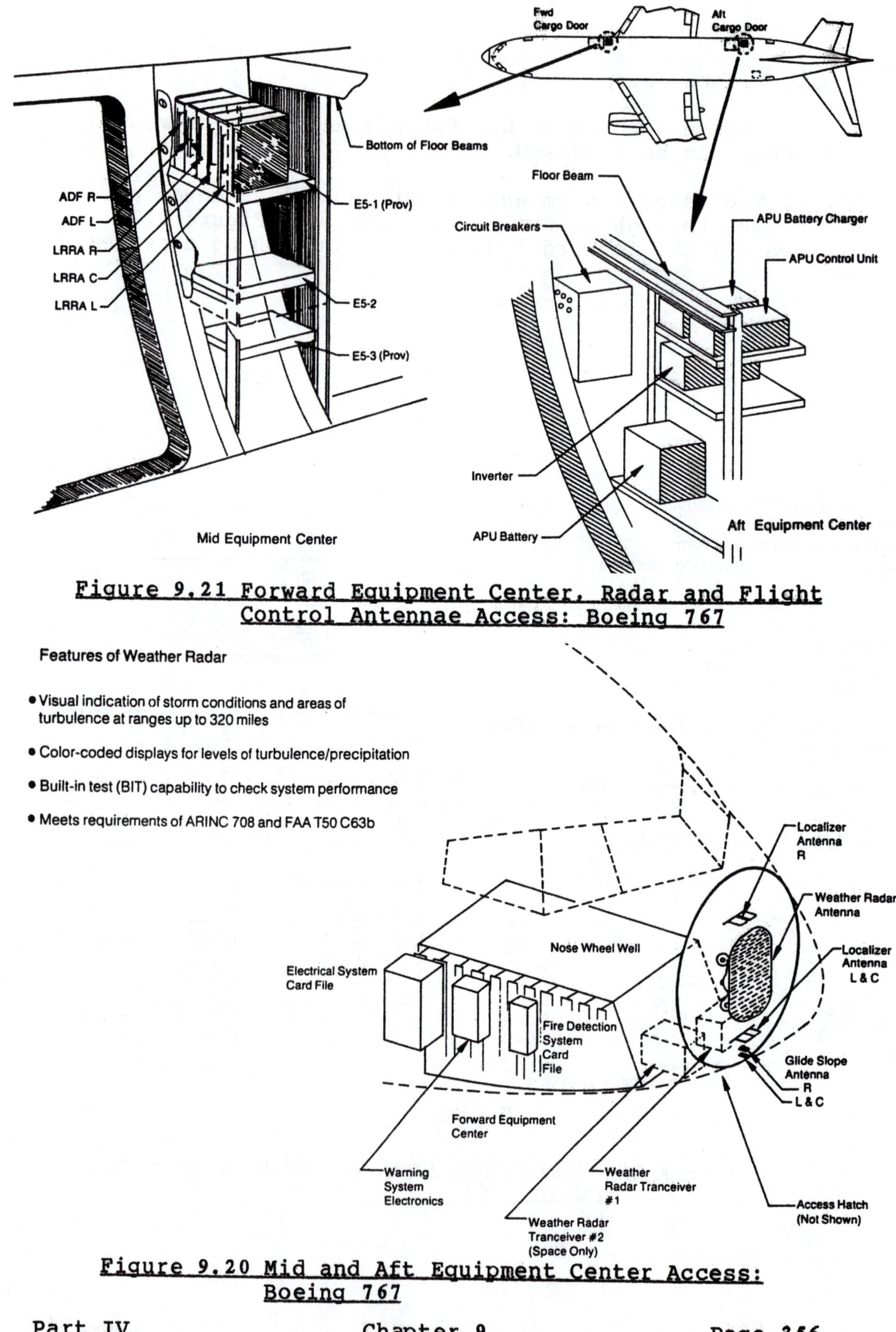

Figure 9.21 Forward Equipment Center, Radar and Flight Control Antennae Access: Boeing 767

Figure 9.20 Mid and Aft Equipment Center Access: Boeing 767

10. DE-ICING, ANTI-ICING, RAIN REMOVAL AND DEFOG SYSTEMS

Whenever an airplane can be expected to be operated into known icing conditions, special systems must be installed to prevent the accumulation of ice and/or to remove ice which has formed.

In addition, when flying through severe rain it is possible that water accumulates on the windshields such that visibility is seriously degraded. A rain removal system must then be available.

Under certain combinations of humidity and temperature fog forms on the windshields, again restricting visibility. A defog system helps to combat this.

Reference 33 contains detailed descriptions of de-icing, anti-icing and rain removal systems.

The material in this chapter is organized as:

10.1 De-icing and anti-icing systems

10.2 Rain removal and defog systems

10.1 DE-ICING AND ANTI-ICING SYSTEMS

Ice formation can have the following consequences:

1. Ice formed on wings and tails can distort the aerodynamic contours such that:

a) the drag increases markedly which may cause the airplane to slow down or to loose climb capability.

b) the lift decreases. Particularly a sharp drop in maximum lift coefficient may lead to earlier than normal stalls when the pilot maneuvers the airplane.

c) the pitching moment changes. This can lead to unexpected trim changes as well as to changes in the stick-force-speed and/or stick-force-per-'g' gradients.

Figure 10.1 illustrates typical ice accumulation shapes which can occur on airplane lifting surfaces.

2. Ice formed in engine inlets can result in serious degradation of engine performance. In addition, when ice breaks loose from the inlet it can cause serious damage to the engine.

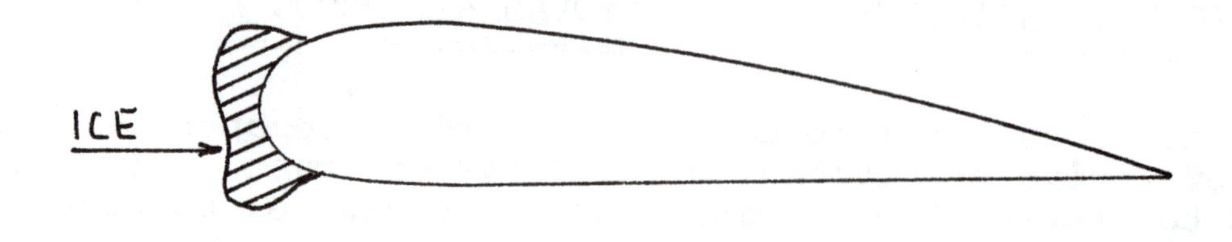

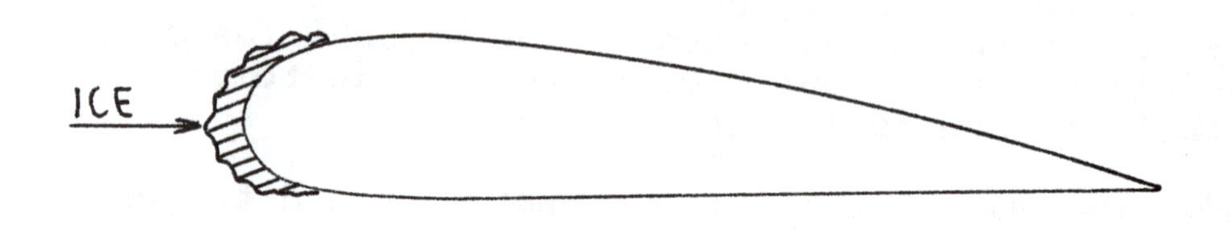

Figure 10.1 Ice Accumulation Shapes

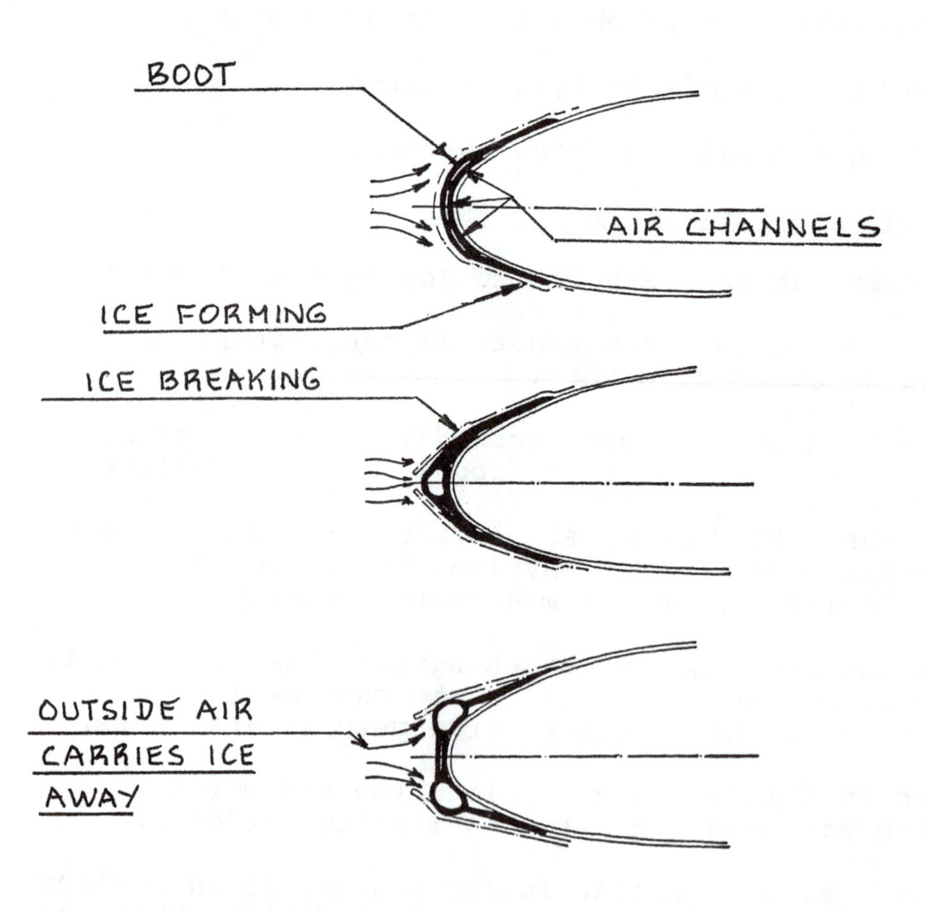

Figure 10.2 Operation of a De-icing Boot

3. Ice formed on pitot inlets, stall vanes or other sensors crucial to the safe operation of an airplane can result in accidents.

The purpose of this section is to discuss the following systems:

10.1.1 De-icing Systems: these are systems which are designed to remove ice which has already formed.

10.1.2 Anti-icing Systems: these are systems designed to prevent the formation of ice.

10.1.1 De-Icing Systems

The following de-icing systems will be discussed:

1. De-icing boots

2. Electro-impulse systems

1. De-icing Boots

De-icing boots consist of a thin rubber bag attached to the leading edge of a lifting surface. Figure 10.2 shows a cross section of a boot installation. Note that three separate air channels are pressurized in an alternating manner. This cracks and breaks the ice away.

Figure 10.3 shows an example of a de-icing boot system installed in a Beech King Air. The system uses engine bleed-air to pressurize the boots.

2. Electro-Impulse Systems

These systems operate by delivering mechanical impulses to the surfaces on which ice has formed. These impulses are delivered by electromagnetic coils installed on these surfaces. Figure 10.4 shows a cross section through a leading edge with an electro-impulse system installed. Ref.37 should be consulted for further details.

NOTE: Airplane configurations must be laid out so that ice which breaks away does not enter jet engines.

10.1.2 Anti-Icing Systems

The idea behind anti-icing systems is to prevent the formation of ice. These systems are to be turned on as soon as the crew suspects that their flight will

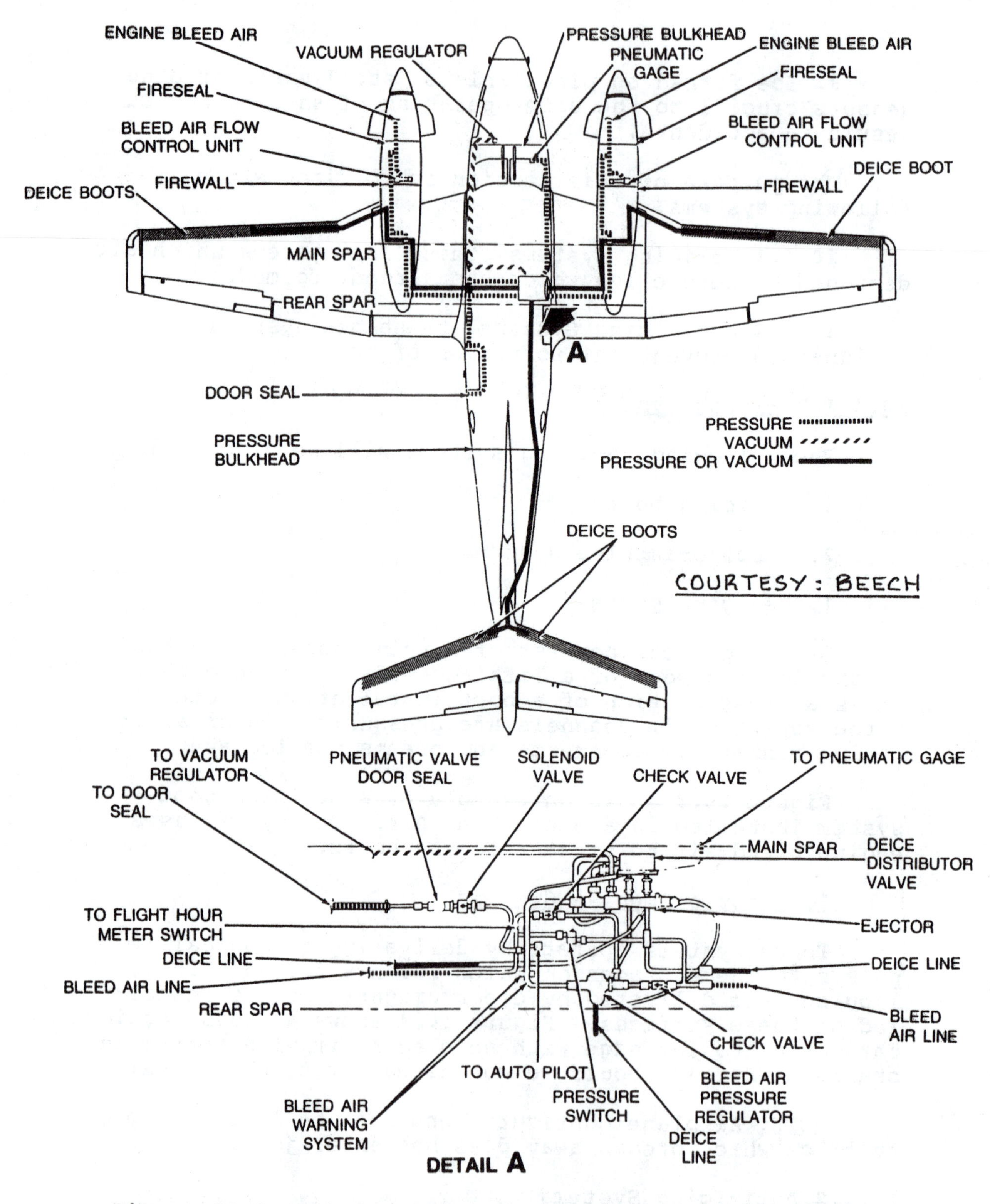

Figure 10.3 De-icing Boot System: Beech King Air F90

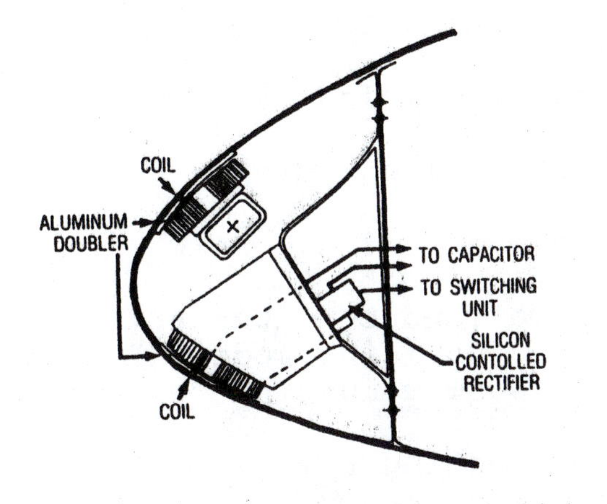

Figure 10.4 Electro-Impulse De-icing System

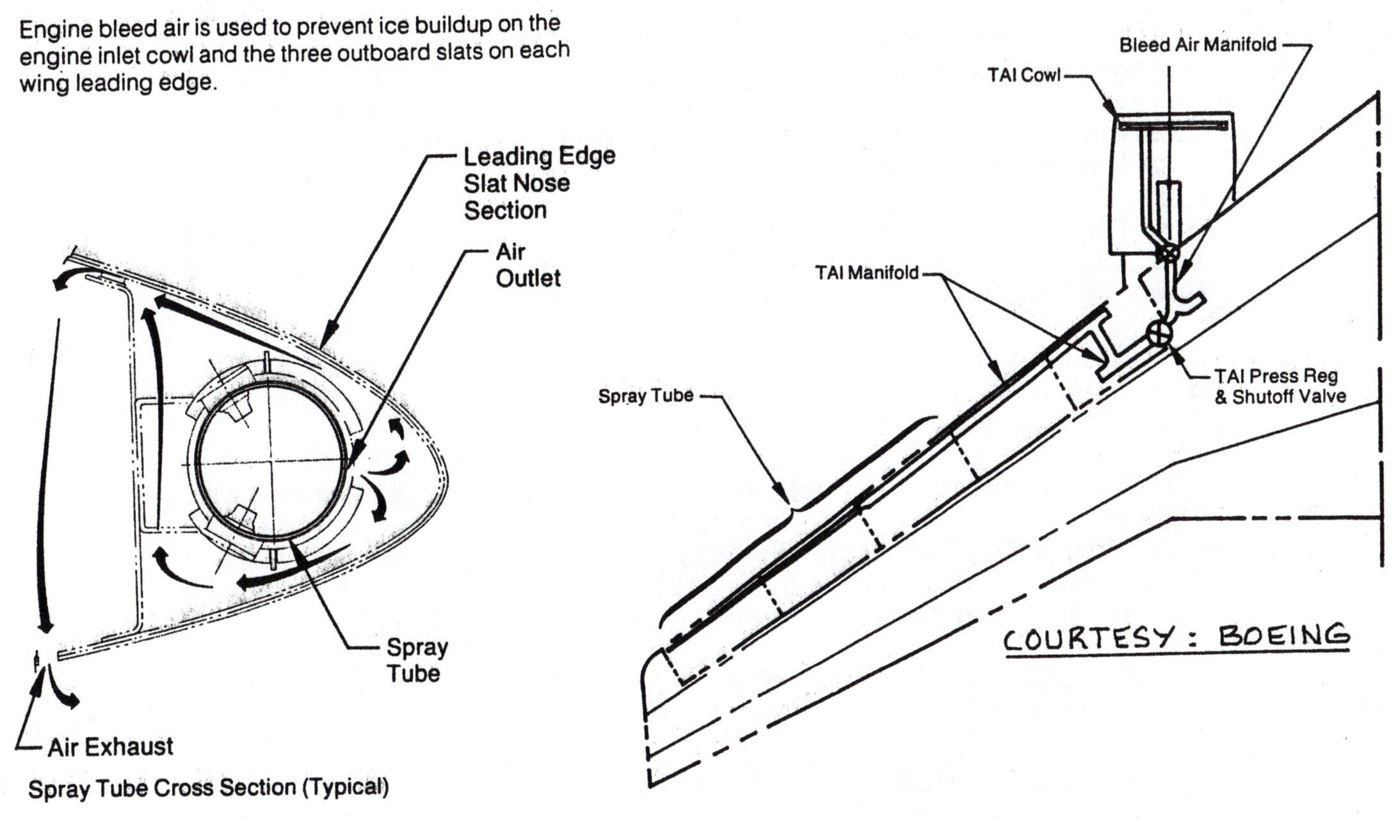

Figure 10.5 Air Heated Anti-Icing System: Boeing 767

encounter conditions favorable to the formation of ice.

The following anti-icing systems will be discussed:

1. Thermal anti-icing system

2. Chemical system

   NOTE: thermal and chemical anti-icing systems can be used in a de-icing mode. If this is done, ice which breaks away must not enter jet engines.

3. Carburetor heating system

4. Inertial anti-icer system

1. Thermal Anti-icing Systems

Two types of thermal anti-ice systems are in use: air heated systems and electrically heated systems.

Air heated anti-icing systems employ hot air to heat surfaces where ice would otherwise form. Thermal anti-icing systems are sometimes used also to de-ice a surface. However, it is far better to prevent ice formation than to get rid of it after it has formed.

Figure 10.5 shows how an air heated system operates. Detailed schematics of air heated anti-icing systems are shown in Figures 10.6 and 10.7.

Air heated systems are also used to protect leading edge slats as well as engine nose cowls (inlet lips). Figure 10.8 shows such a system on the DC-10.

In electrically heated systems electrical resistances are used to heat those surfaces where ice otherwise would form. Electrical heating is used to prevent ice formation on pitot tubes, stall vanes, total air temperature probes, drain masts and engine inlet lips.

Figure 10.9 shows an electric anti-ice system schematic used on the Boeing 767.

2. Chemical Anti-icing System

In this type of system an anti-freeze liquid (for example a glycol base) is 'oozed' through a system of very small holes located in the leading edge. The leading edge is therefore covered with a liquid which does not freeze and thus prevents the formation of ice.

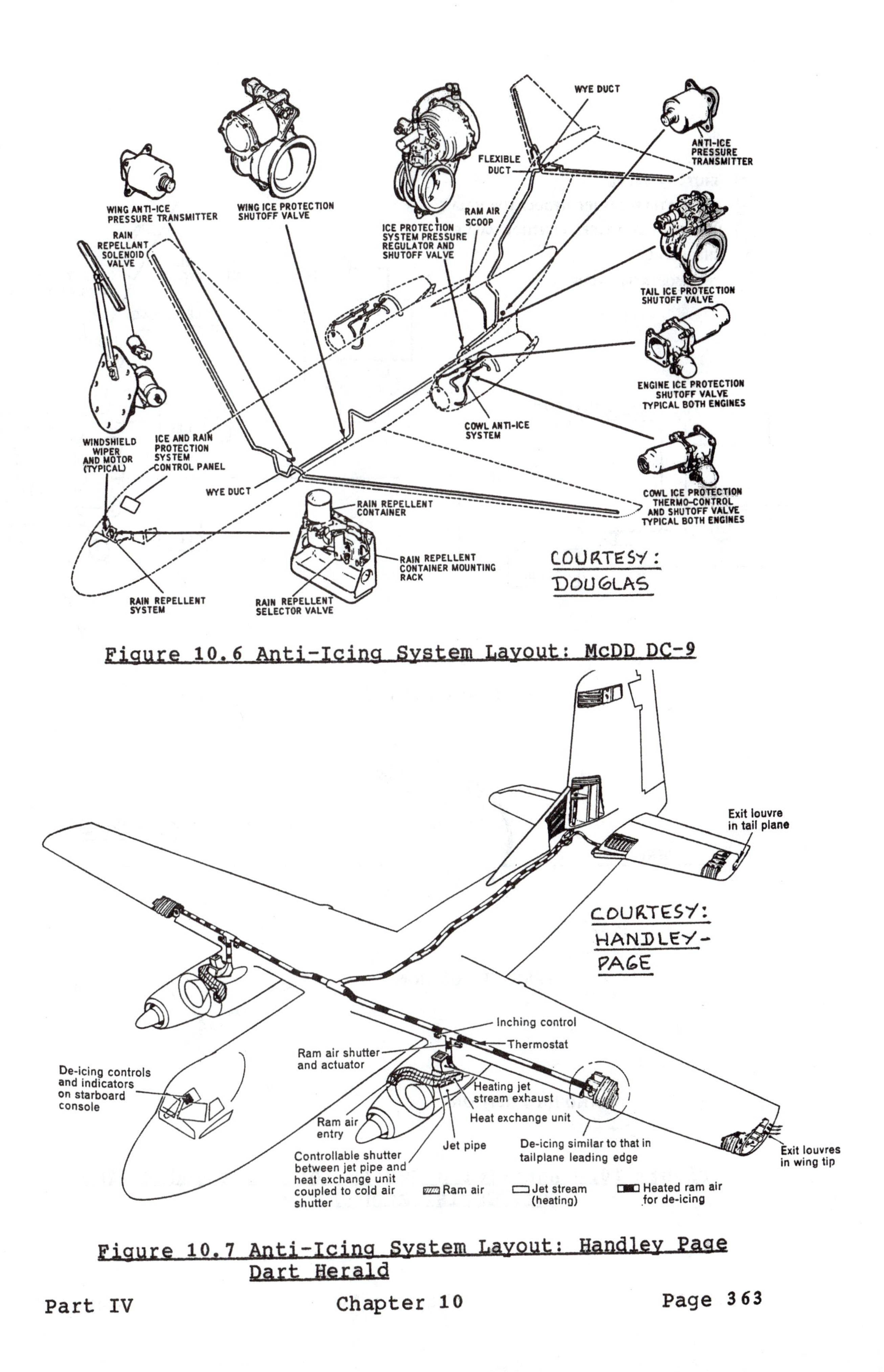

Figure 10.6 Anti-Icing System Layout: McDD DC-9

Figure 10.7 Anti-Icing System Layout: Handley Page Dart Herald

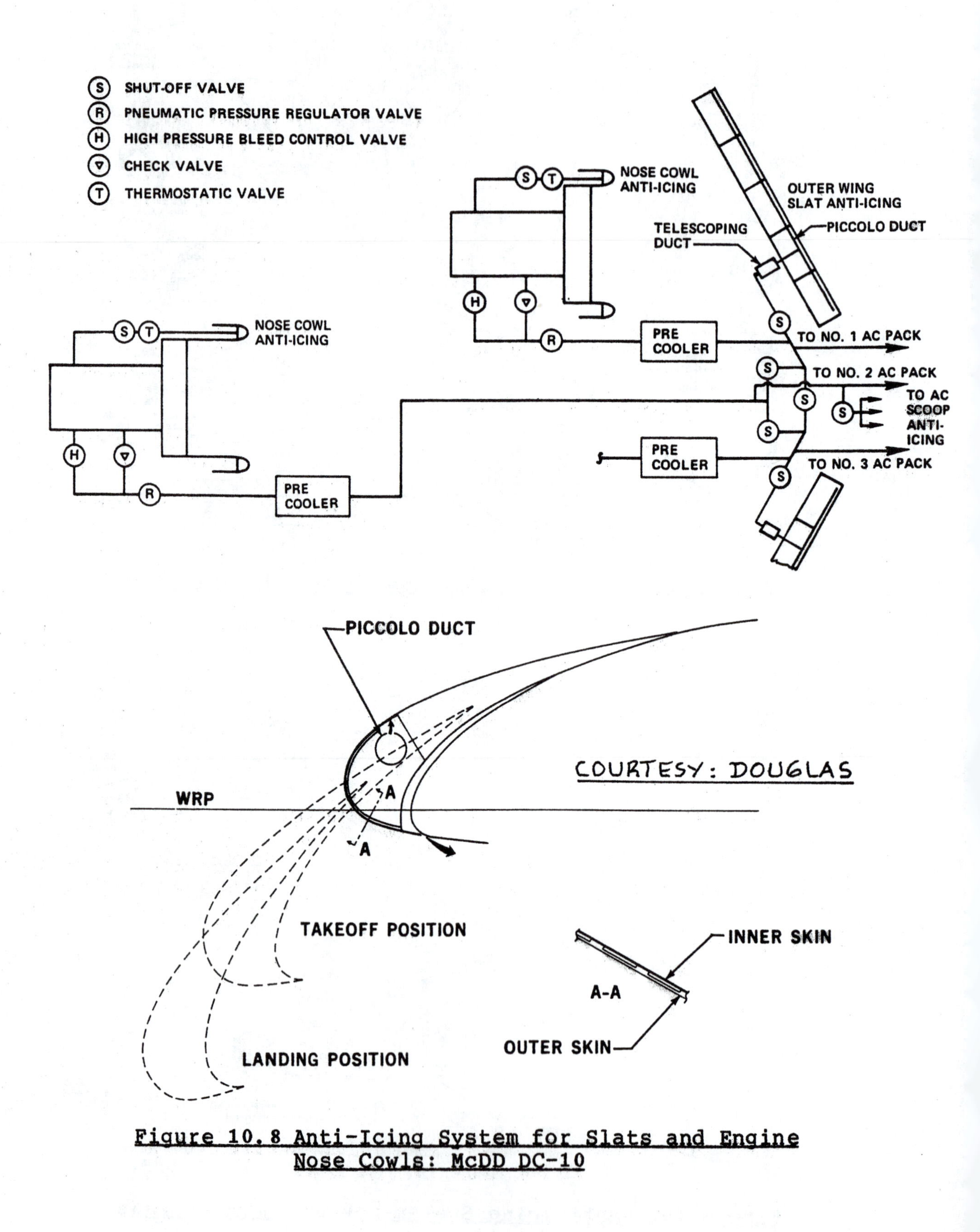

Figure 10.8 Anti-Icing System for Slats and Engine Nose Cowls: McDD DC-10

Electric Heat Anti-Ice

Electric heater icing protection is provided for the angle of attack sensors, the total air temperature sensors, the pitot-static probes, and the drain masts. The two forward flight compartment windshields are electrically heated to antifog those surfaces.

Rain Removal

Dual speed electrically operated windshield wipers are provided for the two forward windshields. A rain repellent solution is available to independently apply fluid to each of the two forward windshields.

COURTESY: BOEING

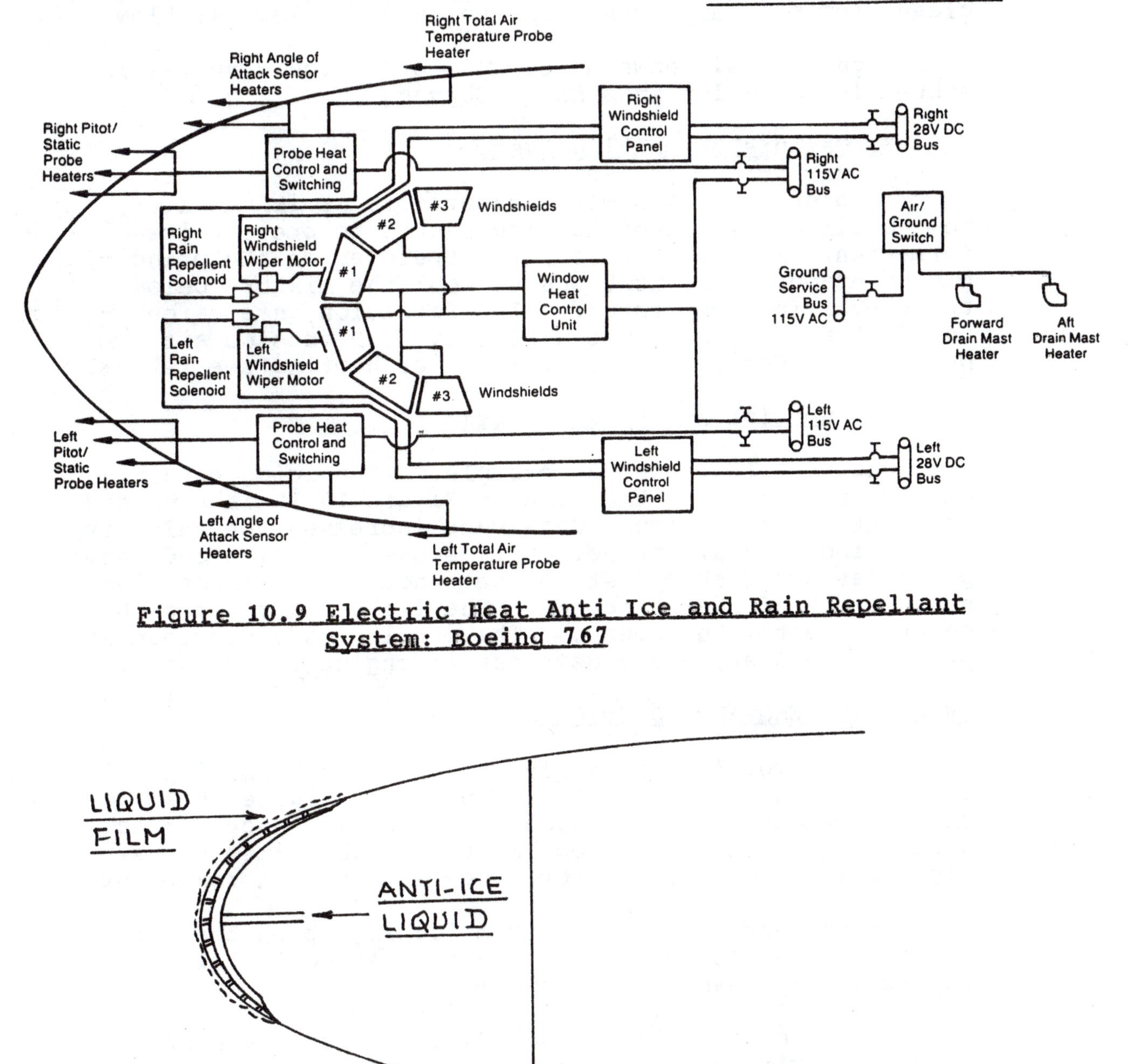

Figure 10.9 Electric Heat Anti Ice and Rain Repellant System: Boeing 767

Figure 10.10 Operation of Chemical Anti-Icing System

Figure 10.10 shows how this system works. A disadvantage of this system is that it operates from a finite supply of anti-freeze. A major design question is: how much liquid should be carried? An advantage of this system is that it can be used to keep the wing leading edge clean from bugs and thus helps to maintain laminar flow.

Figure 10.11 shows an example of a liquid anti-icing system installation in a Short Skyvan.

3. Carburetor Heating Systems

In piston engines with carburetors it is mandatory to install a carburetor heating system to prevent ice formation. Because a carburetor operates on the principle of evaporation, it tends to cool the mixture below the prevailing temperature. Ice formation in a carburetor is therefore possible at fairly high outside air temperatures. For details on carburetor heating see Ref.38.

4. Inertial Anti-icer Systems

Figure 10.12 shows an example. When icing conditions are encountered, the vane in Figure 10.12 is extended into the inlet airflow. This vane increases the velocity of the incoming air by Bernoulli's Law. Any ice and snow particles are rushed past the entrance to the inlet plenum because of their inertia. The lighter air turns the corner to enter the plenum. The penalty is a decrease in pressure recovery and a decrease in engine performance.

10.2 RAIN REMOVAL AND DEFOG SYSTEMS

Rain removal is normally accomplished with windshield wipers similar to those in cars. Figure 10.13 shows an example installation. In most instances it is necessary to add a rain repellant into the wiper paths. Fig.10.14 shows a system for dispersing a rain repellant.

To prevent windshields from fogging up on the inside and/or on the outside a defog system can be installed. Figure 10.15 shows such a system.

Defogging can also be achieved with the help of electrical wiring buried inside the windshield material.

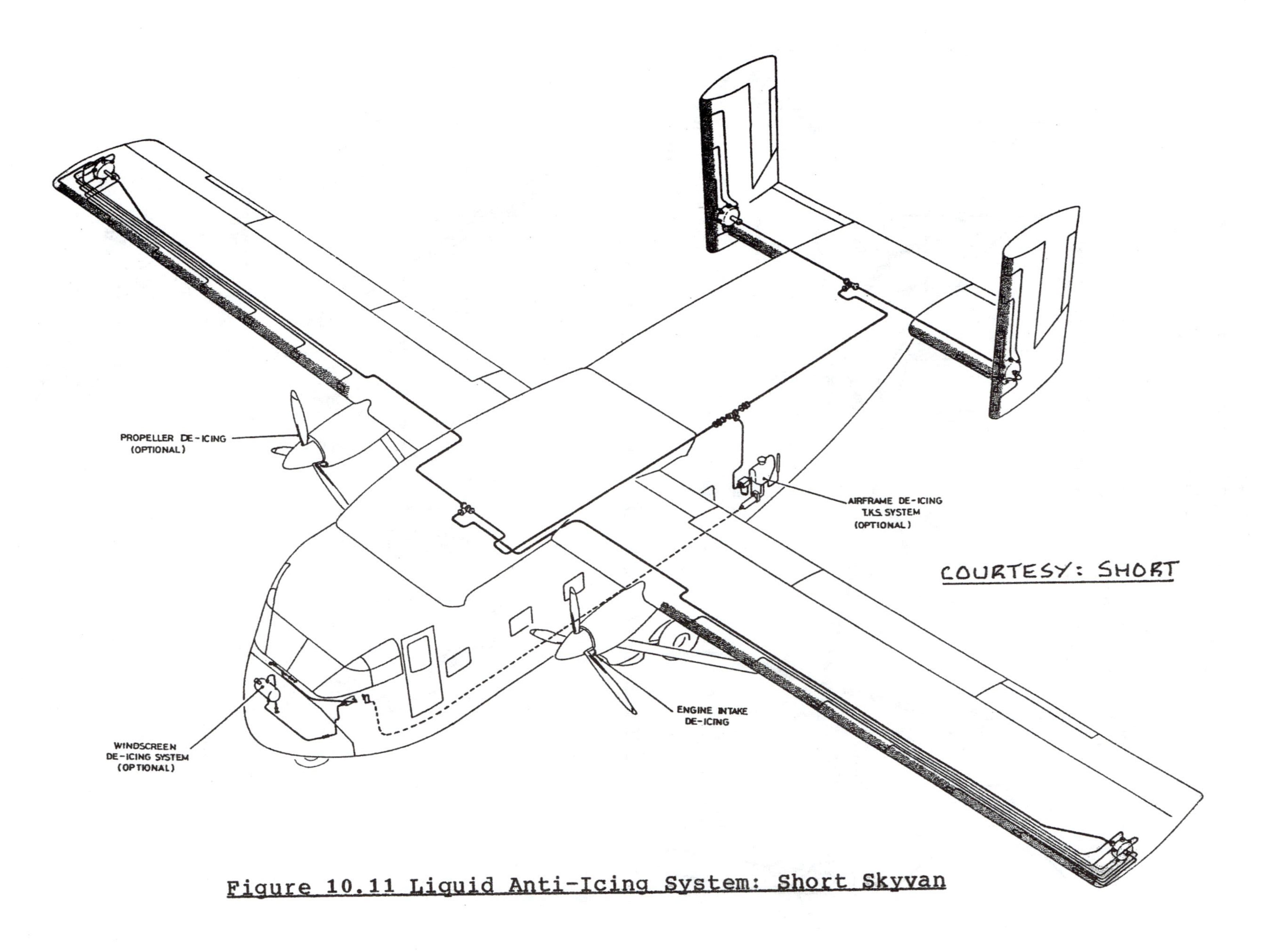

Figure 10.11 Liquid Anti-Icing System: Short Skyvan

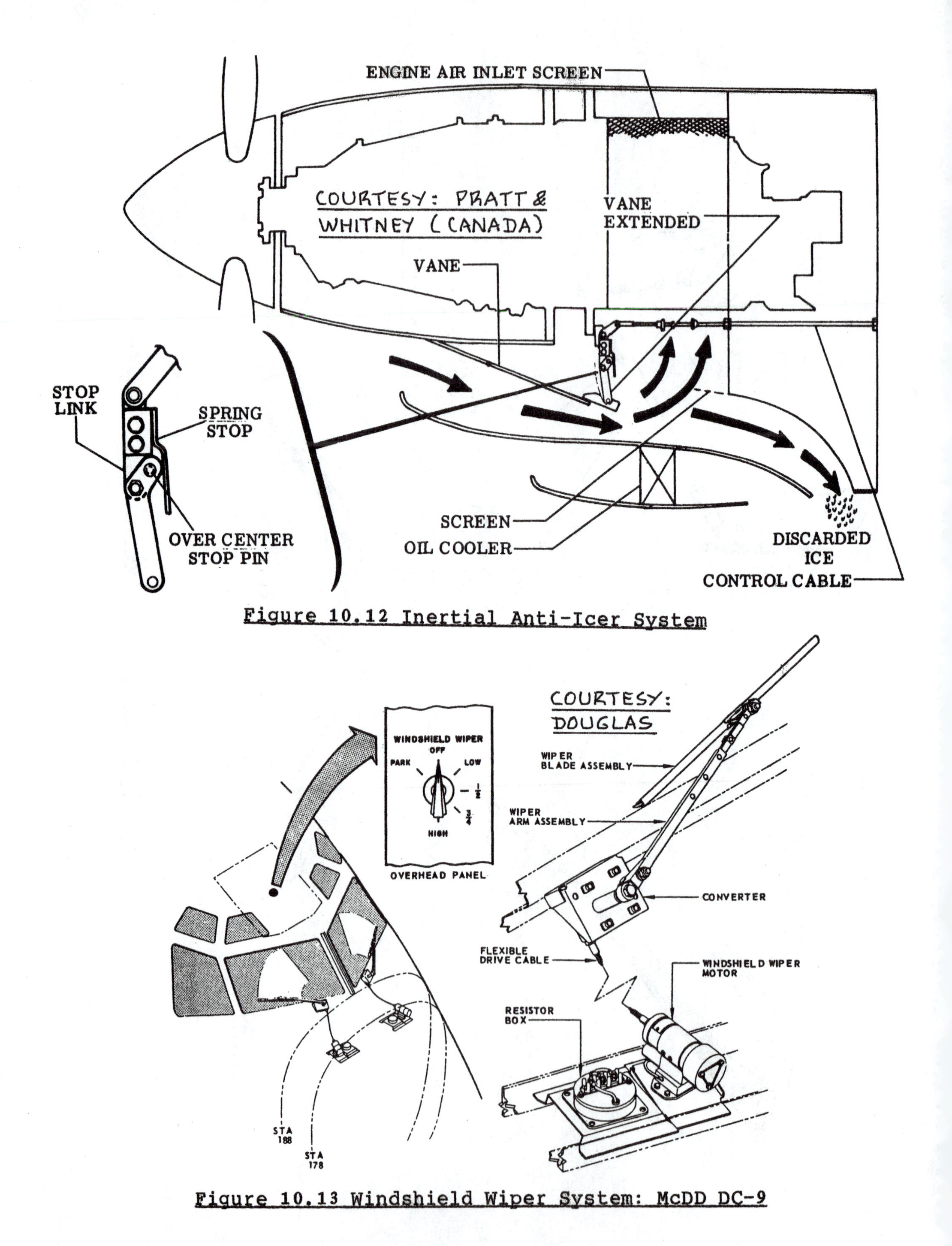

Figure 10.12 Inertial Anti-Icer System

Figure 10.13 Windshield Wiper System: McDD DC-9

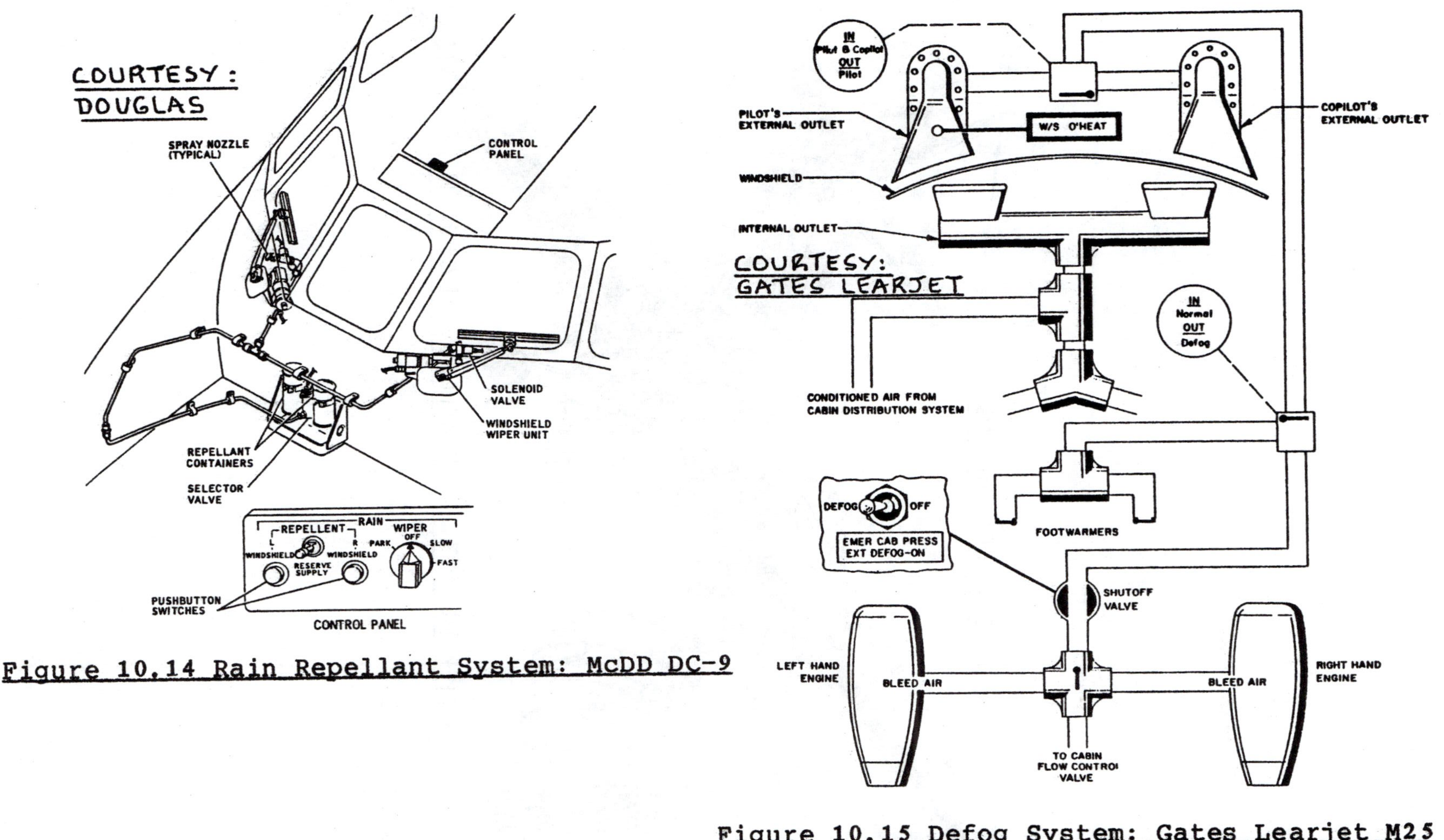

Figure 10.14 Rain Repellant System: McDD DC-9

Figure 10.15 Defog System: Gates Learjet M25

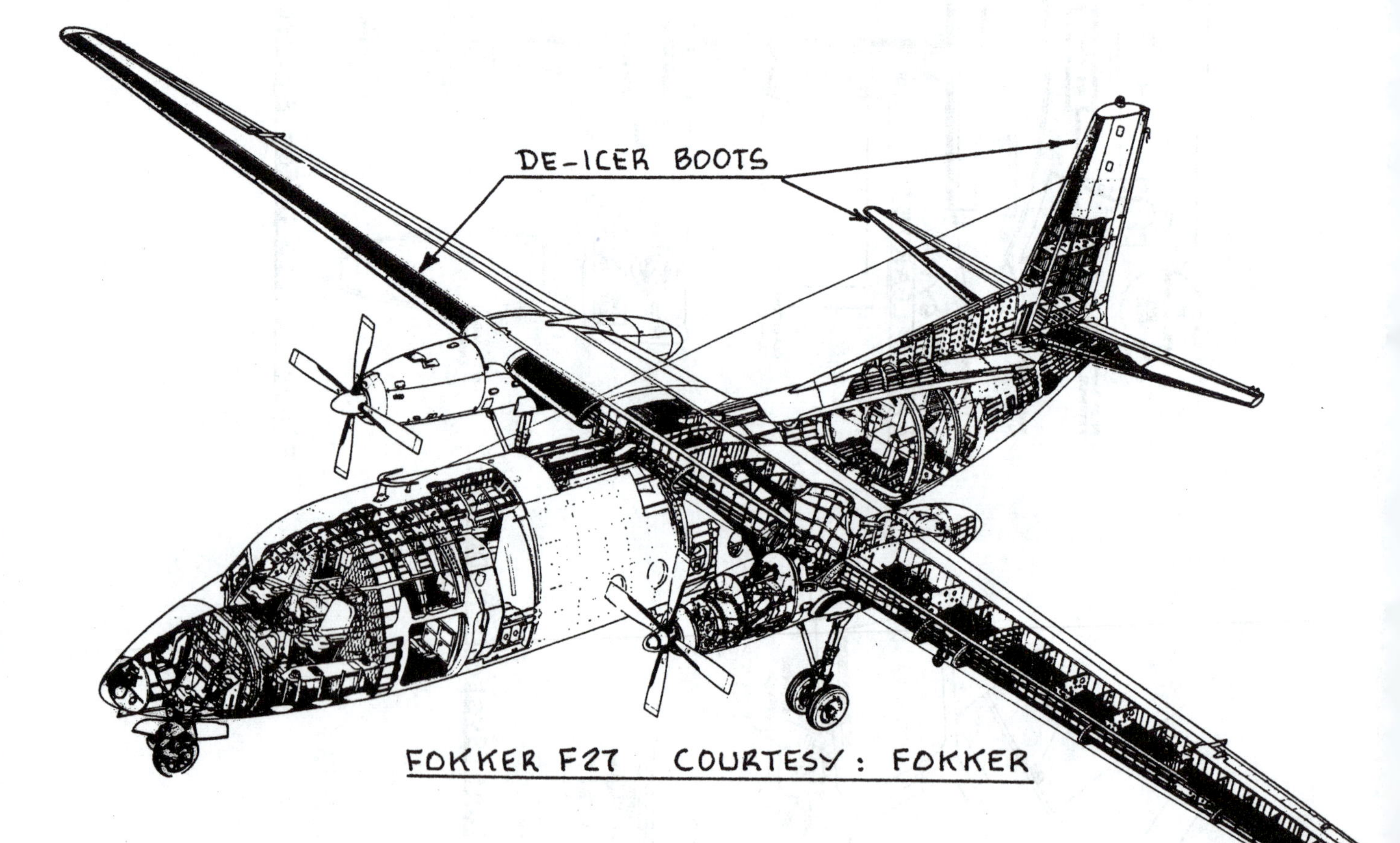

FOKKER F27 COURTESY: FOKKER

11. ESCAPE SYSTEM LAYOUT DESIGN

The purpose of this chapter is to discuss the layout design of emergency exits and escape systems for commercial and for military airplanes.

The material is organized as:

11.1 Emergency exits and escape systems for commercial airplanes.

11.2 Emergency exits and escape systems for military airplanes.

11.1 EMERGENCY EXITS AND ESCAPE SYSTEMS FOR COMMERCIAL AIRPLANES

All passenger airplanes must have reasonable means for passenger egress in emergency situations. Ref.21 contains the pertinent regulations for the determination of size and number of exits required. Ch.3 in Part III (pages 68-72) provides design guidelines for the number, size and type of emergency exits required for FAR 25 certified airplanes.

All exits must be properly marked. They must also be equipped with self illuminating signs. The operation of each emergency exit must be prominently displayed on the exit. Figure 11.1 shows an example escape hatch.

All passenger transports must be equipped with life jackets. These life jackets must be within easy reach of each passenger. For that reason these jackets are normally installed under the seat.

Figure 11.2 shows the emergency system layout of the BAC 111. Note the break-in areas which are marked on the airplane. This is a requirement on military airplanes, not on commercial airplanes.

Large, low-wing commercial transports which sit high on the landing gear must be equipped with escape hatches and escape slides. To prevent passengers from slipping on a smooth wing surface, that part of the wing which can be expected to be used as an escape route must be covered with an anti-skid material. Note that this makes laminar flow over this part of the wing impossible! Figure 11.3 shows an example situation.

Figure 11.4 illustrates the overall deployment of

emergency slides in a Boeing 767.

For overwater flights (in excess of 30 minutes) passenger transports must carry emergency rafts. Fig.11.5 shows an example of a design where the fore and aft emergency slides are in fact configured as rafts. There must be sufficient space in the rafts to carry all passengers and crew members. Figure 11.5 also shows the additional equipment which rafts must include.

11.2 EMERGENCY EXITS AND ESCAPE SYSTEMS FOR MILITARY AIRPLANES

In military transports the requirements for emergency exits are similar to those for commercial transports. In addition, break-in areas must me prominently marked on the outside of the airplane.

Many military airplanes must be equipped with ejection systems. All modern fighters must be equipped with so-called zero-zero ejection seats. Such seats allow the pilot to eject from an airplane which is at standstill on the ground and survive the process.

An ejection seat would be useless if the canopy were not automatically ejected or if the seat could not be propelled straight through the canopy. Many fighters are equipped with systems which have both capabilities.

Figure 11.6 shows two examples of ejection seats. For ejection seat dimensions and ejection seat clearance requirements the reader should see pgs 20-22 of Chapter 2 of Part III.

An example of a detailed ejection seat plus canopy installation is given in Figures 11.7 and 11.8. Note in Figure 11.7 the underwater jettison relief valve. This system allows the pilot to open the canopy under water after allowing the cockpit to flood.

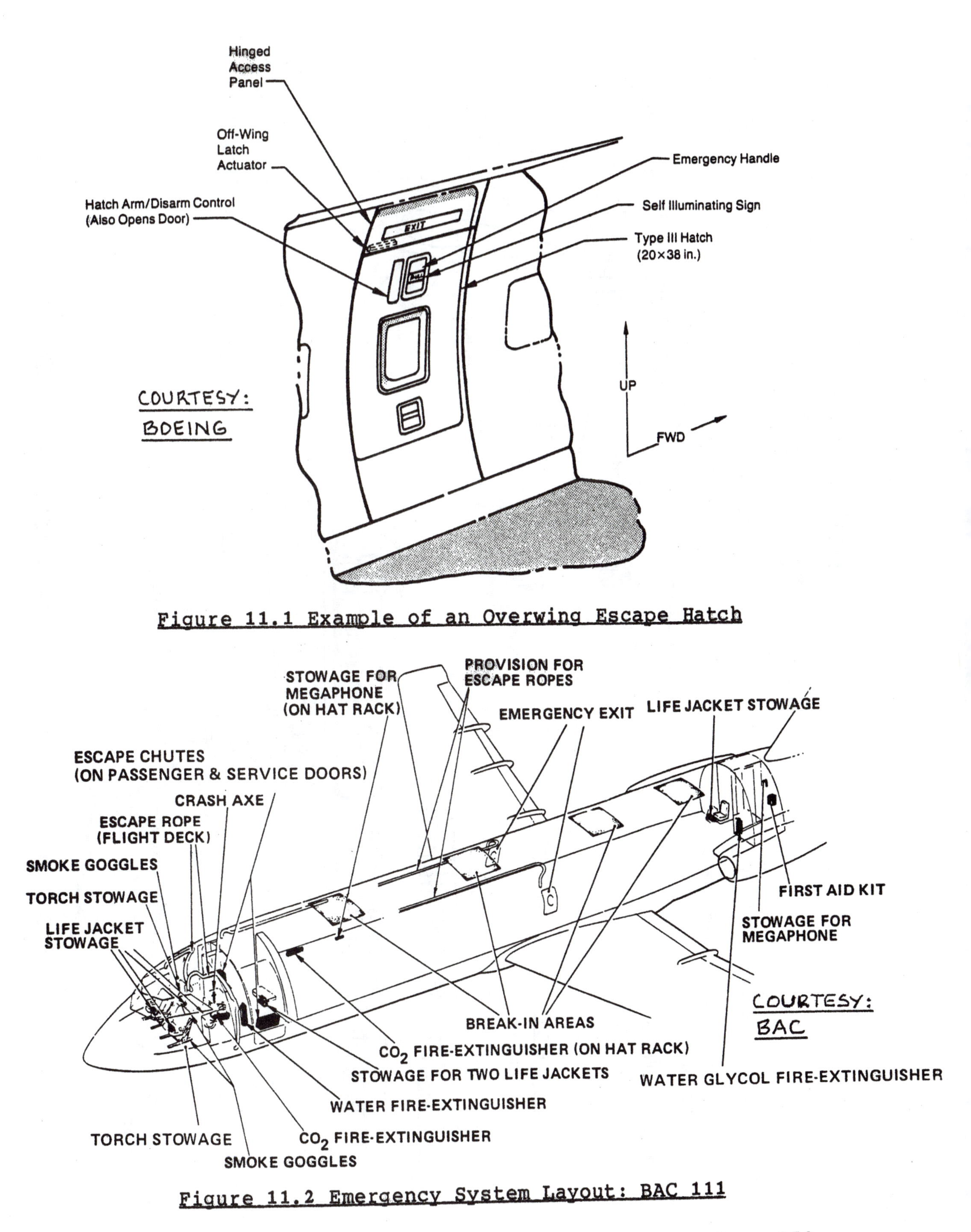

Figure 11.1 Example of an Overwing Escape Hatch

Figure 11.2 Emergency System Layout: BAC 111

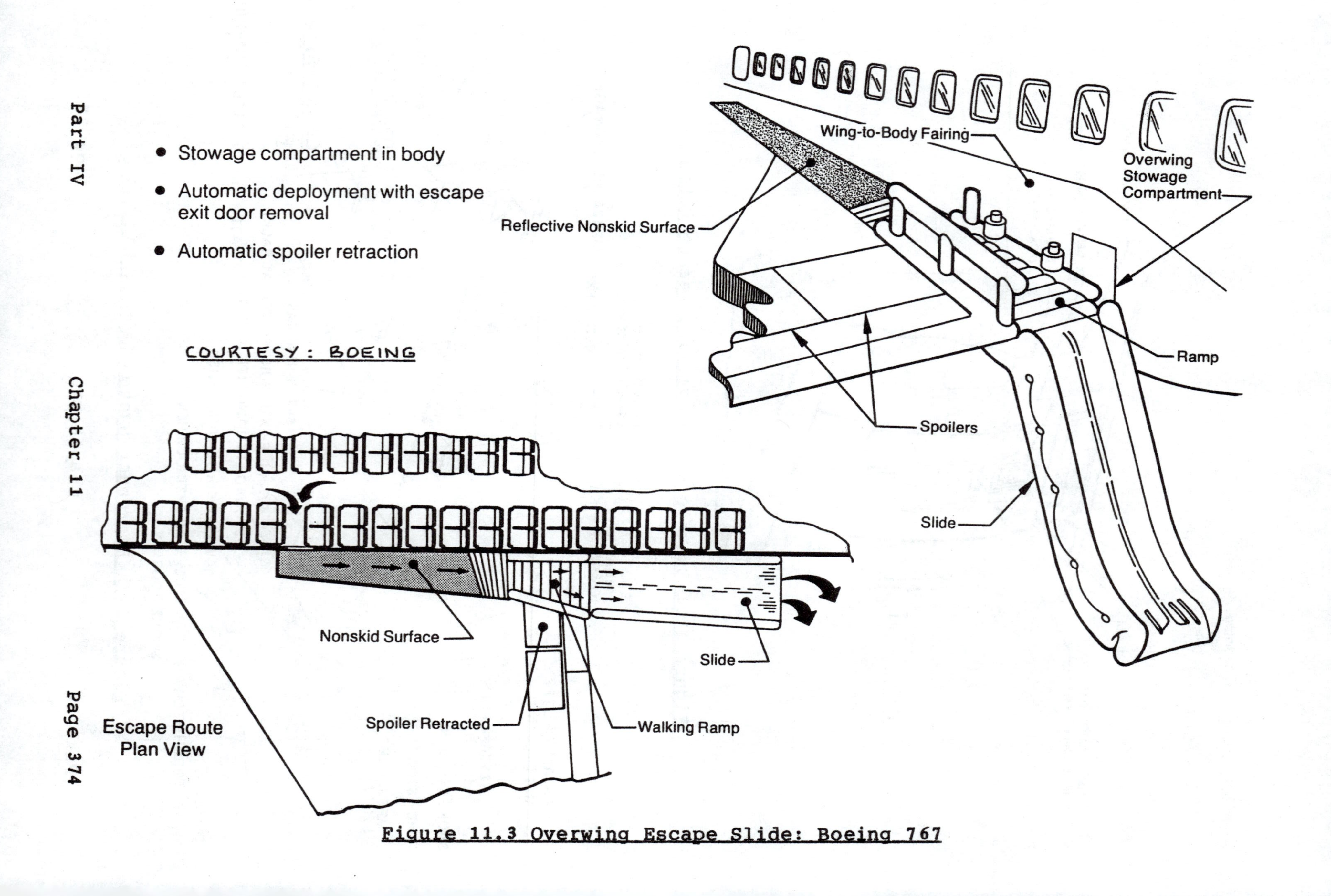

Figure 11.3 Overwing Escape Slide: Boeing 767

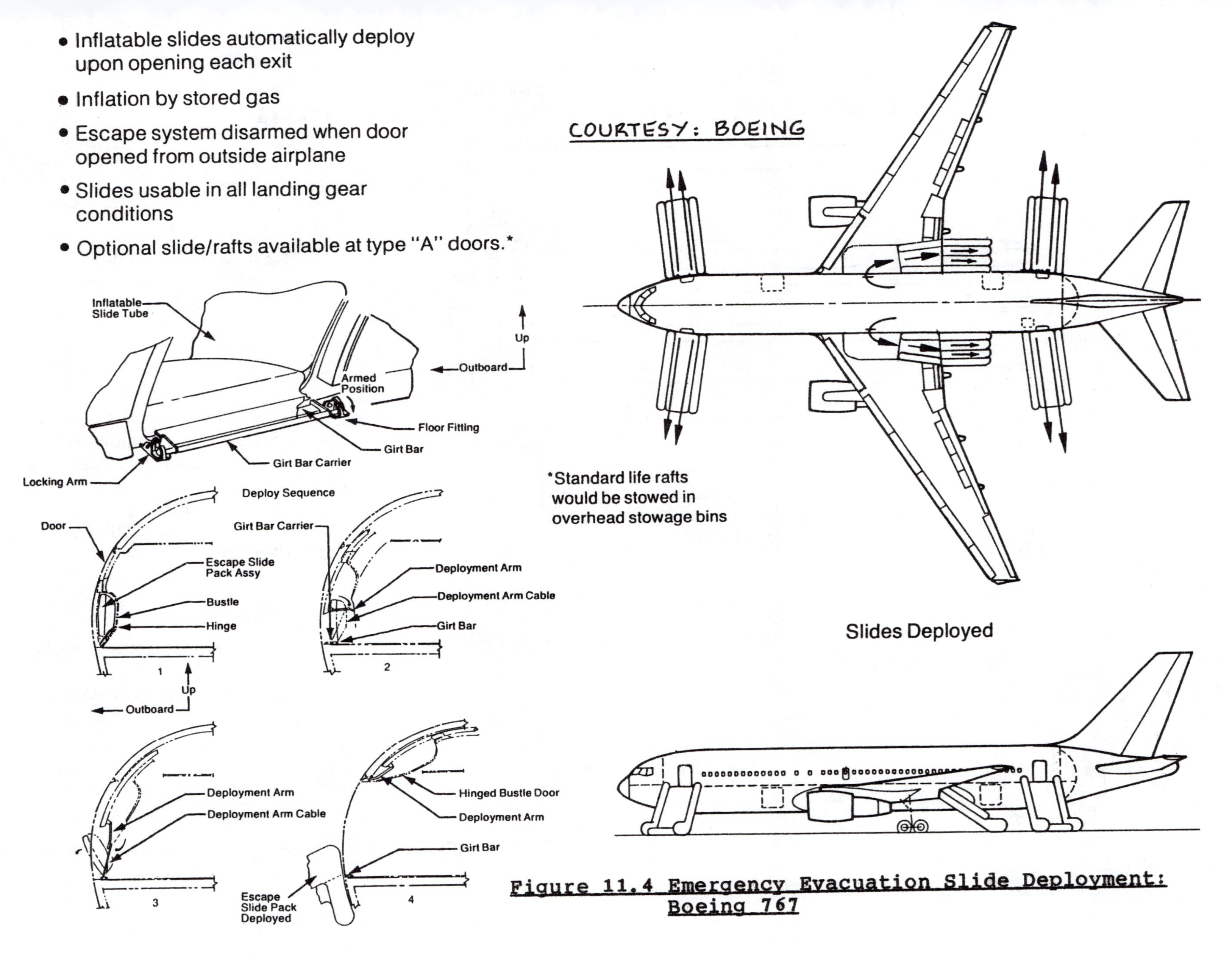

Figure 11.4 Emergency Evacuation Slide Deployment: Boeing 767

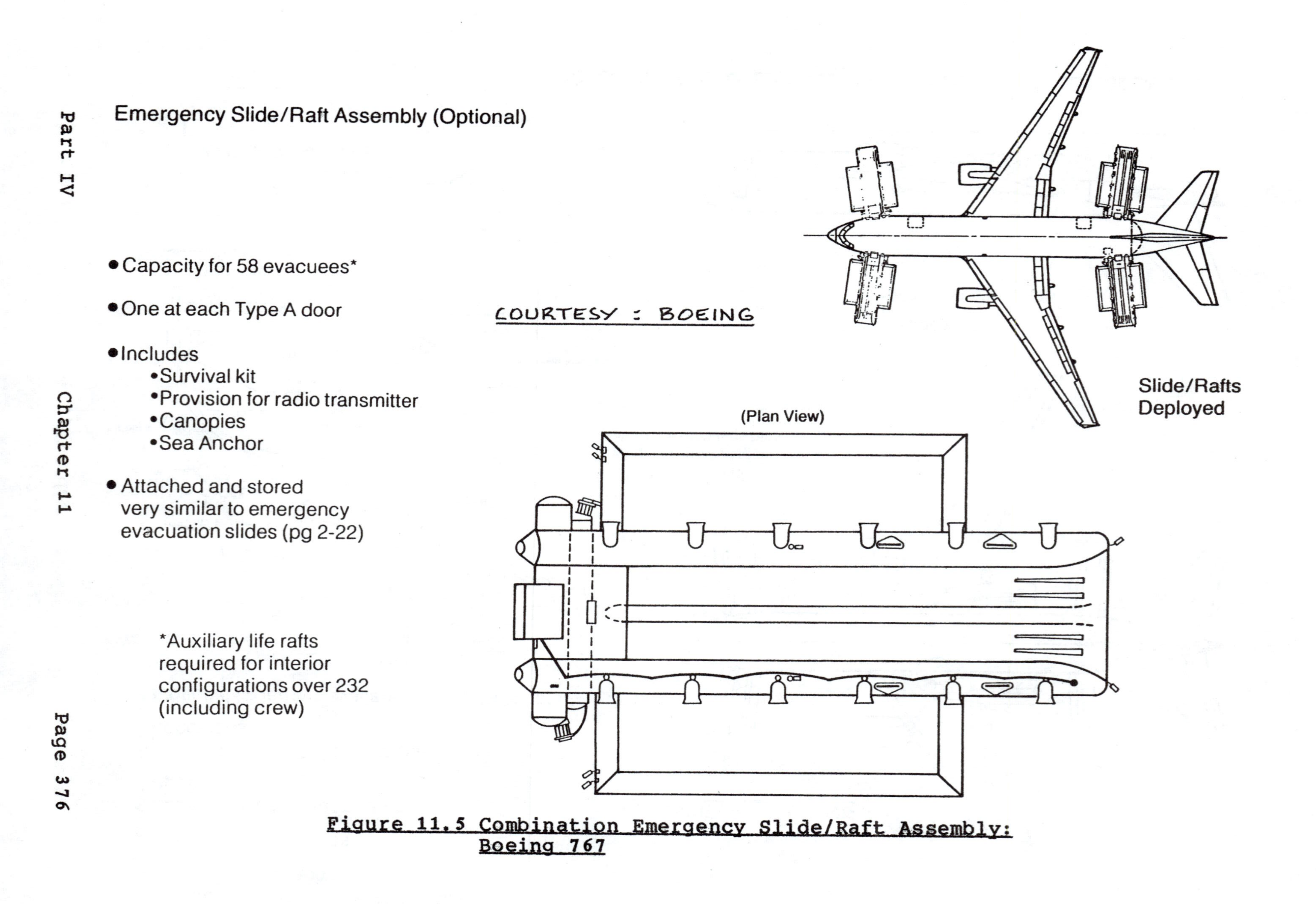

Figure 11.5 Combination Emergency Slide/Raft Assembly: Boeing 767

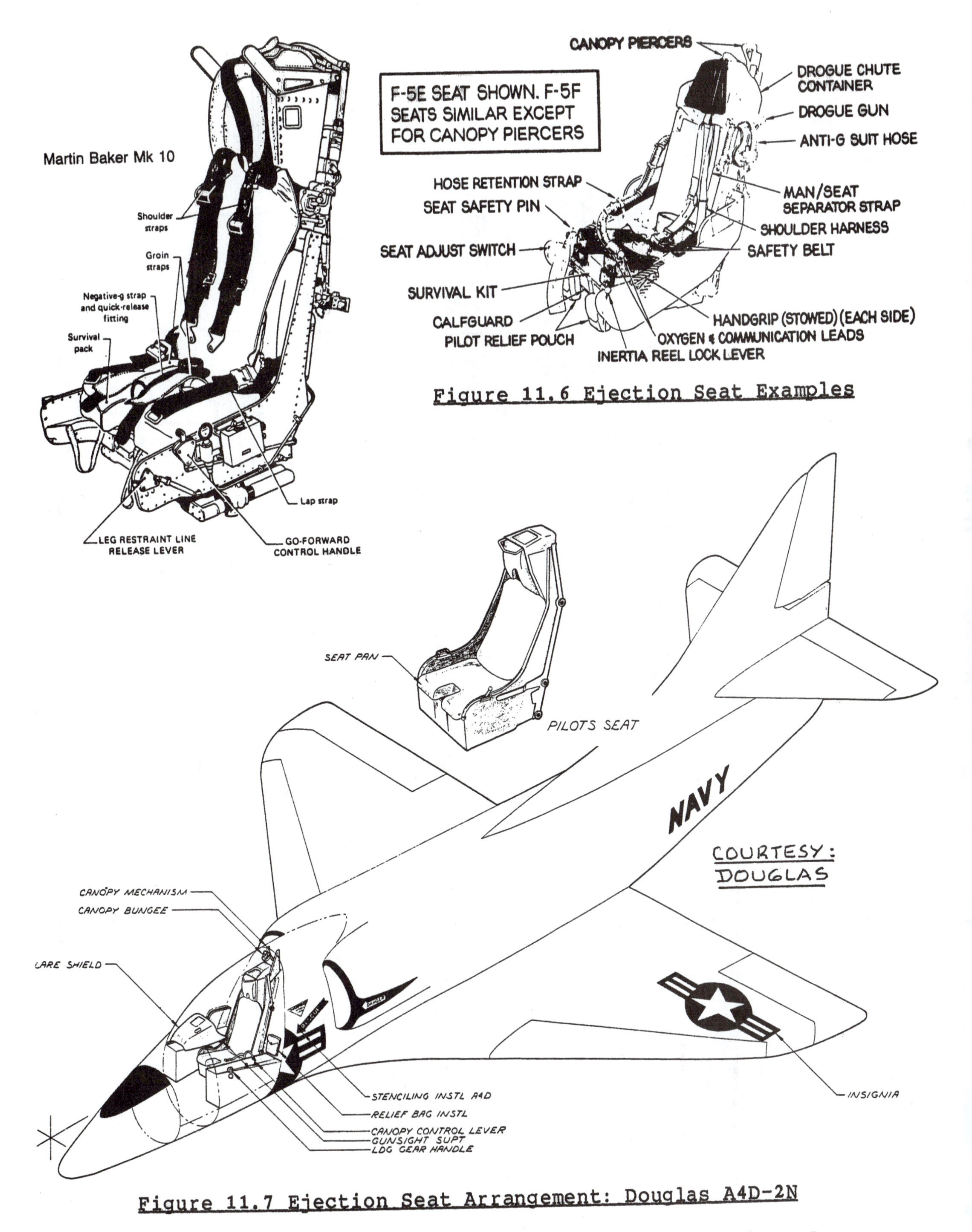

Figure 11.6 Ejection Seat Examples

Figure 11.7 Ejection Seat Arrangement: Douglas A4D-2N

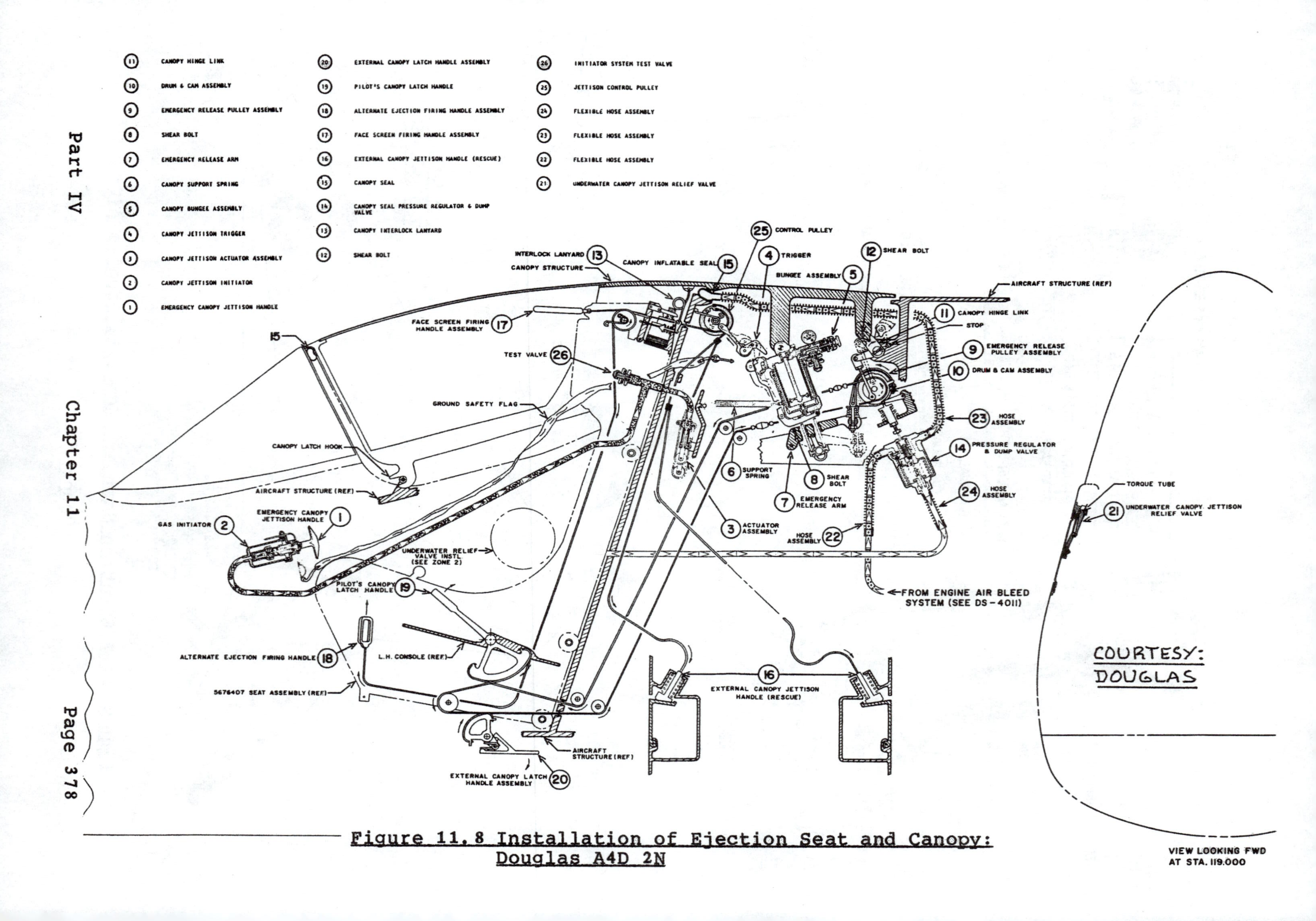

Figure 11.8 Installation of Ejection Seat and Canopy: Douglas A4D 2N

12. LAYOUT DESIGN OF WATER AND WASTE SYSTEMS

Nearly all passenger transports are equipped with water and waste systems. These systems represent a large investment in weight and volume. It is important to include these systems in preliminary design considerations. Section 12.1 presents the fundamental layout design considerations for such systems.

For use in fighting large fires (such as forest fires) airplanes are frequently used. The flying boat and/or the amphibious airplane is ideally suited for this purpose. Section 11.2 addresses the problem of laying out systems to rapidly take on water as well as to disperse the water over a fire.

12.1 WATER AND WASTE SYSTEMS

Reference 33 contains detailed descriptions of water and waste systems.

Water systems in passenger transports are typically sized for 0.3 US gallon per passenger. The system is usually pressurized with air from the airplane pneumatic system (discussed in Chapter 8). Figures 12.1 and 12.2 show potable water systems as used in jet transports.

Note the drain masts in these systems. These drain masts must be heated to prevent freezing. An item of concern in the location of these drain masts is what happens if the drain mast heating system fails. Large blobs of ice can then form. When these ice blobs break away from the airplane they should not be ingested by engines!

In most transports both warm and cold water is available. Warm water is supplied by running cold water through an electrically heated heat exchanger.

Waste systems in passenger transports are self-contained: these systems have waste tanks (collector tanks) and flushing units which mix the waste with chemicals contained in the flushing liquid.

The number of lavatories per passenger varies with the use of the airplane. On the average, 1 lavatory per 30 passengers is deemed sufficient.

Figures 12.3 and 12.4 show examples of waste systems in passenger transports.

Both water and waste systems need to be serviced after each flight. Access to the service panels for these systems should not interfere with access to other services required during loading and unloading of the airplane.

The service drain locations of both water and waste systems are of concern to preliminary designers: if failures occur in the drain valves so that fluids leak out, ice can form. When this ice breaks away from the airplane it should not be ingested by the engines!

12.2 WATER BOMBING SYSTEMS

The system selected for inclusion in this section is that of the Canadair CL215, an airplane specifically designed for the water bombing role.

Figure 12.5 shows the interior arrangement of this airplane. Note the two water tanks in the center of the fuselage.

Figure 12.6a illustrates how water is taken onboard: a bucket is rotated into the water while the airplane is skimming a lake (or another large body of water). When the tanks are filled, the bucket is retracted.

Figure 12.6b shows the water door mechanism. The water doors are opened over the fire as indicated by the door operating sequence in Figure 12.6c. The water system controls in the cockpit are shown in Figure 12.7.

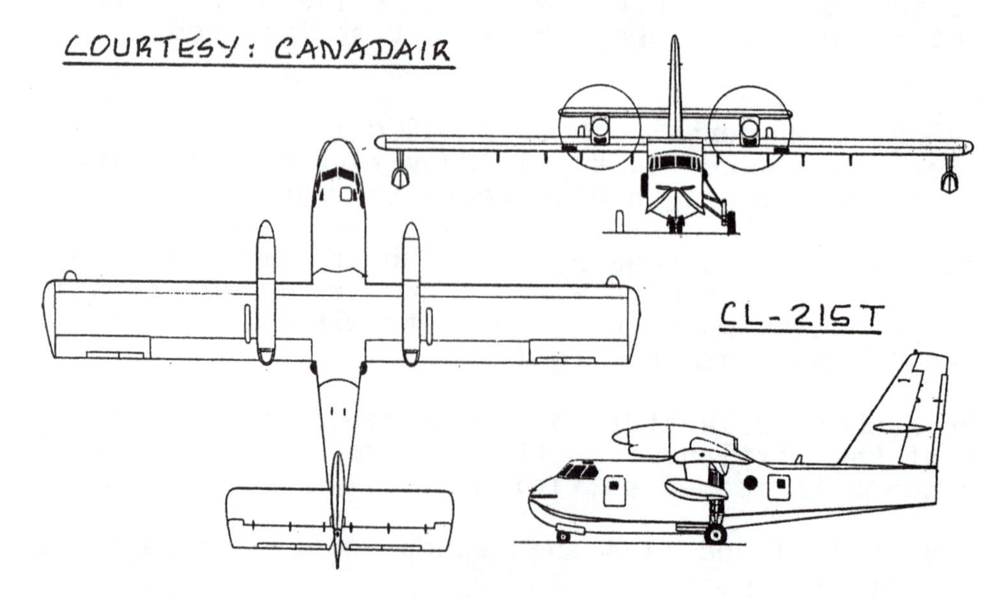

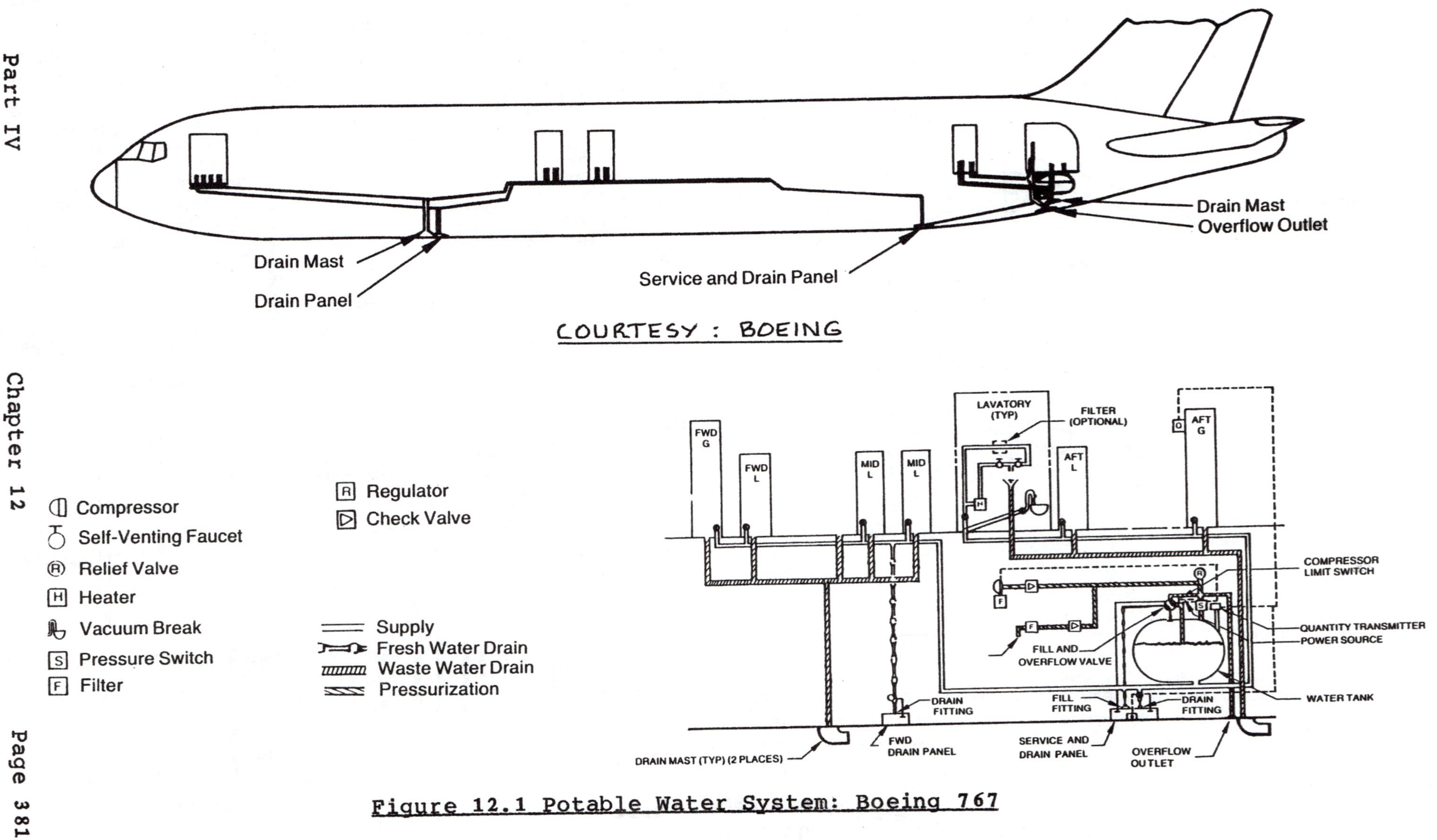

Figure 12.1 Potable Water System: Boeing 767

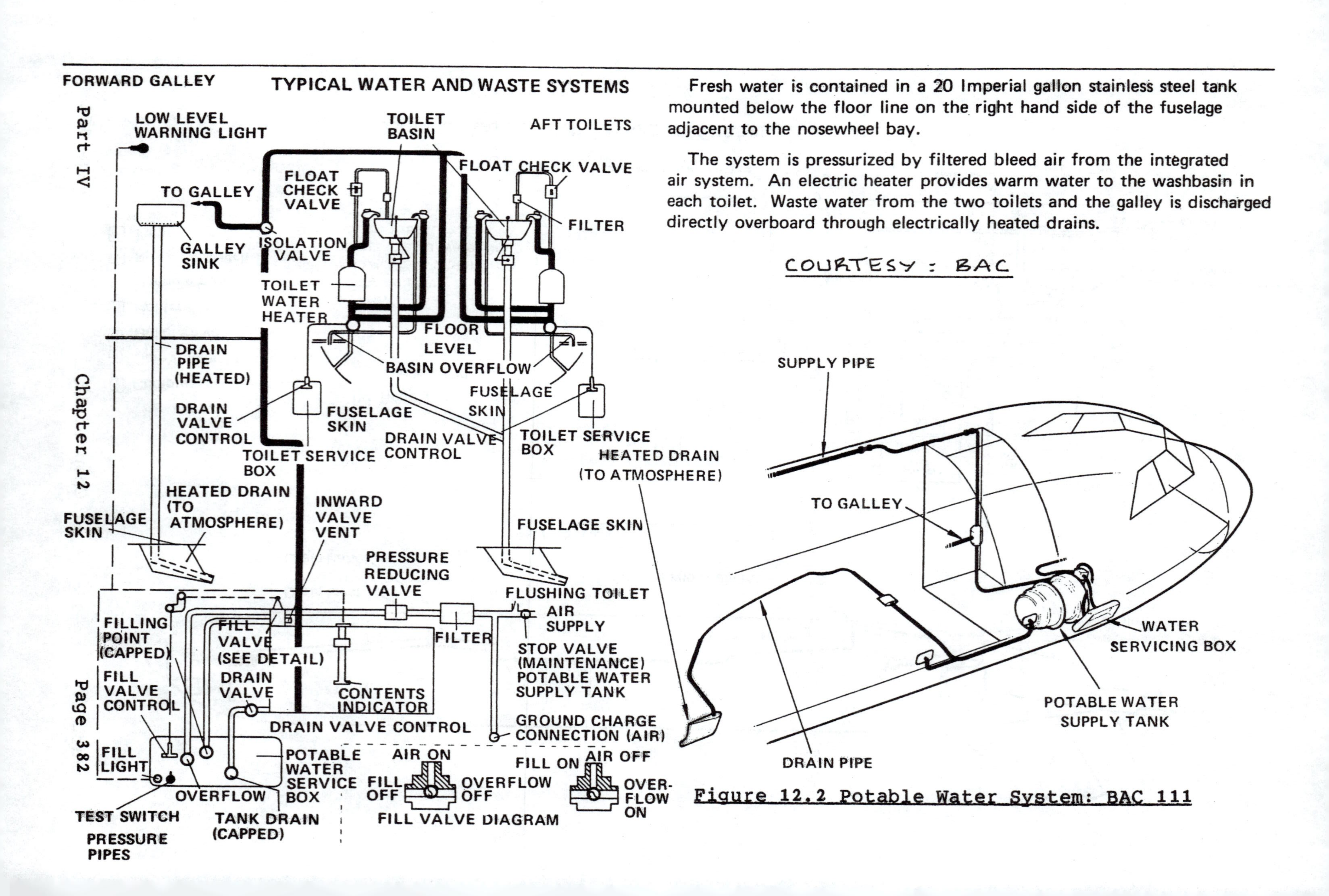

Fresh water is contained in a 20 Imperial gallon stainless steel tank mounted below the floor line on the right hand side of the fuselage adjacent to the nosewheel bay.

The system is pressurized by filtered bleed air from the integrated air system. An electric heater provides warm water to the washbasin in each toilet. Waste water from the two toilets and the galley is discharged directly overboard through electrically heated drains.

COURTESY : BAC

Figure 12.2 Potable Water System: BAC 111

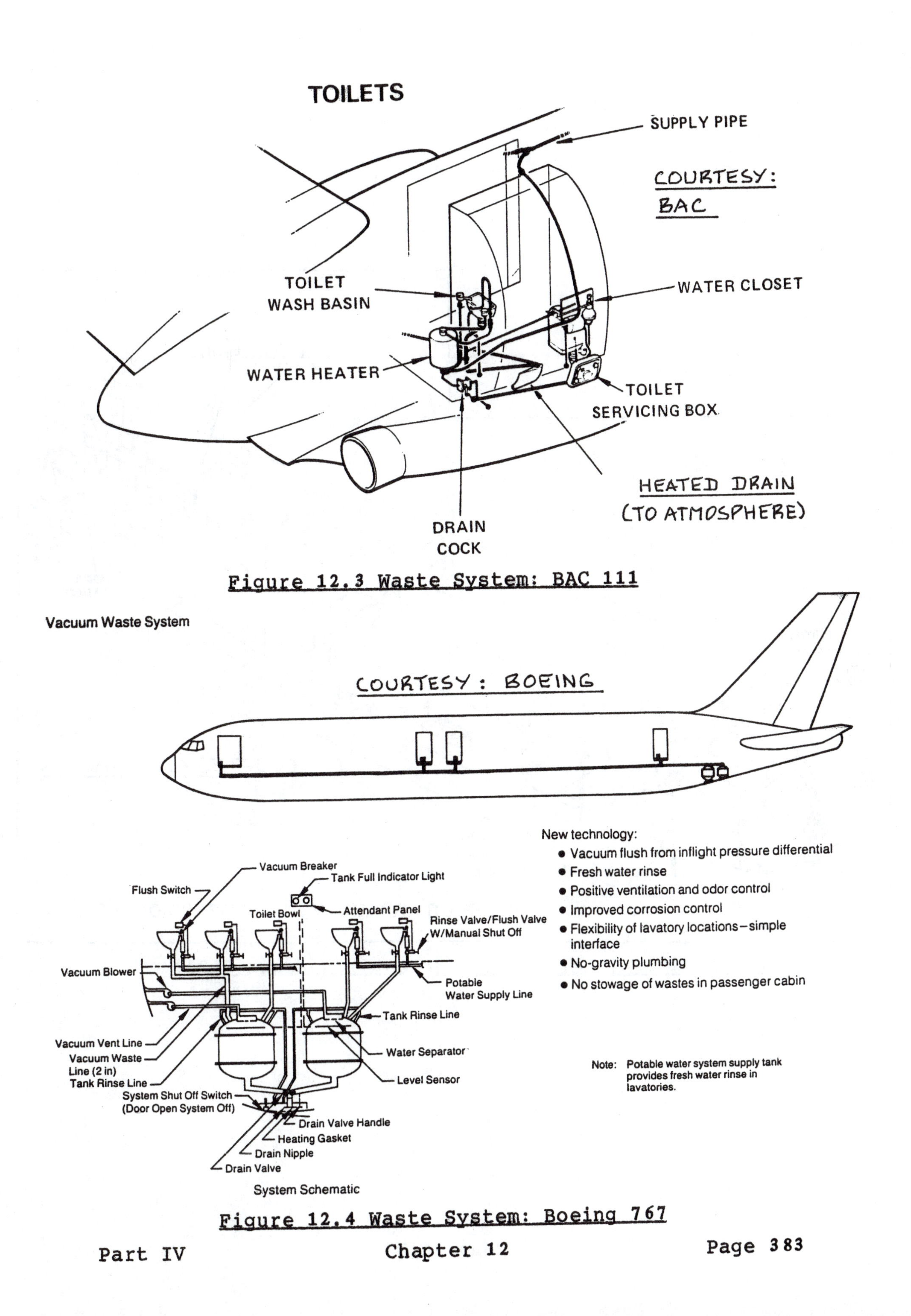

Figure 12.3 Waste System: BAC 111

Figure 12.4 Waste System: Boeing 767

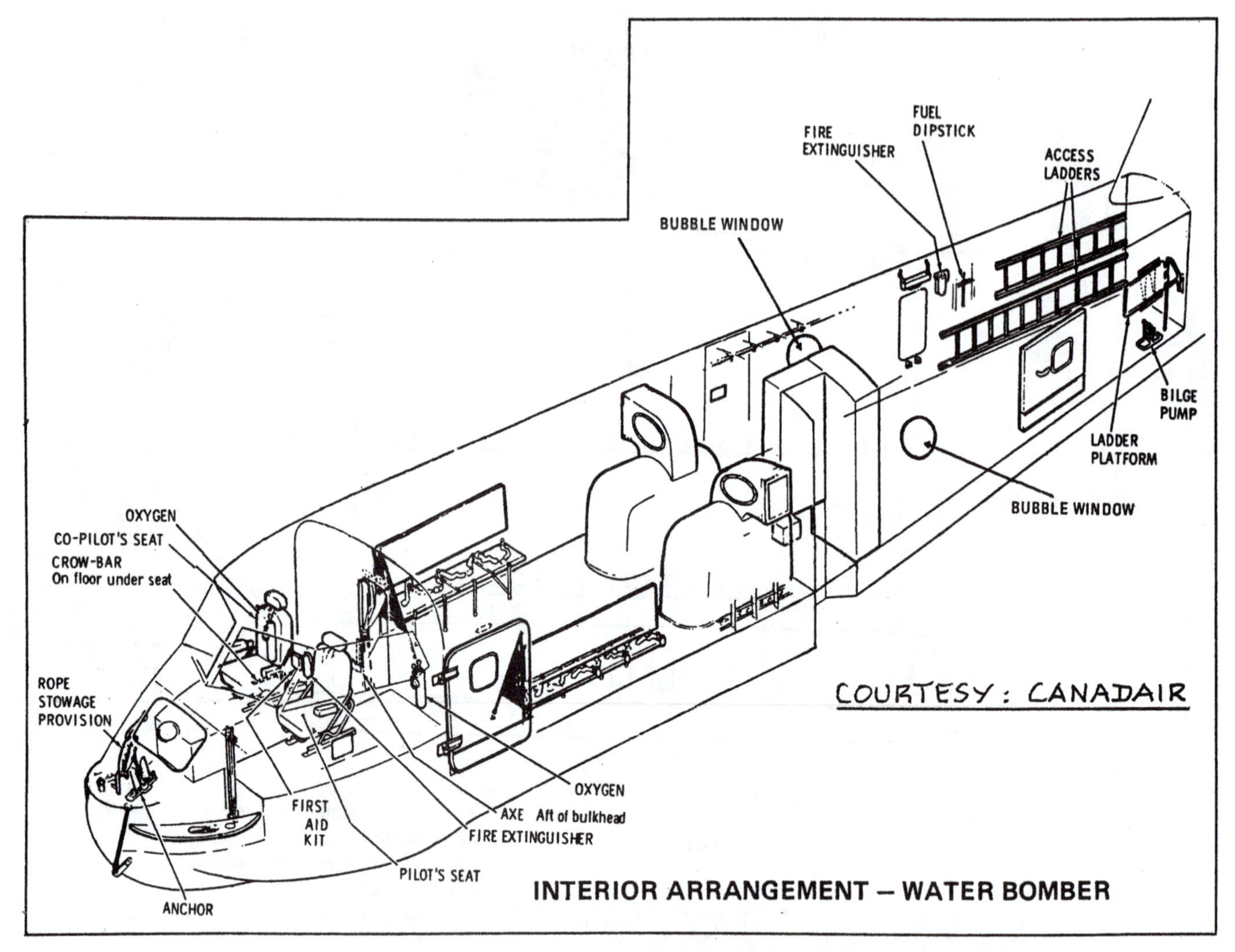

Figure 12.5 Interior Arrangement: Canadair CL215

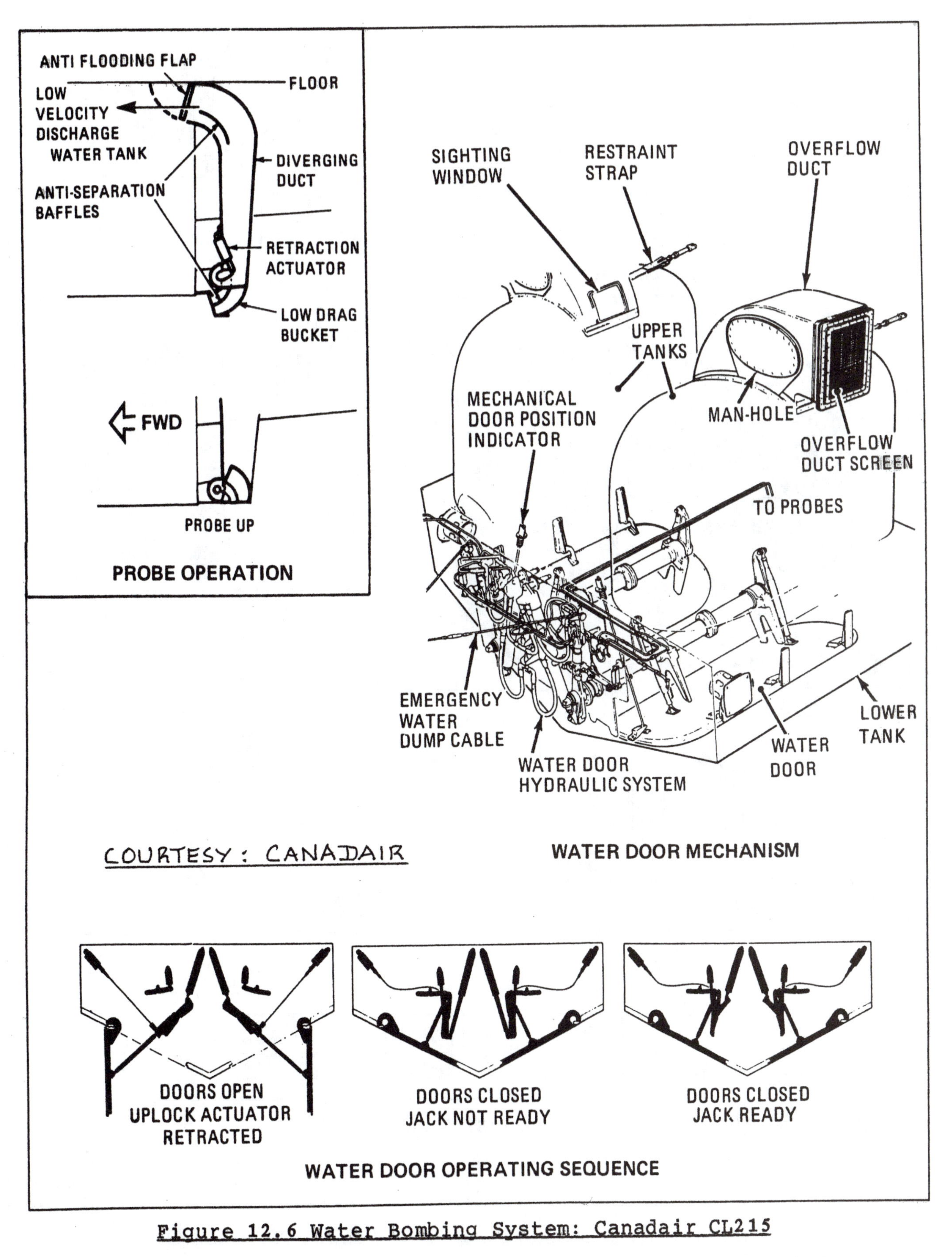

Figure 12.6 Water Bombing System: Canadair CL215

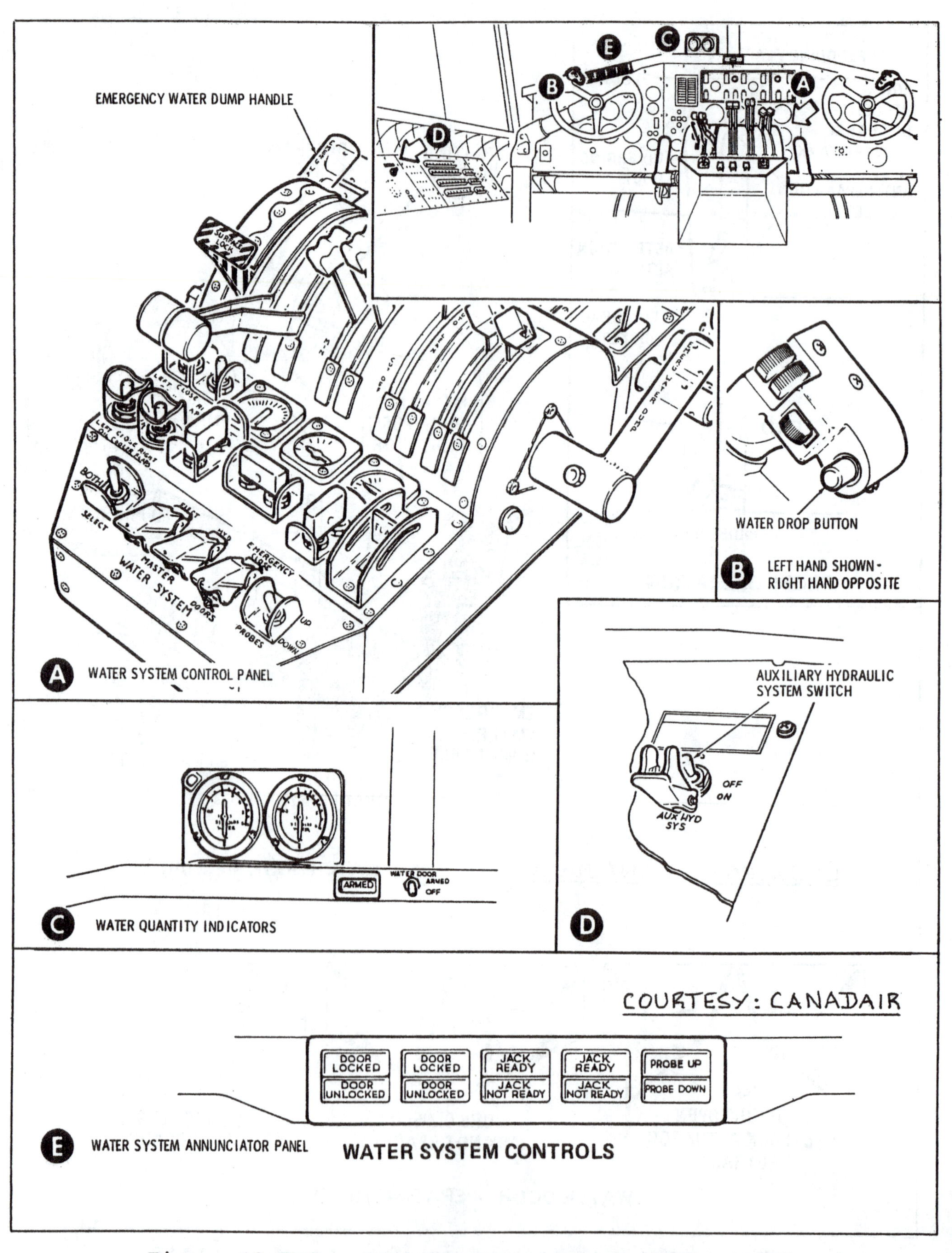

Figure 12.7 Water System Controls: Canadair CL215

13. SAFETY AND SURVIVABILITY

The purpose of this chapter is to provide some insight into design problems associated with safety and survivability considerations. These considerations are important both to commercial and to military airplanes.

The problems of safety and survivability are addressed as follows:

13.1 How safe is safe enough?

13.2 Safety and survivability in commercial airplanes

13.3 Safety and survivability in military airplanes

13.4 The role of the preliminary design engineer and management in creating safe airplanes

13.1 HOW SAFE IS SAFE ENOUGH?

Definitions: 1. An accident is defined as an occurrence which causes the death of at least one passenger or crew member.

2. An incident is defined as an occurrence which falls outside the normal operating events but which causes no fatalities. Non-fatal injuries may be incurred during incidents.

Lemma: It is IMPOSSIBLE to design airplanes such that the probability of fatal accidents is zero.

Accepting this lemma (which is presented here without proof), the question is: how many fatalities are acceptable?

This is a cruel question. The answer to this question will be different, depending on such matters as religion, morality and personal experiences. Neither of these matters are addressed in this text. Instead, the position will be taken that acceptable levels of safety arise as a result of societal trade-offs made between the number of fatalities and the cost incurred to lower them.

Figure 13.1 illustrates how safety (expressed as a relative accident rate) and costs are related. Aviation tends to operate to the left of the minimum cost level.

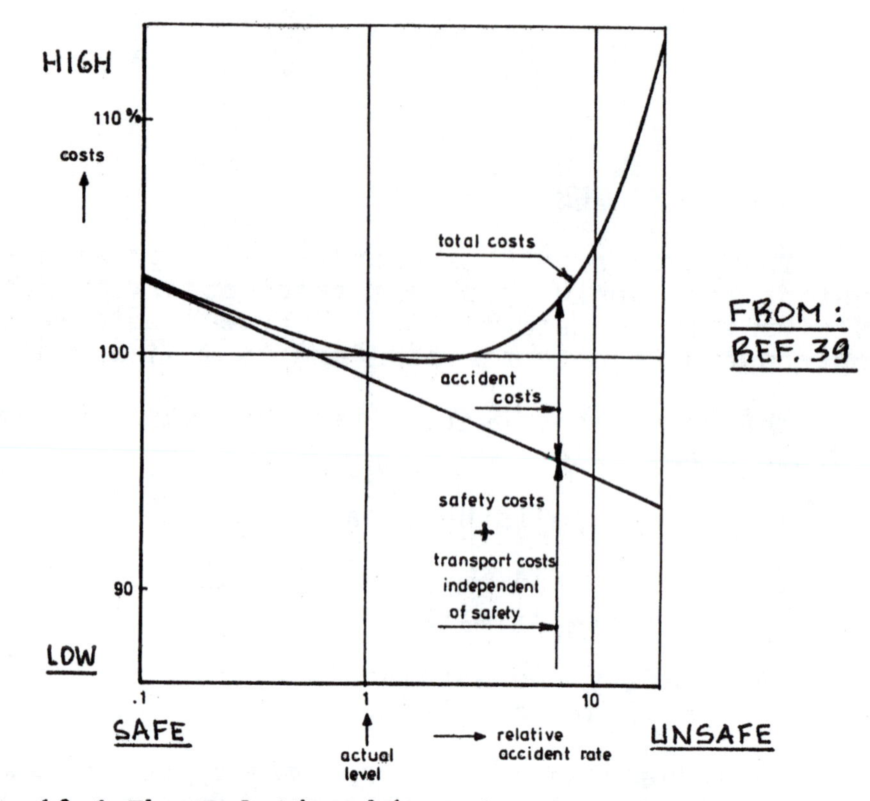

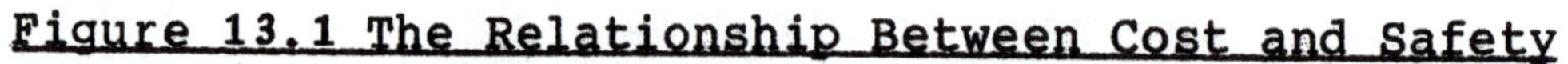

Figure 13.1 The Relationship Between Cost and Safety

Table 13.1 Safety Comparisons for Several Modes of Transportation

Transport mode		Average speed (km/h)	Number of fatalities *) per 10^9 pass.km	Number of fatalities *) per 10^7 pass.hrs
public transport	European Railways, 1968-1975 (Union Int. des Chemin de Fer)	80	.4	.32
	ICAO-world-scheduled air services (excl. USSR, China), 1975-1977	590	.9	5.3
private transport	Road traffic, the Netherlands, 1976 - private car - autocycle - motorcycle/scooter	 70 20 60	 10 92 150	 7 18 90
	USA general aviation, 1975-1977	280	68	190

*) For private transport: driver/pilot included

FROM : REF. 39

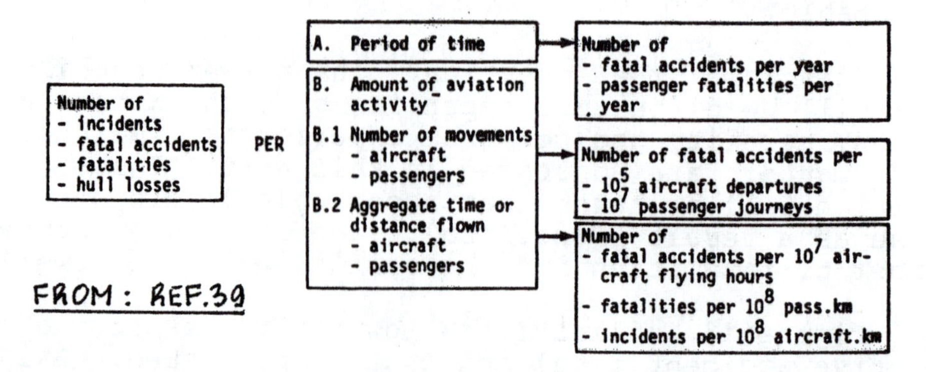

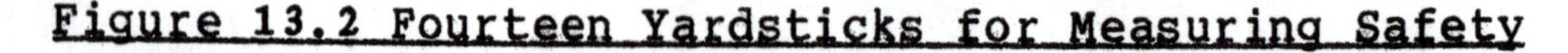

Figure 13.2 Fourteen Yardsticks for Measuring Safety

It is of interest to compare the safety levels associated with various modes of transportation. Table 13.1 presents this comparison. Observe that there are two ways of presenting relative safety:

1. fatalities per passenger mile (or km)

2. fatalities per passenger hour

Note that different conclusions are arrived at depending on which method of comparison is used:

I. on the basis of distance travelled, the safety record of aviation equals that of railway transportation. Note the poor safety record of cars and motorcycles.

II. on the basis of time spent, the safety record of aviation is only slightly better than that of the car. The safety record of railway transportation is an order of magnitude better than that of aviation.

At this point it is useful to point out how the safety yardsticks just mentioned are derived from statistical data on aviation accidents. The derivation has been taken from Reference 39.

The follwowing definitions are required:

P = the number of passenger miles (or km) flown
U = the number of airplane flight hours produced
S = the number of airplane miles (or km) flown
K = the number of passenger fatalities
R = the number of fatal accidents

The fatality rate per passenger mile (or km) is: K/P

The fatal accident rate per flight hour is: R/U

It is possible to write:

$$K/P = (R/U)(K/R)(U/P) =$$

$$= (R/U)\{(K/R)/(P/R)\}(U/S) \qquad (13.1)$$

Introduce the following quantities:

$k = K/R$ = the average number of fatalities per fatal accident

$p = P/S$ = the average number of passengers per airplane

V_B = the average block speed in mph (or km/hr)

Eqn.(13.1) can now be written as:

$$K/P = \{(R/U)(k/p)\}/V_B \tag{13.2}$$

or in words:

$$\frac{\text{passenger fatalities}}{\text{passenger mile}} = \left(\frac{\text{fatal accidents}}{\text{flight hours}}\right) * \left(\frac{\text{fatalities per accident}}{\text{passengers per airplane}}\right) * \left(\frac{1}{\text{block speed}}\right) \tag{13.3}$$

Which yardstick should be used in safety comparisons depends on the objective.

Figure 13.2 presents fourteen yardsticks for measuring safety levels. From an engineering viewpoint, yardsticks B are to be preferred over yardsticks A. Relations between 'the amount of aviation' and 'accidents per amount of aviation' are usable in rational design procedures. <u>However:</u> engineers should not forget that the total numbers of people killed (yardsticks A) are important not only in human terms BUT ALSO in terms of public perception of safety. <u>Never forget</u> that the public ultimately decides on aviation activity levels: it does so in the market place and in the polling place!!!!

Table 13.2 presents aviation accident data using the quantities of Eqn.(13.2). Note that in the time period considered, the fatality rate per passenger mile (or km) has decreased much more than the fatal airplane accident rate per hour. The reason for this is twofold: the block speed has increased due to the introduction of jets and the average number of passengers per airplane has increased due to the introduction of larger airplanes.

One significant point needs to be made: the general public appears to 'accept' these safety levels!

An important question is: what should future aviation safety targets be? Table 13.3 compares aviation fatalities with death rates due to all causes. It is interesting to note that aviation fatality rates (per flight

Table 13.2 Safety Statistics for World Air Travel

year	1945	1950	1955	1960	1965	1970	1975
transport in 10^8 pass.km(P)	80	280	610	1090	1980	3820	5750
total distance flown in 10^8 aircr.km (S)	6	14.5	23	31	41	71	74
total flight time in 10^6 aircr.hrs (U)	2.5	5.0	7.3	8.6	8.8	12.2	12.5
no of fatal accidents (R)	20	27	26	34	25	28	20
no of passenger fatalities(K)	240	551	407	873	684	687	445
no of fatalities per accident (k = K/R)	12	20.4	15.7	25.8	27.4	24.5	22.3
no of pass. per aircraft (p = P/S)	13.3	19.3	26.5	35.2	48.3	53.8	77.7
average block speed in km/h (V_B = S/U)	240	290	315	360	446	582	592
no of aircraft acc. per 10^6 aircr.hrs (R/U)	8.0	5.4	3.6	4.0	2.8	2.3	1.6
no of pass. fatalities per 10^8 pass.km (K/P)	3.00	1.97	.67	.80	.35	.19	.08

PERIOD 73-84 AVE.

15.6

518

FROM: REF. 39

Table 13.3 The Safety Target for the Year 2000

Passenger fatality rates in aviation	Fatalities per 10^9 pass.km	Fatalities per 10^7 pass.hrs
Scheduled air services, 1975-1977 (ICAO-world, Fig. 9)	.9	5.3
Future target for the year 2000	.2	1.2

Average population death rate by all causes (The Netherlands, 1976):
- age group 15-44 years: 1.05 deaths per 10^7 person.hrs
- whole population : 9.5 " " " " "

FROM: REF. 39

hour) appear to be similar to overall death rates (per person hour).

If the demand for public air transportation keeps growing at current rates (about 7 percent per year) a factor 3 increase in fatalities can be expected by the year 2000. The assumption is made here that such an increase in absolute number of fatalities will not be accepted by the public. Instead, the assumption is made that the total fatality level should not increase. However, this implies an increase in relative safety. That target safety level is also shown in Table 13.3.

Such improvements in relative safety level are not going to happen automatically. Remember:

<u>SAFETY IS NO ACCIDENT.</u>

It will take a major effort on the part of airplane designers, airplane design management, airplane air and ground crews and the air traffic control system to make this happen.

Sections 13.2 and 13.3 discuss the safety and survivability aspects of commercial and military airplanes in some detail. The role of the preliminary design engineer and design management in assuring safe and survivable designs is discussed in Section 13.4.

13.2 DESIGN FOR SAFETY AND SURVIVABILITY IN COMMERCIAL AIRPLANES

The FAA (Federal Aviation Administration) is responsible for setting up airworthiness regulations, air carrier operating rules as well as enforcing these rules and regulations. The FAA is also responsible for the operation of the air traffic control system.

Commercial airplane accidents in the USA are investigated by the NTSB (National Transportation and Safety Board). The results of all NTSB investigations are made public, including the recommendations made to the FAA for changes in airworthiness regulations or directives, or for changes in air traffic control procedures.

The author believes that a major reason for the relatively high safety of US commercial aviation is the openness with which the overall business of aviation is conducted AND the fact that rule-making/rule enforcement is separated from investigative efforts.

It is useful to quote from FAR 25, par.1309(b):

'The airplane system and associated components, considered separately and in relation to other systems, must be designed so that the occurrence of any failure condition which would prevent the continued safe flight and landing of the airplane is 'extremely improbable'.'

From a design engineering viewpoint the words 'extremely improbable' are interpreted to mean a probability level of less than $1x10^{-9}$ per 1-hour flight.

Table 13.4 defines the relationship between frequency of occurrence of a failure and its allowable effect on flight safety. Fig.13.3 illustrates this relationship.

Overall, the factors which contribute to safety in aviation are listed in Table 13.5. Accidents are usually classified by causes as defined in Table 13.6. In accident analyses, two types of accidents are considered:

1. Predominantly airworthiness and

2. Predominantly operational.

Table 13.7 shows accident breakdowns according to this classification.

A summary of the statistics in Table 13.7 is given in Table 13.8: note that operational accidents dominate.

A study of the human factor in aviation accidents reveals the data in Table 13.9: over half of all accidents are caused by human factors. A breakdown of the human factor into identifiable deficiencies is given in Table 13.10. Clearly, the contribution of the flight crew to accidents is greater than that of ground crews.

Table 13.11 shows a comparison of commercial accident statistics by region: the USA and Canada have a better safety record than other areas of the world.

An alarming trend is clear from Table 13.11 and from Table 13.12: the airplane hull loss rate per million flight hours is very high: more than two hull losses per million flight hours are registered worldwide. This is approaching two hull losses per month on a world wide basis! If the hull loss rate is not curbed significantly, airplane hull insurance rates will increase out of sight.

Table 13.4 The Airworthiness Code for Airplanes

Frequency of Occurrence			Allowable Effect on Flight Safety		Probability	
Code	FROM: REF.39	Description	Effect	Description	Probability P per hour of flight	Safety Class S
Frequent	Re-current	Occurring from time to time for each individual aircraft	Minor	At most small increase of crew work load and slight degradation of flight characteristics	10^{-0} - 10^{-3}	0-3
Reaso-nably Probable					10^{-3} - 10^{-5}	3-5
Remote		Not likely to occur for an individual aircraft, but may happen a few times during the total operational life of all aircraft of one type	Major	Significant increase of crew work load and considerable change in flight characteris-tics. Emergency procedures may be applied, but safe flight and landing still possible	10^{-5} - 10^{-7}	5-7
Extremely Remote		Unlikely to occur during total opera-tional life of all aircraft from one type, but never-theless possible	Hazar-dous	Dangerous increase of crew work load and serious degra-dation of aircraft performance, handling qualities or aircraft structure. Immediate landing necessary; marginal conditions for occupants/injuries	10^{-7} - 10^{-9}	7-9
Extremely Improbable		So extremely remote that it can be considered not to occur	Cata-strophic	Loss of aircraft and/or human lives	$<10^{-9}$	>9

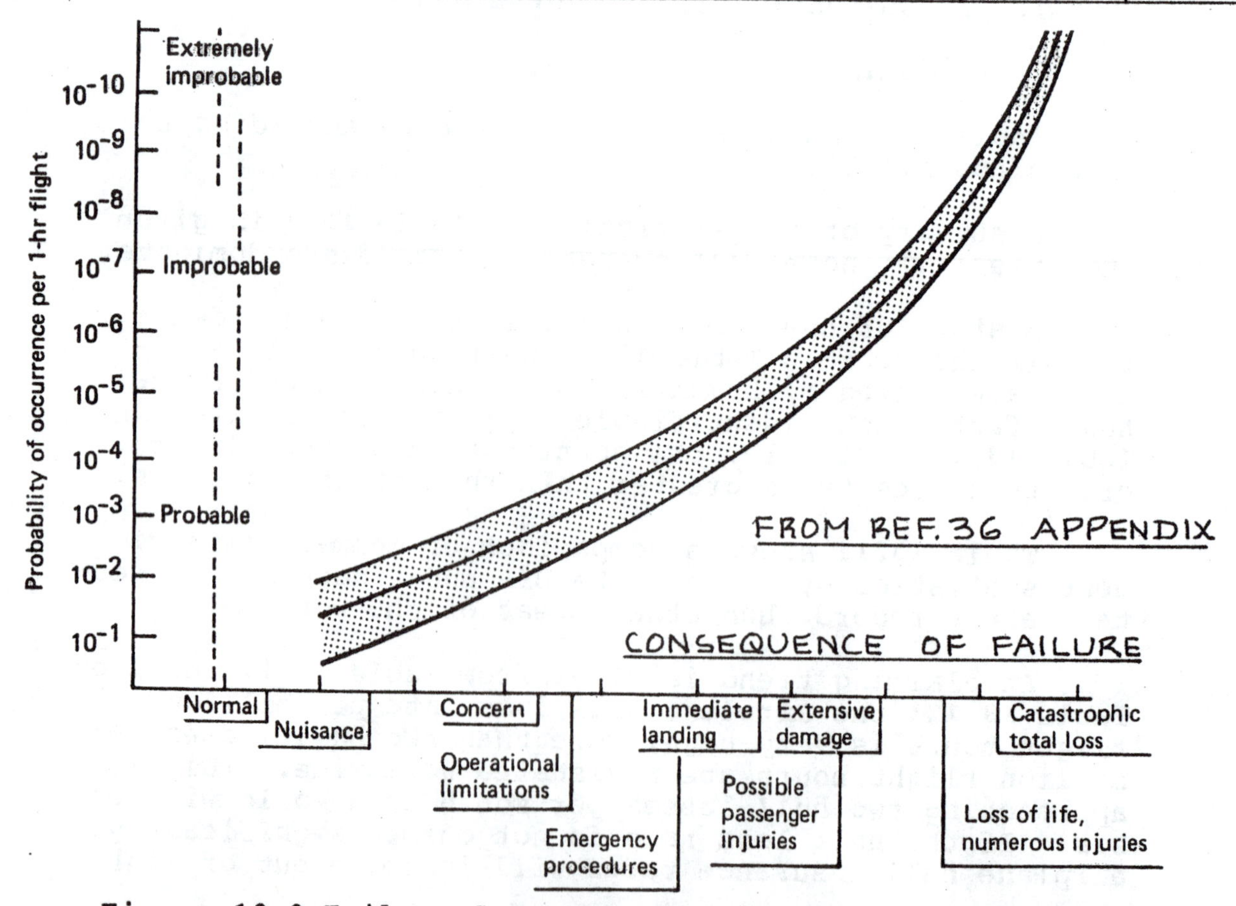

Figure 13.3 Failure Consequence and Failure Probability

Table 13.5 Factors Which Contribute to Aviation Safety

Topic	Contributing Factors
Aircraft design and manufacturing	Airworthiness requirements: - performance and flying qualities - aircraft structure and loads - powerplants - aircraft systems - crashworthiness Aircraft production control
Aircraft flight operations	Flightplanning Air Traffic Control and Air Navigation Airport lay-out and facilities Aircraft maintenance
Personnel	Selection, training and licensing of - operational staff - technical staff
Abnormal events	Occurrence reporting and accident investigation

FROM:
REF. 39

Table 13.6 Classification of Airplane Accidents by Cause

Aircraft condition prior to accident	Basic accident cause	Occurrences, leading to accident
Normal (aircraft under full control, engines running)	Collision with ground, obstacle, birds or other aircraft	Navigational error by flight crew
		Failure nav. equipment of aircraft
		Failure ATC-services
Abnormal flight condition	Loss of control	Mechanical or electrical failures in aircraft systems
		Piloting error with loss of lift, stability and control (stall, dive)
	Loss of engine power	Engine failure
		Mismanagement of engines or fuel system
		Fuel shortage
	Loss of structural integrity	Structural fractures e.g. by fatigue, corrosion
		Outbreak of fire
		Structural failure by excessive loads due to - wrong manoeuvres - extreme bad weather - hard landing - flutter

FROM:
REF. 39

Table 13.7 Analysis of Airplane Accident Types

Accident type (1969 - 1975)		Number of accidents: Total	Fatal	Fatal/Total
Predominantly Airworthiness	Fire (cabin, toilet)	7	2	29%
	Structural failure	21	1	5
	Landing gear mechanism (G)	13	0	0
	Landing gear failure/fire (G)	20	1	5
	Engine failure/fire	58	5	9
	System failure	14	7	50
	Sub-total	133	16	12%
Predominantly Operational	Bird strike	19	0	0
	Weather	18	6	33
	Struck high ground	14	14	100
	Under-shoot	45	23	51
	Overshoot/over-run (G)	28	4	14
	Running off runway (G)	23	0	0
	Heavy landing (G)	16	0	0
	Miscellaneous	42	8	19
	Sub-total	205	55	27%
	Grand-total	338	71	21%

(G) = low-speed accidents at or near ground, few fatal accidents.

FROM: REF. 39

Table 13.8 Airworthiness and Operational Accident Causes

Accident cause predominantly	Number of accidents: Total	Fatal
Airworthiness	39%	23%
Operational	61%	77%
Total	100%	100%
Flying hours: 7.8×10^7 hrs; number of flights: 5×10^7		

FROM: REF. 39

Table 13.9 Airplane Accidents Classified by Causes

Fatal accident causes for commercial aircraft (139 accidents; 47 with jets, 29 with turbo-props, 63 with piston-engined aircraft)	
Human factor, HF	51%
Technical factor, T	14%
Both HF and T	14%
Weather, W	7%
Both HF and W	8%
Sabotage and others	6%
Total	100%

FROM: REF. 39

Table 13.10 The Human Factor in Airplane Accidents

Flight crew	Failed to follow approved procedures (improper IFR-operation included)	34%
	Misjudged speed/altitude/distance (improper level-off included)	19%
	Spatial disorientation	8%
	Failed to see and avoid other aircraft	4%
	Inadequate supervision of flight	5%
	Inadequate preflight preparation/planning	7.5%
	Other crew failures	10%
Ground personnel	Improper maintenance and others	12.5%
Total		100%

FROM: REF. 39

Table 13.11 Fatal Airplane Accidents by Region

Region	Scheduled air services (ICAO, 1970-1976, excl. USSR and China)	Civil jet Aircraft registered per region (1959-Sept.1978)
	Fatal accidents per 10^6 aircr. flight hrs	Hull loss rate per 10^6 aircr. flight hrs.
U.S.A.	} .55	1.31
Canada		2.13
Europe	1.9	2.43
Africa	-	4.61
Central and South America	7.4	8.62
Asia	3.6	4.26
World average (jets only)	2.3 (.85)	2.17

FROM: REF. 39

Table 13.12 Summary of World Jet Hull Loss Rate

WORLD JET HULL LOSS RATE
by generation of aircraft
(Cumulative 1975 — Sept 1980)

Area	1st	2nd	3rd	Total
World	1/385,000	1/767,000	1/1,112,000	1/640,000
Australasia	—	—	1/399,000	1/1,427,000
USA	1/818,000	1/1,468,000	1/941,000	1/1,180,000
Europe	1/426,000	1/599,000	1/1,643,000	1/599,000
Canada	—	1/759,000	—	1/1,297,000
Africa	1/222,000	1/736,000	—	1/413,000
Asia	1/258,000	1/227,000	1/976,000	1/313,000
C&S America	1/182,000	1/544,000	—	1/321,000
World excl. USA	1/301,000	1/497,000	1/1,282,000	1/458,000

FROM: REF. 40 OCT. '80

In designing for safety and survivability the following factors are offered for consideration:

I. Preventive factors:

1. Benign flying qualities.

   This means plenty control power with moderate cockpit control forces, certainly in engine-out emergencies. Changes in flap setting and power setting should be easily controlled. Good damping characteristics are needed.

2. Easy inspectability of the structure for fatigue crack detection and monitoring.

3. Production and materials quality control: know what goes into the airplane.

4. Design systems for ease of operation and prevention of design induced mistakes.

5. Remember: as a general rule, inherent component reliability is better than redundancy.

II. Post crash factors

Crashes will occur despite the best of care in preventive design. For survivabilty: make the cabin environment survivable in a survivable crash.

1. The structure and the seats should not fail in a hazardous manner under g loadings which are survivable by the human body.

2. Prevent fires by safe fuel system design, see Chapter 5.

3. Prevent the use of materials which generate toxic fumes when ignited by fires.

4. Arrange emergency exits so that people really have a chance of getting out.

Review the system design guidelines and design considerations given in Chapters 2 through 12.

13.3 DESIGN FOR SAFETY AND SURVIVABILITY IN MILITARY AIRPLANES

Military accidents are investigated only by military authorities as long as no loss of civilian life has occurred. Since military organizations are not compelled to publish the findings of their accident investigations openly, not much statistical information is available.

Reference 40 publishes listings of military airplane accidents on a regular basis. Aeronautical engineers should study these lists carefully because they often reveal design problems with certain types of airplanes. Selected data from Ref.40 are given in Table 13.13.

Table 13.13, excerpted from Reference 40, is presented here without comment.

Military airplanes are subject to combat damage of a variety of types. Reference 41 provides detailed methods of analysis for combat survivability.

Table 13.14 shows the loss rates sustained in several military conflicts.

Table 13.15 shows the causes of combat inflicted losses in five wars.

Table 13.16 shows the probability of military crews completing a combat tour under different values of assumed loss rate.

It is clear that safety in military aviation is a difficult subject. Even in peace time, by the nature of the military business (need for training in extreme circumstances is one example) it will be virtually impossible to achieve safety levels comparable to those achieved in commercial aviation.

From a design viewpoint, all factors mentioned in Section 13.2 also apply to the design of military airplanes.

13.4 THE ROLE OF THE PRELIMINARY DESIGN ENGINEER AND DESIGN MANAGEMENT IN CREATING SAFE AIRPLANES

The process of airplane sizing (Part I) and preliminary airplane configuration design (Part II) was seen to be dominated by airworthiness requirements (civil or military). Minimum performance requirements as well as stability and control requirements were seen to exert

Table 13.13 Summary of Military Airplane Accidents for 1985, from Reference 40.

Date	Type	Service	Location	F*	I*	S*	Remarks
17.1	F-14A Tomcat	USN	USS Constellation	0	10	2	Clipped nose of A-6 on landing
22.1	C-130 Hercules	USAF	1/2 n.m. off Trujillo, N.Honduras	21			
23.1	VA-3B Skywarrior	USN	Between Atsugi, Japan and Guam	9			Missing at sea
29.1	A-4M Skyhawk	USMC	San Clemente, CA			1	Pilot ejected during post maint. check-flight
?.1	A-10A Th'bolt II	USAF	New York	no data			
7.2	A-7E Corsair	USN	NAS Fallon, Nv			1	departed in IMC climb at low level
7.2	F-16A Ftg'Falcon	USAF	Townsend Range			1	
8.2	F-16A Ftg'Falcon	USAF	La Cesa			1	pilot ejected
9.2	A-10A Th'bolt II	USAF	30 n.m. of Flagstaff, Az.	1			Weapons sortie
10.2	F-16A Ftg'Falcon	USAF	Near Luke AFB, Az	no data			
13.2	TA-4J Skyhawk	USN	NAS Pensacola		1	2	Thrust loss in circuit, crew ejected

* F = fatality I = injured S = survived, no injury

Table 13.14 Airplane Loss Rates in High Intensity Sorties

Data taken from Reference 43.

Operation	Loss Rate in Percent
RAF Bomber Command, Night attacks, January,1942 - June, 1944	> 4
RAF Bomber Command, Battle of Berlin November, 1943 - March, 1944	6.4
U.S. 8th Air Force, January, 1943 - September, 1943 (Airplanes penetrating enemy defenses over Germany only)	8.2
U.S. 20th Bomber Command over Japan, June, 1944 - December, 1944	4
Israeli Air Force, 1973, Golan, days 1-4	4

Table 13.15 Causes of Combat Losses in Several Wars

Data taken from Reference 43.

War	Losses due to:	AAA	SAM	Fighters
U.S., WWII		10,287	---	8,743
U.S., Korea		1,109	---	144
U.S., Vietnam		1,605	196	82
Israel, 1973		31	46	15
Arab Forces, 1973		72	25	334

Table 13.16 Probability of Completion of Combat Tour

Data taken from Reference 43.

Attrition Rate per sortie, %	Probability of completing sorties 25 %	50 %	100 %
0.1	98	95	90
0.5	88	78	61
1.0	78	61	37
2.0	60	36	13
4.0	36	13	2
8.0	12	2	-

much influence on the sizing and the configuration design process of commercial as well as military airplanes.

The preliminary design engineer must therefore be thoroughly familiar with civil and military airworthiness regulations.

The role of the preliminary design engineer in designing airplane systems for safety and for reliability was first hinted at in Part II. In Step 17 of p.d. sequence II (p.18 of Part I) the designer was asked to do the following:

Step 17: List the major systems needed in the airplane. Also: prepare 'ghost' views indicating the general system arrangements and their location in the airframe.

As a follow-up to Step 17, in Step 32 (p.22, p.d. sequence II in Part II) the designer was also asked to:

Step 32: Prepare a preliminary layout drawing for all essential airplane systems, in particular the primary and secondary flight control systems.

Chapters 2 through 12 in this part dealt with the layout design of the most important systems which are normally required by airplanes. What is meant by system layout design in Step 32 is the completion of the following 'what-if' safety and maintenance checklist:

<u>'WHAT-IF' SAFETY AND MAINTENANCE CHECKLIST</u>

1. Check whether all performance requirements which are critical to the airworthiness of the airplane are satisfied. Determine how close the airplane is to NOT satisfying one or more of these requirements and identify the reason. Start thinking about possible design fixes.

2. Check whether all stability and control requirements which are critical to the airworthiness of the airplane have been met. Determine how close the airplane is to NOT satisfying one or more of these requirements and identify the reason. Start thinking about possible design fixes.

3. Prepare the following drawings:

   3a) a system schematic drawing

3b) a system inboard profile
3c) a system 'ghost' view

Table 13.17 provides a key to where in this text examples of schematic drawings, inboard profiles and ghost views of systems can be found.

4. Prepare a functional description of the system.

   This should include a description of all system components with their function.

5. Identify all component attachments to surrounding structure and identify the implications for structural design and analysis.

6. Prepare a list of required redundancies in the system with preliminary reasons explaining the need for the redundancy.

7. Prepare a list of servicing and/or maintenance requirements. This must include an analysis of requirements for accessibility.

8. Play the 'WHAT-IF' game. How to do this is explained below.

One idea behind completing the items in the checklist is to be able to identify possible conflict situations. Conflict situations exist if:

A. two systems or system components apparently occupy the same space.

B. two systems or system components are so close together that in case of system failure or local structural failure other serious failures would result.

With the help of Computer Aided Design (CAD) systems it is relatively easy to spot such conflict situations: various 'ghost' layouts can be overlaid on the computer display. Conflicts are immediately obvious.

At this point it is essential that the designer play the so-called 'WHAT-IF' game. The 'what-if' game consists of asking the question:

<u>What happens if: component X in system Y fails?</u>

The answer to this question must be spelled out in

Table 13.17 Examples of System Schematics, System Inboard Profiles and System Ghost Views

System	Chptr	Schematic		Inboard Profile		Ghost View	
		Fig.	Pg.	Fig.	Pg.	Fig.	Pg.
Landing Gear	2	2.94d	116	2.46	73	2.89b	108
Weapons	3	3.8,	133	3.2	129	3.7	133
Flight Controls	4	4.68	258	4.94	279	4.96	283
Fuel	5	5.15	299	5.13	298	5.2	288
Hydraulic	6	6.4	306	6.11	315	6.5	309
Electrical	7	7.8	327	7.7	325	7.1	318
Environmental	8	8.4	334	8.7	337	8.6	336
Cockpit Instr. and Flight Management	9	9.7	346	9.15	352	9.20	356
De-icing etc.	10	10.3	360	10.12	368	10.11	367
Escape	11	11.8	378	11.3	374	11.7	377
Water and Waste	12	12.2	382	12.1	381	12.3	383

terms of detailed failure and consequence scenarios. Depending on the seriousness of the consequences, modifications of the proposed design layout may be required.

The 'what-if' game is the follow-up to Step 32 in p.d. sequence II as described on p.22 in Part II.

What follows now is a list of rules for designing safe airplanes. Each aeronautical engineering student and each practicing design engineer should always keep this list in mind!

RULES FOR DESIGNING SAFE AIRPLANES:

1. Know the airworthiness regulations!
2. Never be complacent about the safety of any aspect of your design.
3. Pay attention to details: often the 'little' things cause serious accidents
4. Do not become a specialist in any design area. Design engineers should be as broad as possible in their knowledge of the airplane and all its systems.
5. LEARN TO FLY! The author considers it essential that airplane design engineers be licensed pilots. This teaches additional respect and understanding for the products being designed and appreciation for the problems encountered by those who professionally fly and maintain airplanes.
6. READ, READ, READ. The design engineer should spend at least 20 percent of his time reading airplane accident reports and the airplane literature. The author recommends regular reading of the following periodicals:

   1. Aviation Week and Space Technology (USA)
   2. Interavia (Switzerland)
   3. Flight International (England)
   4. AOPA Pilot (USA)
   5. Flying (USA)
   6. Journal of Aircraft (USA)
   7. US Naval and Air Force Safety Reviews
   8. NTSB accident reports and recommendations

   The designer must learn from mistakes made by

others and not repeat these mistakes. There is only one way to prevent making the same mistakes over and over again: READ, READ, READ.

7. An ethical point: whenever the question of the airworthiness of a design arises, the first question the designer should ask is:

   Is it safe enough for myself and/or my family to ride on? Only when the answer to this question is YES, should reference be made to airworthiness specifications. Remember that all airworthiness specifications represent MINIMUM safety requirements. If at all possible the designer should do better than the minimum which is demanded of him.

8. If management fails to alter a design which is judged to be unsafe by the design engineer he should insist on a review of the design by a peer group in the organization. Normally the decision of this group should prevail.

   If management ignores the peer group decision and wants to proceed with the unsafe design, the engineer has only two options:

   A. write a memo outlining the problem, its solution and its status. Address this memo to the appropriate level of management. In extreme cases this may mean the bypassing of immediate supervision!

   B. if design changes are not made, send a copy of the memo to the certifying authority.

In this regard it is sad to observe that certifying agencies not always react with due speed to problems with air safety. Ref.42 contains many examples of cases where certifying agencies as well as corporations failed to react on time in the face of known design deficiencies.

The role of management in the process of creating safe designs is that of monitoring engineering decision making, insisting on meeting at least the minimum requirements AND insisting on ethical conduct of the design, development and certification process.

14. REFERENCES

1. Conway, H.G., Landing Gear Design, Chapman Hall, London, England, 1958.

2. Currey, N.S., Landing Gear Design Handbook, Lockheed Georgia Company, Marietta, Georgia, 30063, 1982.

3. Torenbeek, E., Synthesis of Subsonic Airplane Design, Kluwer Boston Inc., Hingham, Maine, 1982.

4. Stinton, D., The Design of the Aeroplane, Granada Publishing, London, England, 1983.

5. Bingelis, T., The Sportplane Builder, T.Bingelis, 8509 Greenflint Lane, Austin, Texas, 78759, 1979.

6. Taylor, J.W.R., Jane's All The World Aircraft, Published annually by: Jane's Publishing Company, 238 City Road, London EC1V 2PU, England.

7. Bruhn, E.F., Analysis and Design of Flight Vehicle Structures, Tri-State Offset Co., Cincinnati, Ohio, 45202, 1965.

8. Thurston, D.B., Design for Flying, McGraw-Hill Book Co., New York, 1978.

9. Heinemann, H., Rausa, R. and Van Every, K., Aircraft Design, The Nautical Publishing Company of America, Inc, Baltimore, Md, 21201, 1985.

10. Horne, W.B., Yager, T.J. and Taylor, G.R., Recent Research on Ways to Improve Tire Traction on Water, Slush or Ice, De Ingenieur, September, 1966. (Journal of the Royal Netherlands Inst. for Engineers).

11. Anon., Aircraft Carrier Reference Data Manual, NAEC MISC 06900, Naval Air Engineering Center, 1974.

12. Erdman, A.G. and Sandor, G.N., Mechanical Design: Analysis and Synthesis, Volume 1, Prentice Hall, N.J., 1984.

13. Erdman, A.G. and Sandor, G.N., Advanced Mechanical Design: Analysis and Synthesis, Volume 2, Prentice Hall, N.J., 1984.

14. Buzzard, W.C. et al, Tests of the Air Cushion Landing System on the XC-8A, AFFDL-TR-78-61, 1978.

15. Anon., Air Cushion Landing Gear Application Studies, Report D7605-927001, NASA Contract NAS1-15202, Bell Aerospace, Buffalo, N.Y., 1978.

16. Ball, R.E., The Fundamentals of Aircraft Combat Survivability Analysis and Design, AIAA Education Series, AIAA, N.Y., 1985.

17. Fuhs, A.E., Radar Cross Section Lectures, US Naval Postgraduate School, Monterey, CA, 93940, 1984.

18. Sweetman, B., Stealth Aircraft, Motorbooks International, Osceola, Wisconsin, 54028, 1986.

19. Pretty, R.T., Jane's Weapon Systems, 1984-1985. See Reference 6.

20. Roskam, J., Airplane Flight Dynamics and Automatic Flight Controls, 1981, Roskam Aviation and Engineering Corp., Rt 4, Box 274, Ottawa, Kansas, 66067.

21. Anon., Federal Aviation Regulations, Part 23 and Part 25, Department of Transportation, Distribution Requirements Section, M-482.2, Wash., D.C., 20590.

22. Anon., MIL-F-8785-B/C, Military Specification-Flying Qualities of Piloted Airplanes, 1969 and 1982.

23. Scanlan, R.H. and Rosenbaum, R., Aircraft Vibration and Flutter, Dover Publications, N.Y., 1968.

24. Fung, Y.C., An Introduction to the Theory of Aeroelasticity, J.Wiley and Sons, 1955.

25. Bisplinghoff, R.L., Ashley, H. and Halfman, R., Aeroelasticity, Addison-Wesley, 1955.

26. Wood, N.E. et al, Electromechanical Actuation Feasibility Study, AFFDL-TR-76-42, WPAFB, OH, 45433.

27. Anon., MIL-F-9490D, Military Specification-Flight Control Systems-Design, Installation and Test of, Piloted Aircraft, June 1975.

28. Brahney, J.H., Pumps for 8000 psi Hydraulic Systems Examined, Aerospace Engineering, July 1986.

29. McKinley, J.L. and Bent, R.D., Basic Science for Aerospace Vehicles, Northrop University, McGraw Hill Book Co., N.Y., 1965.

30. Casamassa, J.V. and Bent, R.D., Jet Aircraft Power Systems, Northrop University, McGraw Hill Book Co., N.Y., 1965.

31. Marks, L.S., Mechanical Engineer's Handbook, McGraw Hill Book Co., N.Y., 1951.

32. Weinstock, G.L., Lightning Protection on Advanced Fighter Aircraft, SAE/AFAL Lightning and Static Electricity Conference, McDonnell Douglas Paper, MDC 70-035, 1970.

33. McKinley, J.C. and Bent, R.D., Maintenance and Repair of Aerospace Vehicles, Northrop University, McGraw Hill Book Co., N.Y., 1967.

34. McKinley, J.C. and Bent, R.D., Electricity and Electronics for Aerospace Vehicles, Northrop University, McGraw Hill Book Co., N.Y., 1971.

35. Beauchamp, E.D., Opportunities and Challenges for Electric-Drive Systems on Aircraft, SAE Paper 841627, Aerospace Congress and Exposition, 1984.

36. Anon., Integrated Application of Active Controls (IAAC) Technology to an Advanced Subsonic Transport Project, NASA Contractor Report 3880, 1986.

37. Zumwalt, G.W. and Mueller, A.A., Flight and Wind Tunnel Tests of an Electro-Impulse De-icing System, AIAA/NASA General Aviation Technology Conference, AIAA Paper No. 84-2236, Hampton, VA, 1984.

38. Bingelis, T., Firewall Forward, Published by T.Bingelis, 8509 Greenflint Lane, Austin, Texas, 78759.

39. Wittenberg, H., Safety in Aviation, Achievements and Targets, Memorandum M-353, Delft University of Technology, Department of Aerospace Engineering, Delft, The Netherlands, 1979.

40. Flight International, British Aerospace Weekly Magazine, Business Press International USA, 205 East 42nd Street, New York, N.Y., 10017.

41. Ball, R.E., The Fundamentals of Aircraft Combat Survivability Analysis and Design, AIAA Education Series, AIAA, 1633 Broadway, New York, N.Y., 10019.

42. Tench, Bill, Safety is no accident, Collins, 8 Grafton Street, London, W1X 3LA, England.

43. Greene, T.E., Surviving Modern Air Defenses, Aerospace America, August 1986.

15. INDEX

Airplane Design & Analysis Book Descriptions

All books can be ordered from our on-line store at www.darcorp.com.

Airplane Aerodynamics & Performance
C.T. Lan & Jan Roskam

The atmosphere • basic aerodynamic principles and applications • airfoil theory • wing theory • airplane drag • airplane propulsion systems • propeller theory • fundamentals of flight mechanics for steady symmetrical flight • climb performance and speed • take-off and landing performance • range and endurance • maneuvers

Airplane Flight Dynamics & Automatic Flight Controls Part I
Jan Roskam

General steady and perturbed state equations of motion for a rigid airplane • concepts and use of stability & control derivatives • physical and mathematical explanations of stability & control derivatives • solutions and applications of the steady state equations of motion from a viewpoint of airplane analysis and design • emphasis on airplane trim, take-off rotation and engine-out control • open loop transfer functions • analysis of fundamental dynamic modes: phugoid, short period, roll, spiral and dutch roll • equivalent stability derivatives and the relation to automatic control of unstable airplanes • flying qualities and the Cooper-Harper scale: civil and military regulations • extensive numerical data on stability, control and hingemoment derivatives

Airplane Flight Dynamics & Automatic Flight Controls Part II
Jan Roskam

Elastic airplane stability and control coefficients and derivatives • method for determining the equilibrium and manufacturing shape of an elastic airplane • subsonic and supersonic numerical examples of aeroelasticity effects on stability & control derivatives • bode and root-locus plots with open and closed loop airplane applications, and coverage of inverse applications • stability augmentation systems: pitch dampers, yaw dampers and roll dampers • synthesis concepts of automatic flight control modes: control-stick steering, auto-pilot hold, speed control, navigation and automatic landing • digital control systems using classical control theory applications with Z-transforms • applications of classical control theory • human pilot transfer functions

Airplane Design Part I
Preliminary Sizing of Airplanes
Jan Roskam

Estimating take-off gross weight, empty weight and mission fuel weight • sensitivity studies and growth factors • estimating wing area • take-off thrust and maximum clean, take-off and landing lift • sizing to stall speed, take-off distance, landing distance, climb, maneuvering and cruise speed requirements • matching of all performance requirements via performance matching diagrams

Airplane Design Part II
Preliminary Configuration Design and Integration of the Propulsion System
Jan Roskam

Selection of the overall configuration • design of cockpit and fuselage layouts • selection and integration of the propulsion system • Class I method for wing planform design • Class I method for verifying clean airplane maximum lift coefficient and for sizing high lift devices • Class I method for empennage sizing and disposition, control surface sizing and disposition, landing gear sizing and disposition, weight and balance analysis, stability and control analysis and drag polar determination

Airplane Design & Analysis Book Descriptions

All books can be ordered from our on-line store at www.darcorp.com.

Airplane Design Part III
Layout Design of Cockpit, Fuselage, Wing and Empennage: Cutaways and Inboard Profiles
Jan Roskam

Cockpit (or flight deck) layout design • aerodynamic design considerations for the fuselage layout • interior layout design of the fuselage • fuselage structural design considerations • wing aerodynamic and operational design considerations • wing structural design considerations • empennage aerodynamic and operational design considerations • empennage structural and integration design consideration • integration of propulsion system • preliminary structural arrangement, material selection and manufacturing breakdown

Airplane Design Part IV
Layout Design of Landing Gear and Systems
Jan Roskam

Landing gear layout design • weapons integration and weapons data • flight control system layout data • fuel system layout design • hydraulic system design • electrical system layout design • environmental control system layout design • cockpit instrumentation, flight management and avionics system layout design • de-icing and anti-icing system layout design • escape system layout design • water and waste systems layout design • safety and survivability considerations

Airplane Design Part V
Component Weight Estimation
Jan Roskam

Class I methods for estimating airplane component weights and airplane inertias • Class II methods for estimating airplane component weights, structure weight, powerplant weight, fixed equipment weight and airplane inertias • methods for constructing v-n diagrams • Class II weight and balance analysis • locating component centers of gravity

Airplane Design Part VI
Preliminary Calculation of Aerodynamic, Thrust, and Power Characteristics
Jan Roskam

Summary of drag causes and drag modeling • Class II drag polar prediction methods •airplane drag data • installed power and thrust prediction methods • installed power and thrust data • lift and pitching moment prediction methods • airplane high lift data • methods for estimating stability, control and hingemoment derivatives • stability and control derivative data

Airplane Design Part VII
Determination of Stability, Control, and Performance Characteristics: FAR and Military Requirements
Jan Roskam

Controllability, maneuverability and trim • static and dynamic stability • ride and comfort characteristics • performance prediction methods • civil and military airworthiness regulations for airplane performance and stability and control • the airworthiness code and the relationship between failure states, levels of performance and levels of flying qualities

Airplane Design Part VIII
Airplane Cost Estimation: Design, Development, Manufacturing, and Operating
Jan Roskam

Cost definitions and concepts • method for estimating research, development, test and evaluation cost • method for estimating prototyping cost • method for estimating manufacturing and acquisition cost • method for estimating operating cost • example of life cycle cost calculation for a military airplane • airplane design optimization and design-to-cost considerations • factors in airplane program decision making

Design • Analysis • Research
1440 Wakarusa Drive, Suite 500, Lawrence, Kansas 66049, USA - Tel: (785) 832-0434 - Fax: (785) 832-0524
info@darcorp.com – www.darcorp.com